Adrian Smith is Emeritus Professor of Modern History at the University of Southampton, and previously taught at the Royal Military Academy Sandhurst and the University of Kent. An established author, broadcaster and journalist in the fields of modern British political, social and cultural history, his books include *Mountbatten: Apprentice War Lord* and *The City of Coventry: A Twentieth Century Icon* (both I.B.Tauris), *Mick Mannock, Fighter Pilot: Myth, Life and Politics* and *The New Statesman: Portrait of a Political Weekly, 1913–1931*. In researching and writing *The Man Who Built the Swordfish* he worked closely with the family of Sir Richard Fairey and the National Museum of the Royal Navy's Fleet Air Arm Museum.

'British business biographies are few and far and for this reason alone this brilliant account of the life of Sir Richard Fairey will be welcome. Yet it is much more than this – it gives us a unique insight into the history of the aircraft industry and the Air Force, but also the temper of the British political right wing, transatlantic relations in the Second World War, and the private and leisure world of a British tycoon. It gives an astonishing, and exceedingly rare, insight into the nature of the British elite.'

– David Edgerton, Hans Rausing Professor of the History of Science and Technology and Professor of Modern British History, King's College London

'Adrian Smith's spectacular biography breaks the myth of British industrial decline in the twentieth century, highlighting the contribution of a larger than life individual to the development of cutting edge industries. Richard Fairey created aircraft that pushed the limits of range, speed and durability, and helped mobilise American industry to support the British war effort. The durable, war-winning Swordfish, and the fabulous FD2, test bed for Concorde, are his legacy.'

– Andrew Lambert, Laughton Professor of Naval History, King's College London and author of *Crusoe's Island*.

THE
MAN WHO BUILT
THE SWORDFISH

The Life of
Sir Richard Fairey

ADRIAN SMITH

Published in 2018 by
I.B.Tauris & Co. Ltd
London • New York
www.ibtauris.com

Copyright © 2018 Adrian Smith

The right of Adrian Smith to be identified as the author of this work has been asserted by the author in accordance with the Copyright, Designs and Patents Act 1988.

All rights reserved. Except for brief quotations in a review, this book, or any part thereof, may not be reproduced, stored in or introduced into a retrieval system, or transmitted, in any form or by any means, electronic, mechanical, photocopying, recording or otherwise, without the prior written permission of the publisher.

Every attempt has been made to gain permission for the use of the images in this book. Any omissions will be rectified in future editions.

References to websites were correct at the time of writing.

ISBN: 978 1 78831 336 0
eISBN: 978 1 78672 418 2
ePDF: 978 1 78673 418 1

A full CIP record for this book is available from the British Library
A full CIP record is available from the Library of Congress

Library of Congress Catalog Card Number: available

Typeset by Riverside Publishing Solutions, Salisbury, Wiltshire, UK
Printed by CPI Group (UK) Ltd, Croydon CR0 4YY

To Adam John Smith, 1984–2016

Contents

List of Plates	ix
Foreword and Acknowledgements	xi
List of Acronyms	xxi
Chapter 1: Early Life	1
Chapter 2: Aviation Apprenticeship	27
Chapter 3: World War I and the Founding of Fairey Aviation	51
Chapter 4: Great War and Great Expectations	77
Chapter 5: Twenties Bust and Boom	111
Chapter 6: Into the Thirties, 'This Age of Crooners and Safety First'	147
Chapter 7: Through the Thirties, That 'Low Dishonest Decade'	183
Chapter 8: Wartime in Washington – the British Air Commission	221
Chapter 9: Charging to an End, at Supersonic Speed, 1946–56	271
Chapter 10: Conclusion	305
Appendix: My Father, Sir Richard Fairey	321
Notes	327
Bibliography and Filmography	411
Index	427

Plates

1. The Fairey family, late 1890s (Fairey family)
2. CRF with J.W. Dunne's D8 prototype (Fairey family)
3. Dick Fairey with the model aircraft he sold to Gamage's for production, 1911 (Fairey family)
4. Glider model advert, 1911 (Fairey family)
5. Lord Brabazon of Tara, 1908 (Fairey family)
6. Murray Sueter, founding father of the Royal Naval Air Service (Fleet Air Arm Museum – FAAM)
7. Fairey IIIF, mainstay of interwar sales (FAAM)
8. CRF in the 1920s, company founder, chairman, and managing director (FAAM)
9. CRF and fellow yachtsman and aviation industrialist, Louis Charles Breguet (Fairey family)
10. Richard and Esther Fairey, July 1935 (National Portrait Gallery)
11. Fairey Swordfish, Fleet Air Arm icon (FAAM)
12. Lord Halifax chairs a meeting of British and French mission heads in wartime Washington (Fairey family)
13. Postwar shooting party (Richard Fairey behind his father and Lord Brabazon to CRF's left) (Fairey family)
14. Sir Richard and US Ambassador Lew Douglas at Bossington, 1947 or 1948 (Fairey family)
15. Fairey Delta 2, world air speed record breaker, 1956 (FAAM)
16. CRF inspection of FAA Flight703X, April 1954 (FAAM)

Foreword and Acknowledgements

March 1956 saw the supersonic, super-streamlined Fairey Delta 2 smash the world air speed record, averaging 1,132 miles per hour as it streaked across the south coast of England. Fairey Aviation's final fixed-wing aircraft was a stunning design, in both looks and performance. Few of its predecessors were similarly easy on the eye; Fairey as a firm scorning aesthetic considerations in the interest of functionality. The FD2 looked terrific and it flew fast; but the record breaker's future lay as a test bed, and not as the prototype for a high-volume production fighter which would save the company's fortunes. Within four years, Fairey Aviation would cease functioning as an autonomous airframe manufacturer. But what a way to bow out, and how fitting that Fairey's founder, although in poor health, lived long enough to remind the world of his company's key role in the remarkable progress of powered flight across the preceding half century.

Test pilot Peter Twiss, in his account of the FD2's historic flight, described Fairey Aviation as 'not so much a family firm as a benevolent autocracy' ruled for over four decades by the 'formidable figure' of Sir Richard Fairey: 'He was a very awe-inspiring individual, both physically and mentally, and anyone who wished to get his ear was acutely conscious that he was addressing 6'7" of aristocratic aeronautical history.'[1] 'Dick' Fairey was a big man, both in physique and achievement. He was the consummate plane maker, equally at home in the shabbiest shed or the smartest hotel. A typical day might start down on the shop floor dealing with a design problem and end at the Savoy Grill hosting a Royal Aero Club reception. Charles Richard Fairey – universally known as 'CRF' and with the first name discarded – founded a company which in due course would supply air forces from Canada to China, including of course the RAF. Yet, across two world wars and deep into the Cold War, it was the Royal Navy with which Fairey Aviation enjoyed a unique partnership. For 70 years, machines assembled in the firm's hangars at Hayes, Heaton Chapel and Hamble flew off the flight decks of countless aircraft carriers, whether adapted or purpose-built. Unsurprisingly, Fairey – a prominent figure for much of his life – was synonymous with a succession of famous aircraft which bore his name, not least the FD2. Yet Sir Richard Fairey became best known as the father of the Fleet Air Arm's most iconic machine – quite simply, he was the man who built the Swordfish.

On the eve of World War I, Dick Fairey was working for Shorts in north Kent, the epicentre of British aircraft design, development and construction. A qualified engineer, party to the testing and manufacture of the Royal Naval Air Service's state-of-the-art sea planes, CRF founded Fairey Aviation at the Admiralty's behest in 1915.[2] His company survived postwar retrenchment to become for a time Britain's largest airframe manufacturer, supplying the RAF with a succession of front-line aircraft. Supported by a small, ultra-loyal circle of friends and acolytes, Fairey remained the dominant shareholder after his firm was floated in 1929, and his substantial commercial and property interests were scarcely dented by the Depression. The young W.B. Yeats speculated that 'there is some one myth for every man, which, if we but knew it, would make us understand all he did and thought.'[3] For Sir Richard Fairey that one myth was surely the emotional impact of his Edwardian family's fall from suburban prosperity into genteel poverty. If the child is father to the man, then it is scarcely surprising CRF saw the accumulation and the enjoyment of wealth as confirmation that any suitably talented individual can triumph over adversity. Fairey's reputation as a self-made man, and his high-profile position as a figurehead of the aircraft industry, ensured his prominence across the interwar period; especially in the 1930s when success sailing 12-metre and J-class yachts saw him share the spotlight with fellow plane maker Tommy Sopwith in challenging for the America's Cup.

When not exercising hands-on control of a company expanding rapidly in response to rearmament, CRF was pacing the corridors of power, or courting politicians and playmakers on the links, on board his motor yacht or on a weekend shoot at one of his three estates. The last of the latter was Bossington, acquired in 1937, and still the principal home of the Fairey family. Although Fairey, his second wife and their children spent much of World War II in Washington, home was now the Test Valley, albeit with Bermuda as a winter retreat. Ownership of prime beats on arguably England's finest chalk stream meant Fairey took seriously his responsibilities as a keen fly fisherman committed to preserving the integrity of the rural environment.

While CRF's significance as a pioneer of conservation and a champion yachtsman deserves full recognition, clearly his reputation rests on Fairey Aviation's key contribution to British manufacturing industry across the first five decades of the last century. Long before the Delta 2, Fairey Aviation had designed and built cutting-edge experimental aircraft, not least the long-distance record breakers which raced across the Empire from Cranwell to Karachi. Fairey's inventive flair drew on his engineering expertise and his entrepreneurial initiative. Plane maker or helmsman – whatever the role – he had to be the best, his scientific, technological and commercial credentials underpinning eventual success. He was a hard man, whose neo-liberal free-market mentality fuelled a deep suspicion of government, notwithstanding his company's dependence upon state patronage. Dick Fairey was a man of strong opinions, although, unlike several friends, his political prejudices never fuelled a flirtation with fascism.

Ironically, for all his right-wing convictions Fairey was happy to cultivate close contact with Russian designers such as Tupolev and Ilyushin, and, although his company's contribution to the first Five Year Plan never materialised, CRF frequented the Soviet Embassy throughout the 1930s. Having met Stalin's *apparatchiks* in Moscow in 1931, a decade later Fairey found himself in hard bargaining with Roosevelt's New Dealers as he sealed successive procurement contracts for the supply of American aircraft across the Atlantic. His wartime success as head of the British Air Commission earned him a knighthood, but necessitated a change of role when he returned to a very different Fairey Aviation in 1945.

In poor health and with a less interventionist executive role, for much of the time Fairey directed operations from afar, wintering in Bermuda, where he railed against the Attlee government and sought to influence local colonial politics. For all his – in relative terms – low-level activities, Fairey still oversaw the creation of a boatbuilding division, Fairey Marine, and sanctioned his original company's embrace of new rotary technology to develop the revolutionary, if ill-fated, vertical-take-off Fairey Rotodyne. Fairey died before his company's costly project, combined with the failure to build on the success of the Delta 2, confirmed British aircraft manufacturers' inability to invest in jet-age R&D. With no guarantee of extended production runs driving down unit costs, firms like Fairey increasingly lacked the capital necessary to compete in a global, or even a domestic, market.

Yet the story of Sir Richard Fairey can only be told against the backdrop of the rise and fall of the one industry which in the immediate postwar era could still challenge its American rivals on a near equal footing. Fairey's technical education illustrates the fallacy of any suggestion that late Victorian and Edwardian Britain lacked institutions capable of providing a necessary grounding in applied science, while his whole career – across two world wars and a critical interwar period – demonstrates the very real strengths, as well as the inherent weaknesses, of the indigenous aircraft industry. In his day Dick Fairey was a powerful figure, indeed a pivotal figure. His life tells us much about the exercise of power in David Edgerton's 'warfare state', providing an insight into the nature of British manufacturing industry at its wartime peak(s) and on the cusp of its twilight years.[4]

David Edgerton's sustained critique of crude declinism and his analysis of a complex yet dynamic working relationship between applied science, the military and the British state, from the Empire's zenith to its final eclipse, informs each of the succeeding chapters (drafts of which he kindly read and commented upon). The shape and structure of what follows owes a similar debt to the biographer and philosopher Ray Monk. A quietly influential figure, Ray has for over a decade helped shape the intellectual agenda which scholars from disparate disciplines consciously or unconsciously adopt when engaging with that multi-faceted biographical phenomenon we call 'life writing.'[5] An astute and intuitive commentator on matters biographical, Monk's readiness in his life of Oppenheimer to grapple with hard science and yet retain the interest of his reader gives encouragement to any author explaining the theory of flight, let alone the complexities of particle physics.[6] That same sharp intuition and eagerness to embrace and comprehend the previously unfamiliar – however challenging – is equally evident in the writings of the late Alex Danchev. A genuine polymath, in his biographies Danchev vigorously waved the flag of interdisciplinarity, sharing Edgerton's disdain for C.P. Snow's 'two cultures' reductionism.[7] Alex Danchev's influence is less visible than that of Edgerton and Monk, but the man and his intellect were greatly valued during the writing of this book, and both are now greatly missed.

This is the third occasion on which one of Sir Richard's children has initiated a biography of his or her father. Not long after Fairey's death, Richard, the only child of his first marriage, approached the author and journalist James Wentworth Day, an old friend of the family.[8] As described in Chapter 5, Wentworth Day became acquainted with CRF some time in the 1930s. Both men shared a passion for field sports and a deep dislike of the National Government. Yet, unlike several in their circle, a mutual contempt for Stanley Baldwin's 'One Nation Toryism' never translated into open or even tacit admiration of the Duce and the Führer. Instinctively suspicious of the state and fearful of Nazi intent, Dick Fairey saw fascism as a continental aberration, and appeasement of its exponents a craven policy rooted in pacifist naivety. Wentworth Day's fiercely right-wing views were rooted in a passionate love of the English countryside, especially East Anglia. Like Fairey, he was proud of his family's fenland roots, and this constituted a common bond. Wentworth Day was a colourful personality, notorious after the war for his outspoken opinions ('He is of course what you might call Right Wing Conservative and believes that "Socialism and water" is the wrong policy…') and his tabloid-feeding private life – he was briefly married to the unashamedly bisexual film critic Nerina Shute.[9] Remarried, respectable and reliably reactionary in his world view, Jim Wentworth Day was a regular presence at Bossington, especially in the last ten years of Sir Richard's life. He would frequently fish the Test, or on other occasions join CRF and his guests for a day's shoot.

Given the strength of his friendship with Fairey and his reputation as a writer who rarely missed a deadline, Wentworth Day seemed to Richard the obvious chronicler of his father's life. Richard's fellow directors at Fairey Aviation agreed with him that Wentworth Day was best qualified to take on the job and, in April 1957, they paid him 2,000 guineas to research and write a book of 80,000 words. Wentworth Day's proposal for 'Wings over the world, the official life of Sir Richard Fairey' was approved by the board and in due course accepted by the stately, if by now antique, publishing house of George G. Harrap.[10] A manuscript was submitted within the agreed timeframe, but by then directors had more important matters to attend to, namely the survival of their company. Few fretted over a delay in publication, and eventually Harrap concluded that the book was no longer commercially viable. With Richard Fairey dead and copyright now held by Westland, Wentworth Day saw himself unlikely to secure a fresh publisher, and his plans to draft a second manuscript were duly put on hold.[11] As for the original, it must be in Edinburgh, Harrap's last location before the company ceased trading in 1986, or in Wentworth Day's papers, should they have escaped the shredder. More likely the manuscript was discarded and destroyed at some point over the past half-century.

Early in the 1960s a serious-minded Hampstead artist and freelance writer with wartime service in the RAF and an interest in matters mechanical embarked upon a second biography of Sir Richard Fairey.[12] Peter Trippe was an acquaintance of the recently deceased Richard Fairey junior and of his widow and third wife, Atalanta. Through them he agreed to act as ghostwriter for Peter Twiss's account of smashing the world air speed record. The painter turned biographer's life of Sir Richard – 'The plane maker' – was never published because, quite frankly, it is unpublishable.[13] The surviving manuscript runs to hundreds of pages and would challenge even the most dedicated editor. Trippe's epic is wordy, verbose, opinionated, clumsily written, poorly structured, unfocused, on occasion factually incorrect, and too often reliant upon invented dialogue. Nevertheless, for all the omissions and misinterpretations, the writer has a story to tell and,

FOREWORD AND ACKNOWLEDGEMENTS

as the narrative unfolds, it draws upon multiple interviews and written submissions. Trippe talked to everyone still alive who had held a senior post at Fairey Aviation and therefore knew Sir Richard. In consequence, his final text is a gold mine of oral and written evidence, and as such invaluable in the writing of this book. The manuscript contains multiple rewrites and amendments, particularly when Trippe is commenting on CRF's troubled relationship with his first-born son. Thus, although the author's account of Sir Richard's life is unashamedly hagiographic, when family members read the first draft they clearly insisted on a major rewrite. Trippe endeavoured to address these concerns, but in the end his much amended manuscript never left Bossington.

Nearly 50 years later, following John Fairey's death, his sister Jane decided the time was right for a third biography of their father. The passage of time had seen extensive study by reputable scholars of the British aircraft industry before and after two world wars; crucially, a wealth of relevant documentary evidence was now in the public domain. The papers of Fairey Aviation had been donated by Westland Aircraft to the RAF Museum, and, appropriately, Sir Richard's personal papers were held by the Fleet Air Arm Museum. Both archives are well catalogued, with informative indexes accessible online. The availability of this material encouraged the late Jane Tennant (née Fairey) to initiate a fresh life of Sir Richard Fairey. I was invited to take on the task. My home university, Southampton, in partnership with the Fairey family and the Fleet Air Arm Museum, identified a clear and simple outcome for the project: the publication of an informative and suitably well researched biography.

Jane Tennant, her husband David Tennant, and her daughter and son-in-law Esther and Fran Fernandez-Llorente were extraordinarily helpful and hospitable, with visits to Pittleworth Manor in the Test Valley both productive and pleasurable. Working with Jane was very enjoyable, and a highly civilised experience. We never disagreed, not least as Jane displayed a healthy respect for academic integrity and independent judgement. She might correct a factual inaccuracy or offer an alternative viewpoint, but she would never insist on a fresh interpretation. Quite frankly, a biography of CRF is inconceivable without her valuable input. Jane's posthumous view of her father can be found in the affectionate and revealing portrait that forms an appendix to the main text. Further up the Test Valley from Pittleworth is Bossington, the house Sir Richard Fairey loved best, and his final resting place. Although busy running the Bossington estate, Sarah-Jane Fairey endeavoured to be helpful and, as a long-time resident of Lymington, I thank her for showing me an impressive array of action shots taken while her grandfather was racing his J-class yacht in the Solent. I am especially grateful to Charles Fairey, son of Richard, as his

FOREWORD AND ACKNOWLEDGEMENTS

ideas and suggestions at an early stage in the project helped shape my thinking regarding how to proceed.

At the University of Southampton, Anne Curry, Maria Heyward, Sarah Pearce, Mary Hammond and Mike Hammond were all incredibly supportive at a very difficult time. In fact, all colleagues in History and within the Faculty of Humanities as a whole were extremely kind, sensitive and thoughtful; particular mention should be made of Jane McDermid, Tony Kushner and Ray Monk, each of whom combined solicitude with helpful suggestions regarding specific aspects of the book. Southampton Business School's Roy Edwards read my initial account of Fairey Aviation's complex corporate history following World War I and advised accordingly. The head of the Hartley Library's special collections, Karen Robson, agreed to tranches of Fairey's personal papers being held in Southampton. Without the assistance of Karen and her fellow archivists, this book could not have been written. Staff in iSolutions, photocopying and all the other support services at Highfield and Avenue campuses displayed characteristic efficiency and good cheer whenever I sought their assistance – a special mention for the convenor of the two four-hour 'Building a website' training sessions (eight hours!) who realised after ten minutes that the oldest member of the class was already nine minutes behind everyone else; the class was one fewer for the second session and a Fairey website never materialised. A blog, however, was created, and my progress since the inception of the project could – and can – be traced via www.blog.soton.ac.uk/srfb. Beyond the University of Southampton, the person to whom I am most indebted is Barbara Gilbert, chief archivist at the Fleet Air Arm Museum. Barbara's expertise, and her agreement that boxes of Fairey papers could be held in the Hartley Library, negated the need for long and frequent journeys to Yeovilton. Without her agreement I could not have undertaken the research. Barbara even ferried the boxes backwards and forwards herself, which was well beyond the call of duty. She was characteristically helpful in the final stages of the project, seeking out suitable photographs in the Fleet Air Arm Museum's extensive archive – and thank you to the Museum for permission to use these images. Barbara's colleague Dave Morris read the manuscript to check the accuracy of technical detail and my thanks to him for this invaluable service. The Fleet Air Arm Museum is part of the National Museum of the Royal Navy, and various collaborations in recent years with present and former staff – not least Duncan Redford – have been both rewarding and enjoyable.

Working on the early history of Short Brothers and Fairey Aviation opened my eyes to the central role of the Royal Navy in the early history of British aviation. To my shame the Royal Naval Air Service scarcely featured in my biography of RFC/RAF air ace Edward 'Mick' Mannock; today, I would write

FOREWORD AND ACKNOWLEDGEMENTS

a markedly different book.[14] I am grateful to all those, from Oxford to Ypres, who have invited me to talk about aerial warfare 1914–18 and in particular the contribution of the Royal Naval Air Service. The successors of Sueter, Samson and their intrepid comrades invited me to speak at a Taranto Night dinner in November 2012, and my thanks to the Fleet Air Arm wardroom at HMS *Excellent*, Whale Island, for a truly memorable evening.

Complementing Sir Richard's personal papers at the Fleet Air Arm Museum are the archives of Fairey Aviation held at the RAF Museum, and again I am heavily indebted to Peter Elliott, Ross Mahoney, Bryan Legate and their colleagues at Hendon. Other archivists and librarians I must thank are Ian Gerrard (City and Guilds Institute), Andrea King (Ardingly College), Joanne Badrock (Harrow School), Megan Dwyne (National Archives and Records Administration, Washington), Doug Stimson (The Science Museum), C. Fitley (London Metropolitan Archives), Isabelle Chevalott (Guildhall Library), John Wells (Cambridge University Library) and Martin Stromberg (University of Illinois). For their help in providing invaluable information, further thanks are due to Les Hewett, Richard North (IEA), Bob Fisher and Martyn Kemp (RLymYC), Hannah Gay (Imperial College, London), June Barrow-Green (Open University), Mark Womald (Pembroke College, Cambridge), Ivan Rodionov (Warwick University), Matthew Willis (Naval Air History website), Chris Reeves (The Flydressers' Guild), Alastair Robjent (Robjent's Fine Country Pursuits) and Gabriel Gorodetsky (University of Tel Aviv). Charles Lawrence generously gave me a copy of his history of Fairey Marine, a true labour of love, and Gordon Currey kindly lent me the personal papers and photographic collection of his father Charles, managing director of Fairey Marine. The Robert Pooley Trust generously gave permission to use the portrait of Sir Richard Fairey on permanent loan to the Museum of Army Flying. I am grateful to Jane Wood for contacting me at a late stage regarding the identity of Peter Trippe, and for defying my scepticism over whether a north London *flâneur* was qualified to chronicle the life of Sir Richard. Thank you also to Lester Crook, Jo Godfrey and their colleagues at I.B.Tauris for overseeing the editing and production of what is my third book with their publishing house. Inevitably there are omissions from my list of those I should thank, and my apologies to anyone who provided assistance but is not mentioned here.

Particular mention must be made of fellow historians beyond the University of Southampton to whom I am indebted for their help and support: Matt Kelly (University of Northumbria), Frank Cogliano (University of Edinburgh), David Kynaston, Dilwyn Porter and Richard Holt (De Montfort University), Huw Richards (London Metropolitan University), Lord Hennessy (Queen Mary

FOREWORD AND ACKNOWLEDGEMENTS

University of London), Roy Foster (Hertford College, Oxford, now Queen Mary University of London), David Edgerton and Andrew Lambert (King's College London), Mark Connelly (University of Kent), Brad Beaven (University of Portsmouth) and Mark Harrison (Warwick University). Again, this list is by no means definitive and my apologies to anyone who should be mentioned but is a victim of my sexagenarian memory.

The research and writing of this book took considerably longer than I had anticipated. This was partly a consequence of various time-consuming commitments, both before and after retirement from the Faculty of Humanities in 2015. However, the principal reason for the project taking so long concerns my family. *The Man Who Built the Swordfish* would never have been completed were it not for the selflessness of my cousin, Maryrose Bodycote, in helping care for my mother, and for the support in difficult times of close friends in Lymington and further afield, notably Mike and Mary Hammond, Dave and Linda Read, Rob and Vickie Goffee, and Helen Henderson and Simon Fox; and above all, for the devotion of my wife, Mary, and our daughter-in-law, Georgia McDonald, across two dark and desperate years in 2015–16: for nearly two years of a terminal illness, Adam Smith displayed an astonishing degree of dignity and courage, always showing an interest in his father's research when he could justifiably have pleaded more pressing matters to address. For Mary, Georgia and I, Adam was our darling boy – and this book is dedicated to his memory with all my deepest love.

Acronyms

AASF	Advanced Air Striking Force
AEU	Amalgamated Engineering Union
AID	Aeronautical Inspection Division
AMLC	Army Motor Lorries Company
ASW	anti-submarine warfare
BAC	British Air Commission
BBC	British Broadcasting Corporation
BEA	British European Airways
BOAC	British Overseas Airways Corporation
BSC	British Sportsman's Club
BSC	British Supply Council
CAS	Chief of the Air Staff
CAT	Commonwealth Air Training Plan
CID	Committee of Imperial Defence
CRF	Charles Richard Fairey
DFC	Distinguished Flying Cross
DNAD	Director of the Naval Air Division
EPD	Excess Profits Duty
FAAM	Fleet Air Arm Museum
FD2	Fairey Delta 2
IAS	Institute of Aeronautical Sciences
IYRC	International Yacht Racing Union
JAC	Joint Aircraft Committee
MAEE	Marine Aircraft Experimental Establishment
MAP	Ministry of Aircraft Production
MBE	Most Excellent Order of the British Empire
NYYC	New York Yacht Club
R&D	Research and Development

ACRONYMS

RAF	Royal Air Force
RFC	Royal Flying Corps
RIIA	Royal Institute of International Affairs
RLYC	Royal London Yacht Club
RNAS	Royal Naval Air Service
RTYC	Royal Thames Yacht Club
RYS	Royal Yacht Squadron
SBAC	Society of British Aircraft Constructors
TIFA	Test and Itchen Fishing Association
TsAGI	Ventral State Aero-Hydrodynamic Institute
USAAC	US Army Air Corps
USAAF	US Army Air Forces
USN	US Navy

CHAPTER 1
Early Life

Prelude: contrasting family fortunes

In late 1931 the President of the Royal Aeronautical Society received a letter from one of the British Empire's more isolated outposts: Hagley, a small rural community in northern Tasmania. As soon as mail could be carried by aeroplane all the way from Australia to England, the Reverend Leslie Fairey became eager to correspond with distant relatives in the mother country. His first, and most obvious choice was the founder of Fairey Aviation, a company with an international reputation reaching deep into the southern hemisphere. Charles Richard Fairey, familiar to close friends and family as Dick, and known always by his second name, had in the course of the 1920s become one of Britain's best known plane makers.[1] His workforce knew their chairman and managing director as 'CRF', a man driven by the need to be number one, not least when competing with Hawker's Tommy Sopwith to provide a potent strike force for the RAF and the Fleet Air Arm. By November 1931 sales of machines such as the Firefly and the Fox had confirmed Fairey Aviation's global status. As evidence of the founder's overseas ambitions, a subsidiary company began production in Belgium around the time Reverend Fairey was putting pen to paper.

The Presbyterian preacher knew who he was addressing, as his local newspaper regularly reported sightings of the latest Fairey aircraft in the skies above Launceston. Equally newsworthy were Fairey's experimental monoplanes, intended to accelerate imperial communication. These futuristic-looking alternatives to flying boats pioneered non-stop transcontinental flights, with Australia the ultimate destination. Tasmania sat on the edge of the Empire, and yet, after a mere quarter century of manned flight, the prospect of travelling there in days, not months, was the realisable vision of aircraft designers 12,000 miles away. These were the Edwardian pioneers whose engineering and entrepreneurial skills had combined to facilitate their fledgling companies' rapid response to the unprecedented demands of the nation's war effort. The quality and quantity of aircraft produced in the course of World War I constituted a dramatic transformation in the fortunes of the British aircraft industry. The commercial nous of founding fathers like Fairey allowed a select number of enterprises to survive a crippling slump in aircraft sales after the Armistice and to grow again once fresh opportunities arose. In building up his business from its inception in 1915, Richard Fairey experienced his fair share of setbacks, albeit not in such testing physical conditions as cousin Leslie. He certainly knew all about cultivating cutting-edge technology in harsh and unfriendly environments: Fairey's career as an aircraft manufacturer had begun on the bleak foreshore of Sheppey, the island in the Thames estuary from which convicts once departed for Van Diemen's Land.

The considerable wealth acquired as a result of floating his company in 1929 had left the founder of Fairey Aviation largely unaffected by that year's sharp downturn in the economy. For CRF the years following the Great Crash were a period of adjustment, his doctor having advised a more leisurely lifestyle. In practice, competitive sailing generated the same response as running a company, and from the late 1920s Fairey aggressively raced his 12-metre yacht. The landlocked Reverend Fairey could never imagine that the majestic *Flica* marked an early indication of how one day the family name would be as familiar on the water as in the sky. Clearly, the Australian had little idea of the level of affluence enjoyed by Richard Fairey, but he took pride in the family's most industrious member achieving 'such a summit of success.' Nor could the minister envisage the size and scale of Richard Fairey's business interests at the start of the 1930s, as with charming naivety he insisted that, 'If I could be of any assistance to you in any way out here just let me know, and I should be glad to help.' This was an offer made by someone desperate to find work for his eldest boy. Australia in 1931 was a country ravaged by the worst effects of the Depression, and Tasmania as the smallest and poorest state was especially hard hit. Leslie Fairey's brothers had served at Gallipoli and in France, and his surviving sibling now struggled to make ends meet in the face of acute rural poverty. Without a hint of bitterness, Reverend Fairey contrasted his children's bleak prospects with the athletic and motoring achievements of CRF's son Richard, still at Harrow but following in his father's footsteps by learning to fly.

Of this short-lived correspondence only one letter – probably the only letter – from Richard Fairey survives. Typed, to the point, and with a tactless error in the recipient's initials, here is a polite yet revealing reply to Leslie's inaugural missive. CRF displays paternal pride in listing the unscholarly

Richard's accomplishments when away from the classroom, and corporate pride in pointing out that he heads one of the world's leading aviation companies, laying claim to 'the largest output of any.' For a man maintaining scant regard for organised religion, CRF's refusal to acknowledge Leslie Fairey's pastoral status is scarcely surprising. Richard had followed his father's example in refusing to uphold the Fairey family's staunchly nonconformist beliefs, and, although on occasion vaguely acknowledging a deity, he was for all intents and purposes an atheist.[2] Yet for three hundred years, since the yeomanry of East Anglia consolidated their Puritan credentials in the decades leading up to the English Civil War, Fairey kith and kin in and around St Neots had been both evangelical and evangelistic. Previously coach builders and parchment makers, by the Victorian era the Huntingdonshire family's income derived from drapery. Yet, every Sabbath, Richard's god-fearing grandfather set aside his curtain sales to preach in the chapel. The intensity and inflexibility of Charles Fairey's fundamentalist Christianity forced his son – the first Richard Fairey – from mild disenchantment to total disbelief.[3]

Seventy years later, on the far side of the world, Leslie Fairey maintained the family's evangelical mission. Day after day he rode a remarkably robust Ariel motorcycle the length and breadth of his sprawling parish. Yet, with interwar Tasmania home to scarcely 13,000 Presbyterians and Hagley's total population fewer than 200, Reverend Fairey's congregation on a Sunday must have been tiny.[4] The minister's strict Calvinism encouraged him to believe that he was one of God's chosen few, but the irony of such a determinist dogma is that mere mortals are never certain of their place in the afterlife. Richard Fairey, like his similarly apostate father, focused solely upon life on earth, firmly believing in the achievement of material success as a consequence of the individual maintaining a firm grip upon his or her personal destiny. In breaking free of genteel poverty at the age of fifteen and through relentless hard work creating 'one of the leading Aviation companies in the world', Fairey applauded the triumph of human endeavour, not the benevolence of a supreme being. It is scarcely surprising, therefore, that he celebrated being born at the height of a second industrial revolution. Scientific discovery and philosophical inquiry were seen as challenging all the old orthodoxies, and anticipating an era of material progress and personal discovery: 'Darwin, Huxley, Spencer and many others had turned upside down the convictions of centuries and released human minds to a new perspective of past and future.'[5] What for a captain of industry signalled liberation, for his Bible-believing distant cousin constituted the triumph of the godless. Yet both men, living out their lives at opposite ends of the newly designated British Empire and Commonwealth, were in their very different fashions emblematic of empire. Their contrasting fortunes, and their coming together in late 1931 at a moment of severe economic and political crisis for Britain and its dominions, demonstrates how the life of Sir Richard Fairey can be seen as an imperial story: it began in May 1887 with the Queen-Empress in celebration of her golden jubilee and in timely fashion ended on 30 September 1956 with the Suez crisis moving inexorably towards its humiliating climax.

The rise and fall, rise and fall, of Richard Fairey senior

The first Richard Fairey must have been a great disappointment to his father. He scarcely disguised his indifference to organised religion and displayed no interest in drapery. In the mid-nineteenth century, St Neots was a small, quietly prosperous community scarcely affected by the rapid urbanisation of Victorian England until the arrival of the railway. Here was a market town where life seemed timeless, with successive generations' daily routines still moulded by the surrounding fenland's seasonal cycle. This was no rural idyll, but town and country were bound together by a common culture and tradition, rooted firmly in the unique environment of the fens. Although his grandfather died when he was only four, CRF would in later life recall with affection staying at the family home in Eaton Socon, with his Aunt Ethel singled out as a warm and affectionate woman who was especially supportive after Richard senior died. Today a suburb of St Neots, Eaton Socon was at that time a small, self-reliant village where men and women still worked the land – and the water. Here was an ideal location for a growing lad to learn the finer points of fowling, angling and traversing the 'waterland' by punt or pattens.[6] Family holidays in Eaton Socon gave Dick Fairey a broad knowledge of, a residual respect for, and a deep pride in the English countryside, its rural traditions and its people. Given the second Richard's lifelong love of shooting, fly fishing and sailing – and in later life his commitment to preservation of the wild – here was a clear case of the child being father to the man.[7] Richard Fairey senior rarely displayed his son's affection for a rural way of life, from which he had escaped at the earliest opportunity. Yet Fairey's childhood outside the classroom and chapel generated the resilience and resourcefulness necessary to survive a rollercoaster career very different from the secure, uneventful and yet wholly unfulfilling life he would have led as head of the family business.

Like his son, the first Richard Fairey was a big man with big ambitions. Tall and broadly built, he must have stood out among the diminutive clerks of Morgan, Gellibrand and Co., a City-based timber company with interests in broking and a suitably Dickensian moniker.[8] This is certainly a story with echoes of *David Copperfield* and *Little Dorrit*, but George Gissing seems a more suitable chronicler of the older Fairey's changing fortunes. As a timber merchant wheeler-dealing within the biggest market in the world, Fairey earned the healthy respect of his employers. To be warmly appreciated and suitably compensated was gratifying, but greater reward lay in establishing one's own business. By 1892 Fairey was a married man in his mid-thirties supporting a devoted wife, three young children, a large house in north London and an array of domestic staff. His outgoings must have been substantial, not least as he incurred sizeable expenses serving as a borough councillor.[9] Yet there seemed scant prospect of further advancement at Morgan, Gellibrand, a company with only two directors. However, having worked in the commodity markets for so long the entrepreneurial-minded Fairey could draw upon a network of contacts within the City and the docks, as well as contractors across the Empire, most notably in eastern Canada. With the global recession easing and the economy at home again growing rapidly, Fairey judged the time was right to set up his own company. His incentive was access to extensive timber concessions along the Miramichi River in eastern New Brunswick. He knew the province well, having visited there as a young man with his two brothers. James Fairey ultimately crossed the border and headed for the west coast, but Benjamin remained with Richard until his return to London, after which he moved to Manitoba. Writing to the Reverend Fairey in 1931, CRF speculated that his grandmother and aunt had also spent time in New Brunswick, where their house had been destroyed by the famous fire that engulfed the port city of St John on 20 June 1877.[10]

It is conceivable that widow and supportive daughter might have been resident in the province at the time. However, in an era when journeys across the Atlantic for those of modest income were normally one-way, Canada does seem a very distant destination if the intention was always to return home. For an elderly lady to leave the comfort and security of St Neots in order to spend time with her offspring in rural Canada suggests the same spirit of adventure that saw her grandson take to the skies four decades later. The same applies to Frances Jackson who, after marrying Richard Fairey, similarly swapped East Anglia for New Brunswick. Convinced she would never survive a second winter in Canada, the young Mrs Fairey was relieved to learn from her husband that his career necessitated an early return to England.[11] As Richard progressed through the ranks of Morgan, Gellibrand, Frances Jackson found herself setting up home in a succession of increasingly salubrious residences, until finally the couple settled in the suburban splendour of 'Ray House', a striking double-bayed villa in Hendon. Domestic bliss on Finchley Lane was very different from keeping house on the banks of the Miramichi. Thus, Frances was delighted to discover that her ambitious husband's New Brunswick enterprise would be run from offices in Bishopsgate, at the heart of the City.

Unfortunately, by remaining in London, Richard Fairey was dependent upon colleagues in Canada, one of whom acquired a substantial financial stake in the new company. Reliance upon an overseas business partner was bound to entail risk, but for six years the Canadian operation ran smoothly. The volume of trade grew, as did the profit margin. With steam succeeding sail, the process of importing timber had speeded up. Wholesalers inspected samples from North America or the Baltic on the dockside, before suppliers like Fairey processed the orders and orchestrated delivery. This historic system of supply was built on trust, and Richard Fairey enjoyed a reputation for being both efficient and honest. However, in the early months of 1898, run-of-the-mill problems with delivery became serious delays, and confidence in Fairey's company appeared on the verge of collapse. The final blow came with a major order's failure to arrive. A friend just back from North America warned Fairey that to save his firm he should catch the next steamship bound for St John. Richard left that same night, but before embarking he hastened home to warn Frances that the family faced financial ruin. Upon arrival in New Brunswick, Fairey's worst fears were fulfilled, as his eldest daughter later recalled: 'When he got to Canada his partner had gone. He had gone to the Argentine and had taken every asset. He hadn't only not delivered, but he had cut the timber and sold it. There was no extradition and they couldn't do a thing. My father was absolutely – well, everything was gone.'[12]

The business had collapsed, but Fairey refused to declare himself bankrupt. The impact upon his family was instant and brutal – both house and servants were gone in days, not weeks. Yet remarkably, by July 1898 all debts had been cleared. Fairey's success in paying off his creditors so speedily reinforced the residual goodwill felt towards him in the City and within the trade. Ironically, by honouring his debts Richard Fairey confirmed for investors that he was a safe risk, enabling him to raise fresh capital. That summer he set up a second timber-importing business in Bishopsgate, close to the offices of his former employer in St Helen's Place. The omens were good, and the family confidently anticipated a rapid restoration of their fortunes – if not a return to Ray House, then rebuilding their lives in a north London home of similar size and position. Sadly, a triumphant but chastened *paterfamilias* became seriously ill. Nursed by the ever faithful Frances and with his children packed off to a family friend in Huntingdonshire, Richard Fairey senior died on 10 September 1898. The cause of death was tubercular laryngitis, a terminal condition which at that time was identifiable only in its late stages. He was 41, and his widow would outlive him by almost a quarter of a century. The return to Hendon was posthumous as Fairey was buried in the graveyard of the parish church; among the mourners were his two eldest children, bemused, bereft and justifiably apprehensive.[13] Still a village at the turn of the century, Hendon would soon be home to England's earliest 'aerodrome' and thus a place of pilgrimage for the adolescent Dick Fairey. Sir Richard's close connection with north London's least loved suburb continued after his death: his father lies buried in the local cemetery, and his company's archives reside in the RAF Museum.

Life with and without Mr Fairey

In early 1938 the BBC's Home Service broadcast C.R. Fairey's account of his early years in aviation. Wireless listeners learnt of the speaker's 'somewhat pampered childhood as the only son in a very large late Victorian family and, believe me, that is a very soft job.'[14] This recollection was not strictly true in that Dick Fairey had a younger brother, Frederick, who was born in 1897 but died of whooping cough at the age of four. Whether the juvenile CRF was pampered is open to debate, but for his first 11 years he undoubtedly enjoyed the creature comforts of domestic life in an aspirational middle-class household. As well as two younger siblings – Margery and Phyllis – he had an older sister, Frances Ethel. Born two years before Charles Richard, 'Effie' Fairey was the girl closest to him in age. Both children slept in the nursery, and as toddlers they forged a lifelong bond. In terms of who was the dominant partner, age made no difference, with the juvenile Dick dogged and determined in getting what he wanted, if necessary at Effie's expense. Big for his age and clearly bright, here was a younger brother who invariably took the lead in an assortment of infant escapades, setting a pattern for later years when the teenage model maker assumed Effie's readiness to fulfil the role of technical assistant.[15]

Needless to say, Dick's enrolment at Hendon Preparatory School in the spring of 1893 scarcely affected his affection for Effie: he found new friends to play with, but not at the expense of his big sister. It was at prep school that the fascination with flight first became obvious, witness young Richard constructing ever more elaborate paper darts and bringing his classmates home to mess about with wood, string and glue.[16] He clearly stood out in the classroom, his scholarly endeavours being rewarded each speech day with a suitably worthy tome. In mathematics Dick was always top of the class, an aptitude for figures attributed to his enthusiasm being fired at a very early age by an inspirational teacher: Mrs Curry maintained a dame school near the Faireys' home and she fired her young protégé's imagination whenever they tackled sums together. The dame school left a lasting impression and CRF never forgot its mistress. His gratitude was such that later in life he bought Mrs Curry a house, where she lived until her death aged 90 years.[17]

In the autumn of 1897 Richard Fairey junior became a pupil at Merchant Taylors' School, with the family judging him well suited to join the Royal Navy. Naval training was at that time in a state of flux, with Osborne College not opening until 1903 and the Royal Naval College's shore establishment dating from 1905. Thus, training for a commission at the turn of the last century still took place on board Dartmouth's two redundant cruisers. In such circumstances the future naval cadet would have gained immeasurably from extending his education at one of London's most prestigious day schools. Merchant Taylors' had moved out of the City to Clerkenwell 20 years earlier. With the Northern Line not reaching Hendon until the 1920s, the daily journey to and from school must have been tedious and time-consuming. Given Merchant Taylors' close proximity to Bishopsgate, Dick presumably shared his father's hansom cab.[18] Not that the daily commute lasted long, as the collapse of Mr Fairey's business meant he could no longer afford the fees. Having sold their home, by April 1898 the family had moved into 78 Goldsmith Avenue, a newly built terraced house in Acton, on the then furthest fringe of the capital. Late Victorian London's remorseless spread westwards was now well beyond Hammersmith, or even the White City, home to the Olympics a decade later. Acton was virgin territory, an especially dreary suburb where builders and land developers exploited loose planning regulations to maximise accommodation for the burgeoning lower middle classes. Rightly anticipating a trend towards fewer children, the speculator and architect who conceived of Goldsmith Avenue made no provision for a family of seven. Effie was still too young to leave home, but with help from relatives Dick could attend a minor public school less expensive than Merchant Taylors', but providing accommodation. After Richard senior died, a financial contribution from the Jackson and Fairey clans back in St Neots became imperative.[19] It transpired that to pay off lingering debts and relaunch the business, Richard Fairey had mortgaged his life insurance, scarcely anticipating that within months he would be dead. In the autumn of 1898 the Fairey family was for all intents and purposes destitute. The trauma of the past ten months had taken its toll on Frances's health, but the need to make money meant working long hours as a needlewoman. Once she turned 14, Effie took a job as the book-keeper in a local laundry, after which she taught in a convent school, before finally travelling to India as governess for a couple reluctant to leave their children at home. Frederick spent the second half of his short life in Wanstead Orphanage School, while the two youngest sisters found themselves in similar refuges for the bereaved children of the petite bourgeoisie. Such a dramatic change of fortune was both tragic and catastrophic, made that much worse by a mid-summer glimpse of happier times ahead. This implosion of domesticity placed the heaviest burden upon the two eldest children. After her father's death, Frances Ethel proved older than her years. With the passage of time she filled a vacuum. Sensitive to Mrs Fairey's feelings, and with Dick away at boarding school, Effie combined work at the laundry and later the school with ensuring her scattered family could always look on 78 Goldsmith Avenue as home. In a society obsessed with class and social hierarchy, impoverished members of the middle class experienced a daily struggle to maintain a veneer of non-proletarian respectability, hiding the harsh reality of deprivation and lost income from the outside world: keep your curtains clean, your doorstep scrubbed and your darning invisible. Frances Fairey and her eldest daughter proved adept at maintaining appearances, the tacit message to neighbours of being down but by no means out reinforced by the absence of young Richard at boarding school in Sussex. A discreet subsidy from cousins in Huntingdonshire enabled Dick to study at St Saviour's School – today's Ardingly College – until the age of 15, at which point 'I found that I had to fend for myself.'[20]

St Saviour's School

Richard Fairey arrived at St Saviour's in early September 1898 and stayed for four years. His father had yet to pass away, as Mr Fairey's occupation is duly recorded in the school ledger, along with other men of modest income eager for their sons to secure a higher rung on the ladder of material success and social advancement.[21] Enhanced prosperity was not a priority of the founder 40 years earlier, the Reverend Nathaniel Woodward seeing St Saviour's as 'a home where the sons of persons of very small means may be boarded and thoroughly educated and instructed in the subjects necessary for their station in life at an expense little more than the cost of food.' Not surprisingly, religious instruction and daily worship in the Anglo-Catholic tradition was deemed vital to the life of the school. Even in Fairey's day the headmaster and most of his staff were clergymen (somewhat surprisingly, the chaplain hailed from New Zealand). However, the curriculum placed a strong emphasis upon mathematics, both pure and applied, and this played to Dick's strengths; as did the presence of technical drawing on Reverend Woodward's prescribed timetable. While 'English and Latin

Grammar' were deemed core subjects and all boys were instructed in history and geography, notable by its absence was the study of a modern language. Clearly, 'the sons of persons of very small means' were not intended to travel abroad, or to converse with foreigners in anything but the Queen's English.[22]

Ironically, Dick's final year prize at St Saviour's was the published journal of someone eager to visit exotic locations and to interrogate the local inhabitants, the young Charles Darwin. An earlier volume bestowed on the high achiever was *Mafeking* by Major F.D. Baillie, no doubt chosen by the headmaster as suitably inspirational for any sixth former contemplating imperial service on the battlefield or the *veldt*, or indeed both.[23] One can only assume that the disproportionate amount of time devoted to the Almighty, and the minimal attention given to pure and applied science, prompted CRF's recollection of how 'a public school education was very poor equipment, at least in the technical sense, with which to face the world…' As if to illustrate the point, it was an intellectually myopic master at St Saviour's who punished Fairey for building a model aeroplane out of cane and brown paper, sneeringly dismissing his pupil's obsession with the possibility of manned flight: 'But the hiding failed to cure me, and the subject remained my chief obsession. I read eagerly anything on aviation that came my way.'[24]

Presumably a dutiful son, he wrote regularly to his mother and eldest sister, but between the ages of 11 and 15 his contact with home was limited. St Saviour's boys enjoyed just two holidays a year, totalling four weeks. Only in the summer of 1902 did Dick at last become a permanent resident in the Acton household. By this time he was no longer the man of the house, as his mother had recently remarried someone very different from her first husband. Richard Fairey had been tall, dark and whiskered, whereas Henry Hall was stocky, fair and clean shaven. The former had a lively personality and a determination to make his own way in the world, while the latter was a quiet man content to earn a decent salary as transport manager for Carter Paterson, one of the capital's best known haulage contractors. Henry Clayton Hall deserves rescuing from the 'condescension of posterity' because he was evidently a decent and generous person whose love for Frances extended to supporting her offspring in any way that he could. His new wife was by no means robust, physically or emotionally, her second son having died less than a year earlier. The surviving children still recalled fondly their home in Hendon and at first were bound to be wary of their stepfather. Yet Hall built up a close and loving relationship with his adopted family and sought to give the younger siblings a half-decent start in life. No-one was more grateful than Dick, whose apprenticeship and night-school tuition were subsidised by his mother's second husband.[25]

Henry Hall could in no way be considered a wealthy man, but he held a position of considerable responsibility and was thus suitably rewarded. Not only could he ensure his stepson secured a trade, but he also facilitated the family's move to more convenient and commodious accommodation. The return to north London finally came about at the start of 1905. As the address suggests, 27 Station Road was adjacent to a railway line and only a short walk from what today is Finchley Central. The slightly larger houses opposite backed onto the railway cutting and in due course a desire for greater living space would see the family cross the road. For the moment, however, no. 27 marked another small step towards restoring family fortunes: the semi-detached house was much roomier than its predecessor in Acton, and first-floor accommodation included a small side bedroom in which Dick could sleep, study and experiment.[26]

Apprentice and student

Given his fascination with science and technology, Richard Fairey could recognise the huge gaps in his education resulting from St Saviour's well-intentioned but hopelessly unbalanced curriculum. Whatever his acquired skills in computation, geometry and draughtsmanship, he had received only basic tuition in physics and mechanics. Thus, almost all knowledge of electricity and its industrial application had been acquired through initiative not instruction. A young man contemplating a career in engineering had to engage with the relevant theory, while at the same time absorbing the most recent advances in chemistry. The biological sciences so beloved of H.G. Wells, Dick's literary hero, could only be studied at leisure. A priority was to comprehend the nature of the technology driving successive waves of industrialisation, and for this Fairey had to immerse himself in the theoretical *and* the practical: 'He realised his lack of knowledge and was impatient to gain a practical training in order that he might play a full part in the great technical revolution that was taking place around.'[27]

In Edwardian Britain this necessitated 'getting a trade': undertaking a premium apprenticeship, which provided a comprehensive training on the job and a complementary technical education.[28] But this came at a cost, in the form of employer and college expenses being offset by only a nominal wage. This was where Henry Hall's financial support proved so vital to his stepson's eventual success. Equally supportive was Adrian Jones, managing director of the Jandus Electric Company. This was a factory of around 250 workers producing and installing arc lamps, a form of electrical illumination that has been obsolete for almost a century.[29] The company employed Dick Fairey as a premium apprentice and paid him the princely sum of five shillings a week. However, Jones waived his firm's entitlement to the regular fee for registration and training. It would be tempting to view working on the shop floor in Holloway as Sir Richard Fairey taking the first deliberate step towards fulfilling his destiny, and yet the man himself viewed his employment at Jandus with a grim pragmatism: 'My chief recollection of those times is a feeling of resentment at my changed circumstances and an overwhelming desire to return to my previous condition. This was the road back.'[30]

Fairey's half brother, Geoffrey, born to Mr and Mrs Hall in 1906, served later in life as a confidant to his role model, patron, employer and eventual co-director.[31] Despite a 19-year difference in age, the two men were always very close, their relationship rooted in mutual trust and a shared memory of the family's testing circumstances at the start of the new century. Thus, whenever Geoffrey Hall commented upon his mentor's early life, he spoke with authority, enjoying a unique insight into CRF's thinking. Looking back on the time after he left school, Fairey always remembered how he felt himself to be taking on the mantle of family provider. He was driven by a need for personal and professional fulfilment, but also by an undisguised desire to achieve material success, thereby ensuring his family could enjoy a standard of living far beyond that enjoyed at Ray House, let alone the cramped households of Acton and Finchley. This was ambition not arrogance, with young Richard in no way seeking to usurp his stepfather, who had after all rescued the Fairey family from near destitution. None knew this better than Dick, whose advancement depended so heavily upon Henry Hall's modest munificence.[32]

Fairey worked at Jandus for less than three years, but in that time he gained invaluable experience in the drawing office before moving on to that part of the factory where the arc lamps were tested prior to despatch or installation. For large contracts the company would take direct responsibility for wiring up a lighting system, and it is testimony to CRF's rapid acquisition of the necessary technical expertise that he was given responsibility for illumination of the docks and warehouses in the northwest port of Heysham. In early 1904 the Midland Railway Company was preparing to open Heysham Harbour as a ferry terminal for services to and from Belfast and the Isle of Man. Aged only 17, Fairey was surely nervous at the prospect of overseeing such a substantial project. Working so far from home must have been a huge culture shock as CRF had never ventured beyond the Ouse, let alone the Trent or the Humber. The industrial north of England in the Edwardian era was a starkly different environment from anything with which he was familiar. Furthermore, the Lancashire working man's dialect at the time would have been extremely difficult for a young lad from down south to understand. First encounters were no doubt rooted in mutual incomprehension, and Fairey's stay in Heysham must have been a deeply challenging and formative experience.[33]

In terms of exercising authority it probably helped that, given the average height of Britain's pre-war industrial proletariat, Fairey towered above his workmates by nearly a foot. Lanky and still to fill out, by 1904 Dick stood at over six feet five inches. Within a few years he would, by the standards of the day, be a very big man. Such a sizeable person could scarcely be ignored, and the future Sir Richard used his physical presence to powerful effect. Not surprisingly, in every group photograph he looms large, instantly attracting the viewer's attention. Even in youth he looks like a man who knows what he wants and intends to get it. Clearly, here was someone who did not suffer fools gladly. Photographs reveal him as someone who, like any aspirational young man in Edwardian England, dressed smartly, displaying the quiet confidence that comes with a well-cut suit, a stiff collar and a full pipe. Erstwhile biographer Peter Trippe displayed rare insight when he described the young Richard Fairey as, 'already rock hard, both physically and mentally. He needed to be. The streak of Victorian sentimentality, which never really left him, was kept rigidly under control. No sentimentalism was allowed where he planned to go – to the top. Money. Wealth. Power. These were his criteria. The point has to be made quite clear.'[34]

In February 1938, C.R. Fairey reminded Home Service listeners that 'workshop experience is less than half the education of an engineer. He needs, much more importantly, theoretical and technical training … In the evening schools of British cities a man may be trained at a negligible cost in every branch of applied science…' CRF delighted in having attended Finsbury Technical College, which, by the time he broadcast on the BBC, had been shut for over a decade. From 1893 to the mid-1920s, Finsbury cultivated a reputation for academic excellence, ironically paying the price when its staff strayed into areas of scientific inquiry which the high-flying Imperial College considered its preserve. Finsbury was the flagship college of the City and Guilds Institute for the Advancement of Technical Education, a vocational body founded by the City's liveried companies in 1878 and still thriving today. When Finsbury closed in 1926, rival technical colleges across London filled the gap, and one of the capital's most distinguished institutions was soon a distant memory for all but its appreciative alumni.[35] Finsbury before World War I boasted a faculty that any civic university would have been proud to employ. Well-qualified lecturers applying for a position at the college were inspired to do so by the eminent electrical engineer and mathematician Silvanus Thompson, a fellow of the Royal

Society. A lifelong Quaker whose reputation as a researcher was exceeded only by his achievements as an educationalist, in 1885 Thompson's enthusiasm for promoting technical education saw him swap a chair at University College Bristol for the dual post of principal and professor of physics at Finsbury Technical College. Although physicists would look first to Imperial College and then to University College as pre-war London's principal centres of research, Silvanus Thompson and colleagues like Raphael Meldola and later E.G. Coker boosted Finsbury's credentials as a hothouse of scientific investigation. The college's high point came in 1909 when Coker, the then professor of mechanical engineering, developed 'photoelasticity', a system of harnessing polarised light to identify stress patterns.[36]

Silvanus Thompson was a polymath, with a quite astonishing range of interests, but his industry as a textbook writer, biographer, consultant, educationalist, translator, campaigner and ubiquitous committee man reflected a strong streak of entrepreneurship. This insistence that scientific knowledge, however obscure, must in the final analysis be put to good use was a fundamental principle underpinning the college's curriculum and modus operandi. The same message was implicit and explicit in the ten or so lectures Thompson delivered each week to either day or evening students. When visitors to the college observed its similarity to a state-of-the-art factory, with shirt-sleeved students operating sophisticated machinery, the Principal interpreted their remarks as complimentary.[37] Thompson envisaged that, for British manufacturers, a vocational award from Finsbury Technical College would carry more weight than a university degree, and 'From this directness of purpose we have never swerved.' He had ambitious plans for a 'prescribed systematic three years' course of study', which with continuing education and workplace experience would promote a City and Guilds associate to a prestigious fellowship within five years of leaving Finsbury. This remarkable vision was never realised, but if widely adopted would have seen successive generations of highly skilled, non-graduate engineers across a broad spectrum of manufacturing industry more than match the invention and ingenuity of their German counterparts.[38]

Although the Principal Thompson still supervised students in the laboratory: as both a compelling lecturer and a demanding tutor he left a deep impression upon Dick Fairey, registered as an evening student from the autumn of 1902 to the spring of 1904. CRF claimed that he studied four subjects each term, necessitating him to attend Finsbury every night of the week. This explains why he completed his studies in two years, enabling him then to undertake work away from home, notably the illumination of Heysham harbour. In term time, Fairey followed a gruelling schedule, having only an hour after work to go home for his tea and then catch the cable tram to Finsbury. At college he would pursue his studies for up to three hours before at last making his way back to Finchley, snatching a few hours of precious sleep before daybreak and another hard day's graft at Jandus.[39]

In the circumstances it is scarcely surprising that Fairey's academic record was, based upon surviving records, mixed. Given the excellence of his later achievements it is easy to assume scholarly prowess from an early age, but this was scarcely the case. Dick Fairey was clearly a bright boy at school, witness the prizes for mathematics and science, but he was in no way exceptional – that came later, on the factory floor and not in the classroom. The poverty of his schooling at St Saviour's was cruelly exposed in the four constituent parts of the entrance examination he sat in September 1902 to secure a place at Finsbury: he scored a total of 110 out of a possible 400, with the ablest entrant achieving a remarkable 356. Nevertheless, CRF was one of the 114 candidates accepted onto the Electrical Engineering programme.[40] He clearly learnt fast, witness his second place on both the first year Physics and Electrotechnics courses. In his second year he studied Electric Technology under the great man himself, Thompson awarding him a mark of 62, which placed Fairey third in a class of twenty. This modest success compensated for mid-table rankings on that year's Mechanics courses, his overall performance brought down by a mark of 31 for the examination he sat in May 1904. The star students were W.P.C. Russell, with 92 for the exam, and W.P. Buchan, with a remarkable 98 for his lab work, and yet neither of them for all their scholarly credentials went on to found a world-famous aircraft company. Puzzlingly, Fairey's name does not appear in the 1902–3 and 1903–4 records for Mechanical Engineering, Practical Geometry and Mechanical Drawing.[41]

Strangely the name of C.R. Fairey does not appear on the relevant roll of matriculation, and yet in the summer of 1904 he completed the requisite number of courses for award of a certificate of association in Electrical Engineering. Fairey's hard-won status as a Fellow of the City and Guilds of London Institute qualified him for part-time employment by Middlesex County Council's Technical Education Committee.[42] It would have been almost impossible for him to undertake any municipal part-time teaching without a formal qualification. In his wireless talk, CRF recalled where he taught and, while mistaken in the name of one college ('Great North Roads'), his mention of Tottenham Polytechnic makes sense: the antecedent of today's Tottenham Centre was a vocational college located until 1936 at Grove House, on the south side of Tottenham Green. Given the Polytechnic's curriculum and its relative proximity to Finchley, where by 1906 Fairey was both living and working, his claim to have taught there seems credible. It would be extraordinary for such a highly respected public figure to describe to his listening audience a biographical detail which simply was not true. So if, with the encouragement and endorsement of Silvanus Thompson, Dick Fairey supplemented his income by teaching at night school, then hidden somewhere in the archives of Finsbury Technical College is documentary confirmation of how he acquired his final certificate.[43]

If Fairey encountered any problems with matriculation then the answer might lie in serious injuries sustained at some point in the mid-1900s. Quite often his children heard him talk of the long period he spent in hospital, but CRF never indicated precisely when this was. Geoffrey Hall had only the vaguest of recollections and was thus unable to pinpoint the occasion on which his half brother nearly died. It transpired that CRF had borrowed a motorcycle from his friend and fellow student at Finsbury, Clifford Milligan. Throughout his life Richard Fairey drove and rode at high speed, as one would expect of someone whose whole career consisted of making machinery perform to their optimum: here, after all, was a man who in the final months of his life took great pleasure in seeing an aircraft bearing his name be the first to fly faster than 1,000 mph. Half a century earlier the young man on a motorcycle hurtling down a narrow lane in north Kent had little experience of handling such a fragile machine. Furthermore, he wore no suitable clothing, let alone protection for his head. The front fork of the bike snapped, sending Fairey over the handlebars and into a flooded ditch. There he lay for an estimated 12 hours until spotted by a little girl travelling in her father's motor car. Dick was extraordinarily lucky not to have drowned and to have been spotted by the juvenile passenger of what remained a rare sight deep in the Garden of England. He was unconscious, but clearly carrying some form of identification, as following emergency care Fairey was taken by ambulance to Finchley Hospital.[44]

The patient remained in a coma for several days, with a suspected fractured skull and extensive facial injuries. Over the ensuing weeks Fairey's nose was reset and efforts were made to minimise scar tissue. This primitive skin graft proved surprisingly successful and it was only when he 'was tired or very hot, say after a strenuous day of tennis, you could see this white line round the side of his face.'[45] More serious was his amnesia, as when CRF finally regained consciousness 'his memory had entirely gone. Even the basic things ... he never could remember what happened to about six months before the accident. He told me that this was one of the most tiresome things of all. Because he had been doing a lot of studying in those six months and the entire fruits of all his studies had gone completely from his mind.'[46] John Fairey was talking to Peter Trippe, who in his speculative account of the accident locates the 'lost' six months long after CRF left Finsbury Technical College. Trippe may have been right, but the Fairey family could provide no evidence to support his assumption. While still a student, Dick Fairey was just old enough to ride a motorcycle, albeit with a high element of risk. The younger the rider the more reckless the riding and thus the graver the consequences should the machine let him down. Yes, Fairey eventually made a full recovery, but he himself recognised the motorcycle accident as a genuinely traumatic event, with profound consequences. Family members recalled the intensity of his renewed studies: '…he would have sat down and done it all over again and got it right … All his life he had that sense of application. Nothing would defeat him. He had tremendous powers of concentration…'[47]

Trippe believed Fairey's commitment to continuing education was on his own initiative.[48] While CRF clearly was the lower middle-class autodidact so familiar from the late Victorian and Edwardian era, there are nevertheless degrees of studying. If these events occurred sometime after finishing at Finsbury then spending every spare hour immersed in technical literature was scarcely conceivable given the time Fairey now devoted to building ever more sophisticated model aircraft. With a full-time job and night-school teaching, reading was surely rationed. Fairey needed to maximise the time he spent on a design project seen as the means by which one day he would work on the assembly of real aeroplanes. This suggests that in fact his serious accident occurred several years earlier, while he was still a student. If so, then it might explain puzzling gaps in Finsbury College's official record of Dick Fairey's progression to City and Guilds certification.

Silvanus Thompson's academy of applied science reinforced in the minds of its students a keen awareness that they were standing on the shoulders of giants – the Principal had been personally acquainted with several of the late Victorian era's most eminent scientists, in some cases sharing their experiments or chronicling their achievements.[49] Within a remarkably short space of time their inventions had become absorbed into middle-class lives – those stratas of an advanced industrial society which could afford the various appliances arising out of a revolution in energy provision. The generation of power on an unprecedented scale paralleled profound changes in telecommunications and transcontinental travel, hastening the emergence of a global economy and the credit and exchange mechanisms conducive to volume production and free trade. Relative to growth levels in Germany and the United States, Britain's economic recovery in the aftermath of the late nineteenth century's 'long depression' gave cause for concern. Furthermore, early setbacks in the Boer War encouraged the view of imperial pessimists such as Kipling and Joseph Chamberlain that the Empire must embrace radical measures if the mother country was to remain a truly great power.

Yet, despite elite soul-searching and deep foreboding, for young men and women like Dick Fairey there was a keen sense of optimism, buoyed by a belief that the momentum for scientific invention had scarcely waned since Faraday first revealed the full potential of electricity nearly a century before. The development of the internal combustion engine, and the transition from coal to diesel in powering turbines, reinforced the view that second, third and fourth waves of industrialisation were scarcely distinguishable: in a parallel with today's digital-based knowledge economy, this was not so much a paradigm shift as a permanent revolution. Like his father, Fairey recognised the power of science-based secularism's challenge to organised religion since publication of Darwin's *On the Origin of Species* in 1859. The veil had been lifted, and in a new age of technocratic enlightenment, the focus was on solving material problems in this life rather than suffering in silence ahead of the next. H.G. Wells could conjure up mildly disturbing apocalyptic scenarios, but conflagration on the scale of the Great War seemed scarcely conceivable. In 1938 Fairey recalled an era of innocence rooted in ignorance, where progress was relentless and for the benefit of all: 'we were on the threshold of a new world from which the curse of Adam had been lifted. The fear of the machine had not then been born.'[50]

Power station maintenance man and model-maker extraordinaire

No machine generated greater fear in the course of the last century than the aeroplane. Yet, in 1900, 'Aviation was the one great field of conquest left; and some were saying that the air would never be conquered.'[51] As we have seen, the schoolboy Dick Fairey keenly anticipated the advent of powered flight. Studying at Finsbury Technical College deepened his acquaintance with the abstract calculation and practical experimentation undertaken across the course of the past hundred years. While respectful of the Wright brothers' triumph at Kitty Hawk on 17 December 1903, he viewed their brief forays into the air as the consequence of earlier work in France and Britain to comprehend first principles of flight and forward thrust. Thus, pioneers of the previous century like Sir George Cayley, John Stringfellow and the recently deceased Percy Pilcher were not eccentric amateurs indifferent to scientific method. For Fairey, these men were the visionaries who provided a necessary foundation upon which Orville and Wilbur Wright could translate their long hours of experimental gliding into mechanised lift-off.⁵² Perhaps insouciance explains why his elder sister had no impression of the news from Ohio generating great excitement in the Fairey household, and his widow and children retained no memory of CRF recalling how he learnt that at last a machine had taken to the skies.⁵³ In a 1930 lecture to the Royal Aeronautical Society, Fairey traced what he saw as an evolutionary process heavily indebted to British invention. At the end of his talk the Wright brothers were damned with faint praise: although undoubtedly 'great pioneers', at Kitty Hawk they had simply 'put the finishing touch on the accomplishment of their predecessors.' Critically, Orville and Wilbur's machines were dependent for lateral stability on continuous control by the pilot. This inherent instability was a fundamental technological flaw, which Boer War veteran and aviation enthusiast John William Dunne set out to rectify. Before long, Charles Richard Fairey would find himself assisting Captain Dunne in his expensive endeavour to jump a generation and build an aeroplane that freed the pilot from the need to preserve balance.⁵⁴

Five years before the two men finally met, CRF's fledgling career as an engineer received a setback when the arrogance of youth left him jobless. As he himself put it, early promotion at Jandus, and the subsequent realisation that he could take on a complex project like Heysham Harbour, gave him 'a swollen head and a few months afterwards the sack.' Luckily, before he had worked out his notice, a superior post arose, conveniently close to home. Finchley Power Station was the other side of the railway line from Station Road, and Dick Fairey was appointed as the manager's assistant. For the foreseeable future he was a public servant, as an employee of Finchley Urban District Council. Fairey assumed a position of responsibility that belied his years. His flexibility and scientific competence were confirmed by the ease with which he fulfilled the duties of an analytical chemist. Monitoring the quality of boiler feed water and of coal or coke, and analysing the purity of flue gas, together constituted a time-consuming operation, but it was in no way intellectually testing. Fairey quickly got to grips with the job, relying upon a bicycle to speed up the process of regularly touring the plant. He thus maximised the time available for alternative activities, whether tutorial work, reading with the intensity of a lifelong autodidact, or building ever more elaborate model aircraft.⁵⁵

Effective time management only went so far. CRF's tardiness compounded the tension between a traditionalist manager of advancing years and a young buck full of new ideas and with a relaxed view towards punctuality. The manager thoroughly disapproved of Fairey's dependence on a bike to minimise the time spent on what were routine tasks for someone with such a low boredom threshold. Admonishment short of dismissal was not uncommon, but Fairey's boss never worked out how, from the winter of 1909–10, someone with such poor timekeeping now arrived at work ahead of him. The answer lay in the Hall/Fairey family having moved across the road into a slightly larger house: 52 Station Road boasted an entrance hall, a front parlour and a living room with French windows which opened out onto a narrow garden leading directly down to the railway line. Each morning Dick's sister Effie kept a lookout from the bay window of the front bedroom, warning him when his boss had turned into Station Road. Fairey knew that it took the manager eight minutes from passing his house to reaching the power station, during which time he finished his breakfast and took the direct route to work via the back garden and the railway cutting. On occasion he was still in bed when Effie sounded the alarm, which suggests that, like many working men at the time, he shaved last thing at night and not in the morning.[56]

Late in life Effie recalled how her brother worked late into the night on the design and construction of model aeroplanes. The reason for his remaining in bed long after the alarm clock rang was because he was so tired. However long his working day, Fairey would labour until the early hours building elastic-powered aircraft so delicate and sophisticated that they rarely survived a handful of flights. His bedroom housed a production line, except that no two models were identical. Any available material was utilised for the construction of scaled-down aeroplanes too readily dismissed by the rest of the family as toys. Of course Mrs Hulme had the benefit of knowing what Dick went on to achieve, but she presented herself as her sibling's dutiful acolyte, always on hand to defend his hobby, provide necessary scraps of fabric (and even strands of her hair), retrieve the crumpled prototypes, and clean the bedroom/workshop ('I was the only person … because he said I never moved things.'). Fairey's room at 52 Station Road was large enough to accommodate a workbench as well as a bed. Purpose-built box shelves covered every wall except for the end, where he needed easy access to launch aeroplanes out of the window. Whereas previously Dick and Effie flew their models on open ground, after moving they could trial them in the back garden. Only when the trim had been suitably adjusted and the latest model was effortlessly gliding its way from the bedroom window to the edge of the railway cutting did they take their 'toy' to the park for its maiden powered flight.[57]

All this was a source of endless fascination to their half brother, now of an age that he could be entrusted with carrying the model aeroplanes back upstairs for further modification. Presumably young Geoffrey was not allowed near the power station's cooling pond when CRF tested his single-propeller float planes. There can have been few model-makers at the time prepared to risk their fragile, painstakingly constructed creations on water, and yet Fairey was unknowingly signalling where his future lay. Especially prescient was the use of asymmetrical floats made out of recycled tin. The larger of the two floats boasted greater buoyancy, thereby reducing the impact of the torque caused once the propeller began to rotate. The less the weight of the wing and airframe – and these were extremely light artefacts – the greater the capacity of the propeller to turn or twist the model as a whole. Without the torque effect of pushing down the opposite float, the aircraft remained on an even keel. This meant the suitably balanced float plane adhered to a straight line as the taut elastic band unwound, generating sufficient revolutions of the propeller to allow take-off from calm water.[58]

What Fairey's family failed to appreciate was that he never regarded himself as a model-maker per se. Building larger and more sophisticated model aircraft, and then entering them in fiercely contested competitions, was a means to an end, namely entry into the still small circle of enthusiasts who, spurred on by the Wright brothers' achievement, were pioneering powered flight this side of the Atlantic. CRF was still watching from the side line, although he did on occasion get his hands dirty helping out at Hendon aerodrome. Furthermore, he began to meet designers and engineers who would soon become well known, for example, Alliott Verdon Roe, who at Brooklands on 8 June 1908 became the first British aviator to fly an aeroplane built entirely in England.[59] A.V. Roe would soon establish a company in his own name, and this enterprise eventually became Avro. Similarly entrepreneurial was the polymath Frederick Handley Page, who had attended Finsbury Technical College as a day student when Dick Fairey was there in the evenings. It would appear that the two men never met. This was probably a good thing, as CRF's self-confidence would have suffered from an encounter with Handley Page, a larger than life and extraordinarily well-read young man with big ideas for building big aeroplanes.[60]

Richard Fairey's career since leaving St Saviour's had enabled him to consolidate his practical skills while at the same time securing a transferable technical expertise: the physics and mechanics studied at Finsbury Technical College or absorbed in late-night reading at home provided a sound theoretical base for applying the theory of flight to sophisticated aircraft design. The sophistication lay in adhering to the Wright brothers' principle of a forward elevator – a type commonly known as the canard – while at the same time enhancing performance through the combined thrust of not one but two 'pusher' propellers. Most model builders preferred a single engine, in order to minimise the effect of torque on airframes, which were ultra-light and susceptible to roll. Fairey's solution was to have the propellers rotating in opposite directions, so that the torque effect on one side was counter-balanced by that on the other. The principle of using contra-rotating propellers to negate torque dated back 50 years, but Fairey was the first to apply it to model aircraft. Similarly, his familiarity with the latest aerodynamic research ensured the adoption of a simple wing design that allowed air to flow uninterrupted across a wide flat surface.[61]

CRF had a strong preference at this time for monoplanes, particularly when building pure gliders. Trim was crucial for sustained gliding, with powered models intended to descend from a considerable height once the heavy-duty elastic bands had ceased unwinding and the propellers stopped turning: ideally the aircraft would land gracefully on its skis, with the centre of gravity such that the airframe remained parallel to the ground. Just as the idea of asymmetrical floats was revived in the 1920s when record-breaking seaplanes experienced excessive torque on take-off, so the principle of contra-operating propellers was resurrected when Fairey Aviation developed its P24 dual power unit from the mid-1930s: using a common crankcase but opposite-rotating crankshafts, two separate 12-cylinder supercharged engines each drove a three-blade co-axial propeller. A decade later, the Fairey Spearfish, a carrier strike aircraft that never entered into production, was a variant on the same principle; as was the company's last front-line aircraft for the Fleet Air Arm, the turboprop Gannet.[62]

Richard Fairey junior had a lifelong fascination with propellers, from dashing off exquisite designs of technical precision and aesthetic perfection through to carving and moulding a simple sculpture of minimum weight and maximum robustness. Similarly, he could construct an airframe of great delicateness yet remarkable tensile strength. Dick Fairey was always a dab hand with the glue brush, partly through the need to minimise weight but primarily because the contents of his tool box did not extend to a soldering iron. Later, when building real aircraft on the Isle of Sheppey, CRF would make his name as a rigger. Thus the adroit use of piano wire to strengthen his model aeroplanes clearly stood him in good stead.[63] The experience he acquired building ultra-light, high-performance models was quite literally a microcosm of his future career. Fairey still had much to learn, but he clearly had the requisite skills to blossom as a plane maker. As his half brother later acknowledged, 'Right from the outset there's this wonderful perception. And right from those early days, he [CRF] had an eye for good structures, simple structures and proper stressing.'[64]

Fairey's long hours at the workbench and in the park perfecting his model were duly rewarded in the spring and summer of 1910. It is hard to imagine the level of interest in competitive flying at the time, but pre-war Britain enjoyed a boom in model-makers endeavouring to produce the micro machine that could soar the highest, fly the farthest and adhere most strictly to its pre-determined line of flight. The technology employed in building real aircraft was still so simple that, other than the power unit, it could be replicated in miniature form. Indeed, in the hands of an uber-inventor such as Charles Richard Fairey, ideas employed in constructing a model could be adopted for a full-size equivalent. In May 1910 CRF entered what, despite its south London location, was the equivalent of a national championship: the Model Aeroplane Competition, held at Crystal Palace. The omens for his entry performing well in this prestigious contest were not good, given that the model's final test flight had ended in a cross-wind smashing it into a tree. Grim-faced but determined, Dick carried the wreckage home and, with Effie's assistance, spent all night building a replacement. Early the next morning, having had no time to test the substitute model, Fairey set off alone for the competition, leaving his exhausted sister to get ready for work. Mrs Hall and Geoffrey were alone all day in the house, the monotony of daily domestic life interrupted late in the afternoon by delivery of a telegram. It read simply: 'HAVE WON OUTRIGHT.' This was a stunning and comprehensive triumph in the shadow of Sir Joseph Paxton's great glass edifice. Such an inventive architect would surely have applauded the complete victory of Fairey's demonstrably innovative design: not only did his model aeroplane win the Challenge Cup outright, but it received a gold medal for steering the straightest line and for the stability necessary to fly further than any other entry. This was no single success as the same design was similarly rewarded at a meeting of the Kite and Model Aeroplane Association on 4 June 1910. Fairey then went on to win a silver cup for best model at the summer contest of the Aero Models Association on 13 August.[65]

The emergence of an engineer *and* an entrepreneur: thus the future beckons

For someone with such a strong business instinct, Fairey must have sensed immediately that the victory at Crystal Palace had a clear commercial implication. As with kite sales, the purchase of model aircraft by both children and grown-ups reflected a nation fascinated with flight. The largest and best-known store selling model kits was Gamage's on High Holborn. A.W. Gamage & Co. complemented selling directly to its London customers with a mail-order catalogue that was required reading in every corner of the Empire. For Dick Fairey to persuade Mr Benjamin Varlars, manager of the nation's most famous toy shop, to sell his model in large numbers would be a major coup and a first clear step on the path to fame and fortune.

Fairey wrote to Varlars and received a speedy reply inviting him to bring his model into the store and provide a demonstration flight. According to family legend, CRF took a taxi from Finchley to Gamage's, telling the cabbie to wait while he met the store manager. His gamble was that Varlars would request an immediate demonstration in Hyde Park and when they arrived there offer to pay the taxi fare. With the meter running uninterrupted but not enough cash to cover the cost of the cab, Fairey could have been severely embarrassed. However, having calculated that his host would cover all necessary expenses, to his considerable relief he was proved right. Once inside the park, using his impressive height and strength, CRF hurled the assembled model into the air and was relieved to see it fly the necessary distance of 100 yards. Duly impressed by the display, Gamage's chief executive agreed terms on royalties of sales, paying around £300 to secure sole right of purchase. Fairey knew exactly how much he wanted, and, given his relative youth, had proved himself to be an uncompromising negotiator. Given that Varlars had made his new acquaintance such a generous offer, he felt justified in demanding confirmation of the model aircraft's attributes. Again, the 100 yards marker was passed and all was well. However, a cautious CRF felt that it was unlikely he could achieve a third flawless flight if asked. His companion did indeed request a further demonstration. With suitable calmness the quick thinking but in his way quietly ruthless young designer pretended to trip up when retrieving his creation. Fairey stretched out his hand as if to save himself from falling heavily on the ground, and in so doing, crushed the beautifully crafted model aircraft.[66]

A mortified Benjamin Varlars regretted that he might in any way be responsible for the accident, swiftly dismissing CRF's disingenuous offer to rebuild the prototype and stage a third demonstration. The manager hailed a cab back to Holborn, and the model-maker caught a bus up the Edgware Road. Once home in Finchley he began drafting an instruction manual for the assembly of 'Fairey's Miniature Aeroplane.' Nearly half a century later, Geoffrey Hall commented upon his half brother's cunning response to the request for a third flight: 'There was genuine pity on the part of the Manager but – the successful punter knows when to stop. Both men were gambling and neither lost. Fairey got his price and the store its profit. The little booklet of instructions bears witness today to the thoroughness with which Fairey kept his side of the bargain. The money earned by the down payment and licence fees was the material reward for years of hard work, but it would have to be put to good use.'[67]

The commercial arrangement with Gamage's proved highly remunerative, providing Fairey with the capital to draw upon when incurring necessary expenses in setting up his company five years later. He subcontracted construction and packaging of his production model to a small firm in Tooting called The Aerial Engineering Company. The latter therefore assumed responsibility for printing of the instructional leaflet. By early November 1910, an assembly line in Balham had already produced 250 boxes for delivery to Gamage's. The gross price for each box was 30 pence in decimal currency (six shillings at the time), and with the subcontractor keeping unit costs low there was considerable potential for profit at every level of the manufacturing and distributive process. Fairey's Miniature Aeroplane was, in the parlance of the day, a nice little earner.[68] J.W. Dunne would soon exercise a keen interest in a product that unintentionally drew upon his innovative approach to aircraft design, but this never translated into demanding a slice of the profit.[69]

The second half of 1910 was one of several critical moments in Sir Richard Fairey's life. He was still working as an assistant manager at a municipal power station, with a modicum of night-school teaching to bolster his weekly wage. Yet what family and friends had previously dismissed as a hobby verging on an obsession could now be seen as a lucrative side line – why not design and construct a second prototype and then find another receptive retailer? This of course was a myopic view of CRF's success in feeding his fascination with flight while at the same time earning a healthy profit. The commercial application of Fairey's forward-thinking design was certainly an achievement for one still relatively young, but it was simply a signpost for the future. Fairey rightly recognised his talent for translating new ideas into actuality *and* his ability to exploit a marketing opportunity. Here was a tyro designer with a shrewd eye for business, but the key element in Fairey's singularly tough mental make up was patience. Building an elastic-powered model aeroplane was one thing, but designing and assembling the wings and airframe of a full-size machine was a set of skills scarcely acquired overnight. The years of study and toil since leaving St Saviour's had left Dick Fairey with an impressive breadth of technical expertise, underpinned by a readiness to engage with fundamental principles of applied science. This was someone enthused by what might loosely be termed the new technology, and yet, although experimental powered flight in Britain had scarcely progressed beyond the embryonic stage, Fairey still had much to learn.

The ready student needed an enlightened patron prepared to let him learn on the job, while at the same time discovering whether ideas brilliantly realised in the bedroom and on the common remained valid when designing actual aeroplanes. Furthermore, that same student required a suitably intense environment in which he could engage with those of like mind and shared enthusiasm, both learning from and contributing to the debate surrounding how British engineers could generate critical mass, creating the basis for what might credibly be called an aircraft industry. From the spring of 1911 nothing would ever be the same again, even if at the start of the year Dick Fairey was unaware just how much his life was about to change. His patron would prove to be the most unlikely of mentors. Furthermore, the locale in which he would hone his craft was a corner of Kent made famous by Charles Dickens but scarcely conceivable as Edwardian England's hothouse of aerial invention.

CHAPTER 2

Aviation Apprenticeship

Prelude: an afternoon in Edwardian England – Eastchurch, 31 July 1911

The cradle of British aviation in the Edwardian era was an unlikely location, a consequence of chance not design. Only a short distance down-river from the capital, the north shore of the Isle of Sheppey appeared a convenient site from which to fly machines that drew directly upon the designs of Orville and Wilbur Wright. The reality, as the brothers would discover on a brief visit in May 1909, was a bleak, unforgiving marshland that fostered a sense of remoteness redolent of *Great Expectations*. Beyond the future seaside resort of Leysdown was Mussell Manor, an outpost of the fledgling Aero Club. This modest farmhouse stood on a road that seemed to go nowhere, other than into the sea, but in fact led to a factory boasting 80 or more employees. Here was surely the most isolated workforce across the south of England, an impression scarcely diminished by the now Royal Aero Club's move early in 1910 to the neighbouring village of Eastchurch, and fields more forgiving of the fragile prototypes flown by pioneers who, should they survive, were to shape the British aircraft industry across the ensuing half century. One such pioneer, aged 24 and employed at Eastchurch in the summer of 1911, was Charles Richard Fairey.

Weekdays and often weekends that summer would see Dick Fairey working in the sheds and hangars of Stonepitts Farm, today swallowed up by a sprawling prison complex. A grand distraction at the start of July 1911 was the Royal Aero Club's hosting of the James Gordon Bennett Cup, by its third year already a prestigious and hotly contested race. Astonishingly, as many as 10,000 spectators travelled to Sheppey. Once on the island they braved non-metal roads and a malfunctioning light railway to descend upon Eastchurch and witness the United States retrieve the trophy. Fairey was a bystander that weekend, distracted perhaps by news from San Francisco of a devastating earthquake and from Paris of the German gun-boat *Panther*'s provocative presence in the port of Agadir.

By the time CRF arrived for work on the last day of the month, a settlement of the second Moroccan crisis seemed likely. A palpable sense of relief across the Home Fleet may help explain that Monday's staging of a flying display, which had lasting consequences for all concerned. Fine weather on 31 July belied north Kent's reputation for being invariably cold, windy and drizzly, and it facilitated an excursion to Eastchurch by the family of Vice-Admiral Prince Louis Battenberg, who within 18 months would be appointed First Sea Lord. For Fairey, an ambitious young man whose company would be uniquely associated with the rise of naval air power in an era of two world wars, the key protagonists in his early career were present that day: the engineers and industrialists who cultivated a problem-solving mentality and an instinctive entrepreneurialism, and the Admiralty modernisers who fostered technological initiative and commercial innovation.

Prince Louis Battenberg, the Home Fleet's most senior admiral and Commander-in-Chief (C-in-C), The Nore, was fascinated by flying. His earlier visits had been formal affairs, inspecting the progress of the Royal Navy's first four trainee fliers, but on this occasion Victoria's son-in-law arrived with two of his children. Father, son and daughter all flew as passengers, each precariously sat astride the petrol tank behind the pilot with only the struts on which to hold. Princess Louise's box camera ensured a permanent record of all the flights, including those of her 11-year-old brother: in later life CRF did on occasion encounter Admiral Lord Mountbatten, but this marked his first sight of the then Prince Louis.[1] However high the risk, the late Queen's grandchildren had taken to the skies, their father publicly demonstrating his faith in the new technology.

That afternoon, Fairey saw visible confirmation of how seriously the Royal Navy viewed manned flight as a means of waging war. Furthermore, the aeroplanes flown that day proved to him that, however exciting the experimental project he was presently engaged upon, rival designers at Eastchurch were already able to translate fresh thinking into volume production. CRF was project supervisor for the mercurial J.W. Dunne. The first generation of Royal Naval Air Service (RNAS) pilots were trained by the Short brothers – Horace, Eustace and Oswald – who also built their aeroplanes. Dunne provided Richard Fairey with the opportunity to experiment and to learn, Short Brothers taught him how to assemble aircraft in large numbers, and the Admiralty encouraged him to set up his own company. All were formative influences and all warrant closer inspection in order to understand why, on that hot summer's afternoon in 1911, a young man could survey the feverish activity around him and envisage a time when he would be masterminding events, translating ambitious plans into reality. With war still a distant prospect, Fairey Aviation may only have been a vision, but it was a realisable vision.

John Dunne and Richard Fairey: a meeting of minds

J.W. Dunne arrived at the Eastchurch airfield, south of the village, late in 1909. The site had been gifted by Frank McClean, the unsung hero of aeronautical activity in the period immediately prior to the Great War.[2] It had been the future Sir Francis, owner of as many as 16 aeroplanes, who donated Mussell Manor and the adjoining Shellbeach to the Aero Club. Indirectly, McClean would prove an important influence upon Richard Fairey's future career through his patronage of Short Brothers and sponsorship of the Royal Navy's first trainee pilots. As a fellow Aero Club member he knew Dunne well, but this was scarcely surprising, given that the paucity of pioneer aviators in Britain enabled a rapid fusion of critical mass, with Sheppey in many respects the core. McClean, a one-time water engineer in India, and Dunne, a Boer War veteran twice invalided out of the Army, were ten years or so older than Fairey's generation of plane makers. Indeed at the age of 24, Fairey could be seen as a latecomer when he joined Dunne on Sheppey in the spring of 1911: Thomas Sopwith, a year younger, had already won the Baron de Forest prize for the longest flight from England to the Continent, taking off from Eastchurch the previous December. Albeit extraordinarily talented, Sopwith took full advantage of his comfortable background. Yet, unlike privileged enthusiasts such as Charles S. Rolls or John Moore-Brabazon, as a student he sought a solid grounding in applied science.[3] The first generation of aircraft manufacturers – not least those like Tommy Sopwith, Dick Fairey and Frederick Handley-Page, who more than met the unprecedented demands of 'industrial war' and then survived the rude shock of peacetime retrenchment – were young men reaching their creative peak at the very moment aeronautics accelerated away from the rudimentary technology that had lifted the Wright brothers off the ground in December 1903.

These were designers *and* entrepreneurs, very much products of a late Victorian middle class that placed a premium on manufacturing and on commerce. On the whole they eschewed a conventional university education, however impressive the faculties of engineering at Cambridge, Manchester, Glasgow or London. They were mechanical polymaths, stripping down and rebuilding cars and motorcycles before moving on to balloons and aeroplanes. Resisting the narrow specialism of the varsity graduate, the likes of Geoffrey de

Havilland or CRF looked to municipal technical colleges and polytechnics for a thorough grounding in all aspects of applied science and mathematics, not least mechanics. These men were comfortable with physics and unfazed by scientific theory, but by dint of training and direct experience were highly practical. Whether at the drawing board or on the shop floor they were quintessential problem-solvers, as adept with a torque wrench as a slide-rule. Ironically, none of this applied to 'John Willy' Dunne, which was why he needed Dick Fairey. Where Dunne, veteran of South Africa and son of a general, shared common ground with his younger colleagues was in forcing Whitehall policy-makers to appreciate fully the military potential of the aeroplane.[4] By the summer of 1911, only the Admiralty's most unreconstructed disciples of big ship orthodoxy remained hostile to aircraft and airships, but within the War Office too many generals grudgingly accepted the reconnaissance value of balloons while dismissing all other forms of flight. At the very highest level of decision-making, the Committee of Imperial Defence (CID) had reversed its earlier dismissal of state-sponsored research and development, but not before ordering the Balloon Factory at Farnborough to dispense with the services of 'Colonel' Samuel Cody and Lieutenant J.W. Dunne.[5]

Dunne had made too many enemies at Farnborough.[6] He left the soon-to-be-renamed Royal Aircraft Factory in 1909, a victim of the CID's original decision to accept a sub-committee recommendation that government funding focus upon non-rigid airships, leaving aircraft R&D to private enterprise. The rapid advances in aircraft technology across continental Europe and in the United States, plus growing anxiety over the threat from Germany at sea and in the sky, prompted a harsh reappraisal of Britain's capacity to defend itself. The Secretary of State for War, R.B. Haldane, overhauled a creaking administration at Farnborough and established an authoritative Advisory Committee for Aeronautics. It was a well-intentioned reform, but reversing earlier policy by placing so much emphasis upon government-sponsored R&D discouraged private investment and stifled corporate initiative.[7]

Furthermore, the Army failed to respond with the same energy and enterprise as the Royal Navy, its newly established Air Department eager to collaborate with inventive designers such as Sopwith and the Short brothers. Having relocated to Sheppey, Dunne sought Short Brothers' assistance in building an aircraft which would at last vindicate a theory of flight and control advanced for nigh on a decade. Demobilised and demoralised, Dunne appeared at first sight yesterday's man. The experimental flights of his gliders and powered aircraft on the Duke of Argyll's Blair Atholl estate in the summers of 1907 and 1908 had proved a costly failure. Dunne's machines suffered from a debilitating power–weight ratio as a consequence of his ambition to build a technically stable aircraft. He knew what he wanted, but he lacked the training and technical expertise to convert pure theory into practice. In consequence Dunne was heavily dependent upon his mechanics. Fortunately, in late 1910 he came across a student of flight with an insight into what he was seeking to achieve, one Richard Fairey.[8]

From adolescence Fairey greatly admired H.G. Wells, later adopting from *The History of Mr Polly* his inspirational family motto, 'If the world does not please you, you can change it.' CRF's library at Bossington boasted the complete works of H.G. Wells, a signed set of the limited Atlantic Edition including his wartime satire *Bealby*.[9] In this lightweight picaresque novel, generously dedicated to Haldane, Wells drew heavily upon J.W. Dunne for his hero: Captain Alan Douglas, a charming, fair-minded and highly unorthodox subaltern boasting 'a sinister strain of intelligence and inventiveness and lively curiosity', sacrifices the love of his life for an obsession with manned flight and its military potential. A thinly disguised Dunne is the intrepid inventor in several of Wells' short stories. The imaginative lieutenant's premonition of the tank inspired Wells' 'The Land Ironclads' and his vision of the future influenced Wells' prescient and apocalyptic pre-war novel *The War in the Air*. Keen to promote his protégé, Wells made a point of lobbying scientifically minded ministers such as Richard Haldane and munitions manufacturers such as Sir Hiram Maxim. All three men in their different ways supported the sickly Dunne, as did the array of aristocrats who invested around £4,000 in the company he set up at Eastchurch: the Blair Atholl Syndicate. Wells regularly inspected Dunne's prototypes at Farnborough and later on at Sheppey, although unsurprisingly his first flight came courtesy of a more successful rival. The novelist's continued belief in Dunne as the solitary genius who could transform the infant science of aeronautics meant that at some point in the spring and summer of 1911 CRF for the first time came face to face with his literary hero.[10]

Dunne attracted an abundance of capital and goodwill from his admirers because he articulated his ideas with such passion and conviction, building aeroplanes fundamentally different from machines indebted to the Wright brothers' original design. With crescent-shaped swept wings and no tails, Dunne's prototypes visibly embraced modernity. Their inventor sought to build stable machines that could endure trying conditions and maintain their equilibrium even when the pilot surrendered control in order to undertake other tasks, presumably of a military nature. The pitch of a Wright brothers' aircraft was exaggerated, with the pilot fully occupied because the forward elevator required constant adjustment in order to maintain level flight. Dunne's rejection of the Wright model was instinctive and dependent upon practical experiment rather than mathematical calculation.[11] Furthermore, his conceptual approach had to take account of mediocre power units, a perennial problem for British plane makers until wartime necessity saw the birth of an indigenous aero-engine industry.

The War Office's later reluctance to support expensive R&D was partly a consequence of the 1907 and 1908 Highland tests: at considerable cost to the Army a succession of gliders and powered machines had failed to take to the skies, calling into question Dunne's alternative mode of aircraft design. One consequence was that the ex-yeomanry man at last immersed himself in applied maths. His basic thesis was correct and by 1910 even the Wright brothers had rejected the concept of a front elevator. Yet Dunne remained insistent that swept-back wings accentuated stability in pitch, and he dismissed any suggestion that his designs were fundamentally flawed. He saw the costly excursions to Scotland as serving simply to confirm the inadequacy of his power units and the fragility of his aircraft. Future machines should be sufficiently rugged as to tolerate his unorthodox design and to support the increased horse power necessary for sustaining lift, thrust and direction.[12]

Eastchurch provided a suitable base for Dunne's fresh initiative, the D5. This machine, first flown by its inventor on 27 May 1910, appeared at first sight a very simple design, not that dissimilar from earlier tailless prototypes. However, closer inspection revealed wing-tips and side-curtains connecting the upper and lower wings. Independently operated elevons (trailing-edge flaps) at each end of the upper wing allowed the pilot to pitch and roll courtesy of two levers to his left and right; if the pilot let go of either lever an automatic locking device enabled level flight free from direct control, and sustained as a consequence of inherent stability. A system of chains and pulleys enabled a British-built 60-horsepower engine to drive twin seven-feet-diameter 'pusher' propellers mounted on outriggers behind the pilot. This arrangement, plus the large undercarriage required to support it, added to the weight, and compounded the problem of an aerodynamically unsound wing design. Nevertheless, the D5 did fly, and Dunne's later machines owed much to a prototype conceived at Farnborough but born on Sheppey. The aeroplane's real significance lies in who took his crude drawings and turned these latest ideas into a credible invention.[13]

Having relocated to Eastchurch, Dunne needed a qualified project manager and a contractor who together could realise the concept of 'inherent stability.' Short Brothers were, as we shall see, both convenient and reliable, but Dunne's priority was an engineer mechanic with the knowledge and self-confidence to liaise effectively between the contractor and himself. 'John Willy' gambled on Dick Fairey to revive his fortunes, and the revolutionary D5 biplane vindicated its owner's belief in a young man who for the first but by no means the last time demonstrated his capacity to drive through a complex and demanding project. A serendipitous meeting of the two men arose as a consequence of CRF's all-conquering model aircraft, from the autumn of 1910 available in commercial form via Gamage's. It transpired that Fairey's design infringed patents registered in the name of J.W. Dunne. CRF had allowed for varying incidence of the wing to boost lift, which was unusual given the governing principle that a wing would be symmetrical across its length and at a fixed angle to the parent fuselage. Furthermore the model's wing-tips were gently swept back in a manner familiar to any close observer of a full-size Dunne glider.

Although unaware of it, Fairey was thinking along similar lines to Dunne, who had taken the precaution of patenting his radical alternative to orthodox wing construction. Dunne took the infringement in the right spirit. When they met at the United Services Club he asked only that his approval be properly acknowledged in the packaging of Fairey's production model. The younger man readily agreed and a strong professional and personal bond was swiftly formed. Wells may have idealised Dunne in *Bealby*, and it is true that his harsher critics saw in him a deeply flawed personality, but he was no fool and he had an eye for talent. Fairey had scant experience of aircraft construction, but someone so intelligent, single-minded and quick to learn had obvious managerial skills which, when combined with his engineering expertise, rendered him an obvious candidate to realise Dunne's latest ambition – construction of the D5.[14] Dunne offered him a job, to the consternation of Dick Fairey's family.

CRF's first instinct was to seize the opportunity, but he had underestimated the fierce reaction of his mother and two youngest sisters. National Insurance had arrived only a few months earlier, but unemployment remained the great fear of the Edwardian lower middle class, the strata of a sharply differentiated society to which the Fairey/Hall family now belonged. For the industrial working class, loss of employment was a harsh and all too frequent reality. The absence of a regular income fostered a deep fear of penury and dependence upon charity, leaving only affluent workers of the 'labour aristocracy' to agonise over status: a proletarian priority was survival, not downward social mobility. This contrasted with members of England's burgeoning middle class, many of whom were deeply sensitive to their place and position within the local community. However, Frances Fairey's mixed fortunes over the past 15 years left her less fearful of a further fall in the family's suburban standing and far more concerned regarding Richard's readiness to surrender a secure job with prospects. Management of a municipal utility, however dull, was still some years off, but promotion when it finally came would provide a salary sufficient to support a comfortable married life, with a modest pension in old age. Meanwhile, in his present job, and still single, young Dick's weekly contribution significantly boosted the household budget, ensuring a quality of life not enjoyed for nearly two decades. Thus the family lamented a potential drop in their standard of living, even if an absent CRF set aside a proportion of the slender sum Dunne intended to pay him. Disapproving relatives confidently anticipated a second catastrophic collapse of fortune.[15]

Given the fledgling condition of the British aircraft industry in 1911 and Dunne's record to date, one can hardly blame Fairey's mother for maintaining that a move to Sheppey was both rash and short-sighted. They insisted that working for Dunne would be poorly paid and brief. An array of relatives was called upon by the errant engineer's mother to dissuade him from ruining his career. Even five-year-old Geoffrey sensed that, for all the familiar gloom of the Sabbath, this was no ordinary Sunday. In later life he recalled 'this semi-darkness, the aspidistra plant, lots of Victorian furniture with hardly room for anyone to stand, and a lot of people in the room, a lot of people talking ... Dick standing there and being lectured by everyone.' No-one present was prepared to see his point of view and his eldest sister was sorely missed: Effie's job as a governess kept her from pleading her brother's case, having 'always realised at the back of my mind that Dick had tremendous possibilities ... if he could break away and get the chance...' After what had been a weekend of endless wrangling, CRF took the heat out of the situation by announcing that he would go upstairs to reflect upon what had been said. In reality his mind was made up, and the following day he resigned his post as Finchley Power Station's analytical chemist. At this point, in the early months of 1911, CRF found himself with no immediate income and drawing heavily upon his savings. Money was already going out to the subcontractors responsible for the model aeroplane, with Gamage's yet to commence payment for the finished product. Furthermore, Fairey was forced to postpone commencing his new position as he was yet to regain his full strength after being struck down by peritonitis, still a serious condition today, but a century ago life-threatening. By the time he was fully fit, Dick Fairey had been forced to spend the £100 saved in the bank, but yet again he had demonstrated a remarkable robustness and speed of recovery. At last he set off from Hendon on a not unfamiliar journey to north Kent, leaving behind a family reconciled to the great adventure if still apprehensive as to its outcome. Sir Richard Fairey's career as an aircraft designer and manufacturer had belatedly begun.[16]

Fairey was no stranger to Eastchurch, his fascination with flying having prompted visits on a borrowed motorcycle or by train. The circuitous railway journey, from London Victoria, via Chatham, Sittingbourne and Sheerness, was only possible after the Sheppey Light Railway opened in 1901. An airfield and factory at the eastern end of the island would have been unthinkable without the rail link and the growing availability of motorcars. A motorcycle ride from north London to Eastchurch was a major – and exhausting – excursion, experienced only by the truly committed.[17] No wonder that once Fairey was resident on Sheppey he avoided brief trips up to London: the roads were poor and no doubt he feared a second serious accident if travelling by motorcycle.

Visits to Eastchurch across the previous 12 months had been solely to watch and to learn. However, his half brother was insistent that CRF and a friend once hijacked the machine of someone seen as singularly incapable of taking to the air: they succeeded in doing so before a rapid return to earth necessitated discrete emergency repairs. Fatalities were not uncommon among pioneer aviators, but Fairey would later insist that these early aeroplanes were not dangerous. Nevertheless, he was ultra-cautious in control of an aircraft, and realised early on that he was not a natural pilot. Nor, ironically, did he in later life enjoy being a passenger. This unhappy initial experience also fed a lingering sense of guilt (he convinced himself the patched-up propeller was acceptable given that it would never be tested aloft).[18] If one ignores this unfortunate incident, then prior to 1911 Fairey's only direct involvement with a full-size aircraft had been at the future Hendon aerodrome when he helped build a monoplane for Everett Edgcumbe, a family firm of instrument makers based in neighbouring Colindale.[19]

In his first year at Eastchurch, CRF would see through to completion no less than four prototypes, including Dunne's most famous machine, the D8, a much modified D5 which in 1913 proved capable of reaching Paris. Such an achievement was only possible because Fairey removed excess weight by a systematic identification of key stress points across each machine, which in turn ensured the precise deployment of struts and wire, to optimal effect. Intuitive 'stress men' were few and far between, and were thus much in demand. They were usually maths graduates, like Farnborough's 'brilliantly clever' Edward Busk, whose own experimentation in inherent stability was cut short by a fatal air crash. Nevertheless, Fairey put to good use his short time at Finsbury Technical College. He added to his basic knowledge by acquiring the published calculations of Harris Booth, another graduate of Cambridge's Mechanical Sciences Tripos, first taught in 1894. Booth was at that time engaged in theoretical work on stress at the National Physical Laboratory in south London. Like Fairey, he would progress to designing his own aircraft, but in his case on the Isle of Wight. CRF successfully exploded the ideas and measurements first employed in building model aircraft and applied the same thinking to a full-size machine capable of sustained flight. At this early stage of aeronautics a competent designer could successfully scale upwards given the commonality of principles, and nowhere more so than in the science of countering and minimising stress.[20]

It comes as no surprise therefore that Fairey rose to the challenge of fulfilling Dunne's dream, recalling his time at Eastchurch as 'the most blissful time of my life.' He always maintained that pre-war aircraft could never be matched for thrills and excitement. In 1938 CRF nostalgically recalled how 'The sensation was rather like tight-rope walking or balancing on the top of a flag pole. A little too far in any direction and over you went, but we were learning a lot and learning quickly.'[21] Dunne delegated an unusual degree of responsibility to his protégé, but Fairey was nevertheless serving an apprenticeship. He had much to learn and yet he needed to exercise firmness and confidence when overseeing construction of Dunne's aeroplanes. Above all, he had to exercise discretion when discussing technical matters with Dunne's prime contractor, Short Brothers.[22] Fairey's initial success is measured by the airworthy machines built for Dunne and by his recruitment as Short Brothers' chief stress man towards the end of 1912. In the course of the next 30 months Fairey would progress through the company to become works manager and chief engineer. Given the high calibre of Short Brothers' senior employees and the ambitious commissions the company undertook before and after the commencement of war, his rise was meteoric. No wonder CRF maintained that it was while working for the remarkable triumvirate of Horace, Eustace and Oswald Short, 'I learnt my trade.'[23]

Short Brothers: the ideal company for Richard Fairey to join

Mussell Manor is today Muswell Manor, surrounded by mobile homes but still able to recall former glories. Visitors may be surprised to find a modest exhibition in the bar and a memorial in front of the house. Yet most striking is the bronze statue that stands across the road from the main entrance, its three 'magnificent makers of flying machines' standing in line, their arms aloft in a bold life-affirming statement that anything is possible. The trio are of course Horace, Eustace and Oswald Short, the sculpture unveiled by a distant descendant in May 2013, four years after an even grander tribute to the brothers was installed on the edge of Eastchurch: a biplane-inspired sculpture marked 100 years since Moore-Brabazon had flown the Short No. 2 around a mile circuit, winning the *Daily Mail*'s £1,000 prize for being the first in Britain to do so.[24]

In addition, the Short brothers dominate Eastchurch's very grand aviation memorial, laid down in 1955 following a six-year fundraising campaign. Given the eminence and wealth of the patrons and subscribers, it is surprising how long the organising committee took to pay for what was admittedly an ambitious project. Sole surviving brother Oswald Short was a patron, the others being Lord Brabazon, Sir Francis McClean and Sir Winston Churchill. All three had of course been present at the birth of the British aircraft industry, as were many of the listed subscribers. Longevity was a marked feature of those pioneers who survived an era when fatal flights were all too common, although McClean died shortly after the art deco memorial's unveiling and Fairey just a year later. Brabazon and Short were both octogenarians when they passed away, while Sopwith was an astonishing 101 when he died. All these men lived well into the jet age and all saw their achievements applauded at Eastchurch, with the notable exception of CRF. The memorial booklet, published to mark completion of a monument that drew heavily upon Greek mythology, noted regretfully Sir Richard's absence from the roll of honour; no doubt words had been said.[25]

In 1955 Oswald Short remained a bitter man, having been forced during World War II to leave the company he co-founded. Short Brothers shared with Fairey Aviation the ignominy of the Ministry of Aircraft Production insisting upon a change of managing director, but the firm was unique in being state-owned from March 1943 until the end of the war.[26] Unlike Sir Richard Fairey, Oswald Short boasted no honours, other than his fellowship of the Royal Aeronautical Society, but he could claim pride of place on Eastchurch's memorial, alongside his siblings. Back at the turn of the century, Oswald and his younger brother Eustace had forged a reputation for themselves as first-class balloon-makers. Combining inventiveness and energy with entrepreneurial flair, and boosted by government contracts, they employed a growing number of craftsmen at their Battersea factory. By 1908 the quality of the balloons bought by wealthy members such as Moore-Brabazon, Rolls and Sopwith ensured a close working relationship between the Aero Club and the two brothers. Impressed by Wilbur Wright's demonstration flights in France, Oswald and Eustace were keen to convert tyro aviators' crude drawings into actual aeroplanes. Their credentials as constructors were obvious to all, but they lacked a theoretical base upon which to translate vague ideas into airworthy designs. A breakthrough came at Le Mans in November 1908 when Wright subcontracted the newly incorporated Short Brothers to build six of his demonstration machine, the Flyer. With the original designs unavailable, older brother Horace, by now a partner in the firm, travelled to Pau in February 1909. There he made detailed drawings of Wilbur Wright's aeroplane.[27]

By this time Horace was already at work on a glider for Rolls and a proto-Flyer for McClean, Short No. 1. He continued work on these two projects while overseeing the world's first aircraft assembly line at the recently opened factory on Sheppey. All six Short–Wright biplanes were built on budget and inside 12 months, but Aero Club customers such as Charles Rolls and Alec Ogilvie had to wait until French engines reached Leysdown in the second half of 1909. The most famous client was John Moore-Brabazon, securing his prize from the *Daily Mail* on 30 October 1909 before going on to claim a succession of records and rewards. Horace Short was crucial to the success of the contract, as Wilbur and Orville Wright confirmed on a visit to Mussell Manor in May 1909. The creation of Short Brothers as a bona fide family business, with an initial working capital of £600, had convinced Horace he should leave his secure and lucrative job with Charles Parsons, one of Tyneside's biggest employers.

As the inventor of the steam turbine and a primary contractor for the Royal Navy, Parsons headed one of the most strategically vital companies in Britain. He must have been loath to lose Horace Short, whose reputation as a visionary engineer had been forged across the Atlantic in South America. Short was in CRF's mind a genius, a mentor and 'the greatest engineer I have ever met', while the future Lord Brabazon relished the company of a 'remarkable man', with 'an extraordinary grasp of every scientific department of life.' Moore-Brabazon formed an unlikely partnership with Horace Short, both men recognising that the Flyer contract was simply a prelude to Short Brothers embracing the latest design technology. Only then could the company create machines that were genuinely original, exploiting to the full Horace's theoretical knowledge of flight and aerodynamics. The oldest Short brother needed more than most to display his intellectual brilliance. Horace suffered from the cephalic disorder known as brachycephaly, or more commonly flat head syndrome. The nature of his health challenge was scarcely understood at the time, with various explanations given for his disproportionately large skull – the most generous was that he had two brains and Short's unique capacity for problem-solving gave every reason for believing this to be true. Horace Short is rightly first in line on the plinth opposite Muswell Manor and to her credit the sculptor toned down his unusual cranial features.[28]

For a short period in 1909–10 Horace found himself working closely with Moore-Brabazon and Charles Rolls, the three men exploiting French aviators' readiness to adapt or alter the Wright brothers' original model. They saw design technology as evolving, albeit at speed. This approach contrasted with that of Dunne and Fairey, who were building machines fundamentally different from the prototype first flown at Kitty Hawk six years earlier. Thanks to Frank

McClean's generosity, in May 1910 Short Brothers joined the Royal Aero Club at Stamford Hill, south of Eastchurch. Now visitors could inspect both projects: Dunne's small team endeavouring to build a machine fundamentally different from anything flown to date and the three brothers' workforce busy constructing hybrid biplanes based on Horace's adaptation of French airframe and engine design. The first of these Farman-derived aeroplanes, Short Nos S.26 and S.27, flew in June 1910. Short Brothers' second series, produced in volume and subject to relentless modification and improvement, proved critical in sealing the company's reputation as a balloon *and* aircraft manufacturer, and in establishing the credibility of the RNAS inside the Admiralty. Regrettably, the earliest, still highly experimental versions resulted in Charles Rolls' death and Moore-Brabazon's consequent withdrawal from competitive flying. Overnight, Horace lost the two pilots whose technical input he most valued, even if adventurers like McClean and Ogilvie remained keen to test new ideas, and crucially, within a year the Sheppey factory would be benefiting enormously from the first generation of naval fliers. Nearly 40 years later, Lord Brabazon puzzled over why, given their closeness later in life, he had no memory of meeting Dick Fairey at Eastchurch. CRF was by then liaising with Short Brothers on Dunne's behalf, but his move to the company occurred 18 months after a shocked aeronaut had acceded to his wife's wishes and sold No. S.28 to Frank McClean. Only after the war did the Hon. J.T. Moore-Brabazon, Royal Flying Corps (RFC) veteran and now Tory MP for Chatham, belatedly encounter company chairman Richard Fairey, the two men forging a lifelong friendship.[29]

Of the series, Nos S.26–S.81, built by Short Brothers between 1910 and 1914, no two machines were identical. Basic airframe design was uniform, as was the 'pusher' configuration of propulsion: aircraft with power units mounted behind the pilot were known generically in Britain as 'Farmans' given the French model's pre-war ubiquity. 'Tractor' aircraft, with the propeller mounted in front of the pilot, experienced less drag, but the adoption of the nacelle, which by 1914 could carry a machine gun, saw Short Brothers find virtue in repeatedly reworking Henri Farman's original design. Just a cursory glance at company photographs reveals a radical progression from the first machines built through to the final variations on the eve of war. The latter conform far more to what we popularly assume an aircraft should look like, and it is not so surprising that subcontractors continued building No. 38 as a trainer until as late as 1916. The latter reflects a key early development, as provision for a passenger was followed quickly by a facility for dual controls. Slightly larger engines facilitated an improving power–weight ratio, with No. S.80 carrying as many as four passengers in November 1913. Four months later, McClean and Ogilvie proved the same machine's versatility with long-distance flights the length of the River Nile. This expedition was only possible because McClean, Short's de facto test pilot, had pioneered the use of twin floats in the spring and summer of 1912. On 10 August that year he hit the headlines by flying his seaplane through Tower Bridge. Landing and taking off from the Thames marked the climax of experiments Horace Short had initiated on the relatively calm waters of the Swale, which separates Sheppey from the north Kent shoreline: a rigorous inductive process, rooted in complex computation, had produced the necessary float construction and settings.[30]

Working so closely with pilots like Frank McClean and Cecil Grace left Horace Short bereft if any of his friends and colleagues were killed in a crash. As a result, Short endeavoured to build safer, more rugged aircraft, especially once he came to focus upon seaplanes. This was one reason why he was attracted to Richard Fairey, who when building a wing seemed to have a sixth sense as to what would and would not work.[31]

This was developmental progression, rooted in hard science. Yet the rate of progress remains astonishing, and here the derring-do of the aeronauts acted as a spur: Frank McClean's imagination fused with Horace Short's readiness to translate speculative notion into solid invention. Short Brothers, like their pre-war rivals, could only enjoy modest growth, as the absence of long production runs deterred serious capital investment. Every client's order was unique, and yet by working from a Farman template some economies of scale were secured. Horace's variations on a theme extended as far as a 'triple twin' and a 'tandem twin' biplane, the former driven by two propellers at the front and another to the rear.[32] Cecil Grace's loss over the Channel left Short haunted by inadequate engine capacity. His fantastical multi-engine machines highlighted domestic manufacturers' inability to secure anything greater than 80 horsepower. The failings of pre-war machine tool and aero-engine builders threatened the development in Britain of larger, high-performance, demonstrably modern aircraft.[33]

An adaptation which more than any other proclaimed the modernity of Short's machines was when McClean for the first time covered a fuselage with fabric and aluminium sheeting. The aircraft in question was one of a sequence of 'tractor' biplanes built in 1911–12 around the same time as two experimental monoplanes. The latter led nowhere, but the former saw civilian aviators for the first time work closely with naval personnel in reconfiguring an expensive recreational pursuit as a potent means of waging war. Shorts' tractor-biplanes provided the experimental base on which were built the Admiralty seaplanes flown throughout World War I.[34] In this respect they were in the vanguard of naval air power, and when CRF joined Short Brothers as chief stress man in late 1912, for the first time he found himself in direct contact with the Royal Navy. For all Sir Richard Fairey's dealings with the RAF across the years, it would be the Senior Service with which he would share the closest and most profitable – if not always the easiest – working relationship.

Richard Fairey at Short Brothers: witnessing the birth of naval air power

In 1909–10 both the CID and the Treasury gave the Royal Navy every encouragement to take to the skies. The CID displayed all the enthusiasm of the convert, and the Exchequer accepted the Admiralty argument that aerial reconnaissance was a cheap and reliable means of observing the enemy. An immediate result was Vickers' construction of the ill-fated No. 1 Rigid Naval Airship, popularly known as the 'Mayfly.' This disastrous project, heavily over budget and beset by problems of design and construction, ended with the airship's accidental break-up in September 1911, providing the Admiralty with a convenient reason to terminate its exclusive agreement with Vickers. Arguably the whole sorry story had been overtaken by events, in that by late 1910 the Royal Navy had ignored the CID subcommittee's concern for the safety of aeroplanes operating at sea and sanctioned a scheme to train pilots. Ironically, the officer who initiated the training programme also sought to save the 'Mayfly' project, as one would expect of someone freshly appointed as Inspecting Captain of Airships. The officer concerned was Captain Murray Sueter, whose expertise in naval ordnance, notably torpedo design, encouraged an early interest in the offensive potential of aircraft. Vain, temperamental and easily offended, Sueter was not an especially attractive individual. Yet he was energetic and enthusiastic about every project he undertook, hence his reluctance to abandon construction of rigid airships while at the same time insisting that Short Brothers could fulfil the clear potential of naval air power.[35] In this respect Sueter and Horace Short were visionaries, each exercising a considerable influence, whether direct or indirect, upon the early career of Richard Fairey.

If Murray Sueter's superiors needed confirmation of the aeroplane's capacity to observe and attack then it came courtesy of Cecil Grace and Claude Graham-Wright, who both flew courtesy calls above the Home Fleet in July 1910. Later that year Grace was the preferred instructor when Frank McClean offered the Admiralty two of his Short biplanes. Maclean had persuaded fellow members to sponsor a naval training programme for award of the Royal Aero Club pilot's certificate. Taken together, this was an extraordinarily generous gesture, with the

Club's chief instructor George Cockburn agreeing to replace Cecil Grace after his death in December 1910. As well as learning to fly, trainees would receive technical tuition from Horace Short and his senior mechanics. For this invaluable grounding in aeronautics the Admiralty paid Short Brothers a flat fee of £20 per head.[36]

The level of understanding required for the long, detailed and intellectually challenging lectures delivered by Horace Short in the first fortnight of the course was equivalent to the demands of a second-year physics degree. On the evening of their arrival the fledgling pilots could expect a lengthy treatise on air content and pressure, with particular reference to wind. The second lecture, contained in 36 pages of the speaker's dense prose, focused upon 'the primary problems involved when dealing with mechanical flight.' No doubt to the relief of the exhausted trainees there was a mid-week break before the Thursday lecture offered an astonishingly sophisticated discussion of aircraft design and propulsion, illustrating how far Short's ideas were in advance of what his workers were capable of building at the time. Access to all six of Horace Short's lectures – his notes at a crude estimate totalling around 100,000 words – offers a valuable insight into the degree of expertise expected of his closest aides.[37] Short Brothers would not have employed Richard Fairey unless he could easily digest its chief designer's heavily theoretical reflections upon the theory of flight and then transfer those ideas to the drawing board. CRF could readily comprehend the abstract, on the basis of which he approached complex design problems with an open mind.

More than 200 navy and marine officers stationed on the Medway put their names forward for six months leave of absence on full pay. Those chosen would all vindicate their selection before, during and after World War I. A mark of their commitment was that the four pilots qualified within six weeks of their arrival at Eastchurch on 16 March 1911. The senior officer was Lieutenant C.R. Sansom, whose subsequent career rendered him 'a key figure in the history of naval aviation. An ally of Sueter and protégé of Battenberg, Charles Sansom was a striking figure. He wore an expensively tailored uniform, studiously waxed his twisted moustache, and modelled his beard on that of Sir Francis Drake, and yet, for all his flamboyance, Sansom was no lightweight. He submitted weekly reports on a course which continued until the start of September, boasted a succession of inaugural test flights and duration records, and embraced visits to admire the latest advances in French engineering. The less colourful Arthur Longmore gained his certificate the same day as Sansom and in time would progress from jolly jaunts aloft with the Battenberg family to wartime command of the RAF in the desert. Churchill would sack the allegedly risk-averse air chief marshal in May 1941, which was ironic, as 30 years earlier the then First Lord of the Admiralty praised Longmore for risking his life with a test landing at sea.[38] Once qualified, Sansom, Longmore and their companions stressed to the C-in-C The Nore and to Their Lordships in London the significance for the Royal Navy of Short Brothers' relentless inventiveness. The pioneer pilots' final report, in mid-August 1911, confirmed a role for 'scouting' aircraft, mapped out a procurement programme, urged an expanded training scheme, and recommended acquisition of the Royal Aero Club's lease.[39]

Battenberg's support, Sueter's energetic lobbying, and above all Churchill's arrival at the Admiralty in October 1911, together ensured implementation of the lieutenants' master plan for exploiting the opportunity granted them by Maclean. In consequence, four biplanes were purchased, a steady stream of trainees gained their pilot's certificates, and a dozen senior ratings set about constructing a Naval Air Station. Sansom was the obvious choice to command what in February 1912 was renamed the Eastchurch Naval Flying School. Not that Sueter envisaged Sheppey as solely a training establishment: his August 1912 paper to the newly constituted Air Committee saw north Kent as vital for maritime reconnaissance and aerial defence of naval dockyards and land bases. The Royal Navy would quickly exercise a dominant presence and yet, given its continued dependence upon Short Brothers and the Royal Aero Club, experimentation at Eastchurch on the eve of war confirmed the maintenance of a unique civil–military partnership. In the meantime Sansom saw himself promoted to commander and ranked the most senior officer in the Naval Wing of the RFC, forerunner of the RNAS. Sueter had few friends inside the Admiralty, but creation of an Air Department in the summer of 1912 determined his immediate future. Independent-minded, reluctant to account for his actions, and suspicious of his immediate superiors, Murray Sueter served as Director for a tumultuous three years, exploiting to the full Battenberg's forward thinking and Churchill's boyish enthusiasm for action and ingenuity.[40]

The First Lord of the Admiralty understood the need to keep a close eye on 'the enterprising and energetic' Sueter. In December 1912 he recommended that the Director of the Air Department (later Division) report directly to Prince Louis Battenberg, by then Second Sea Lord, although in practice the minister's hands-on approach to the 18-month gestation of the RNAS blurred the lines of accountability.[41] The RNAS was very much Winston Churchill's creation. He promoted naval aviation at the highest level in Whitehall, appeased or silenced Senior Service sceptics, and in contemporary parlance micro-managed the entire grand project. Crucially for the Short brothers, Churchill understood the need for uniformity of design, with Navy Wing/RNAS squadrons ideally flying variations of the same machine, sharing a standard engine and wireless facility. In seeking a 25 per cent rise in the Naval Estimates from 1910 to 1914, the First Lord identified aircraft procurement as a priority, boasting of his achievement a decade later in *The World Crisis 1911–14*. Churchill enthusiastically promoted Sueter's blueprint for extending coastal air defence across the south and east of England, but he recognised Eastchurch's unique contribution. Short Brothers was seen as playing a vital experimental role, and endorsement of the company's activities came with the minister's frequent visits to Sheppey.[42]

While he never secured his pilot's certificate or even flew solo, Churchill defied conventional opinion as to the maximum age for taking to the skies and received regular instruction from a succession of service personnel. His favourite instructor was Gilbert Wildman-Lushington, one of the Royal Marines' earliest trainees at Eastchurch. The hazardous nature of flight at this time rendered a fatality on Sheppey near certain, and on 2 December 1913 the young captain died after crash landing. Wildman-Lushington's death was clear evidence to Churchill's wife and friends that he was taking far too many risks. All were insistent that, having suffered only one mishap in two years of flying, his luck was surely running out: the Liberal Government's least sedentary member had been aloft 'nearly 140 times, with many pilots, & all kinds of machines' by the time he stopped flying in June 1914.[43]

Given its relative proximity to London and the commodious accommodation available in Sheerness, Churchill viewed Eastchurch as an ideal location for regular gatherings of his cronies. Especially welcome were key opinion formers in Fleet Street. All visitors were offered the opportunity to fly, with Sansom happy to oblige if the outcome was a favourable view of the Royal Navy's newest initiative. Similarly, guests were invited to meet Horace Short, 'a good man, but terrible to look at.' Churchill courted the press at other locations, from Calshot to Cromarty Firth, but Sheppey was a unique location in that he could show off Commander Sansom's 'War Squadron' while at the same demonstrate the scale on which Short Brothers were operating once the full potential of their aircraft had been realised.[44] As a senior member of the workforce Dick Fairey was meeting an array of personalities, although given the depth of partisan division at Westminster there was a notable absence of Unionist politicians accepting the hospitality of a Liberal firebrand most Tories considered a traitor to his class and to his former party. Single-minded and unashamedly interventionist, Churchill appeared to be establishing a personal fiefdom within the world of naval aeronautics. Not surprisingly, this infuriated those retired or serving flag officers suspicious of the new technology. Their claims of ministerial malpractice encouraged an Opposition insistent that Asquith's administration lacked a coherent policy.[45]

It was certainly true that the Army and the Royal Navy were heading in different directions, and when the Central Air Office was set up in Sheerness at the end of 1913 it largely ignored development work at the Royal Aircraft Factory and dealt directly with entrepreneurial aviators such as Tommy Sopwith and the Short brothers. These proto-industrialists focused upon development work suited to naval requirements and largely ignored the needs of the Army. Thus there was an absence of anything akin to knowledge transfer, with communication largely informal and rooted in personal connections. To take two examples from the summer of 1912: the Royal Navy Aeronautical Branch ignored a War Office ban on flying monoplanes after two fatal accidents; and Short Brothers displayed scant interest in British & Colonial's testing of a 'hydro-seaplane', despite Sueter having authorised and funded the Bristol company's research into hydrofoil technology.[46]

The absence of volume production meant most manufacturers saw little point in learning from their ostensible rivals, unlike in France or even Germany. Only in wartime conditions did collaborative R&D evolve, and then largely as a consequence of subcontracting. Similarly, it took the onset of war for plane makers to rely upon exclusively British designs and to reduce significantly their dependence upon French airframe and engine technology. Edwardian aviation was still a small world, and nowhere was this more obvious than in meeting the needs of the Royal Navy. At executive level the minister went well beyond constitutional propriety in determining future policy, especially when advised by a compliant and sympathetic Chief of Staff, and at operational level a relatively junior officer exceeded his authority when liaising directly with contractors and determining the size and shape of an embryonic strike force. Churchill and Captain Sueter shared a common agenda, realised with the establishment of the RNAS on the eve of World War I (at which point the Director was finally promoted to commodore). Long before this, however, the Navy Wing was paying only token respect to inter-service cooperation, with Sueter encouraging favoured companies to focus upon the threat across the North Sea and leave Farnborough to accommodate the needs of the Army.[47]

Short Brothers needed little encouragement from Whitehall or Sheerness to test its biplanes to the limit. Increasingly the company's de facto test pilots were overshadowed by the Navy's first fliers, particularly when Maclean was out of the country. The 12 months prior to Fairey's arrival saw a succession of genuinely historic flights by the Royal Navy's newly acquired aircraft. Once Longmore had confirmed the reliability of Oswald Short's flotation bags, Sansom successfully launched his S.38 biplane from a platform built above the bows of a battleship. At the Portland Naval Review in May 1912, both forms of take-off suitably impressed George V when the Navy Wing staged an inaugural demonstration, adroitly orchestrated by its Commanding Officer in tandem with Horace Short.[48] Similar displays were staged in July at the Portsmouth Naval Review, by which time the Admiralty War Staff had agreed a strategy for deployment of both land- and ship-based aircraft that heralded Sueter's offensive ambitions. Sansom and his fellow pilots had proved such a strategy could work, and by the autumn of 1912 the Navy Wing's complement of aircraft had doubled in size. While Short Brothers was building easily modifiable 'pusher' or 'tractor' machines, the Admiralty's reluctance to become dependent upon a single manufacturer saw Armand Deperdussin supply one of his monoplanes. The futuristic Deperdussin Monocoque was such a revolutionary design that it must surely have influenced Horace Short when he sought to enhance the performance of the Navy's S.41s by reducing wing weight, adapting Geoffrey de Havilland's double-action ailerons, and streamlining the floats. Regrettably, speed was sacrificed by enlarging the cockpit to accommodate an observer and navigator. When Fairey saw the first blueprints of the Swordfish at some point in the early 1930s he must surely have recalled Short's attempt to balance engineering capability against naval need.[49]

By the spring of 1913 the Navy Wing was operating 34 aeroplanes out of five air stations, with the ageing cruiser HMS *Hermes* modified to serve as a support vessel. A year later the embryonic RNAS would boast nearly three times as many machines, with 50 more on order and HMS *Ark Royal* soon to enter service as a mother ship for seaplanes on active service. Such growth in so short a time appears impressive, although one should note the claim that on the eve of the war only around 50 of the Royal Navy's aircraft could be considered to have genuine offensive capability, drawing upon the rapid advances in armament and ordnance undertaken in the course of the previous 12 months, and allegedly half of these machines were out of action at any one time. Volume production for both the Army and the Navy from 1912 to the start of World War I was small by comparison with German output, as was the number of firms with adequate manufacturing capacity. The War Office depended heavily upon the Royal Aircraft Factory, with leading arms manufacturers working to Farnborough's basic design. The Central Air Office enjoyed greater flexibility, with access to a wider range of models, but the Royal Navy was nevertheless dependent upon one or two key companies. Short Brothers was a trusted partner, and first among equals, but the emergence of Sopwith Aviation signalled an important new player, eager to exploit a rising demand for its technologically sophisticated aircraft.⁵⁰ This then was the environment in which Dick Fairey made his mark as both a mechanic and a manager, capable of assuming ever greater responsibility within the Short Brothers hierarchy.

Coda: an afternoon in Edwardian England – Manchester, 31 July 1911

Dick Fairey saw out July 1911 admiring Prince Louis Battenberg's insouciance as his children flew for the first time, their lives in the hands of two young naval officers still consolidating the knowledge acquired since their arrival at Eastchurch fewer than five months before. Fairey was dedicating his life to the pursuit of manned flight at precisely the moment a slightly younger Ludwig Wittgenstein was abandoning his commitment to the science of aeronautics and embracing analytical philosophy. The two men would seem to have had little in common, yet both eschewed university by studying programmes of applied science at technical colleges, and both benefited from late Victorian Britain's municipal flowering of higher mathematics. Admittedly, Imperial College rather than its junior partner in Finsbury provided a more obvious parallel with Charlottenberg's Technische Hochshule, but the mathematicians and physicists at Manchester University were contemporaries of Silvanus Thompson, with the same commitment to cultivating engineers whose work rested on a firm theoretical basis. For this reason, in 1908 Wittgenstein's father sent him to Manchester, where eminent professors such as Horace Lamb and Arthur Schuster sought answers to precisely the problems Horace Short highlighted in his lectures to the Royal Navy's trainee pilots. An authority on hydrodynamics, Lamb advised Farnborough on how best to model seaplane performance at take-off and landing. The physicist Joseph Ernest Petavel combined a chair in engineering with a pilot's certificate, suitable credentials for later in his career rebuilding the National Physical Laboratory's first wind tunnels.[51]

Initially, Wittgenstein flew kites with Schuster and Petavel in an ambitious programme of experimental aerodynamics. Manchester's preoccupation with wing design was rooted in advanced mathematical calculation, with the equations fully tested prior to laboratory-based empirical investigation. Horace Short's approach was not that dissimilar, but his can-do mentality had more in common with cerebral pioneers like Geoffrey de Havilland and CRF, who understood the dynamics of flight but saw *ab initio* research as an inductive and applied process requiring hard graft on the runway and in the engine shed. Rivals of Short Brothers like Manchester's own A.V. Roe, from whom in

1913 the Admiralty ordered for Eastchurch six Avro 500 biplanes, bemoaned his company's dependence on French engines. An applied science department at a civic university would enthusiastically accept the challenge of designing a light, fuel-efficient power unit, which is why towards the end of 1909 Wittgenstein set aside his work on kites to address a problem more worthy of his intellect.[52]

Contrary to popular myth, Wittgenstein never designed a jet engine akin to Frank Whittle's model of compressed air mixing with high octane fuel in a combustion chamber such that the consequent force drove a turbine and provided necessary thrust. The outcome of costly and haphazard experiments was a patent application in November 1910, with submission of the full specification seven months later. On 17 August 1911 the Patent Office registered Wittgenstein's 'Improvements in Propellers applicable for Aerial Machines.'[53] The idea is simple, in that air and vaporised fuel are fed through the hub of a radial motor with centrifugal force pushing the inflammable mixture to compression chambers at the tip of a revolving hollow propeller where the compressed gas is ignited. The patent drawing displays a twin-blade propeller, with combustion chambers tangentially mounted on each tip, but the technical description specifies the possibility of multiple-blade propellers with tip-jets. The concept was highly problematic, with no provision for regulating air and gas flow, or for ignition in the twin combustion chambers. Crucially, propeller construction in the early decades of aviation was based on the principle of bonding laminated wood, with a metal sheath on the tip of each blade: how could the universal principle of solid propellers accommodate the blade-length piping of a compressed fuel mixture? Wittgenstein provided no answer, and yet his capacity to comprehend and extend complex mathematical analysis arose out of his absorption in the now abundant literature on the aerodynamics of airscrew design.[54]

This was a field similarly absorbed Fairey, even if he would have struggled to comprehend the level of abstraction at which Lamb operated and which Wittgenstein exceeded when his absorption in Bertrand Russell's *The Principles of Mathematics* began a two-year journey away from aeronautics and towards the study of mathematical logic at Cambridge.[55] As already noted, Fairey was fascinated by propeller design, and he would doodle on the back of menu cards or memoranda neatly drafted technical drawings of twin blades pitched at varying angles. That fascination fed into a succession of cutting-edge piston-engined aircraft across ensuing decades, and reached its apogee in adapting Armstrong-Siddeley's dual propeller system to the Gannet, the Fleet Air Arm's last purpose-built reconnaissance aircraft.[56]

In the summer of 1911, when Wittgenstein remained uncertain as to whether he should relinquish his studentship and wholeheartedly embrace philosophy, CRF naturally had no notion that a wholly unorthodox approach to aircraft propulsion had been patented. Had the opportunity arisen he would have relished rendering real Wittgenstein's invention, and with the passage of time he – or to be more accurate, his designers – did. If Whittle applied the concept of centrifugal flow compression, Fairey Aviation developed the principle of tip-jet technology, rediscovered in the 1940s by Austrian designers unaware of their exiled compatriot's experiments 30 years earlier. In 1955–6, Fairey's final year, his company tested the Jet Gyrodyne: 'pusher' propellers on stub wings provided propulsion, with lift courtesy of compressors mounted together under the main pylon forcing air up to the tip burners at the ends of a twin-blade rotor. The same principle was applied to the Fairey company's final prototype, the hybrid passenger-carrying Rotodyne. With two turboprop engines, an auxiliary air compressor and four tip-jet combustion chambers, this combination of helicopter and aeroplane could never overcome the dual problem of weight and noise. Although the project staggered on until February 1962, the Rotodyne fell foul of Fairey Aviation's merger with Westland two years earlier. Yet this was the flying machine that came closest to fulfilling Wittgenstein's dream. Remarkably, by the time of its cancellation the Rotodyne programme was 11 years old. This means that Richard Fairey was aware of jet-tip technology's commercial potential while Wittgenstein was still alive (the philosopher died of cancer in April 1951).[57]

During his life Wittgenstein produced just one full-length publication: the English translation of *Tractatus Logico-Philosophicus* appeared in 1922. The book was dedicated to the mathematician David Pinsent, with whom one autumn afternoon ten years earlier Wittgenstein had studied intently the Royal Navy seaplanes flying out of Rosyth. Late in the war the tyro philosopher's travelling companion put his expertise in aerodynamics to good use, compiling reports for the National Physical Laboratory. The test flights at Farnborough became ever more hazardous, and Pinsent was killed on 8 May 1918. He had been a late recruit to the team of mathematicians and scientists assembled at the Royal Aircraft Factory from the spring of 1915.[58] Future policy-makers Henry Tizard and Frederick Lindemann forged their reputations as physicists adept at addressing the technological imperative of 'industrial war.' This was evident from the speed with which they gained their pilot's certificates once civilians were allowed to fly in August 1916. The future Lord Cherwell's exploits, most famously his systematic spinning of notoriously unstable aeroplanes, may have been exaggerated. Nevertheless, Lindemann's reports on auto-rotation were invaluable and speedily transmitted to manufacturers such as the Short brothers, Sopwith and Richard Fairey. The contrast with relaxed attitudes pre-war to the sharing of information was stark.[59]

While Pinsent paid the ultimate price, senior colleagues like Lindemann and Tizard survived their tenure as test pilots, both men becoming rival power-brokers in the course of World War II. Their power and influence within the 'warfare state', contrasting so starkly with Pinsent's abrupt denial of intellectual fulfilment, fuelled Wittgenstein's post-Hiroshima exposure of applied science as a false god.[60] In the course of his life Wittgenstein moved from a fascination with the infinite prospects for manned flight to a denunciation of aircraft as the most visible symbol of humanity's delusionary faith in scientific progress. Such thinking was of course an anathema to Sir Richard Fairey at the start of the jet age: CRF could celebrate and embrace Wittgenstein, the prophet of jet-tip technology, while at the same time be appalled by the Austrian's readiness to turn his back on scientific endeavour not once but twice.

CHAPTER 3

World War I and the Founding of Fairey Aviation

Short Brothers and Richard Fairey: consolidating one company and inspiring another

Richard Fairey had quit the sickly Captain Dunne in late 1912, but his loyalty and gratitude extended to advising on his former employer's final and most futuristic prototype. The D9's failure saw the invalid designer surrender his patents to Armstrong Whitworth, seen by some as Dunne's secret sponsor. In the late 1940s the same company revived memories of the Blair Atholl Syndicate when testing its AW52 'flying wing'; ironically, this revolutionary aircraft ended its days at Dunne's *bête noir*, the Royal Aircraft Establishment.[1] 'Willy' Dunne sought solace in popular psychology, but Dick Fairey was on a sharp trajectory from the moment he joined Short Brothers. The 30 months he spent with the company were, in his own words, 'the best years of my life and my real beginnings in aviation.' He considered Horace Short his mentor, but this was by no means a one-way relationship, and from the outset CRF enjoyed considerable responsibility and authority.[2]

That Fairey fulfilled the trust placed in him by the Short brothers is evident in his rapid elevation to chief engineer of a company thriving on its status as the Admiralty's prime aero contractor. For the new recruit, building aeroplanes was now a serious business, with commercial considerations dictating output, costs, productivity, and above all, the nature of the product: Short Brothers sought to satisfy Royal Navy requirements, and crucially, to anticipate them. Thus Horace Short's ambitious programme of R&D prioritised the means by which naval aircraft might strike at the enemy in time of war. Keen to minimise their outgoings, Horace and his brothers had no time for procedural niceties: they drew scant distinction between uniformed and civilian experimentation, insisting that Gordon Bell, their regular test pilot, work closely with Charles Sansom and his fellow officers. Fairey found himself part of an expanding workforce, with labour costs dominating a budget based on second-guessing how much the Navy's Central Air Office was prepared to pay per machine. Short Brothers was one of the earliest firms to embrace volume production, and yet few aircraft were uniform in specification: each new machine incorporated the latest design features, and CRF soon learnt how to reconcile fresh modifications with economies of scale.

Fairey was earning his spurs as an engineer, but his entrepreneurial skills had been scarcely tested. Marketing model aircraft was scarcely the same as selling the real thing; but not long after joining his new employer he gained an insight into the competitive nature of Britain's fledgling aviation industry. On 14 February 1913 London's largest exhibition centre unveiled its fourth and much anticipated Aero Show. Olympia boasted a comprehensive gathering of aviation pioneers, with the Short Brothers stand manned by key personnel in the company's brief but successful history. Frank Maclean was given the honour of explaining to George V how the seaplane ('hydro-aeroplane') on display embodied Horace's enthusiastic embrace of alloy tubing, comprehensive waterproofing, robust floats, fuselage-mounted lower wings and cockpit-based fuel control and ignition. This was the technology which, within two years, would enable the RNAS to cross the North Sea and attack German Zeppelin sheds. With no show in 1912, the King had been obliged to wait two years before reacquainting himself with the ingenuity and imagination of Britain's premier aircraft designers. Hastening past the eccentric and unproven, His Majesty was intent on meeting the key personalities and the credible plane makers.[3] The reliably jingoistic C.G. Grey, founder-editor of *The Aeroplane* and later one of Fairey's closest friends, rejoiced that the nation's most modern machines now matched their French rivals. Yet this 'assertive man, who felt none of the intellectual or class diffidences of his contemporaries', also observed that half of the aeroplanes on display were built across the Channel. Even aircraft as sophisticated as the Short Brothers' 'tractor' seaplane remained dependent upon engines shipped across from France. Manning the company stand at Olympia, as yet too junior to make the King's acquaintance, Dick Fairey could scarcely ignore British engine builders' failure to match the inventiveness of an A.V. Roe, a Frederick Handley Page or, indeed, a Horace Short.[4]

The 1913 Aero Show signalled recognition of the seaplane's martial credentials, with Short Admiralty No. 2 testimony to Murray Sueter's belief that here was an ambitious company capable of translating extensive testing at sea into the manufacture of reliable and robust machines. Dick Fairey was crucial to that process of workers maintaining a keen sense of craftsmanship when engaged in assembly-line construction. Short Brothers enjoyed an early competitive advantage, but neither siblings nor senior employees could ignore the impact of a budding rival. The Sopwith Aviation stand failed to interest George V, but fellow exhibitors acknowledged the potential of the firm's Three-Seater, which within months was being delivered to both the Navy and the Army at a rate of one a week. The sensation of the show was Tommy Sopwith's single-float 'Bat Boat', based on revolutionary methods of hull design and intended to fly from both sea and land. Britain's first genuine flying boat went on to win assorted aviation prizes. Ironically, in the light of their interwar achievements, the Short brothers were deeply sceptical of Sopwith's and partner S.E. Saunders' technological breakthrough; but tellingly Churchill authorised Sueter to purchase the prototype. Another notable customer was the Imperial German Navy, although presumably Sopwith's acute cash-flow problems never prompted a demand for advance payment such as the plea made to the Admiralty in late 1912.[5]

Tommy Sopwith was familiar to Fairey from excursions in and out of Eastchurch, but the Olympia exhibition revealed the size and scope of his Kingston-based operation. Crucially, the show highlighted how Sopwith had gathered around him a design team capable of constructing machines that could match Short seaplanes in terms of functionality and originality, but were also easy on the eye: his shrewdly selected cohort of hands-on mechanics and varsity theorists saw speed and style as synonymous.[6] Looks ranked low on Horace Short's list of design priorities, as evident in his reliance upon large rectangular radiators to maximise engine cooling capacity.[7] Arguably the same would prove true of Richard Fairey, but in February 1913 Tommy Sopwith's keen sense of the aesthetic was the last thing on his mind. What surely made a deep impression was how quickly a suitably qualified student of aeronautics could create a viable company if driven by single-minded ambition and entrepreneurial flair. In CRF's mind, Tommy Sopwith was a role model. Creating one's own company was by no means inconceivable, and yet so early into his stay with Short Brothers the notion of going alone must have seemed a distant prospect.

For the moment Fairey was content overseeing seaplane production and supporting the test programme Short and Sansom saw as vital if the Navy Wing could establish its credentials as a credible strike force. Ceaseless invention ranged from Horace's cockpit-controlled folding and self-locking wings, seen as crucial for maximising deck stowage on board the future *Ark Royal*, through to his quick-release mechanism for launching heavy ordnance torpedoes.[8] By the spring of 1914 the tireless Short was dividing his time between Sheppey and the seaplane stations on the Isle of Grain and on Southampton Water at

Calshot. The proximity of the former meant Fairey frequently flew across the Medway estuary, serving as Gordon Bell's observer while he tested or delivered the latest addition to the Royal Navy's varied assembly of aircraft.[9] It was at Calshot, on 27 July 1914, that Bell successfully completed the RNAS's first torpedo firing. Both Horace and Oswald were in attendance, but Fairey was almost certainly carrying out his regular duties in north Kent. Those duties now included overseeing the work necessary to equip four seaplanes with torpedo gear and Gnome's most powerful – yet most unreliable – engine. Ironically, although completion of these aircraft coincided with the start of World War I, none of them launched a torpedo in anger. The same could not be said of their successors, seven of which participated in the Christmas Day raid on the Zeppelin base at Cuxhaven; the sole survivor of this expedition and the last of Horace Short's prewar seaplanes fell victim to Turkish gunfire in the aftermath of the Gallipoli landings.[10]

This advance guard of maritime air power was built at Eastchurch, with each machine then transported along narrow lanes to Sheerness for final rigging and initial test flight. The next generation of seaplanes built on Sheppey would take to the air from Queenborough Pier, its steam crane commandeered by the Admiralty for the duration of the war. By this time, however, a new factory had opened at Rochester, upstream from the bridge. In the months following the Olympia exhibition Eastchurch had been working at full stretch, but too many aircraft were being delivered late. Fresh manufacturing capacity was urgently required and construction of the new works began in October 1913. When seaplane production commenced on the right bank of the River Medway in April 1914, Oswald Short assumed responsibility for a plant which by the Armistice would boast five huge 'shops' and dominate much of the shoreline. The decision to build a second, more accessible workplace made sense and was swiftly vindicated. Rochester's close proximity to the dockyard town of Chatham ensured a plentiful source of suitably skilled workers, and the Medway allowed Short's pilots to operate from a far calmer stretch of water. The Sheppey factory was isolated and, despite the convenience of working in tandem with the naval establishment, the environment was no longer conducive to operating on the scale envisaged by all three siblings. Eustace, mastermind of Short's still buoyant balloon business, was eager for an overall expansion of the company. Yet aircraft continued being built at Eastchurch until late 1917, when Oswald, following Horace's death, handed the whole site over to the RNAS.[11]

CRF enjoyed working on Sheppey so much that in 1914 he was reluctant to become works engineer at the Medway plant. This was surprising given the significance of his promotion. Furthermore, London was more easily accessible from

Rochester, a cathedral city closely associated with Charles Dickens, and offering a quality of life markedly superior to the harsh reality of Edwardian Sheerness. The journey time back to Finchley was significantly shorter, especially since Fairey could borrow a Model T Ford. With Effie away in India and Margery newly married, 52 Station Road was now quieter and less crowded. Geoffrey had started school, easing the pressure on his increasingly frail mother: the former Frances Fairey had over a decade still to live, but she suffered from ever present asthma. Household duties, a young child, and anticipation of weekend visits from her eldest son, together stopped her thinking like an invalid. Crowded days left no time for metaphorically, let alone literally, taking to one's bed. Dick's Sunday visits were the high points of Mrs Hall's week, for six hours rescuing her from the drab reality of suburban tedium. An increasingly busy social life, and the availability of the Ford, dictated how often CRF drove up to north London, but Geoffrey Hall recalled a family-orientated half brother, with a keen sense of filial duty.[12]

Two years of 'perfect bliss' saw the opportunity to work hard and play hard. Ironically, it was in this remote spot that Dick Fairey for the first time became acquainted with another world, one of cosmopolitan affluence and sophistication – a world which eventual wealth and social status would ensure he never left, and in which he thrived as a consequence of ambition, authority and an adamant refusal to be in awe of anyone. CRF gained an insight into the reality of upward social mobility when he became acquainted with three Cambridge undergraduates some six years younger than himself. At their respective public schools Maurice Wright, F.G.T. Dawson and Vincent Nicholl had all pursued mathematics to an advanced level.[13] Yet as engineering students under the supervision of Bertram Hopkinson, a true renaissance man, they also secured a solid grounding in technical design and assembly. Engineering at Cambridge was an unusually hands-on degree programme, focused on the well-equipped, turn-of-the-century laboratory in Free School Lane. Drawn from a sharply contrasting social background and undergoing an education markedly different from that of Dick Fairey, these new friends nevertheless shared the older man's capacity to root practical experimentation in rigorous calculation and computation. The three students' qualifications aptly demonstrate how mistaken Correlli Barnett was to downgrade the calibre of science and technology tuition in British universities, both civic and ancient, when applauding comparable centres of excellence in Wilhelmine Germany.[14]

No doubt encouraged by Professor Hopkinson, himself a future wartime aviator, several students eager to build gliders arrived at Eastchurch in the summer of 1911. Yet only Wright, Dawson and Nicholl stayed for any length of time, and later came back. While working for Dunne, CRF had the time to cultivate their acquaintance and to share their ambition of being able 'to make aeroplanes one day on a grand scale, and our ambition even soared to such heights as the possibility of making twelve aeroplanes in a year.'[15] Fairey retained close contact with the three students after he began working for Short Brothers. He clearly enjoyed each man's company, partly through shared interests and ambition, but primarily because a young blood at varsity before World War I could gently buck social convention, endure only mild sanction, and at no time endanger his privileged place within Edwardian England's carefully delineated social structure. Hence the previously buttoned-up Dick Fairey could enjoy japes and jollities with his new-found chums, secure in the knowledge that their social status allowed him to make mischief in a manner that otherwise would be unthinkable. Thus privilege brought protection when engaging in hearty activities deemed unacceptable to the Kent constabulary. All forms of authority could be challenged, not least music hall managers in the Medway towns. The four friends, by CRF's own admission, regularly disrupted stage shows, even though he himself always had a deep affection for vaudeville.[16]

Clearly Fairey felt empowered to act in a manner he would normally consider inappropriate, and the irregular use of motorcycles and Nicholl's Model T suggests scenarios worthy of Kenneth Grahame.[17] However, P.G. Wodehouse remains a more obvious literary guide, not courtesy of Bertie Wooster but via the mishaps and misadventures of Mike Jackson in the Psmith stories – comic elegies to an era fast disappearing even without the threat of war.[18] Just as Wodehouse's young companions survive a series of scrapes courtesy of a shrewd and sympathetic man of the world, so Fairey and his friends relied heavily upon Charles Crisp, a young solicitor in the City whose knowledge of company law, while unsuited to appeasing local bobbies, would one day prove invaluable.[19]

When war finally came, Wright, Dawson and Nicholl were together commissioned in the RNAS as flight sub-lieutenants, and at Eastchurch on 8 October 1914 all three men were certified by the Royal Aero Club as trained pilots. First among equals was Vincent Nicholl DSO and DFC, who in the course of the war rose to the rank of lieutenant colonel, combining the leadership skills of a station commander with the intuitive qualities of a natural flier. At the Armistice, Nicholl was serving as a test pilot, in which capacity he was recruited by Richard Fairey, competing for the 1919 Schneider Trophy on his old friend's behalf. The former colonel played a valuable role in the early postwar years as a demonstrator and salesman for successive variants of the Fairey III. As we shall see, the commercial success of the Fairey III series, particularly its seaplane version, vindicated CRF's decision to establish his own company. When in 1921

Fairey Aviation arose from the ashes of its wartime predecessor, Nicholl became a director, continuing to serve as chief test pilot for a further two years. Ironically, after risking his life so many times in the cockpit, Nicholl died in 1927 aged only 34. Like his friend and colleague, Vincent Nicholl was unusually tall, but where the two men differed was in bulk. Richard Fairey was in every sense of the term a big man, looming large in group portraits with his three Cambridge companions. The wiry, slimly built Nicholl entered the war fit and strong, but the ordeal of three days adrift in the North Sea left a physical if not a mental scar. CRF was deeply distressed that such an impressive character should die so young.[20]

While Vincent Nicholl was tall and thin, Maurice Wright was of medium height and stocky – by today's standards he would be deemed short, and Dick Fairey towered over him. M.E.A. Wright AFC earned his spurs flying in the Middle East before progressing to be one of the Admiralty's most senior test pilots and then chief technical officer at the Marine Aircraft Experimental Establishment in Felixstowe. He transferred to the newly established RAF in April 1918, retiring seven years later as a squadron leader. This surprisingly modest rank for someone who in uniform had achieved so much was quickly forgotten once the now veteran pilot accepted Dick Fairey's invitation to join the board. Maurice Wright would prove a stalwart of the company, fulfilling a variety of roles up until his death in 1957. World War II would prove especially stressful: Wright agreed to become deputy managing director in 1939, and he assumed full executive responsibility a year later when CRF joined the British Air Commission in Washington. Here was a man Fairey trusted implicitly, particularly when dealing with overseas customers. Beneath the hail fellow well met persona lay the natural linguist's love of travel. The polylingual, cosmopolitan Wright was invaluable between the wars when Fairey Aviation established itself as a truly international enterprise.[21]

The post-1918 experiences of both Vincent Nicholl and Maurice Wright illustrate the premium CRF placed upon loyalty: on Sheppey in the final years of peace he had forged friendships stronger than he could possibly have imagined. Here were men made privy to the future plane maker's wildest dreams, responding enthusiastically and fostering his idealism. Their refusal to display scepticism and scorn saw both men duly rewarded. Not that this was a one-way process, in that Nicholl and Wright proved proactive and energetic company servants, each man endeavouring to justify his generous salary and shareholding.

Although supportive of Dick Fairey's desire to leave Short Brothers and launch his own enterprise, neither Nicholl nor Wright boasted sufficient capital to risk a sizeable investment in the new venture. However, the third young officer,

F.G.T. Dawson, was in a position to provide significant financial backing.[22] His contribution to the founding of the company was crucial. Wounded at Gallipoli, Lieutenant Dawson returned home to London in the summer of 1915.[23] He was thus easily accessible whenever money was required, or legal documents demanded his signature. He was a senior partner in Fairey Aviation's wartime incarnation, and held a seat on the board from 1921 until his retirement in April 1933. Richard Fairey retained a deep personal debt to 'Wuffy' Dawson, who, despite early retirement and long periods out of the country, remained a close friend. The two men saw more of each other after Sir Richard purchased a large villa on the island of Bermuda. Back home in Hampshire, Fairey doubled the size of his Bossington estate when Dawson, intent on remaining in Canada, sold land in the Test Valley to his long-time friend. Their respective families were always close, especially in the final months of World War II when the Caribbean offered a welcome retreat from the roughhouse rigours of wartime Washington: an exhausted Fairey took his family to Bermuda after a heart attack obliged him to resign as head of the British Air Commission. Back in 1940, when CRF's second wife and their two children found themselves penniless in Canada because British assets had been frozen, Dawson provided necessary financial support. 'Wuffy' Dawson was the least demonstrative, most quietly mannered member of the Cambridge troika, content for his colleagues to lead from the front. A thoroughly decent man, he was nevertheless financially astute. His shrewd investment at a moment of maximum risk was, in the fullness of time, suitably rewarded.[24]

Dawson and Fairey were unusually close, but all four friends were bound by a striking blend of camaraderie and commercial endeavour. They trusted each other totally, with CRF the unquestioned leader by dint of energy, enterprise, engineering brilliance and sheer force of character. With a shared purpose and a common interest in maximising personal and corporate profit, they formed the bedrock upon which Fairey Aviation prospered in the second half of the 1920s and through to the advent of World War II. Letters from the interwar period confirm the unique bond they shared, displaying a familiarity and intimacy absent from all but CRF's correspondence with his immediate family. For such a tightly knit group the premature loss of Vincent Nicholl must have been devastating.

Fairey's weekend high jinks were a welcome relief from the pressure he was under as Oswald Short's number two. In later years the two men seem to have had little regular contact, and no record survives of CRF referring warmly to Oswald in the same way that he publicly lauded Horace's character and achievements. One can only speculate as to whether the demands of work put pressure upon his personal relationship with the younger sibling, and this was certainly Geoffrey Hall's opinion.[25] Admiralty orders enabled Short Brothers to raise capital and invest in a new factory, but the size of each contract, and the design changes that affected every machine built, eroded all attempts to secure economies of scale. Monitoring assembly, and enforcing quality control, were only two of the tasks for which Dick Fairey assumed responsibility when production commenced at Rochester in April 1914. Not that he entirely abandoned Eastchurch as he continued to rent a room at the village pub, The Crooked Billet. While the Sheppey workforce focused on Short's larger seaplane, known within the Royal Navy as the Type 166, Rochester built a smaller version designated the Type 830 or Type 827, dependent upon the engine employed. The Type 827 became the RNAS's standard equipment from 1915, ubiquitous on operational duties in the Mediterranean, the Persian Gulf and East Africa, as well as the North Sea. While Petters (later Westland) assumed responsibility for building the final batch of Type 166 seaplanes, the Type 827 signalled the onset of wartime subcontracting, with Rochester responsible for just a third of all machines built.[26]

The success of the Type 827 – Admiralty Type 184 – rested on a breakthrough in domestic engine design and manufacture. At the 1914 Aero Show, 19 British prototype engines were on display, but none of the exhibitors proved capable of volume production. The onset of war saw the Royal Navy look to Rolls Royce for an answer to Britain's aero-engine difficulties, although the first order for what became known as the Eagle was not placed until January 1915. A temporary solution, courtesy of motor racing, was the Sunbeam V-12, tested by Horace and able to power his smaller seaplane up to 225 horsepower. The Type 827/184 only became the RNAS's preferred model in July 1915, but a year earlier its potential was already obvious. Dick Fairey was not alone in recognising that *if* Short Brothers could secure a steady supply of Sunbeam engines, then the company's favoured status with the Admiralty was secure. Furthermore, if Rolls Royce could bring its planned prototype into production, then Rochester might struggle to ensure that output matched engine availability. In actual fact British aviation suffered from a bottleneck in engine manufacturing throughout World War I, for the first two years remaining reliant upon France. After 1916 the Lloyd George coalition singularly failed to improve productivity and enforce standardisation. By the war's end Fairey would be all too familiar with the inability of companies like Sunbeam and Rolls Royce to meet an ambitious government target of 7,000 engines a month.[27]

In 1914 such a figure was scarcely conceivable, with Sunbeam unique in providing the power unit necessary for Horace's torpedo-carrying seaplanes to fulfil their full potential. The works manager at Rochester was sufficiently astute as to ask how, given the length of time it took to build each machine, his employer would cope with a burgeoning order book. The onset of war in August 1914 rendered that question even more pertinent, but evidence suggests CRF had already considered how Short's success might be the means by which he set up his own business.[28] Partly by ability, partly by good fortune, at the start of World War I Richard Fairey found himself in the ideal location from which to launch his own company. Britain had, in a short space of time, spawned numerous aspirational plane makers, some of whom were already starting to realise their ambitions.[29]

Yet, among the first generation of pioneers, only Short Brothers operated on a scale in any way comparable to what was already the norm in France, Germany and the United States. Size is of course relative, and, while the Admiralty Air Department was sufficiently ahead of its RFC counterpart as to have 418 aircraft on the order books by December 1914, 300 of these machines were land-based. Some of the latter were modified seaplanes but, with the likes of Sopwith and Avro rising to the challenge, Shorts faced healthy competition. Nevertheless, for Fairey the company was a role model. It demonstrated that, however much individual genius still mattered, team work was vital, with senior staff providing an example for the wider workforce. Rochester is where Fairey learnt how to build up an operation and also how to pre-empt problems: healthy recruitment in the Medway towns tempered a shortage of male labour when a fifth of Britain's skilled workers joined up in the autumn of 1914. As works manager Fairey recognised the unpopularity of dilution in the early years of the war, with a growing trade union presence reinforcing men's reluctance to accept unskilled, especially female, labour.[30] If Short Brothers was forced to subcontract, which even prior to the war seemed increasingly likely, then regular naval contractors would lack the knowledge necessary to realise Horace's complex designs. Richard Fairey had neither factory nor workforce, but he clearly had the expertise. Assuming he remained a civilian then he could auction his services to successful subcontractors, *or* he could seek to become one himself, exploiting the contacts made within the still small world of naval aviation. Given the man he was and his career to date, only one course of action lay open to him.

The onset of the Great War: C.R. Fairey intends to build *his* aircraft

There seems little doubt that Dick Fairey followed the example of his Cambridge friends and, despite his dislike of flying, sought to join the Royal Navy. A staunch patriot all his life, it would have been astonishing had he not felt the need to do his duty once Britain declared war on Germany, and then on Austria-Hungary and Turkey. By December 1914 the RNAS had four times as many officers as its peacetime complement, but C.R. Fairey was not one of them. With the newly promoted Commodore Sueter taking a keen interest, the Air Department rightly vetoed recruitment of Short's most senior engineer. Geoffrey Hall recalled his half-brother then seeking to join the RFC, and soon after being informed in no uncertain terms that he could best contribute to the war effort by staying out of uniform and building aircraft. However, Hall then claimed CRF received his 'calling-up papers', thereby prompting a renewed effort to join the RNAS. Clearly a failure of communication within the administration responsible for naval recruitment led to Fairey receiving instruction on when and where to report for duty. However, this was not conscription as Hall's use of the term 'calling-up papers' suggests. The notion of CRF eventually being compelled to join up makes for a better story, but is of course wholly erroneous given that military service remained voluntary for a further two years.

Oswald Short was clearly aware of Fairey's eagerness to serve his country, and, while by now there was no love lost between the two men, the chance of securing an equally competent works manager was remote. At Short's behest Sir Oswald Swann, RN Director of Personnel, summoned CRF to the Admiralty. There he was brusquely informed his place was on the shop floor not the parade ground. Fairey's feelings of frustration and annoyance were surely compounded when, on the train journey to Victoria, a young woman in his carriage presented him with a white feather. This was the final insult, and one reason why after a few months CRF again sought a commission in the Royal Navy. This time he secured a meeting with Murray Sueter. The interview proved a painful experience until Sueter inquired of Fairey how such a talented aircraft builder might realise his ambitions given the Navy's insistence that he remain a civilian. Famously, Fairey replied by bluntly asking for the chance to build his own aircraft, at which point Sueter indicated that he would be awarded a contract should he set up on his own. When asked how many machines he could build, a fired-up Fairey promised 12 seaplanes, at which point a contract was speedily drawn up.[31]

Taking place at some point in the autumn of 1914, this clearly was a historic meeting. Here is a powerful and attractive creation myth: tyro plane maker arrives in Whitehall resentful of his employer and of the Royal Navy, but departs with an unexpected opportunity to make his mark in the world. However, that opportunity was not unexpected, given that Dick Fairey had already resolved to leave Short Brothers at an opportune moment in order to launch his own business. Furthermore, Sueter's assurance of work was by no means definite.

The Royal Navy's decision to order over 100 Type 827s was delayed until June 1915, following satisfactory trials on board HMS *Campania*, a converted Cunard liner later used to pioneer deck take-offs and landings. The Rochester factory could build no more than 30 seaplanes by the Air Department's deadline and three sizeable companies in Bristol and the Midlands (Parnall, Brush and Sunbeam) were each subcontracted to build 20 additional machines.[32] The remaining 12 aircraft were assigned to the fledgling Fairey Aviation. Murray Sueter was true to his word, but given the embryonic state of Dick Fairey's enterprise this was a sizeable gamble. One suggestion is that Alec Ogilvie, Sueter's wartime appointment as head of technical design, harboured a grudge against the Short brothers and therefore encouraged the Fairey breakaway. Yes, Ogilvie was a shareholder in the British Wright Company, which Horace Short saw as increasingly irrelevant, but in early 1914 both men had joined Francis McClean on his trail-blazing flights across Egypt. After leaving the Admiralty the entrepreneurial Ogilvie would keep in close contact with Shorts, showing no obvious sign of antipathy.[33]

More likely is that Sueter saw the spring of 1915 as a final opportunity to exercise his unique powers of patronage, using his privileged position to assist the trail-blazing aviators he had cultivated in the final years of peace: major engineering firms were now bidding for contracts, and, at a time when total output was still a modest 200 machines a month, rising stars like Tommy Sopwith – or a fresh arrival like Richard Fairey – could easily be squeezed out. The Director of the Air Department owed such men a debt, but with neither Battenberg nor Churchill to champion the RNAS, Sueter must have known his days were numbered. In September 1915 he would be made accountable to a rear admiral, and in effect demoted to the post of 'Superintendent of Air Construction.' Awarding Dick Fairey the contract to build a dozen of Horace Short's breakthrough seaplane marked Murray Sueter's last hurrah. Ironically, when finally the maverick commodore exhausted Their Lordships' patience he was posted to the Adriatic, where the Type 827s flying out of Taranto were among those built for the station commander two years before.[34]

At some point in 1913 or early 1914 Dick Fairey became acquainted with Frank Rees. 'F.H.C. Rees, Gentleman, Hampstead, London' was, contrary to the description on his business card, a minor business man in East Finchley plagued by a residual cash-flow problem. Through his correspondence Rees comes across as a creation of H.G. Wells; a character worthy of inclusion in *Tono-Bungay* or *Kipps*. Here was an aspiring entrepreneur of vaunting ambition, but scant means of realising his dreams – a venture capitalist with no capital. For both investment and expertise – and the odd fiver to oil the wheels of company law – Rees looked to Fairey and his friends, particularly 'Wuffy' Dawson. In the early summer of 1914, Rees and Fairey drew up plans for a syndicate which would pool capital and thereby secure a privileged position when the time came to float the 'X Aviation Company Limited.' By late July, Rees had drafted a document making clear that the vital expertise of 'Mr X' justified his majority shareholding and budgetary control: all references to 'X' saw the name 'Fairey' inked out.[35] A month later, Rees, aware of Fairey's desire to join up, conceived the idea that partnership in the syndicate be secured via a deposit of one shilling for each £1 share. Thus the shell of a company could be formed, with flotation once the war was over and its prize asset had returned to civilian life.[36]

Frank Rees was not alone in anticipating an early end to the war. As the nature of the conflict became clearer, and Fairey's unsuitability to serve more obvious, Rees hit on the idea of a dual company. The intention now was to build naval aircraft *and* to satisfy an unprecedented military demand for motor vehicles: the Army would doubtless appreciate a ready supply of 'light cars' easily adaptable to peacetime motoring. To fund this much grander scheme Rees sought Fairey's help in raising no less than £32,500.[37] CRF must have pointed out the absurdity of such a request, but Rees never lost sight of his dream to transform Pott Brothers, a south London firm building 'light cars', into the visionary 'D.E.P. Company.' While Fairey soon saw building seaplanes as the sole purpose of his new enterprise, Rees continued to promote a joint venture: he anticipated building as many as 300 military versions of the Potts' 'D.P. Car.' As late as 1916 Rees commissioned a draft prospectus and a vehicle catalogue. He and Fairey had already created the 'Facroldaw Syndicate', to complement the embryonic aircraft business and provide an alternative vehicle for raising capital.[38] Rees's partner could see the value of this initiative, but in April 1915 he had refused to invest £10,000, or a month later £5,000, in vehicle assembly. Apart from the fact that Fairey had no access to such large sums of money, he only wanted to build aircraft.[39] Cleverly, he never distanced himself from Rees's ever more complex financial involvement with Francis and Herbert Pott, and the mysterious and wonderfully named L.F. De Peyrecave. Until 1917 Dick Fairey needed his East Finchley wheeler-dealer, and he saw no point in unnecessarily alienating someone so assiduous and energetic when it came to raising money: let Frank Rees dream of countless cars pouring off the Potts' assembly lines, so long as he secured the capital and plant necessary for the mass manufacture of naval aircraft. When the moment was right Rees could be disabused of the notion that the managing director in any way favoured building land vehicles; only then would he also find out that CRF saw no permanent role for 'F.H.C. Rees, Gentleman' once Fairey Aviation was firmly established.

By late January 1915 Rees was projecting a healthy profit margin of as much as £650 per machine if 12 seaplanes were built in the first year of the syndicate's existence. A guaranteed return, a potentially prolonged conflict and a plane maker far from the front line meant 'The Fairey Aviation Company Limited' might be established earlier than anticipated.[40] Not that Fairey himself relished his name being widely circulated, remaining wary of how the Short brothers would react to such a trusted employee striking out on his own, and at the Admiralty's behest building their aircraft. The works manager was uniquely qualified to serve as a subcontractor, yet he was right to assume the Short siblings' sensitivity to his establishing a firm intent on moving swiftly from dependence to rivalry.[41] Fairey's insistence on anonymity created problems for Rees when drafting company documentation, as only he and Dawson knew who 'F.C. Richards' was. How, Rees asked Fairey in April 1915, would he prove that this was his pseudonym should both partners be killed? Given Dawson's posting to the Dardanelles, this was a not wholly implausible scenario, and the accelerating course of events soon saw Fairey and his company properly named.[42]

While CRF continued overseeing operations in Rochester, Rees laboured long and hard in London creating an initial company structure which would boast investment capital of £60,000 and working capital of £20,000, but absorb a start-up cost of only £7,640. He and Fairey would together be just short of majority shareholders, but, with support from Dawson (plus Nicholl and the anonymous Maurice Wright), enjoy full control of the company. Rees conceived a complicated but lucrative corporate arrangement, such that by 1916 a 30 shillings share value would boost Fairey's £6,000 investment three-fold. In return for making his partner a rich man, Rees awarded himself the title of joint managing director and an annual salary only £100 less than CRF's £600. Rees sought to raise the equivalent of over £5 million at today's prices because, as well as wanting to build cars, he envisaged a takeover of the Edinburgh Engine Company, thereby securing Fairey Aviation the competitive advantage of building all-British seaplanes. Having despatched his draft prospectus to Fairey's Medway lodgings, Rees sent a postscript claiming to have secured an overseas order for 50 seaplanes. Furthermore, a company lawyer was convinced that, 'we ought to make an absolutely stunning business of it … so I hope you will realise that the occasion demands both in your own and our interests a special effort on your behalf to pull this through with Sueter and also get Dawson to put up the Syndicate money at once.' The latter request was prompted by the £50 Rees claimed to have spent to date, and the urgent need for Fairey or his friend to 'raise a fiver' so he could pay off an impatient solicitor.[43]

Rees clearly failed to see the incongruity of his plea, as confirmed by muted complaints throughout 1915 concerning the amount spent on advancing the project. Surviving correspondence suggests Fairey's readiness to let Rees bear all financial costs, and a marked reluctance to meet his ostensible partner whenever in London. Thus negotiations took place largely by post, with CRF making clear from the outset that the opportunity to invest in a potentially profitable enterprise constituted a generous payment for Rees's services; the latter could not anticipate appointment as joint managing director, nor a salary of £500 per year. Fairey was rightly sceptical of Rees's inflated figures, and decidedly uneasy over the naming of two RNAS officers given that their financial involvement clearly flouted King's Regulations.[44] Dawson was sufficiently concerned as to ensure the final prospectus stated that his association with the company was suspended pending demobilisation. He was equally insistent that Rees had no privileged status in the new set-up, recognising Fairey's claim to make all final decisions.[45]

Rees largely ignored CRF's belittling of his role, while repeatedly badgering him for money, *and* for affidavits confirming the appropriateness of expenditure to date. He clearly anticipated recouping the cost of launching the company, not least the expenses incurred cultivating merchant bankers and government agents. At one moment Rees was the hard-headed business man, urging Fairey to pass on insider information about Shorts in order to sway potential City investors, and the next he was totally delusional, seriously proposing the Potts 'light car' project, the Scottish aero-engine takeover, the volume sales to 'one of the leading Continental Governments', and craziest of all, the sister company in Japan.[46]

Despite the distraction of 'DEP Cars', across the spring of 1915 there emerged a viable company structure, based upon realistic shareholdings and sensible salaries. Thus a credible syndicate agreement was in place by the time Rees met Fairey at the Charing Cross Hotel in early July. Later that month, on 23 July 1915, the two men signed off a document confirming Rees's demotion to junior partner in the new venture. Likewise, flaky investors like de Peyrecave were removed from the list of syndicate members. Also dropped was one R. Paterson, recruited by Rees as works manager under an extraordinarily generous share and salary deal. F.G.T. Dawson held the largest stake, and via a 'cash consideration' of £3,500 regained the £490 lent to Rees and Fairey in order to launch the company. Respective cash injections of £1,500 and £2,000 allowed each man to repay Dawson, secure 250 preferred shares, and avoid penury until such time as Fairey Aviation could support salary payments.[47] The source of this 'cash consideration' is unclear from the syndicate agreement, but the future economist and businessman Edward Beddington Behrens, self-proclaimed 'Fairey Godfather', supposedly took a keen interest in the enterprise. Given that he did not join the Royal Field Artillery until September 1915, Beddington Behrens could conceivably have fulfilled the role of patron. As heir to a considerable fortune he certainly had the money to support such a venture, but he was only 18 at the time, and as an officer cadet unlikely to have encountered Rees or Fairey.[48] Yet whoever provided this cash injection gave Dick Fairey a lifeline at a critical moment when he had left Short Brothers but as yet had neither plant nor workforce to satisfy the Admiralty's ambition of maximising manufacture of Type 827 seaplanes.

On 6 August 1915 Charles Grey's magazine noted the recent registration of 'Fairey Aviation Company Limited. Capital £35,500 in 25,500 ten per cent preference shares of £1 each and 200,000 deferred shares of one shilling each. First Director C.R. Fairey.'[49] This announcement, plus Rees's briefings to the trade press on Fairey Aviation's lofty ambitions, significantly inflated the value of the company. Future director and works manager Wilf Broadbent began working for Fairey in 1917, but was reliably informed that two years earlier no more than £15,000 had been available as working capital. Even that figure seems generous given how much Rees and CRF looked to 'Wuffy' Dawson for financial support. For the moment Lieutenant Dawson remained a silent partner, albeit one wholly supportive of the chairman and managing director. The equally reliant Charles Crisp joined the board as its legal advisor, while A.A. Amos served as company secretary. Within a year Dawson had been invalided out of the Navy and able to commence his duties as a working director.[50] Arthur Amos's appointment signalled the young Fairey's resistance to a near ubiquitous prejudice among the English middle classes in the early decades of the past century. As time passed, CRF acquired several Jewish friends, among them the City stockbroker Maurice Myers. Grateful for the lucrative investments made on his behalf, in June 1937 Fairey agreed to second 'Mossy' Myers' nomination to join the Royal London Yacht Club (RLYC). To his distress and discomfort, membership was discreetly stalled.[51]

Fairey Aviation's company headquarters and drawing office boasted the central London address of 175 Piccadilly. The choice of location may have been influenced by its close proximity to the editorial offices of *The Aeroplane*, but more likely was Frank Rees's preference for the grand statement. Throughout his life Richard Fairey had an aversion to half-measures, and his letters show him absorbing the axiom that you should start as you mean to go on. Wartime rent was lower than might be expected, with the Zeppelin threat depressing landlords' hopes of exploiting an increased demand for commercial property.

More pressing was the question of where to locate the firm's first factory, with the original intention being to concentrate all activities on the eastern shore of Southampton Water. In May 1915 the Royal Navy had found Rees a site at the mouth of the River Hamble, and for an outlay of little more than £3,000 he envisaged enough space to support every stage of production. These plans were scaled back when Fairey pointed out that Southampton boasted no fewer than three plane makers, all of whom were working around the clock and hampered by a shortage of skilled labour.[52] Thus work at Hamble Spit would be restricted to a small team assembling and testing seaplanes built inland, with available Admiralty sheds for the moment remaining empty. A wooden slipway was laid down for aircraft to be lowered by crane into the water, and concrete stilts were sunk to raise the assembly shop above undrained marshland. The facilities at Hamble were very basic; yet significant change came within 18 months when the formidable Brice G. Slater was instructed to transform the subsidiary works into a multi-purpose operation, including the manufacture of floats and wings. Henceforth all available property was fully utilised.[53]

Slater would prove a shrewd appointment, as was the recruitment of A.C. Barlow, a uniquely talented draughtsman whom Fairey had persuaded to leave Short Brothers.[54] The loss of such a competent designer no doubt infuriated Horace and Oswald, but there was little they could do to dissuade Barlow from setting up his drawing board in the crowded office of 175 Piccadilly. His task was twofold: to modify the Type 827 design to meet current Admiralty specifications, and to translate his employer's drawings into blueprints. Barlow proved a prize asset, and within a year he was given direct responsibility for all experimental work. To replace him Fred Wright was brought in as chief draughtsman. Able, conscientious, but lacking any great passion for aeroplanes, Wright went back to Kodak, his former employer, once the war was over. Fairey had so much respect for Horace Short that there must have been pangs of guilt over poaching Barlow, but he was determined to secure the best in the field. Like Broadbent and Slater, Barlow devoted the rest of his long working life to Fairey Aviation, proving extraordinarily loyal to its founder. From the outset CRF was a hard taskmaster, but he paid well, and he took a keen interest in the well-being of his closest associates and their families. He rightly valued loyalty and trust, hence the premium placed on his relationship with Dawson, Nicholl and Wright. Fairey especially valued the four friends' attorney in spats, C.O. Crisp, 'the first Director who laid the basis of their financial fortunes.'[55] In London on 22 December 1949 the company arranged a dinner for all employees with over a quarter of a century's service, and of the 282 guests no fewer than 32 had been with Fairey since World War I: sitting beside Sir Richard on the top table was A.C. Barlow. Proposing the toast that night was Brice Slater, insistent that he, and not Barlow, was Fairey Aviation's longest serving employee. Slater's organisational skills were so highly prized that, when finally he did retire, CRF promptly despatched him to Hampshire as manager of the Bossington estate.[56]

Fairey and Rees could now boast an office in central London and a subsidiary works under construction at the mouth of the Hamble, but their urgent requirement was an easily accessible factory. The ideal location would be south of London, and well served by road and rail: the seaplanes' final stage of assembly would require transit to the south coast. Just as the Admiralty had guided Rees in the direction of Southampton Water, so the Ministry of Munitions identified firms with spare factory space. West of the capital in Hayes the Army Motor Lorries Company (AMLC), on Clayton Road, was using only half its requisitioned factory to recondition military vehicles. Fairey and Rees successful negotiated to rent all surplus floor space. At the same time they noted that, across the railway line, on North Hyde Road, their new landlord owned a large disused shed nearly 300 feet long and 60 feet wide. The building was light and airy, and by no means beyond repair: the cost of conversion to manufacturing use could be offset by the low rent. Skilled workers on modest wages were still available as local industry was not yet sufficiently dense as to drive up labour costs in line with wartime inflation. However, the situation was changing fast, as in late 1915 the 200 acre Munitions Filling Factory No. 7 was nearing completion either side of North Hyde Road: a conjunction of two major highways, the Great Western Railway and the Grand Central (later Union) Canal rendered Hayes an ideal place to produce shells. By 1918 the ordnance factory would employ around 10,000 women and 2,000 men, but three years earlier the competition for workers was not so fierce as to force Fairey to seek an alternative site. Nor was he disinclined to buy up adjacent land further along the North Hyde Road. This was an astute investment, as late in the war Fairey Aviation would secure a Treasury loan to build a large new production plant. The factory at Hayes and Harlington would become the hub of the company's worldwide activities and remain so for the next 40 years.[57] For the moment the refurbished shed plus an ancillary two-storey building would allow Barlow to relocate and recruit a team of ancillary designers, and Fairey to set aside space for research and development. By late 1916 these interim facilities were all in place, and a company photograph from the time depicts a drawing office staffed by no fewer than 20 draughtsmen.[58]

Many of the machinists and mechanics Dick Fairey hired to work at Clayton Road had only recently arrived in Middlesex. These were Belgian refugees, despatched from the dispersal centre at Earls Court to be billeted in Hayes.

On 23 October 1914, councillors and philanthropists had responded enthusiastically to the Local Government Board's request that Hayes create a relief fund for exiled Belgians. Within a year the charity was supporting over a hundred displaced families, several of whom had fled the historic university town of Louvain, devastated by German troops at the start of the war. Given Belgium's reputation as a powerhouse of manufacturing industry, those male refugees currently repairing vehicles boasted skills easily adaptable to building aircraft. Furthermore, they were keen on finding fresh employment. The AMLC was a wartime initiative largely run by expatriates. Their compatriots on the shop floor suffered low wages and harsh working conditions, yet any complaints were answered by threats of conscription into the Belgian army.[59] Here, on hand, was the core of a credible workforce, and thus began Fairey Aviation's long and close association with the people of Belgium, a bifurcated nation quietly obsessed with aeroplanes.

Had Fairey delayed setting up his business any later, then he would have experienced great difficulty securing Royal Navy approval. Murray Sueter's intervention was timely, but so too was the War Office's reluctant decision to draw upon the tendering experience of the Admiralty. In October 1915 both departments bowed to pressure from the manufacturers: all future contracts would be negotiated not on a fixed-cost basis but on the principle of cost-plus. This switch offered Fairey a safety-net in that a cost-plus contract guaranteed the supplier a profit, with the state absorbing any undue rise in the price of parts and labour. The Ministry of Munitions noted that a growing dependence upon subcontracting favoured generous returns at each stage of the manufacturing process, but as yet it lacked the power to monitor or restrict excess profits. Only with the creation of Lloyd George's coalition in December 1916 did the former Minister of Munitions place the whole of the aviation industry under the control of what by now was wartime Britain's supreme procurement agency.[60] Fairey Aviation further benefited from a government decision in late 1915 to advance payment on contracts of up to 20 per cent of the original estimated value (with cost-plus this would invariably rise): 'Further advances of 60 per cent of the value of each machine (or up to 80 per cent if the full 20 per cent had not been paid earlier) were then available as soon as the machine was ready to test.'[61] Both pride and profit made Richard Fairey eager to design and build his own aircraft, but this indirect subsidy of R&D rendered the experimental process far less of a commercial risk. The second half of 1915 saw Lloyd George forcing his party, its leader and their unwanted coalition partners to accept the state apparatus's unprecedented role in waging total war. For politicians of every persuasion this was a painful period, their mantra of 'business as usual' brutally exposed as pathetic self-delusion. However, for a prescient, single-minded engineer and entrepreneur, Britain one year into its first 'industrial war' offered a unique set of circumstances within which to prosper.

Eighteen months into the Great War: creating a company intended to last

By the autumn of 1915 the fledgling Fairey Aviation faced a triple challenge: to build on time and on budget the 12 Type 827s, to secure follow-up contracts, and to initiate design of its own aircraft. Successful pursuit of the second and third objectives was clearly dependent upon achievement of the first. Through to June 1916, seaplanes F4-15 were under construction in Middlesex prior to their final assembly and testing on the River Hamble. Sydney Pickles was the experienced pilot hired to sign off the 827s. CRF saw the value of having such a highly regarded flier testing his early aircraft: approval by Pickles assured the Admiralty that Fairey could match Short Brothers in terms of quality control. The Australian proved a key player in the new enterprise's early years, and yet Fairey made it clear Vincent Nicholl would be his chief test pilot once the war was over. Pickles recognised where he stood in the company pecking order, and yet as the years passed he and his boss became firm friends. Even as the first seaplanes were being built the Air Department authorised Fairey Aviation to initiate design of a large twin-engine, three-seat aeroplane intended to serve a variety of purposes, from long-range fighter to light bomber. Development of the prototype F2 was tied up with expansion of the Hayes and Harlington site later in the war, but the project is highlighted here as evidence of how from the outset the Admiralty acknowledged Fairey as an ambitious company capable of fulfilling its potential.[62] To keep the fledgling firm viable the Air Department fed it lucrative repair and conversion contracts. This carried on even after Fairey's assembly work for Sopwith gained momentum and its early prototypes came off the drawing board. Thus, the company's close interwar relationship with Curtiss, the US pioneering sea-plane manufacturer, had its roots in a repair contract from late 1916.[63]

Tasked since September 1914 with sole responsibility for the air defence of the United Kingdom, the Royal Navy had enjoyed mixed results in aggressively pursuing a forward defensive strategy against the Zeppelin threat. One possible role for the F2 was to intercept those airships successful in crossing the North Sea, but all that was in the future; another was to realise Sueter's ambition of bombing Germany. The scale on which, 18 months into the war, the RNAS was operating – at home, on mainland Europe and in the Mediterranean – demanded increased output for those aircraft already available and tested in battle.⁶⁴ The Admiralty had its well-established working relationship with Short Brothers, of which Fairey was now a beneficiary. However, increasingly attention focused upon Sopwith Aviation, not least because the company boasted a winning combination of imaginative design and increasing productivity. In 1915 the Air Department responded immediately when Tommy Sopwith demonstrated his adjustable tailplane, and yet for no obvious reason the RNAS failed to pass on reports of early production models lacking adequate horsepower to handle choppy conditions at sea.⁶⁵

Nevertheless, by the winter of 1915–16 the Royal Navy was cultivating an ever closer working relationship with Sopwith, displaying a proprietorial interest in three distinct projects at advanced stages of development: a triplane; a single-seat scout (fighter) soon to find fame as the Pup; and a two-seat fighter/bomber about to enter production, the 1½ Strutter. Only 152 Sopwith Triplanes were built from the summer of 1916, but British success in downing German aircraft left a deep impression on the enemy's premier designer, Anthony Fokker. The Sopwith Pup's firepower, manoeuvrability, adaptability and efficient power–weight ratio rendered it a vital transitional fighter in the middle years of the war, hence its value to both the RNAS and the RFC. Together the Royal Navy and the Army ordered no fewer than 1,847 machines. This volume of manufacturing was unprecedented, and Sopwith's Kingston factories could only produce a fraction of total output. Even more remarkable was the 1½ Strutter, of which 5,466 were built, the majority under licence in France. For all of the Pup's popularity among pilots, and the awe in which fliers on both sides held the Camel, it was the Strutter which signalled Sopwith Aviation's arrival as a major force in Allied aircraft procurement. When 1½ Strutters appeared above the trenches in the spring of 1916, French observers grudgingly acknowledged *les rosbifs*' success in seizing the technological initiative. The parent company built in total 246 aircraft, after which production levels at the Kingston plant rose exponentially until 1919. The RNAS gained a multi-purpose machine with radio communication and the potential to be launched from modified gun turrets at low speed, and the RFC acquired a scout with the capacity to fire through the propeller arc and enjoy access to auxiliary airfields courtesy of engineer Fred Sigrist's freshly patented airbrakes. Thus, in terms of both quantity and quality, the 1½ Strutter can be seen as Tommy Sopwith's breakthrough aircraft. On a more modest scale, it served the same role for Dick Fairey.⁶⁶

Attention in Britain was focused upon the first great test of Kitchener's New Army when Contract 87A499 for 100 Sopwith two-seaters, with an airframe price of £892, was awarded to Fairey Aviation on 6 July 1916. None had reached the Western Front before the Battle of the Somme ground to a bloody halt in mid-November, but 26 of the subcontracted Strutters had been delivered before the year was out. Creating the means to fulfil an initial contract of 12 seaplanes was a notable achievement, however familiar the model, but scaling up the assembly process to build nearly ten times as many machines constituted a rite of passage. Sopwith provided Fairey with the necessary blueprints, an airframe to serve as a pattern, and a promise of assistance if requested; pressure on the subcontractor to effect early deliveries came not from Kingston but Whitehall.[67]

Government intervention came most directly via the Aeronautical Inspection Division (AID), organised under the auspices of the War Office from 1912 to 1917, after which it became one of the Ministry of Munitions' key monitors of quality control. An AID inspector had been on site scrutinising assembly of the Type 827s, but once CRF had secured the Sopwith subcontract he persuaded this man, by the name of Rose, to leave government service and become production manager. This was a shrewd move in that Rose could second-guess any issues his successor might raise, thereby speeding up the production process.[68]

The down side was the arrival of one Lieutenant Cox RNVR, a resident Admiralty Inspector. Even allowing for the managing director's deep loathing of the man, Cox was clearly the bureaucrat from hell, relishing the power he wielded while at the same time taking full advantage of a cushy posting with rich pickings. Cox enjoyed a host of privileges paid for by Fairey, most notably access to a company car with a full tank of petrol. The 'temporary gentleman' took particular pleasure in undermining senior management's authority and reputation, with CRF convinced a deterioration in labour relations was attributable to the man from the ministry. He was fobbed off when visiting the Air Board to complain about the agent provocateur in naval uniform, but soon after drafted a lengthy indictment. The letter leaves Fairey open to charges of pomposity, although it does give an insight into the hands-on fashion with which he ran *his* company. Yet any charge of arrogance is offset by admiration that he tolerated such an obnoxious meddler for so long, particularly as Cox demonstrably held up delivery of urgently needed machines. The demand for change clearly worked, as company records make no further mention of a resident Admiralty Inspector: Lieutenant Cox, along with his petty officer and personal chauffeur, were clearly posted a long way from Hayes, ideally to a destroyer flotilla out on the Dover Patrol.[69]

As well as relocating the drawing office from Piccadilly to the refurbished shed on the Harlington side of the railway track, Fairey leased additional factory space from the AMLC. He now had the necessary plant but he still required a sizeable area of open ground suitable for testing the Strutters. Hamble was clearly inappropriate, and the field behind the Harlington shed too small. Finally, a field was secured in neighbouring Kingsbury, on the west side of Watling Street. Two miles north on the old Roman road, De Havilland would take over the Stag Lane aerodrome in 1920, assembling aircraft on site until the company relocated to Hatfield 14 years later. De Havilland's arrival confirmed how in the course of the war the British aviation industry's epicentre had shifted from the Thames estuary to west of the capital.[70]

Building the 1½ Strutters began in late October 1916 and continued for over a year. The workforce was significantly expanded, with more Belgians recruited if they possessed the skills necessary to build an aeroplane demonstrably more sophisticated than the Type 827. More employees ensured trade union interest, with the young company disputing the claim that it was party to the engineering employers' London-wide agreement on wages and conditions. Fairey insisted that his absence from negotiations undertaken in the autumn of 1916 was because his company was located outside the capital. Technically this was not the case, but a government-sponsored arbitration report found in Fairey's favour, albeit at the cost to the company of longer paid breaks and a weekly bonus of five per cent of the total hourly rate for the job, including overtime.[71] Ernest Page QC's recommendations increased labour costs almost as much as if he had judged Fairey Aviation to be a London-based enterprise. Given the firm's fragile financial position, its managing director's later advocacy of payment based on results and not by the hour can be traced back to this first unhappy encounter with organised labour. The same can be said for CRF's deep loathing of trade unionism, tempered by a pragmatic recognition that collective bargaining was a reality: wherever possible shop stewards should be isolated by appealing directly to the workforce and vetoing closed shops, and yet it was naive to ignore the disruptive potential of powerful trade unions such as the AEU.[72]

Full production required use of almost the entire Clayton Road factory, and yet Fairey Aviation remained a subtenant, subject to the bureaucratic whims of AMLC managers with time on their hands: all matters relating to the plant were passed on to invisible directors, who only took action when it was in their interest to do so. To compound Fairey's frustration, component suppliers confused as to who constituted the principal tenant more than once charged his firm for materials ordered by the AMLC. Initially, mischarging was generously seen as evidence of maladministration not malpractice. That magnanimity disappeared once the AMLC management and board employed control of the lease as a bargaining chip to secure a shareholding in Fairey Aviation. Relations between respective companies broke down, giving CRF an excuse to act.

A marked feature of Richard Fairey's career was that, while he rarely acted impulsively, he invariably acted decisively; once embarked upon a course of action he saw it through to a successful conclusion. In this instance he unilaterally took control of the whole factory, symbolically occupying the managing director's office and reversing the power structure such that Fairey and not AMLC ran the plant. Once the new order was in place CRF relied on Charles Crisp to see off the inevitable injunction. In purely legal terms his coup de main was indefensible, but this was wartime and inside the Ministry of Munitions there could only be one winner. The AMLC was a busted flush run largely by foreigners; Fairey Aviation was busy building 100 state-of-the-art fighter aircraft, with further contracts in the pipe line. The matter was settled out of court, with Whitehall sending Fairey a strong signal that future plans for expansion of the plant could anticipate full government support.[73]

Richard Fairey now had the confidence and experience to grow, establishing volume production as the norm. This would be no wartime contingency, but the achievement of critical mass, establishing the plant and apparatus necessary to absorb sizeable contracts and exploit economies of scale. At least four companies this side of the Channel which bid to build a similar number of Strutters, or even considerably more, either closed or abandoned aircraft manufacture after 1918. They exemplified a peculiarly British talent for improvisation and mass mobilisation, their embrace of plane making a temporary expedient.[74] Fairey Aviation, as its name proclaimed, was a very different animal, exploiting wartime opportunity, but with eyes fixed on survival and prosperity beyond the return of peace. Yet to forge a clear identity and ensure its long-term survival, the company's priority was to design and sell its own aircraft. In the meantime it maintained a symbiotic relationship with Sopwith Aviation, subcontracted to build the 1½ Strutters and eager to take on any other work beyond the capacity of the Kingston factories.

As yet, Dick Fairey remained in the shadow of T.O.M. Sopwith, a man whose weight within the industry ensured his company's presence on the council of the Society of British Aircraft Constructors. The Society was founded in April 1916, the same month in which Tommy Sopwith joined Charles Sansom and their fellow committee members in initiating a sizeable – and symbolic – expansion of the Royal Aero Club.[75] While Fairey Aviation joined the SBAC immediately, and its managing director would be elected president in 1922, at this point CRF was still a peripheral figure, with no obvious power base and, in the eyes of less enlightened RAeC members, a ground crew image. Within a social order scarcely changed by war, the brave new world of aeronautics was unusually meritocratic, but any shift in status and position demanded a visible display of success. Richard Fairey wanted – and would enjoy – the trappings of wealth, but above all he sought recognition, not least recognition by his peers within an industry that, driven by the demands of total war, was experiencing an extraordinary rate of growth. Fairey Aviation was a direct result of that wartime growth, but the company – and its founder – remained for the moment minor players. With no notion of when the Great War would end, the case for relentless expansion was irrefutable.

CHAPTER 4

Great War and Great Expectations

Dick Fairey's company

The growth of the British aircraft industry in the course of World War I is truly remarkable, not least the dramatic acceleration in production across the final two years of the conflict: monthly output at the start of 1917 was still only 122 machines and yet, by the time of the Armistice, a workforce of around 300,000 had boosted that figure to a remarkable 2,688. The RAF lost as many as 7,000 aircraft in the last ten months of the war, and yet operational squadrons enjoyed a steady stream of replacements. Long-serving ground crews found the supply of spares equally reliable, enabling frontline serviceability above 85 per cent.[1] Production on this scale powerfully demonstrated Britain's belated embrace of 'industrial war', with large, suitably skilled design teams facilitating a vital balance of quantity and quality. Unsurprisingly, poorly performing machines still somehow survived the prototype stage and went into production, feeding a voracious appetite for combat aircraft. Despite these death traps reaching the front line, a Darwinian process of procurement prioritised the production of planes tested in the air war taking place day after day in the skies above Picardy, Pavia and Palestine – this was in every sense a global conflagration, with RFC and RNAS squadrons deployed en masse far beyond the Western Front. Aircraft like the Sopwith Camel and SE5a were proven killing machines, especially when flown by pilots like Mick Mannock and James McCudden. These were the ace tacticians of the embryonic RAF whose engineering credentials enabled them to grasp the implications for aerial warfare of what, by comparison with only four years earlier, constituted a new and strikingly sophisticated technology. Both men earned their spurs in the RFC, surviving above the trenches far longer than those pilots who regarded the combat readiness of their machines the preserve of deferential mechanics.[2]

In contrast, the RNAS, born out of Horace Short immersing Charles Sansom and his fellow fliers in the hard science of aeronautics, always placed a premium on the engineering expertise of its pilots and observers, even as the Air Service expanded in the course of the war. The calibre of air crew and the unique qualities of Short and Sopwith seaplanes in 1914–15 had allowed the RNAS to take the initiative from the outset, laying down a marker for offensive operations on a scale scarcely conceivable to their military counterparts in the opening phases of the war. Thus, even before naval aviators launched their first raids on Zeppelin sheds across the North Sea, the Admiralty directed Handley Page to design a twin-engined heavy bomber. Murray Sueter envisaged a naval strike force directed towards Germany's industrial infrastructure; a strategic initiative finally realised two years later, albeit with only modest results.[3]

The exercise of strategic air power was a deeply divisive issue, crystallising every element of interservice rivalry. The Army complained of the Admiralty's reluctance to divulge plans, share scarce resources and adopt a common procurement programme. The RFC recognised the Royal Navy's unique relationship with the likes of Oswald Short and Tommy Sopwith, but deemed this inappropriate in time of war. The War Office saw naval obstructionism unaffected after Churchill resigned as First Lord in 1915, and Sueter became more accountable for his actions. If anything differences over procurement intensified, with fierce competition for British-built aero-engines. These remained in short supply, and even the most efficient power plants rarely proved reliable. The disconnect between aircraft design and engine efficiency in the middle years of the war was critical for both the RFC and the RNAS. Increasingly, both wings were flying the same marques, which compounded the competition for proven combat aircraft such as the 1½ Strutter.[4]

Murray Sueter was a genuine visionary, and, as in the earliest days of Fairey Aviation, the company's expansion in the second half of the war was a consequence of his prescience and patronage. Even after Sueter was exiled to Taranto in February 1917, his influence across the British aircraft industry remained palpable, not least in the thinking of Dick Fairey and his closest associates.[5]

Fairey Aviation was by no means the only dedicated aircraft manufacturer established in the course of World War I, with Westland similarly founded in 1915. Over 40 years later the two companies would share a shotgun marriage, but 18 months into the Great War neither could imagine the size and scale of their respective operations at the moment of merger. In 1959 both Fairey and Westland would look back on how rapid wartime expansion established an operational model for the ensuing decades: ambitious R&D, ideally funded by government; a range of projects to prevent over-dependence upon a single prototype; multi-plant volume output facilitated by a fusion of skilled assembly workers and modern production methods; the capacity to transport and test finished aircraft; reliable components suppliers locked in on favourable supply terms; aggressive marketing, promotion and lobbying to secure contracts; and a hands-on approach to management, with senior staff maintaining a presence in the design office and on the shop floor.

These were paternalistic business organisations, the reputations of their key players as engineers *and* entrepreneurs consolidated in the course of the war. Plane making was by no means unique in this respect. The aero-engine, machine tool and even motor manufacturing industries all maintained the power of the founding father long after individual enterprises had ostensibly dropped an outmoded business structure and become public joint stock companies.[6] Industrialists like Sir Richard Fairey were adroit at retaining a tight control of their businesses in an era of 'personal capitalism' when individual and institutional shareholders simply sought to secure a reasonable return upon their investment. Thus the emphasis was upon annual performance and dividend, with a reluctance to support major cash injections and thereby maximise deferred profits.[7]

Looking ahead, in the years after 1945 the failure of Fairey's institutional shareholders to think strategically would cause severe problems for an aerospace company pursuing several costly and technologically challenging development projects without adequate government support. Such behaviour was of course consistent with a century of British financial institutions failing to emulate their German counterparts and thus denying hi-tech enterprises hands-on, long-term capital investment.[8]

In a paternalist (and indeed patriarchal) organisation, the passivity of shareholders was reinforced by boardroom support for the chairman and managing director – often, as in the case of Fairey Aviation, one and the same. In most cases the board was largely made up of loyal executive directors answerable on a day-to-day basis to their boss. For Richard Fairey, personal and corporate loyalty was reinforced by a network of close friends, their support for each other dating back to high jinks in the Medway towns before World War I. Dominant, high-profile figures like Frederick Handley Page or John Siddeley retained full authority after their companies were floated on the Stock Exchange, by dint of a controlling interest or compliant shareholders. CRF was unusually adept at maintaining a tight grip on corporate affairs, both in the 1920s when his company was repeatedly restructured to generate working capital, maximise returns and minimise the tax burden, and in the decade after Fairey Aviation went public in March 1929. An unchallenged position, within an enterprise which after all bore his name, was secured only two years after Fairey left Short

Brothers, encouraged to do so by the unlikely combination of Commodore Sueter and Frank Rees.

As Fairey Aviation's first deputy managing director, F.H.C. Rees was living on borrowed time. Wounds sustained on the Dardanelles had necessitated an extended convalescence for 'Wuffy' Dawson. Yet by Christmas 1916 he had hung up his naval uniform and, as planned, taken a senior post at Fairey. The company's principal investor was now operating in tandem with his old friend from Sheppey, and almost immediately both men moved to ease out Rees. Extraordinary general meetings held early in 1917 saw the company's articles of association amended to allow consolidation of control if one of the three directors' shareholding dropped below 7,500. This may already have occurred because within weeks Fairey and Dawson secured the Ministry of Munitions' permission to buy out Frank Rees, as well as any surviving small-scale investors. Securing dual ownership of the company, while retaining an option to grant Nicholl, Wright and Crisp directorships as and when, cost the not inconsiderable sum of £8,632. Dawson and Fairey shared the expense equally, although the former may well have lent his friend some or all of the necessary capital. To seal the deal, CRF persuaded Rees to sign a release form, binding him to make no future 'claims and demands' upon his former partners.[9]

Both men were in an enviable position once wartime cost-plus contracts became of sufficient size to maximise economies of scale, thereby securing a healthy profit margin on every machine built. The key here was early payment, as from the start of the war Whitehall's procurement agencies were notoriously indifferent to the cash-flow problems of new firms like Fairey Aviation.[10] The problem was compounded by the harsh consequences of the Government charging an Excess Profits Duty (EPD) between 1915 and 1921. The EPD accounted for a quarter of the Government's wartime tax revenue. It was imposed on all manufacturers' profits above a prescribed prewar standard, with the rate reaching a peak of 80 per cent in 1917. Fairey and Dawson felt unduly penalised, as any profit made was considered 'excess' owing to their company not existing before August 1914. Not surprisingly, CRF was from the outset an enthusiastic member of the Society of British Aircraft Constructors (SBAC), launched on 29 March 1916 to lobby Westminster and Whitehall. His faith in the embryonic SBAC was vindicated when government accountants acknowledged that young firms like Fairey deserved special consideration. Treasury goodwill was markedly absent once the war was over, and the Society lobbied for suspension of its members' tax liability.[11]

In any case, the de facto co-owners did enjoy a financial safety net: in early 1917 the articles of association had been further amended to allow directors to raise their salaries at the company's annual general meeting. The managing director was now earning £3,000 a year, while Dawson took home £2,500. Arrangements for their remuneration were in stark contrast to the shady system maintained by Rees across the winter of 1916–17. Because the bank vetoed directors drawing salary cheques on their own behalf, CRF had agreed to a delay in listing these weekly payments in the company records. However, in October 1917 Fairey insisted that the accounts be amended, and to his horror he found his former partner had for ten weeks at the start of the year debited salary outgoings under assumed names.[12]

Although for ever indebted to the great wheeler-dealer, from the summer of 1917 Fairey moved swiftly to distance himself from Frank Rees. Indeed he was extraordinarily successful in ensuring Rees's absence from any history of Fairey Aviation. A quarter of a century would pass before Fairey again showed interest in Rees, noting the one-time deputy managing director's death in a letter to Charles Crisp: 'He wrote to me about two years ago when his son was lost. Incidentally, did he [Rees] die a natural death, or did someone murder him?'[13] The question could conceivably concern the child, but a lifetime of dodgy dealing in the wilder reaches of the City suggests it was the fate of Rees senior which intrigued a newly knighted Dick Fairey.

As the preceding chapter made clear, Fairey never envisaged a long-term working relationship with Rees, and what probably sealed the latter's fate was a lengthy, complicated and costly civil case that went as far as the Chancery Division of the High Court of Justice. When acquiring and furnishing the Hamble site Rees had made a series of rash promises to Henry Beazley, a local building contractor. As a result, Beazley believed himself entitled to receive 2,500 shares in Fairey Aviation. He aggressively pursued his claim through the courts, ceasing only when Crisp offered a settlement of £3,000, to be paid in three annual instalments from June 1918. Paying off the litigious builder, and settling Ashurst, Morris, Crisp and Co.'s solicitor's charges, was altogether an expensive business, and it could so easily have been avoided. No wonder that later in life Fairey could imagine someone so irate with Frank Rees that he or she resorted to violence.[14]

If Rees left no lasting legacy the same could scarcely be true of Murray Sueter, who even at a distance maintained a keen personal interest in the fortunes of Dick Fairey's fledgling company. The creation of Fairey Aviation had coincided with the Director of the Air Department's effective demotion and the RNAS's closer integration into the Royal Navy – Sueter and Churchill had run it like a personal fiefdom.[15] Although now technically number two, Sueter remained a live wire; it was never advisable to cross him and yet established favourites relished his friendship. Serial courting of controversy culminated in exile to the Mediterranean, not long after Sir John Jellicoe arrived from the Grand Fleet to become head of the Navy. Early in 1917 Jellicoe appointed a Fifth Sea Lord to oversee the RNAS. With the First Sea Lord an enthusiast for flying and a dynamic Director of Air Services freshly installed, Commodore Sueter's departure created few problems for firms like Fairey, which were dependent on Admiralty contracts but at the same time sympathetic to the Ministry of Munitions assuming control of all aircraft supply.[16]

Dick Fairey's designs

Although he had little or no say in the Sopwith subcontract, Murray Sueter had already given Fairey advance notice of a Royal Navy competition to design and build 'Twin Tractor Armoured Fighting Aeroplanes', one a heavy bomber and the other a fighter.[17] Fairey set out to produce two distinct aircraft, but the end result was a hybrid. His twin-engined biplane carried a crew of three, with a rugged undercarriage to facilitate night-time landings, and its folding wings and opposite-rotating propellers drew upon CRF's seaplane expertise.[18] The aircraft was neither one thing nor the other. It was huge and heavy, but without the capacity to carry a large body of ordnance. Although well armed, as a fighter it was too slow to compete with the speedy single-seat scouts busy redefining the air war above the trenches. In other words, the F2, while an impressive early effort in terms of design and manufacture, was already obsolete by the time Sydney Pickles tested it for the first time on 17 May 1917, over a year after the Royal Navy's original deadline for delivery of respective prototypes. Its predecessor, the F1, a purpose-built bomber, had never even left the drawing board.[19]

Even while the war was still on, Fairey Aviation was pointing to the F2 as evidence that back in 1914 it could design and build machines ahead of their time. This was of course pure fiction, and yet the company had spun the story to friendly editors like C.G. Grey, circulating declassified photographs of 'a peculiarly British contribution to aviation.' Both *The Aeroplane* and *The Graphic* were suitably impressed by the F2's folding wings. Captions noted that at least one British company had pre-empted the Germans in anticipating the significance of aerial bombardment.[20] Such reporting must have grated with Frederick Handley Page, not least as Fairey's 1916 design was for a three-man fighter. As we shall see, by the late 1920s and early 1930s, CRF enjoyed the status and standing of a major industrialist, holding high office in his industry's most distinguished bodies. A frequent lecturer and after-dinner speaker, his mantra was that manufacturers not civil servants always knew best, one lesson from the last war being that Admiralty bureaucrats missed a trick in not seeing the F2 as a means of bringing down Zeppelins, even though the RNAS had surrendered responsibility for the nation's aerial defence as early as March 1916.[21] For Fairey Aviation and its embryonic publicity machine, the F2 was sadly the game-changer that never was, neither taking on German raiders over London nor wreaking revenge above the Ruhr.[22]

These pioneering prototypes, the F1 and F2, generated extensive correspondence between Fairey and officials in Sueter's technical department. The latter were clearly unimpressed by the fact that CRF had the ear of someone so antagonistic towards both senior and junior staff. In their bids, contractors had been 'given an absolutely free hand with regard to design and construction', but presumed capable of satisfying the Royal Navy's exacting technical specifications. The former Director admired Fairey's efforts to meet service requirements, rendered ever more demanding as the Allies struggled to match Anthony Fokker's innovations in fighter design. However, by September 1916 Sueter was advising his protégé to play to his strengths, namely the construction of high-performance seaplanes.[23] If Fairey Aviation were to build state-of-the-art scouts then the blueprints would originate elsewhere, namely in Tommy Sopwith's design office at Kingston.

The last 1½ Strutter flew out of the Kingsbury airfield in September 1917. The assembly line at Clayton Road had been in full production for 11 months, on more than one occasion building a complete machine within the working day. This initial success meant that Fairey Aviation's role as a subcontractor for Sopwith was by no means over. Larger factory space ensured the company now had the capacity to assemble 100 Camels, although none were built before the contract was cancelled in late 1918. A.C. Barlow's duties were by now so wide-ranging that the design team was placed under a new man, Fred Duncanson, accountable directly to Fairey. Duncanson found his draughtsmen could not come up with a fighter aircraft comparable to the Camel, but in the summer of 1917 they seized the chance to make Sopwith's best known seaplane fly faster and higher: the Baby was given larger floats, a bigger engine, a fresh tail unit, and a revolutionary wing design to maximise lift. The RNAS took delivery of 50 machines in the middle years of the war, and they saw action in the Western Approaches and the Mediterranean.[24] The Fairey Hamble Baby, as it was rather clumsily known, marked a double first for CRF. It sent a strong signal to fellow subcontractors, notably Blackburn and Parnall, that their new rival had the capacity to be genuinely inventive *and* build aircraft in ever larger numbers. At the same time Tommy Sopwith and his chief engineer Fred Sigrist could not fail to be impressed by the fact that their original design now flew at almost 100 mph and carried a far heavier load, including the fuel necessary for long periods out on patrol. All these gains in the seaplane's performance were thanks to CRF patenting the variable camber gear; his invention was crucial to Fairey Aviation's good fortune in the era of austerity and retrenchment that followed the return of peace.[25]

From the outset, Fairey and his design assistants sought multiple patents, each submission boasting superbly executed sectional drawings and a carefully crafted explicatory text. Archives at both the RAF and the Fleet Air Arm museums contain boxes bulging with dense and detailed patent requests.[26] The earliest applications to the Patent Office set a precedent, and Fairey Aviation was consistent in securing even the most modest innovation protection from plagiarism. Dick Fairey submitted a patent application for the variable camber gear on 19 May 1916. In due course his invention would become a still fragile company's most valuable asset. The speed with which his invention was formally recognised reflected the input of the Air Department. Sueter's technical advisers were anxious to develop high-lift wings that enabled heavily loaded seaplanes to take off and land at lower speeds, thereby saving fuel.[27]

Squadron Commander John Seddon was already testing aerofoils at the Marine Experimental Depot on the Thames estuary. A physicist in uniform, he drew on data collected by colleagues working in the National Physical Laboratory's unique wind tunnel facility. Endeavouring to translate theory into practice, Seddon tested a standard Sopwith with heavier but steeply cambered wings. The outcome was deeply frustrating, as the load increased significantly, but at the expense of speed. CRF's reputation among the aviators of north Kent made him an obvious choice when Seddon sought collaboration with an experienced, hands-on engineer. He briefed Fairey on his own work, as well as abortive experiments with wing flaps at the Royal Aircraft Factory, and efforts by the Varioplane Company to create a flexible wing surface.[28]

Flexing wings on any aircraft, let alone a biplane, sounds the sort of left-field thinking J.W. Dunne would have enthused over. However, his disciple, ever the hard-headed engineer, sought a more functional solution, devising a cable and pulley system to connect cockpit and wing-flaps. As the years passed, successive patents rendered this system more sophisticated, with the pilot adjusting the trim fore and aft via a steering wheel. The trailing edge of each wing was hinged so it could be lowered to increase the lift on take-off before reverting to its in-flight function as an aileron, with the reverse operation as the seaplane slowed down to land: 'By thus providing adjustable camber some, at least, of both slow-speed and high-speed requirements could be met.'[29] Throughout the 1920s, promotional material for Fairey Aviation invariably highlighted the 'Patent Variable Camber Gear' as clear evidence that this was a company synonymous with cutting-edge technology. The attention of aviation enthusiasts was drawn to patents covering ever more complex flap gear. Naturally Fairey was wary of patent infringement. In November 1919 he demanded financial compensation from the Ministry of Munitions, having charged Farnborough with illicit use of his 1916 Admiralty paper on 'high lift wings.' With his company strapped for cash, CRF offered to sell HM Government the variable camber gear patent for £10,000; luckily for him, the offer was turned down. The offer to sell was not so surprising in that two years earlier the Ministry of Munitions had purchased three patents pertaining to Fairey Aviation's first production aircraft: the firm received £35,000 in instalments, but Fairey himself secured £15,000 in a lump sum. This was a significant sum of money, and design of the Campania seaplane made Dick Fairey a wealthy man.[30]

CRF plugged his mechanical invention in a succession of articles and speeches, arguing that it allowed the Air Ministry to prioritise seaplanes for commercial operations: Fairey Aviation's use of the Thames at Isleworth signalled the potential for establishing short take-off and landing sites on all major rivers. This aggressive promotion of the variable camber gear as vital to the seaplane's bright future reached its apogee in Fairey's address to the Air Ministry's 1923 conference on the future of British aviation, and in the dense technical paper which fleshed out his remarks.[31] Inspired perhaps by Frank Rees's talent for showmanship, the Fairey company's advertising was always impressive, skilfully combining design flair with a serious message. Interwar promotion fused art deco and modernity, while early adverts insisted that 'the British Government buys Fairey Seaplanes ... from experience of test results on standardised figures and not from verbal claims and experimental demonstrations.'[32] These then were aeroplanes tried and trusted, enjoying HM Government's wartime imprimatur: the impression Fairey sought to create in ensuing years was that its variable camber gear bore a quasi-official seal of approval, which indeed it did.

Yet, however ingenious Fairey's attempt to reconcile load and speed, the variable camber gear was heavy and cumbersome, challenging the manual dexterity of even the most experienced pilot. There is no doubt that for Fairey Aviation the flap system proved hugely profitable, courtesy of the best-selling Type III, and yet high-lift devices were rarely employed by rival manufacturers in the postwar years. Fellow designers' ambivalence towards Fairey's invention explains the delay in acknowledging his achievement. He received the Royal Aeronautical Society's Wakefield Gold Medal as late as 1936, the award in part an appreciation of all his work on behalf of the aviation industry.[33]

The precursor of the Fairey III series, and the first factory design to be assembled in sizeable numbers, was the Campania, a two-man seaplane for maritime patrol. Of the Royal Navy's original order for 100 machines, half were built at Hayes and then reassembled on the south coast.[34] Twelve more were made in Scotland on the Clyde, but Fairey's first subcontractor ceased production once the Armistice was signed. Two prototypes had been tested by Sydney Pickles at Hamble in early 1917, and then handed over to the RNAS. The Royal Navy promptly agreed to Maurice Wright flying its first Campania non-stop from the Nore to Scapa Flow, home of the Grand Fleet. The aviation-minded Sir David Beatty, Jellicoe's successor as C-in-C, was surely impressed, while Dick Fairey applauded an initiative in which he no doubt had a hand. The record-breaking flight laid down a marker and strengthened the symbiotic relationship between the Admiralty's Air Department and Fairey Aviation: the upstart enterprise was now shown the same healthy respect as Short Brothers, Sopwith and Supermarine. Like Shorts, Fairey was receptive to the Navy's needs, with variants of the Campania tailored to specific requirements, most notably extremes of temperature.[35]

Admiral Beatty would have appreciated the irony of greeting an aircraft named after the carrier which missed the Battle of Jutland, thereby denying the Royal Navy vital aerial reconnaissance. HMS *Campania* was a converted passenger liner. She had been purchased from Cunard early in the war because she cruised fast enough to maintain station when the Grand Fleet was at sea.[36] Like her companion carriers, *Campania* hoisted her complement of seaplanes in and out of the water, but a series of modifications enabled aircraft to take off from a flight deck 200 feet long. The planes were launched using a retrievable four-wheel trolley, and successful trials with larger machines in the summer of 1916 encouraged the Admiralty to commission from Fairey a suitably rugged machine.[37] Meanwhile, naval aviators experimented with platforms for take-off *and* landing.[38] Only 14 months after a Sopwith Pup's first landing on a modified cruiser, HMS *Furious*, the Royal Navy was fitting out *Argus*, its first aircraft carrier with a ship-length flight deck. Even after the Armistice was signed the Admiralty anticipated a 'Flying Squadron' of ten adapted or purpose-built seaplane or land plane carriers. Technical and operational experimentation throughout the war saw the Royal Navy render maritime air power a reality. At the time, Sopwith Aviation was an obvious beneficiary, its Camels ideally suited to carrier-based attack. When taking the long view, however, Fairey proved the aircraft manufacturer most indebted to the Senior Service's vanguard role.[39]

Like its namesake, the Campania was a transitional model, with its successor, the Type III, a more revolutionary design, but for Richard Fairey and his colleagues this was an aircraft of huge significance. With each machine costing £2,950, the Air Department's order gave Fairey Aviation every reason to expand its production facilities.[40] The company convinced the Ministry of Munitions that any such expansion was only feasible if properly financed, and here the role of Lloyd George's 'warfare state' was vital.[41] By the spring of 1917 the case for Whitehall's newest but most powerful department of state to assume responsibility for aircraft supply was overwhelming: unplanned growth under competing procurement agencies had resulted in chronic duplication with too many manufacturers producing too many models in too few numbers. This failure to rationalise and secure economies of scale was matched by the aero-engine industry. When compared with their French equivalents, British engines were too often heavy, under-powered and built to a poor standard. A new generation of water-cooled engines, reaching between 200 and 400 horsepower, had yet to enter production.[42]

CRF was only too well aware of this, switching between Sunbeam and Rolls Royce power plants to maximise the Campania's overall performance. Fitting seaplanes with a high-lift device was Seddon and Fairey's response to the absence of a fuel-efficient engine and a tolerable power–weight ratio. It is not an exaggeration to say that, as with the air war across the Channel that spring, British aviation at the start of 1917 was in crisis.[43] The turnaround in the course of the next 18 months was remarkable, and the Ministry of Munitions was crucial to the industry's dramatic improvement in output, productivity and quality control. Fairey Aviation clearly benefited from the new regime, acquiring factory space crucial to the firm's survival and prosperity in the aftermath of war. However, as we shall see, expansion of the plant came at considerable cost, threatening the survival of the company while the war was still on.

From the outset, Fred Duncanson and his boss forged a productive partnership. Together they masterminded what, when finally built, were the biggest flying boats in the world, weighing in at over 30,000 lb.[44] Fairey Aviation had no experience of designing, let alone building, big four-engined biplanes. These were flying boats with a crew of five and the capacity to patrol for up to nine hours in the harshest of maritime conditions. It is testimony to the inventiveness of the Fairey–Duncanson partnership that early in 1918 the Admiralty ordered three prototypes from a firm which in actual fact lacked adequate floor space to construct aircraft of this size. With the company already working at maximum output, Fairey was forced to contract out assembly of the first and third machine. All three hulls could only be built by specialist boat builders, with two on the Solent and one over 300 miles away on the Clyde. With no fewer than six subcontractors, production and transportation was a major administrative and logistical exercise which entailed extensive delays. The inaugural machine, named *Atalanta*, finally flew on 4 July 1923. The maiden flight of its successor, *Titania*, was not for another two years. The third of the series was destined never to fly, even though it was built by a Bradford company called Phoenix Dynamo.[45]

Barlow masterminded the operation and it is scarcely surprising that he suffered a breakdown. The pressures upon him were enormous, as was the case for C.F. Bray, the charge hand promoted to chief of Fairey's experimental department. Neither man boasted Silvanus Thompson as his tutor, and both were intellectually challenged by the performance graphs and detailed engineering notes CRF generated day after day. Dick Fairey lacked the theoretical underpinning three years at Cambridge could provide, but he was the empiricist *par excellence* – test, test and test again. Armed only with a slide rule and a set of tables he absorbed, assimilated and analysed a vast body of data. He expected Barlow to follow his example and at the same time run a factory. Duncanson clearly thought along the same lines as his employer. He was a trained engineer and not merely the brightest mechanic on the shop floor. CRF had confidence in his new man's capacity to translate the grand conception or the neat idea into the necessary blueprint.[46]

That confidence was reinforced by Duncanson's ability to work with Fairey Aviation's chief technician. In his fifties and already a father-figure in the design office, P.A. Ralli had been born in Greece, but later studied and then taught engineering science and pure mathematics at the Sorbonne. He had fled Belgium in 1914 and, via Earls Court, finally found himself in Hayes. Fairey had been quick to realise his good fortune in gaining a genial workaholic who, 'just lived for the job and delighted in finding elegant mathematical solutions to avoid long tedious calculations.'[47] Ralli's organisational skills left a lot to be desired, but his powers of computation were formidable, not least when designing propellers for high-performance aircraft such as Supermarine's Schneider Trophy-winning S5. The cosmopolitan and colourful Ralli was by no means unique. Works manager F.M. Charles, of whom more later, was similarly marinated in Cartesian scholarship the far side of the Channel, and in 1920 CRF persuaded Belgian designer Marcel Lobelle to join Fairey Aviation rather than return home. Once night classes in Finsbury had polished up his linguistic and computational skills, Duncanson and Ralli found their new colleague invaluable: Lobelle had volunteered for the Belgian Army at the start of the war, sustained serious wounds, and on recovery been sent to build aircraft in England. His presence reinforced and sustained Fairey's close connection with Belgium, even after hostilities ended. Four years after Lobelle arrived he took control of the design team when

Duncanson had a furious row with his boss, and left to work for Blackburn. A charismatic if somewhat eccentric character, the Belgian was popular and much admired, but denied official acknowledgement as chief designer. Fairey reserved that title for himself, remaining silent whenever colleagues labelled Lobelle's blueprints the work of a genius. Such hyperbole was understandable, for Marcel Lobelle was the father of the Swordfish.[48]

An overstretched young company should never have received the flying boat commission, and yet its chief executive had demonstrated that he had a team capable of designing large, complex aircraft. The firm was already confirming its capacity for bringing in big orders on budget and on time, so it is not surprising that the sad story of the N4 series did not dent the Ministry of Munitions' confidence in Fairey Aviation. Yet a lesson had been learnt – future seaplanes might be of an unprecedented size, but flying boats were best left to rivals like Shorts. CRF had a unique knowledge of float technology, but could exercise only scant control over traditional methods of laying down stressed-skin wooden hulls. The methods were traditional, but the principle of flexible hulls was new, with Fairey acquiring the patent from Linton Hope, a yacht designer working for the Admiralty since 1915. The seeds were sown for what by the end of the 1920s would be a keen interest in ship science and boat building. The hull of the flying boat that never flew was conceived by Charles Nicholson, of Southampton yacht designers and builders Camper and Nicholsons. This was CRF's first acquaintance with the Itchen boatyard, which would feature so prominently in his life before and after World War II.[49] The N4 episode has a further, albeit tenuous link, with another of Richard Fairey's passions in middle age. All three machines landed up the property of the Marine Aircraft Experimental Establishment (MAEE) on the Isle of Grain. Not surprisingly, Fairey would often be back on familiar territory consulting with naval engineers. However, he continued to visit the MAEE on a regular basis after it moved to Felixstowe in 1924 – on an initial foray he discovered a suitably commodious seaside hotel *and* the delights of tackling 18 holes in the teeth of a sharp easterly blowing across the dunes.[50]

Dick Fairey family man

While the 1920s saw Richard Fairey become an extremely rich man, it was the second half of the Great War which witnessed a visible change in his fortunes. Driven by the desire to establish Fairey Aviation as a pioneering, profitable and well-regarded enterprise, Fairey was yet to acquire a hinterland: in both thought and deed the company came first, second and third. As if the pressure to succeed was not enough, he was endeavouring to survive and prosper against the backdrop of a nation mobilised for war on an unprecedented scale. This was a level and intensity of conflict few could have imagined, and yet ironically it enabled Fairey's company to grow at a rate inconceivable in peacetime. The ultimate test would be whether the young firm, having grown so fast, could survive the collapse in demand once the guns fell silent. The return of peace would test Fairey Aviation's commercial viability, with clear implications for each director's personal finances. But all that was in the future, and, compared with his financial situation at the start of the war, Dick Fairey began 1918 boasting a healthy salary, a considerable sum of money in the bank, and a large investment with deferred dividend potential. From the outset he and his fellow directors were adroit at ensuring their accruing incomes were kept at a distance from company accounts. Each of them now had serious money to spend, and, while exercising suitable constraint, they made the most of their newly acquired prosperity.

Pressure of work meant CRF was yet to throw himself into an array of leisure pursuits, all of them passionately pursued – the rod, the iron, the gun and the helm would all have to wait. For the moment motoring was his sole indulgence: long hours were spent at the wheel of an expensive Marlborough saloon, whether commuting to and from the metropolis or speeding Toad-like to the coast down alarmingly narrow country roads. The chairman and managing director worked exceptionally hard, and always led from the front. Yet he was no workaholic, and amid all the demands and pressures of building a company from scratch, Dick Fairey still knew how to enjoy himself. Never was the cliché of 'work hard, play hard' more pertinent. Those heady days in the Medway towns before the war were by no means forgotten. Yet neither was the strong commitment he felt towards his family – memories of the long years of hardship and struggle could not be wiped out by a couple of hefty cheques drawn on the Ministry of Munitions.

Fairey's three sisters were all married and starting families, but his half-brother was not yet a teenager. Geoffrey lived at 52 Station Road with a father who was worthy but dull and a mother who seemed older than her years. For a schoolboy for whom domestic drabness was compounded by suburban London's sober and sombre response to wartime sacrifice, the increasingly sophisticated Dick appeared to inhabit a remote and exciting world young Geoff could scarcely imagine. Here was someone who lived in a fashionable district and who drove a suitably stylish automobile. As time passed, his part-sibling was less a figure of romance and more a role model. The Hall/Fairey family could boast a rising young man, with a company bearing his name: he directed a large workforce in a big factory, which day and night was building the aeroplanes intended to win the war. The high-achieving aviator turned entrepreneur inspired Geoffrey Hall to pursue a parallel career path, similarly endeavouring to secure high job satisfaction and an enviable quality of life. CRF recognised this ambition early on, acting as both patron and mentor to his half-brother as he secured the qualifications and experience vital for executive achievement in manufacturing industry.[51]

Richard could provide professional support, but he relied upon his sisters to give Geoffrey a good home after first his father and then his mother died. The former Mrs Fairey passed away in 1922, but she had been in poor health for the previous five years. During that time she had sought solace in God, to a degree her first husband would have deemed irrational and irresponsible. The same was true of her eldest son, who could scarcely comprehend his mother's conversion to Roman Catholicism, let alone that of his sister Effie. Mrs Hall displayed all the faith and fervour of the convert, dedicating her life and her personal wealth to the archdiocese of Westminster. Fairey found that much of the material support he provided for his mother was gratefully received by the local parish priest – a cleric quite happy to walk off with a family painting until thwarted by a furious CRF. Whatever steps Fairey took to protect filial financial assistance, a Papist conspiracy succeeded in channelling his munificence into the coffers of the Church. This intolerable state of affairs generated distress, frustration and inevitable fury. As a self-declared man of science, proud to come of age in an era of scientific progress and popular enlightenment, Fairey regarded religious belief as a relic of superstition and the supernatural. Solidly atheist, he had previously dismissed organised religion as an irrelevance and a minor irritation. However, his painful encounter with an ultramontane Church of Rome, which for many in Protestant England still seemed deeply alien and threatening, left a lifelong legacy: he now viewed all faiths with contempt, but Catholicism – *not* individual Catholics – he looked on with a deep loathing that even his closest friends would never dare challenge.[52]

Anglicanism, as the established church, was grudgingly accepted as a vehicle for social cohesion. At the insistence of their mother, the children of CRF's second marriage were christened: 'She said it was like water dripping on a stone, and by the time she got her way John had to stand by the font and I [Jane Fairey] was seated upon it.' The purchase of Bossington brought with responsibility for the living of St James and St Mary, a joint parish in the Test Valley. Shortly after World War II CRF enjoyed sherry-fuelled discussions with the Bishop of Winchester as together they chose a new rector, as a result of which 'we ended up with the most marvellous man, who had been in a Japanese Prisoner of War Camp, and one of the finest clergymen I have ever come across.'[53]

By the final year of the Great War, Richard Fairey was a husband and father, enjoying the life of a countryman on a modest rented estate in the Home Counties. His wife was not yet 23, and yet they had already been married for over two years. Mr and Mrs Fairey were very different individuals, so how had they met? However demanding his commitments at Shorts, and then working for himself, CRF ceased every opportunity to enjoy music hall and cabaret. A rising income now allowed him to sample entertainment in and around Leicester Square, away from the rough and tumble vaudeville of the East End. The years around the Great War marked the golden age of musical theatre, and Fairey saw home-grown shows as well as the best of Broadway. He progressed from stalls to circle to box, but his boyish enthusiasm for variety never waned. This tall, imposing, rather stern figure was unlikely to be found mooning over starlets on stage, let alone lingering by stage doors. He scarcely seemed the romantic type, and yet one evening 'up west' early in 1915 he espied a vision of beauty across the footlights.[54]

Thus began a lightning courtship, which led the aspiring entrepreneur to marry a vivacious comedy actress, seven years his junior. The future 'Joan' Fairey was an attractive blonde with a lively character. Queenie Markey's family were all children of the Raj, and she herself had been born in Karachi. Warrant Officer John Markey was bandmaster of the 16th (The Queen's) Lancers, and had met his wife Henrietta while stationed in India. Having returned to England with their father's cavalry regiment in 1904, the Markey children came to live racy lifestyles beyond the imagination of even the most open-minded suburbanites. Elder sister Isabelle wed an aspiring comedian, George Clarke, and they formed a double act. Fresh from touring Australia and New Zealand, George and Isabelle encouraged Queenie to make the most of her lilting voice. As a popular actor and revue artist, both at home and on Broadway, George Clarke had the contacts to set his sister-in-law up as a wartime sweetheart. Soon she was singing in the West End. By this time the Clarkes were on set in Hollywood, returning only after Queenie sacrificed her stage career for the promise of marital bliss.[55]

By agreeing to marry the serious-minded Dick Fairey, with his extended and highly respectable family, 'Joan' entered a different world from the one in which she had been brought up. Dick's closest family thoroughly disapproved of the match. They make no appearance on the marriage certificate and may have been absent from the ceremony itself. Yet already the bride was safely ensconced in Hendon, far distant from either barracks or Bohemia.[56] Soon she would find herself setting up home at 8 Greville Place, Maida Vale, ahead of the anticipated move away from London. No longer Queenie Markey, songstress extraordinaire, now she was on a steep learning curve: expected to manage the household, find fulfilment as a wise and loving mother, and conduct herself in the manner expected of a rising businessman's loyal and supportive consort. Here was her most demanding role to date, and for over a decade the first Mrs Fairey rose to the challenge.[57]

On 21 November 1916 Joan Fairey gave birth to a son, the third Richard in a now firmly established family tradition. The infant's grandfather would have quietly approved the lack of a christening, however unusual its absence at the start of the last century. Baby Richard's father considered himself a man of principle, although as we have seen, his atheism had softened sufficiently by the late 1930s as to allow the two children of his second marriage to be christened.[58]

Drawing on fond memories of fenland adventures, Richard Fairey planned for his children to grow up in the country, albeit with easy access to the city. He sought to acquire a property that was rural in location but not isolated. Ideally this modest estate would be close to the Fairey factory and an easy drive from central London. Conscious of his new status as an industrialist and businessman, CRF anticipated a lengthy guest list whenever he and Joan hosted convivial dinner parties or summer garden parties. Appearance was everything, whether cultivating potential customers, confiding in close associates, or conveying to junior staff a clear message that the company and its chairman sought nothing but the best. At the same time, Fairey was always intent on having a good time. In his daughter's words, he loved to enjoy himself in congenial company, and he was a naturally generous man and now he had the means. 'He was not a snob, nor was he a social climber.' Here was someone brimming with self-confidence and self-belief, who felt inferior to no-one.[59]

'Grove Cottage' in the village of Iver, on the southerly edge of Buckinghamshire, fitted the bill perfectly. Only 20 miles from the heart of the capital and six miles from Clayton Road, the four-acre estate lay on the edge of the Colne valley, its formal gardens merging in to an attractive water meadow with appropriate stream and pond. The lovingly maintained lawn, the carefully cultivated kitchen garden, and the ever popular tennis court were set adjacent to an impressive array of ancillary buildings, including a coach house. The latter quickly became a workshop, with garage space for two cars. The main house boasted no fewer than five bedrooms. Downstairs, as well as a large kitchen and commodious vestibule, there were good sized dining and living rooms, and a suitably spacious studio for the master of the house. Fairey's workroom led directly into a conservatory running the length of one side, allowing him to wander in and out from the garden without anyone interrupting his thoughts.

The rooms were light and airy, which was just as well given that electricity only reached Grove Cottage in 1922, five years after the young couple and their son moved in. When power did at last become available, CRF and Geoffrey Hall took it on themselves to wire the whole of the main building and the outhouses. It must have made perfect sense – who in the county, let alone the village, was better qualified to take on the job?[60]

In and around the village of Iver resided a growing range of facilities: even in wartime the countryside beyond the furthest reaches of west London suburbia was attractive to 'new money', and thus ripe for development. In time the exclusive golf and tennis clubs, smart restaurants, upmarket public houses and heavily panelled *Shell Guide* hotels would together constitute a convenient and fashionable playground for the metropolitan middle classes. In 1935, as the worst ravages of the Depression receded south of the Trent, Iver was to prove the perfect setting for a revival of the British film industry: for many of the nation's aspiring celluloid stars the newly opened Pinewood Studios would be but a brief chauffeured drive from their *bijoux* residences. If they didn't know it in 1917, it was not long before Richard and Joan Fairey came to realise their new home was situated in the most affluent and aspirational corner of the Thames Valley.[61]

Making serious money: the Type III

If Fairey Aviation is synonymous with a single aircraft then it must surely be the Swordfish. Yet it was the Type III that enabled the company to survive postwar retrenchment. It was this series which, in the early 1920s, consolidated Fairey's reputation, transforming an ambitious yet low-key operation into a major, high-profile manufacturer. This was the all-purpose seaplane which, in the first year of peace, contested the Schneider Trophy, delivered the final edition of the *Evening News* in Margate, and flew through Tower Bridge at 120 mph. Further afield, the same all-purpose model pioneered a crossing of the South Atlantic, circumnavigated Australia in four days, and flew in formation from Cairo to the Cape of Good Hope. Later versions of the highly adaptable Type III could be found in every corner of the globe, whether serving as an air ambulance in central America or racing as a King's Cup contender back home.[62]

The Type III series had its origins in a pair of experimental seaplanes, the N9 and N10. These were designed by Fred Duncanson and his employer to meet an Admiralty specification for test aircraft that could be catapulted off converted cruisers. In practice, only the N9 served this purpose, trialled for the last two years of the war in the Thames estuary on the aptly named

HMS *Slinger*. The N10, later designated the first Fairey III, had folding wings and was flown off a variety of seaplane carriers. Both aircraft incorporated all Dick Fairey's innovations to date, not least fitting the variable camber gear across the entire wingspan. Bought back from the Royal Navy in 1919 as two of the earliest machines to be civil registered, these prototypes were subject to constant modification, particularly the N10, which in the course of an eventful life went from seaplane to landplane to amphibian, powered by no fewer than nine engines.[63]

One reason for retrieving the N9 was to re-engineer the original design and enable Sydney Pickles to fly the Atlantic nonstop, thereby winning for Fairey Aviation the prize of £10,000 put up by the *Daily Mail* in March 1919. Thankfully for Pickles and his nominated navigator, the flight never materialised, their challenge being overtaken by events, namely Vickers' success in vaulting from Newfoundland to Ireland. In any case, to CRF's deep irritation, the project had been seriously disrupted by a dispute on the shop floor which led to workers at Hayes walking out. Furthermore, on 20 May Mrs Pickles told reporters that she had persuaded her husband to withdraw. It's noticeable that from this point on Sydney Pickles was less prominent, with Vincent Nicholl the first choice test pilot whenever available.[64]

CRF and his colleagues were keen to maximise publicity for the company and, while they were clearly over-ambitious in hoping to cross the Atlantic, the same was not the case for securing the Jacques Schneider International Trophy. With a Sopwith Tabloid winning for Britain in 1914, the initial postwar event was staged on the Solent in September 1919. Flying the flag in the face of challenges from France and Spain were the test pilots of Supermarine, Sopwith and Avro, along with Lieutenant Colonel Vincent Nicholl in Fairey's radically rebuilt N10. The latter now boasted a 450 horse power Napier engine, but was nevertheless one of the slowest competitors. The renamed Fairey III was, however, the most robust entrant, its designer gambling upon choppy water to guarantee victory. The gamble almost came off, as the plane proved the only contestant capable of regaining its mooring without assistance. Unfortunately, this premature return to the buoy was a consequence of Nicholl abandoning the contest because of poor visibility. The race was a glorious chapter of accidents and, had Nicholl not been handicapped by fog at the Swanage turning point, he might have gained the Schneider Trophy by default. In actual fact no-one did, but, as the Savoia flying boat would have won had it been spotted in the gloom, the Royal Aero Club agreed that Italy be next to stage the event.[65]

The Times saw the whole sorry episode as a vindication of flying boats, provoking a lengthy riposte from CRF, who complained bitterly of inaccurate reporting and faulty conclusions. Charles Grey was swift to endorse Fairey's views. Here, after all, was 'one of the wisest and wiliest of men, besides being a genius as a designer.' *The Aeroplane* saw evidence of that genius in Fairey fencing off his beached seaplane in order to keep the Bournemouth holidaymakers at bay; its editor not alone in noting the Italians' lamentable failure to follow the sharp-thinking Englishman's example.[66]

Fairey generated an astonishing intensity of newspaper coverage at precisely the moment many of its rivals were reeling. The company was generous in purchasing advertising space, expecting the host newspaper or magazine to provide an accompanying article. Aviation supplements in publications ranging from *Aeronautics* to the *Manchester Guardian* carried full-page ads alongside flattering profiles of inventive designer and canny problem solver Mr C.R. Fairey. *Flight* even provided its readers with a full-page portrait of CRF.[67] *The Aeroplane*'s C.G. Grey published a long interview with his close friend in April 1919, and at the start of August a four-page illustrated guide to all Fairey aircraft. In December, Grey clumsily advised his French readers that the inventor of the variable camber gear was 'un genie d'une plus jeune generation.' The *Aeronautics* correspondent's gushing treatment of CRF echoed Charles Grey in portraying 'a big man both physically ... and mentally, with big views.' Here was a dedicated company boss unafraid to speak his mind or take big decisions, and unapologetic for being a hard-headed realist: 'the man who deliberately overlooks obstacles is not an optimist – he is a fool.' It is worth noting that these – and all later – profiles, when describing Fairey's pre-Finsbury education, mentioned only one school – the prestigious and prominent Merchant Taylors'; four years at the more humble St Saviour's had been airbrushed out of the big man's CV.[68]

In the years immediately after the Great War, Fairey Aviation, in the form of its senior management, was appreciative of the chairman's emerging status as a powerbroker in the aviation industry. Yet CRF's growing reputation naturally rested upon the quality of his aircraft. Thus the company looked to the Type III seaplane as the means of making Fairey a household name. Legend has it that CRF accompanied Sydney Pickles on his heart-stopping low-level flight up the Thames and through Tower Bridge on 21 March 1919. Two months later the Australian aviator flew a IIIC emblazoned with his company's name from Isleworth to Westminster. Outside Parliament he collected bundles of Lord Rothermere's *Evening News* for delivery downriver. Having brought the good people of Thanet an up-to-the-hour despatch from the heart of the Empire, Pickles posed for photographers before returning to an enthusiastic reception at Blackfriars.[69] These camera-craving flights were seen as something more than mere stunts, with their instigator seeking to make a serious point: Richard Fairey was an evangelical promoter of the seaplane, insistent that free access to Britain's waterways was the key to maximising private ownership of aircraft. Initially sceptical of civilian demand for light aircraft, he soon came to see flying as the new freedom, providing an unprecedented degree of personal fulfilment. For this reason alone, notwithstanding the obvious benefits for the economy, flying clubs should be established and young men and women urged to reach for the sky. Here were the first stirrings of 'airmindedness', with Fairey scornful of anyone who suggested flying was the preserve of the wealthy and well-connected.[70]

Promotion of the Type III was boosted by Fairey Aviation attaining a creditable third in the Air Ministry's commercial aircraft competition, held at its Felixstowe and Martlesham experimental testing establishments in the late summer of 1920. Faced with a formidable battery of tests, Nicholl and the N10 struggled to compete with rival contestants' flying boats; but as the only entrant in the contest's amphibian section, Fairey attracted extensive coverage in both the specialist and mainstream press: floats boasting state-of-the-art technology and incorporating a retractable undercarriage were seen by admirers like Charles Grey as further evidence of 'Mr Fairey's all-round genius as a designer, business man, and financier.' Plaudits compensated for the paltry third-place prize of £2,000, one-fifth and a quarter of the sums awarded respectively to Supermarine and Vickers.[71]

The money mattered little given the N10 had more than paid for itself, as far back as the spring of 1918 facilitating an order for 50 Type IIIA two-seater land-based bombers. Fittingly, after six years servicing the needs of naval fliers, CRF had secured possibly the RNAS's final commission before its incorporation into the RAF. The Type IIIA flew no missions prior to the Armistice, and was soon declared obsolete, but as an intended successor to the 1½ Strutter it represented continuity and a coming of age for Fairey Aviation, production moving smoothly from one generation of combat aircraft to the next.[72] Through the shortlived Type IIIB, Fairey's multi-purpose machine morphed into the IIIC. The definitive directory of Fairey aircraft describes the Type IIIC as 'the best seaplane designed during the 1914–1918 War.' Although too late for that conflict, of the three dozen aircraft built for the military at least seven accompanied HMS *Pegasus* to Archangel and took part in the Allies' ill-fated expedition against the Bolsheviks. Among the RAF volunteers flying untested seaplanes fresh off the assembly line at Hayes was Massey Hilton. Awarded a Distinguished Flying Cross (DFC) for his counter-revolutionary exploits, Hilton was destined to join the Fairey board, proving a key figure in the company after

World War II. A powerful reason for despatching the IIIC to northern Russia was that its high-performance engine, the well-tried Rolls Royce Eagle VIII, ensured an unusually high power–weight ratio for a float-plane. This enabled larger fuel tanks to be fitted, facilitating long-range patrols. Both airframe and engine were extremely reliable, which reassured pilots flying in a uniquely harsh and dangerous environment.[73] Four Type IIICs were purpose-built for civilian use, signalling a shift of focus: the early years of the new decade would see the all-conquering IIID take on a multitude of roles.[74] Despite its roundels and personnel, this versatile machine would often be deployed on flights of a wholly non-military nature.

The IIID's exploits eclipsed those of its predecessor, but not immediately. However, early in the Type IIID's history, Fairey did attract extensive newspaper coverage when the first of the new aircraft were despatched to the furthest corner of the Empire. This was a big story, reflecting well upon imperial unity at the very moment in August 1921 when the independence struggle in Ireland was approaching a climax. Wartime Australia's larger than life Prime Minister, Billy Hughes, was in London for the first Imperial Conference after the war. However, it was his equally colourful wife, Mary, who hit the headlines when she accepted six Fairey seaplanes on behalf of the newly formed Australian Naval Air Service. A large party of the great and the good travelled on a specially commissioned train to Southampton, from where they were ferried to Hamble. A succession of patriotic, self-congratulatory speeches culminated in Vincent Nicholl announcing that he would take the more adventurous antipodean visitors aloft in the freshly designated ANA1, granting them a spectacular view of Southampton Water to match that of Sydney Harbour. The whole carefully orchestrated event was a corporate publicist's dream, not least because Fleet Street's finest were present in strength. It confirmed the chairman's undoubted talent for promoting his company as aggressively as possible, and it signalled just how much the Type III series was an aeroplane of the Empire. Further confirmation came two years later when ANA3 flew around Australia in 90 hours, earning its crew the 1924 Britannia Trophy.[75]

The British Empire was at its zenith, and in the Type III it possessed a machine capable of conveying proconsuls and path blazers anywhere from Alberta to Zanzibar. This was aptly illustrated when the last of the four civilian IIICs was boxed up and despatched east of Suez to play its part in an early transcontinental expedition. The seaplane had been speedily modified in the spring of 1922 so it could convey veteran flier Norman Macmillan and his two companions (a cameraman and an intrepid reporter for the *Daily News*) on the third stage of their round-the-world flight: from Calcutta to Vancouver.

Rebuilt in India by an RAF ground crew, the Type IIIC suffered engine failure in monsoon conditions, crash landing on a remote island in the Bay of Bengal. Macmillan was lucky to be alive, and in due course to be rescued, but his real luck lay in being stopped from flying a second- or at best third-generation seaplane across the vast expanse of the north Pacific. The great global adventure was aborted, and somewhere in shallow waters south of Chittagong remain the scattered remnants of the first but by no means the last attempt to enter a Fairey machine into the record book.[76]

Ending the war as an instructor, Norman Macmillan had flown a succession of Sopwith scouts in both France and Italy. Looking back a decade later he found the experience of waging war on two fronts, 'a good time and a bad time rolled into one, a picnic and a term of penal servitude, but it was a great and glorious adventure too.' After a brief return to the RAF in 1921 he combined aircraft delivery with freelance work as a test pilot. On various occasions throughout the 1920s he could be found at Northolt or Hamble putting a Fairey prototype through its paces. The resourceful Macmillan displayed an obvious physical and mental toughness, but he was no rough diamond. In fact he was extremely bright, witness the stream of books he wrote across the five decades following his return to civilian life. He flew 163 sorties with 45 Squadron, but away from the airfield would attend La Scala or admire Notre Dame.[77]

Not only was Captain Macmillan MC AFC an accomplished pilot, but he had the technical nous to discover what was wrong, and if necessary repair it. For example, before taking off from the Hugli River in the fated Type IIIC, he had singlehandedly repaired a float damaged in transit. His credentials as a test pilot rested upon combat experience and completion of the RAF's inaugural officer training in advanced aeronautics. Here was someone hard to intimidate, and a year prior to his flight halfway around the world Macmillan submitted a disparaging report on Fairey's first and last fighter–reconnaissance seaplane, the Pintail.[78] Although for some reason the Imperial Japanese Navy ordered three machines, the RAF test pilots endorsed their literary comrade's belief that in this case CRF and his project draughtsman, Marcel Lobelle, had failed to get it right. Vincent Nicholl agreed with Macmillan, so no more needed to be said. Once again attention focused upon the Type III. Unlike the Pintail, which exemplified the Air Ministry's naive belief in a high-performance multi-purpose machine, the Type III was a proven design, and a template for whichever type of aircraft was desired. Macmillan's antipathy towards a universal model, fulfilling multiple roles and carrying all necessary kit, was very much the view of Richard Fairey. Throughout the 1920s he lambasted officials at Adastral House for the false economy of overloading combat aircraft with excess equipment. In the modern era CRF would have been appalled by multi-role aircraft such as the Tornado and the F35, lamenting the impact of cost efficiency upon performance.[79]

A flagship factory, but at a price

Building bigger aircraft necessitated rapid expansion, and a complete factory was constructed on the Harlington site across the winter of 1917–18. While production continued at Clayton Road until the war ended, the new plant came to embody Fairey Aviation as a major presence in British manufacturing industry: outposts would come and go, but between the wars Hayes and Hamble were the heart of Richard Fairey's ever-expanding enterprise. Part and eventually all of Northolt aerodrome was leased for testing land planes, necessitating a hazardous journey of five miles: one of the firm's lorries would carry a set of wings and tow the parent fuselage.[80] With the Thames always in reserve, Hamble and Northolt served as the firm's testing and delivery centres until 1929, when Fairey was forced to find a fresh runway. The Harlington site soon became Fairey Aviation's principal production facility, with six assembly bays, ancillary shops and a spacious construction shed. Such provision for aircraft assembly was impressive by the standards of the day, and yet these buildings would soon be dwarfed by the hangars and offices that followed.[81] The Ministry of Munitions and the Treasury facilitated the new factory, but at a price.

Firstly, the Government had to approve who would manage its investment. Fairey found himself a general manager sure to secure ministry approval: F.M. Charles had been invalided out of the French Army and then worked at Daimler as a designer. The new man boasted the right organisational skills and could deal directly with an increasingly restless refugee workforce. Rose, the internal candidate for the new position, quit in high dudgeon, and so A.C. Barlow took over the Clayton Lane operation. Bray was moved sideways, and the company's experimental section placed in the hands of a young engineer from Yorkshire, Wilfred Broadbent. In the late summer of 1917, when Barlow fell ill from overwork, an untested Wilf Broadbent was given the additional responsibility of running Clayton Road.[82] The tyro works manager's success in rising to the challenge confirmed his boss's skill in spotting talent. CRF had lost an experienced quality inspector in Rose, but his departure was scarcely debilitating. Fairey's founder now had a management team he could delegate authority to, secure in the knowledge that each man would get the job done. At boardroom level Rees's removal was a cause for celebration, with Dawson the trusted confidant every leader needs.

F.M. Charles clearly made a favourable impression, witness friendly references in the works magazine of January 1918. *Fairey Tales* was a cheerful if by no means cheap publication, which, while running to only one issue, proved a precursor of later less naive efforts to promote company pride. The shop floor journalists' compliments seemed genuine, and the ultra-patriotic house journal had a charming naivety. What the magazine reveals is the variety of morale-boosting activities organised in the final 18 months of the war, ranging from tea dances in Ealing Town Hall to garden parties at Gunnersbury Park. Charles's linguistic and managerial skills were no doubt invaluable given the large number of Belgian women who worked at Hayes right through to when the fighting ceased. Mostly unmarried, they were clearly very competitive given their prominence on company sports days: track and novelty events for the ladies were the main attractions, and the predominance of veteran races reflected an overwhelmingly middle-aged male workforce.[83]

Whitehall's interference in Fairey's appointment of a general manager had a happy ending. The same could not be said of the Government's £20,000 capital investment. This not insubstantial sum was authorised by the Treasury on the assumption that increased revenue would facilitate rapid repayment with interest. With turnover of £340,000 in financial year 1917–18, Fairey Aviation had presented a credible business plan, predicated on a steady stream of income generated by substantial subcontract work and sales of its seaplanes, the Campania and the forthcoming Type III. This was a big gamble given the dramatic fluctuation in Fairey's liquid assets across the winter of 1917–18. Into the final year of the war, payment for completed aircraft remained subject to delay, and the company's cash-flow problem only worsened.[84]

By mid-1918 the final cost of the new factory plus interest left Fairey owing the Ministry of Munitions £26,544. The company needed to secure advance payments against a total order of 50 Type III seaplanes in order to keep going, let alone pay EPD and service the Harlington loan. The Ministry of Munitions declined to sanction further lending, advising Fairey to seek assistance from the firm's bank. This hard line was a consequence of the Treasury refusing to sign off pre-delivery bank transfers without partial repayment of the original sum borrowed and a full disclosure of the directors' annual earnings. Senior officials must surely have recalled the generous deal CRF had secured for sale of the Campania patents a year earlier. Heated correspondence was exchanged throughout 1918. That August, CRF detailed the £72,536 he insisted the Government owed his company, but still Whitehall questioned the figures.[85] To complicate matters further, Fairey was at the same time claiming £15,000 compensation for 'the excess cost of developing a seaplane of exceptional performance.' The Treasury refused to accept government liability for development of the Campania, but in July 1918 did authorise a loan of £10,000 pending settlement of CRF's dispute with the Air Board.[86]

In December an exhausted and exasperated Fairey took his case to the relevant minister, Lord Weir. The complainant insisted that he and Dawson were being penalised for success: Treasury obduracy arose out of the Ministry of Munitions' refusal to recognise their status as working directors who earned salaries less than the general manager, and were yet to earn a profit from their entrepreneurialism. Given that in 1918–19 Richard Fairey took home the sum of £1,900 plus a guaranteed £300 in expenses, one wonders if this was a disingenuous comparison, between the directors' net incomes and F.M. Charles's gross salary. Furthermore, while a dividend had yet to be paid, the managing director and chairman had received £15,000 as a generous lump sum payment for the surrender of three patents.[87]

While Fairey's salary dipped in the financial year 1919–20, after that it was on a rising curve. Thus by the spring of 1922 he was earning £3,245 per annum, and held just under £7,460 in his deposit account. The latter sum disguised the sizeable sums going in and out of the account as a result of frequent dealing in shares and gilts – CRF must have been on the telephone to his stockbroker almost daily.[88] All this, plus the comfortable lifestyle at Grove Cottage, suggests considerable wealth, whatever the state of the company. This would be to ignore the demands made upon Fairey for income tax and surtax, the payment of which he did all in his power to delay. By the early 1920s CRF was in a running battle with the Inland Revenue over Fairey Aviation's tax bill, and his own. His resentment in having to pay up is scarcely surprising given that HM Government took nearly one-third of his total income for the three years 1917 to 1920. The MBE awarded in January 1920 for wartime services was probably seen as scant reward for income surrendered to the tax man.[89]

The appeal to Lord Weir was a consequence of continuing losses across the autumn of 1918. Thus, in the week of the Armistice, Fairey Aviation delivered a Type IIIB seaplane and a large body of components, to the value of £6,495. Yet procurement revenue that week totalled £604 and, when set against total unit costs including labour, it left the firm in the red to the sum of just over £6,331. This was on top of an operating deficit of £57,531 and imminent expenses of more than £1,185. The only good news was that Fairey's current account held nearly £3,300, albeit a tiny sum given the size of the company. It is scarcely surprising that F.M. Charles's first task on arrival had been to translate the firm's crude accounts into a weekly briefing for the board on the state of Fairey's cash-flow crisis.[90]

Ideally, a finance officer should have been appointed as, despite Charles's intervention, the company secretary continued to provide back-of-an-envelope calculations: the Ministry of Munitions and the Inland Revenue would have been horrified had they discovered the haphazard fashion in which A.G. Hazell documented Fairey's financial performance. When in February 1920 the Inspector of Taxes lost patience with the failure of Fairey Aviation's auditors to sign off and submit accounts for 1917–18 and 1918–19, a harassed Hazell scribbled down retrospective summaries of income and expenditure which the accountants would hopefully render acceptable to the Inland Revenue. The accountancy firm was clearly unhappy with this state of affairs as three months later the senior partner, backed by Charles Crisp, insisted that the Fairey board could only approve and submit for auditing and tax assessment full and accurate figures for the past three years.[91] To put it bluntly, in the early years of its existence, Fairey Aviation's accounting procedures were chaotic.

The company undoubtedly had the potential to be profitable, even as the Ministry of Munitions exploited its cash-flow problems to drive down tender costs. Fairey's seaplanes sold for just under £3,000 at the end of the war, with upgrading of a Type IIIB to Type IIIC costing around half that price. Despite primitive assembly lines, plane making was labour-intensive, each aircraft calculated as calling on the skills of over 30 employees. This sounds expensive, but in 1919 the average wage was £4 15 shillings (£4.75p), with a weekly bill for the entire workforce of around £4,000. Big commissions, as with the rolling contract for 50 Type IIIs, enabled the company to secure substantial discounts from components suppliers, especially once the war ended and demand dropped. Given the narrow profit margin, big orders and consequent economies of scale were vital, hence CRF's relentless courting and lobbying of Air Ministry personnel. In early 1919 Hazell calculated that the amount owed on 'Harlington New Building' was now not much more than £8,000. Equally encouraging was that revenue had begun to rise, such that a huge unpaid debt of around £95,000 in January 1919 was only £5,800 12 months later. Needless to say, given Hazell's cavalier book-keeping these figures had to be viewed warily.[92]

Yet Fairey Aviation clearly did ride the immediate postwar boom as the Lloyd George coalition cleared its debts to those companies it continued to rely upon. With at least three types of seaplane flying in support of the White Russians, Fairey's short-term survival was deemed vital. Unlike so many of its rivals it was encouraged to keep building aircraft after the Armistice. Only 20 Type IIICs were built across the financial year, but the demand for parts remained high: with a set of floats costing £400, keeping aircraft in the air and on the water was a lucrative business. However, by the spring of 1920 the end of hostilities in northern Russia and an absence of repeat orders left the company carrying a hefty cash deficit of £25,622, and with no reserve to clear any outstanding tax on wartime income.[93]

The Type III series was a lifeline, but demand was intermittent. Like every other aircraft manufacturer, Fairey Aviation required sales of the size secured in the course of the past war. No such orders were forthcoming as the world's largest air force shrank at an astonishing rate. Whereas Sir Frederick Sykes, the second Chief of the Air Staff, had recommended the RAF become a powerful peacetime deterrent, and thus a guarantee of imperial security, his minister – Winston Churchill – sought a less costly option. Sykes found himself sidelined when Sir Hugh Trenchard was reappointed chief of staff. 'The Father of the Royal Air Force' created the institution we know today, working closely with Sir Samuel Hoare, Secretary of State for Air, throughout much of the 1920s. Trenchard favoured a small cadre force, which, in theory at least, adhered to a doctrine of strategic bombing. In 1922 the Home Defence Air Force was formed, envisioning medium and heavy bombers as a credible force of deterrence. This precursor of Bomber Command never reached full strength, with fighter aircraft predominating until 1928, and by the following decade an exaggerated fear of aerial assault saw successive prime ministers prioritise the RAF's defensive role.[94]

Excess capacity brought the aircraft industry to its knees and only a few firms benefited from the newly created Aircraft Disposal Company buying surplus stock for £1 million and a half share in future profits. At the end of 1920 SBAC membership still counted nearly 30 aircraft constructors and around a dozen engine-makers. Yet only a fraction of these firms were working on Air Ministry contracts, with orders so infrequent that even Rolls Royce gave serious consideration to closing its aero-engine division.[95]

Like every other aircraft manufacturer after the war Fairey Aviation had to diversify or go under.[96] This was the advice 'Boom' Trenchard had given Richard Fairey when he met with key manufacturers soon after his reappointment as Chief of the Air Staff. He was quietly confident that at some point in the not too distant future the RAF would have to re-equip, with combat aircraft needed in large numbers. Trenchard's cautious optimism was reinforced after he persuaded Sir Eric Geddes that air power could prove cost effective: when later in 1921 government 'waste' was subject to the notorious 'Geddes Axe' a third was shaved off the RAF's procurement budget, and yet it could have been a lot worse.[97]

Unfortunately, Fairey's alternative to aircraft manufacture rendered a woeful financial situation that much worse. F.M. Charles's previous post had been with the prestigious motor manufacturer Daimler. He remained on good terms with his former employer and negotiated a deal for Fairey to build a unique and expensive body for the Coventry car-maker's top of the range saloon. The original order was for 20 coach bodies, with Fairey Aviation acting as the financial guarantor of a subsidiary specialist company, Fairey and Charles.[98] The project proved a rare false move given its chronic inability to generate a profit on each body built. The workers lacked the necessary skills, and there was scant opportunity to secure economies of scale. The assembly bays were crammed with bulky half-built car bodies, annoying everyone still at work on outstanding aviation contracts. Throughout 1920–21 the parent company's chairman, secretary and solicitor found themselves moving money between bank accounts in order to keep the operation solvent and pay the workforce. In May 1921 Fairey Aviation's principal bank brought the issue to a head: Fairey and Charles would not be forced into liquidation if CRF personally guaranteed that an unsanctioned overdraft would be wiped out within three months. When Hazell and Crisp, under pressure from the Inland Revenue, at last submitted Fairey Aviation's accounts for audit, they frontloaded a £36,000 loss on the Daimler contract, transferred from the sister company. This meant that for 1918–19 the firm turned a healthy profit of £61,750 into the more modest £28,500.[99]

Thus, although hand-built 'laundalette' bodies proved a costly mistake, the Fairey board found a convenient way of avoiding punitive tax on revenue earned from aircraft manufacture up to and beyond the Armistice. In late 1920 the chairman decided Fairey and Charles could risk reneging on the Daimler contract. To the amazement of everyone at Hayes, and indeed throughout the industry, an order had been placed for 50 Type IIIDs. Given conditions at the time this was a huge commission, with the Air Ministry signalling that more contracts might follow. The Type III really was a winner. An exultant CRF needed his factory back, but Daimler, supported by Charles, insisted that production must continue. With echoes of the Clayton Road occupation, Fairey told Wilf Broadbent he needed 'a team of strong and trustworthy chaps' to create necessary space. On a bitterly cold night early in 1921 the chairman and works manager orchestrated the evacuation of every half-built body. The next morning, faced with a fait accompli, Daimler's representatives at Hayes made hasty arrangements for collection. Not long after, monsieur Charles returned to France.[100] Scarred by the Daimler experience, CRF refused any further involvement in car production. Before and after World War II he ignored Moore-Brabazon's suggestion that he buy Lagonda. In March 1945, a month before Munich fell to the Allies, Lord Brabazon was already urging investment in a revival of Frazer Nash's prewar partnership with BMW. Fairey was too ill to respond, but would doubtless have pooh-poohed the idea. Bristol, with its ambitious plans for building luxury limousines after the war, did link up with

Frazer Nash to produce expensive coupés, but a reverse takeover of BMW's car division was always a fanciful idea.[101]

To be fair, the Fairey and Charles initiative was clearly not a deliberate attempt to avoid tax, and managing such a hapless project while trying to keep the parent company solvent placed an enormous strain upon the directors, not least the chairman. In December 1920 Richard Fairey saw a consultant about problems he was having with his throat. The specialist who examined him concluded the patient was physically sound but 'obviously suffering from nervous exhaustion.' Professor Thomson recommended CRF winter abroad, or else there was a strong risk 'he may break down in a more serious way.'[102] There is no evidence that Fairey followed this advice, not least because in the winter of 1920–21 he was preoccupied with sorting out his abortive project's tortured finances. The award of an MBE, as due recognition of wartime service to the Empire, briefly raised the chairman's spirits. Then it was back to bemoaning Lloyd George and his coalition of cost cutters.

Despite offsetting any loss made on the contract with Daimler, Fairey Aviation still faced a sizeable Treasury demand for final payment of its wartime tax bill. The EPD penalised pioneer companies, but its impact upon Fairey was not so severe that it proved terminal, as was the case with Sopwith Aviation. The latter had grown so fast during the war that its tax burden was proportionately higher than that of Fairey, which had only commenced production at the start of 1916. Sopwith Aviation had a workforce in November 1918 of over 5,000, most of whom were speedily laid off. In September 1920, with the Inland Revenue insistent on immediate payment, and a shrewd Tommy Sopwith keen to safeguard creditors' interests, the firm went into receivership. Two months later there arose out of the ashes H.G. Hawker Engineering Company, a more modest enterprise named after its chief test pilot, and the embryo of what over the following four decades became one half of the entire British aviation industry, the Hawker Siddeley Group. It would be another three years before Sopwith saw the first Hawker aeroplane fly, and his design team recruit the incomparable Sydney Camm.[103]

Although the new company was slow to see its aircraft in the sky, Sopwith was still a key figure within the SBAC. He was therefore party to an informal agreement between Whitehall and the Society identifying which constructors could tender for future government commissions. This agreement, while protective of the industry in the short term, led to too many firms securing too many short-run military contracts. An inefficient procurement system bestowed on the interwar RAF too many types of aircraft. With productivity low and costs high, it was hard to make money from manufacturing these machines, leaving almost no surplus capital to fund R&D in the potentially more profitable civil arena. For this reason Geoffrey de Havilland, who in 1920 also relaunched his company, focused upon light aircraft and aerial access to the Empire, while Richard Fairey, who later in the decade did make money out of Air Ministry contracts, complained of a system that discouraged military suppliers from investing on the civil side.[104] Ironically, Fairey spoke as someone who, by dint of his chairing the SBAC between 1922 and 1924, was party to the consolidation of the 'ring', with the Air Ministry clearly identifying its 'family' of increasingly specialised design centres and prime contractors. This informal cartel continued until 1932, after which the SBAC boasted a government guarantee that its members would receive all future contracts for combat aircraft.[105]

The mastermind behind the resurrection of T.O.M. Sopwith as an aircraft manufacturer was City accountant Sir William Peat. W.B. Peat and Co. was a family firm which in recent years had acquired an interest and expertise in the nation's newest industry. One of its clients was Fairey Aviation, and in the winter of 1920–21 the senior partner persuaded Richard Fairey to liquidate the original company and make a fresh start. Thus began a process of restructuring which was to reoccur at critical moments throughout the 1920s.[106] The new enterprise, formally set up on 9 March 1921, was a more complex corporate structure than Frank Rees's original creation. However, its articles of association reinforced the board's tight control over membership, remuneration, size of the dividend and, critically, the sale, scale and ownership of individual shareholdings; a meeting to consider any of the above could be called at any time and was quorate if two of the three inaugural directors – Fairey, Dawson and a token Charles Crisp – were present. Although Nicholl and Wright were not named, provision was made for their joining the board as and when appropriate. Fairey Aviation's complicated corporate history in the 1920s deserves revisiting in a later chapter. Fairey was at last floated as a public joint stock company on 5 March 1929, with a nominal capital of £500,000 in ten-shilling shares. Air Ministry officials believed that by then the firm was in its sixth incarnation. As we shall see, senior civil servants and their legal advisers were remarkably sanguine about these frequent changes.[107]

The technical press was similarly relaxed, with *Flight* reporting the 1921 relaunch as no more than an overdue formality: 'The Fairey works are extremely busy, and are likely to remain so for a long time … the present step is calculated to ensure the continuance of that prosperity.' As if to confirm its editor's prognosis, six months later the magazine carried Fairey's call for qualified draughtsmen to contact the company; the success of the Type III had created a shortage of skilled men in the Hayes design office. *The Aeroplane* saw this as further evidence of Fairey Aviation's 'most remarkable record of progress since its inception', and in November eulogised a company whose founder boasted a unique blend of business acumen and aesthetic appreciation – for Charles Grey, Dick Fairey's genius lay in his ability to build aeroplanes that looked 'eyeable and flyable' *and* made money.[108] Grey was being unduly generous: from the mid-1920s Fairey Aviation certainly did produce aircraft easy on the eye, but for all CRF's love of clean lines his early designs were functional and unattractive.

Driving the Fairey board's serial restructuring was a need to raise fresh capital, minimise the burden of taxation and release equity for board members. The salience of these contributory factors varied at different times, but the overarching principle, even after Fairey Aviation's final flotation, was that a small, intimate circle of colleagues – bound by comradeship, camaraderie and loyalty – together determined every aspect of their common enterprise. The most visible manifestation of that overarching principle was unequivocal allegiance, with Dawson and Crisp – ably supported by Amos, Hazell, Barlow, Broadbent, Charles, Duncanson et al. – rarely if ever questioning the judgement, foresight and practice of someone for whom micro-management and strategic thinking were by no means incompatible. 'Tiny' Fairey exercised, and enjoyed, a remarkable degree of power, whether in the boardroom or on the shop floor. Rees had been deluded from the outset: Fairey Aviation was always its namesake's creation, his authority rooted in a firm foundation of acclamation, approval and personal accessibility. The chairman and managing director had skilfully steered his fledgling firm through the wartime years, and, however tough the task of adapting to a world of austerity and arms cuts, he was still fiercely determined to succeed.

By 1922 Richard Fairey could point with pride to his company's capacity to survive and to the speed with which his standing and status had changed inside the industry – his elevation to head of the SBAC was anticipated, with membership of the Aeronautical Research Council imminent. The recognition of his peers, at the very highest level of executive endeavour, was both pleasing and encouraging. However, professional success was not enough in itself: Fairey now enjoyed a standard of living and a quality of life he could hardly have dreamt of the day his father died and the sky fell in. Furthermore, he had a son who might one dare share his passion for aeroplanes, and the affection of a young and attractive wife. Looking out from the conservatory across Grove Cottage's manicured lawn, not yet in his mid-thirties but already the head of a publicly quoted and reassuringly resilient enterprise, Dick Fairey could scarcely resist reflecting upon his good fortune. The challenge now was not one of complacency – a state of mind CRF could scarcely imagine until the day he died – but of how he could take Fairey Aviation to another level. How, given such an unsympathetic environment, he could drive his company forward, securing its position as a leading, perhaps the leading, aircraft manufacturer in peacetime Britain. For the moment drive and direction relied heavily upon seaplanes, but already the engineer–entrepreneur was reflecting upon the next generation of machines: aircraft, necessarily complex yet rooted in simple design, which could both bind and defend Britain's triumphant yet troubled empire.

CHAPTER 5

Twenties Bust and Boom

The moment of take off

For a man fascinated by mathematics Dick Fairey must have relished the degree to which his aircraft were synonymous with heroic feats of navigation. Given the atrocious conditions in the Atlantic on the evening of 24 May 1941, it is remarkable that Swordfish of 825 Squadron found and crippled the pocket battleship *Bismarck* before returning safely to their home carrier, HMS *Victorious*.[1] It is equally remarkable that in the spring of 1922 a late middle-aged navigator, the Portuguese Navy's Gago Coutinho, could guide the pilot of their seaplane on an epic island-hopping journey from Lisbon to Rio de Janeiro. Not one but three Fairey IIIDs were called on as a shortage of fuel necessitated the designated record-breaker's crash landing beside the Rocks of St Peter and St Paul, a tiny archipelago nearly 1,000 miles east of Brazil. Captain Coutinho's dead reckoning and the IIID's ruggedness ensured the first aerial crossing of the south Atlantic: 4,367 nautical miles in a flying time of just over 60 hours.[2] The Portuguese aviators' determination to overcome the odds and complete this death-defying expedition captured the attention of Fleet Street, with readers repeatedly reminded that these were British planes battling unseasonal weather and storm-tossed seas. Napier and Rolls Royce shared the honours in powering the three machines, but the principal beneficiary was Fairey Aviation. The company enjoyed weeks of free publicity, and the IIID's reputation for versatility and reliability generated further sales at home and abroad.[3]

Whether flown from sea, shore or flight decks, or even catapulted off warships, a grand total of 207 Fairey IIIDs were built between 1921 and 1925. As well as serving front-line squadrons, both RAF and the newly constituted Fleet Air Arm, the IIID fulfilled a variety of colonial tasks, from White Nile taxi in east Africa to riverboat ambulance in British Guiana. Norman Macmillan took advantage of a rare seaside start and finish to contest the 1924 King's

Cup, a round-Britain race usually closed to seaplanes: the Shell/Fairey factory team waiting on Stranraer beach refuelled at a speed worthy of modern-day Silverstone. Again, the press were on hand to report Macmillan's derring-do.[4] Macmillan, like Alan Cobham, was a brilliant self-publicist, pioneering the way for a younger generation of high-profile aviators and aviatrixes to seize the headlines later in the decade. As a test pilot he was brilliant, but high recognition was surely a further factor in Richard Fairey inviting Macmillan to join the company at the start of 1924. A year later Vincent Nicholl handed over the reins of chief test pilot, stepping up to serve as deputy managing director: in running the company 'The Colonel' was accountable only to CRF, who was devastated when his friend died of pneumonia following an operation. In October 1927, aware that his old friend faced imminent surgery, Richard Fairey sent him to stay at Brighton's Hotel Metropole. That weekend the two men spent time together reminiscing and making plans for the future, but within seven days Nicholl was dead. For Macmillan his fellow pilot was 'tall and very thin, a man of charm and irrepressible fun, much consideration for others, but with a body too frail to contain his spirit long.'[5]

Nicholl's elevation to Fairey's number two was the last in a sequence of organisational changes generated by the appointment of the highly experienced Tom Barlow as chief engineer. Major T.M. Barlow, originally commissioned in the RNAS and thus familiar with seaplanes, had a formidable CV. Fairey cultivated him for some time before at last convincing him to leave government service and embrace the wider opportunities (and better money) available in the private sector. As a former factory inspector and technical officer, Barlow's presence at Hayes speeded up the process by which the Aircraft Inspection Department authorised Fairey Aviation to become the first manufacturer with an autonomous inspection procedure.[6]

Epoch journeys in the Fairey IIID could not remain the preserve of the Portuguese and the RAF looked to fulfil Cecil Rhodes' dream of bisecting the 'dark continent.' Thus, in the spring of 1926 an elite formation made its way across Europe to the Mediterranean, and launched an audacious attempt to fly four factory-modified aircraft from Cairo to Cape Town and back.[7] This well-organised effort to fly the flag and extend the RAF's expertise in long-distance flying was a spectacular success. With only modest replacement of major components, and just one complete service, all four aircraft returned to their Solent base in remarkably good condition, and with the same Rolls Royce Lion engines as when they left. They had covered a total distance of 14,000 miles. This was a triumph of service logistics, military coordination, colonial administration, dominion collaboration, post-Locarno cooperation, and above all, British aviation technology. Adastral House would never have sanctioned the whole costly operation without reassurance that the relevant aircraft manufacturer would be heavily involved in both planning and implementation. Thus, the chairman of Fairey Aviation and his colleagues were closely involved in the venture from the start, and they duly shared in the glory.[8]

The Swordfish and the supersonic FD2 may be Sir Richard Fairey's best known aircraft, but it is the success of the IIID which secured his reputation as the driving force behind what soon became a highly profitable company. That success was compounded by Fairey continuing to exploit its monopoly of the variable camber gear.[9] Thus the RAF bought a substantial number of the Fawn, an ugly and ungainly light bomber blighted by the Air Ministry's absurdly unrealistic specification (attack *and* aerial reconnaissance). So many demands were made of this aircraft that its speed was drastically reduced, and overall performance was inferior to the Fawn's predecessor, the DH9A. CRF pointed out that de Havilland had insisted upon full control of designing his wartime bomber, which entered production in a fraction of the time spent on developing the Fawn. The painful experience of meeting Whitehall's unrealistic demands confirmed Fairey in his belief that 'the departmental system of design' ran counter to aeronautics' paramount principle of simplicity: civil servants' excessive interference and demonstrable lack of expertise provided incontrovertible evidence that the company not the ministry should have the final word regarding design. By 1924 CRF saw the dead hand of Whitehall as a key factor in Britain lagging so far behind the United States in the manufacture of high-performance machines. Nowhere was this more obvious than in the lack of light, high-speed engines. A spur to Fairey circumventing ministerial interference and looking to American engine suppliers was the Fawn's painfully slow speed when fully loaded.[10]

The legacy of CRF's hybrid machine was his insistence on retaining ultimate control over future development projects. However, of more immediate significance was the Flycatcher I and II. For a decade this supercharged single-seater served as the Fleet Air Arm's first standard fighter, with nearly 200 machines rolling off the Hayes assembly lines.[11] The Flycatcher was ground-breaking in that its maker forged a close working relationship with the embryonic Fleet Air Arm reminiscent of that previously enjoyed with the RNAS – for the following three decades Fairey Aviation would be closely associated with the Royal Navy, in both popular perception and reality. The robust and speedy Flycatcher sacrificed appearance for performance. It looked squat and stocky, but generated high praise from pilot and ground crew alike: Norman Macmillan, a man rarely satisfied, waxed lyrical over his prototype's 'superb handling skills.' To the surprise of all, the scream of a diving Flycatcher proved popular with audiences at the annual Hendon Air Display.[12]

As we shall see, in the second half of the 1920s CRF was evangelical in promoting clean lines and uncompromisingly aerodynamic aircraft. Nevertheless, prioritising functionality over good looks à la the Flycatcher was by the mid-1930s once again a feature of the Fairey mindset. The firm's marked indifference to the aesthetic dimension of aircraft design chimed with the prevailing *mentalité* of the Fleet Air Arm well into the following decade.[13] Invariably the most attractive aircraft on Royal Navy flight decks were intended originally for use on land, for example, the Supermarine Seafire and, in the jet age, the De Havilland Sea Vixen. Predictably, it was Tommy Sopwith and Sydney Camm who, with the Hawker Sea Fury, belatedly offered a unique blend of beauty and raw power.[14]

In the aftermath of World War I, at a moment of severe retrenchment, Fairey Aviation had defied the odds. Starting with the IIID, the Fawn and the Flycatcher, the firm found itself in the healthy position of selling successive aircraft types in sizeable numbers. Volume output meant economies of scale and large profit margins, more than covering the cost of those prototypes which never went into production. To achieve the magic formula in peacetime was an impressive achievement, but the real test was if a company could secure the same success a second and then a third time, and ultimately establish an enviable track record. Between the wars this was precisely what Fairey Aviation achieved, primarily thanks to a chairman and managing director who led by example, loathed complacency, and looked to maintain a mix of experience and new blood at every stage of the firm's expansion.

Stalking the corridors of power, striding the highways of empire

Fairey Aviation's capacity to buck retrenchment and recession in the aftermath of war made its founder very much the coming man. Awarded the MBE in 1920, Fairey continued to attract plaudits for his stewardship of the company. By now his standing within the aviation industry was on a par with – or even greater than – that of Tommy Sopwith, not least because he carried on building aeroplanes when Hawker was literally still grounded. The accolade of his peers culminated in Fairey's election as the Society of British Aircraft Constructors' chairman from 1922 to 1924. Throughout this period he was also an active member of the august Aeronautical Research Council, while at the same time sitting on various subcommittees of the Royal Aeronautical Society. Elevation to presidency of the Society was only a matter of time, and Fairey duly served two periods in office, 1930–1 and 1932–3. Outside of wartime it would be hard to think of two tougher years in which to defend and promote flying at home and across the Empire. Dealing with the worst effects of the Depression was still some years ahead, but by 1923 Fairey was already preoccupied with aviation as the key to consolidating Britain's imperial domain.

The inclusion of mandate territories, albeit technically under the umbrella of the League of Nations, resulted in the British Empire reaching its zenith between the wars. This vast global network required fast and efficient systems of telecommunications and transport. Reaching out from the metropole to the furthest corners of the King-Emperor's global estate clearly depended upon a technology still only two decades old. Yet advocates of the airship as an effective means of covering these vast distances continued to enjoy Treasury support for both public and private initiatives. Long before the disastrous crash of the state-funded R101 in October 1930, Fairey had dismissed airships as a safe and reliable means of long-distance flight. For several years after World War I he put his faith in seaplanes and flying boats, seeking to prove his point in 1924 when the five-float Fremantle was specially constructed to circle the globe: this mighty machine was a triumph of invention, not least a device to raise the propeller hub in rough seas, but American success rendered it redundant. The role of the US Army and Navy in enabling as many as three Douglas aircraft to fly round the world was a wake-up call for the Air Ministry, and a marker for promoting service-industry partnerships to set records and secure trophies.[15]

While carriers like Pan Am and BOAC would extend the life of the flying boat as a long-haul carrier well into the postwar era, by the second half of the 1920s civil transport increasingly looked to land-based aircraft. Richard Fairey responded accordingly, in 1928 building his long-range monoplane. Nevertheless, he remained insistent that the next generation of flying boats would prove both safe and cost-effective, not least when compared to state-subsidised airships.[16]

However, well into the decade Fairey was still promoting seaplanes as a cheap and practical means of accessing the inaccessible. His argument was advanced courtesy of an extraordinarily dense and highly technical paper delivered on the opening day of the 1923 Air Conference. This was the third such gathering since the war, organised with a clear remit to advise Stanley Baldwin's first administration. Fairey's inordinately long lecture was scheduled immediately after a leisurely lunch in the splendid surroundings of the City of London Guildhall. It would have been remarkable if even the platform party succeeded in remaining awake, let alone pose informed questions, when at last the speaker sat down. An even longer version of this postprandial talk formed part of the conference proceedings, published as a parliamentary paper. CRF was a key contributor in his capacity as chair of the SBAC and thus a member of the Air Conference's planning committee.[17] The event was an unusually high-powered affair, intended to establish a blueprint for Britain becoming the dominant force in civil aviation. No such master plan emerged, although all present were placated by an official assurance that an air mail service to India was already in the pipeline. One tangible result was the appointment of a Civil Air Transport Committee, on the initiative of a secretary of state keen to see his conference generate tangible results.[18]

An enthusiastic and highly motivated Secretary of State for Air was rare indeed, but Sam Hoare was impatient for success. History has not been kind to Sir Samuel Hoare, his ministerial record prior to May 1940 overshadowing his diplomatic success in ensuring that Franco's Spain remained neutral throughout World War II. In the summer of 1940 the widely read, fiercely polemical *Guilty Men* dismissed Hoare as an arch-appeaser, deaf to Churchill's prescient warnings and steadfastly opposed to rearmament.[19] Never mind that the future Viscount Templewood quarrelled with Churchill over India not Germany, or that his record at the Admiralty in the 1930s and the Air Ministry in the 1920s signalled a persuasive and increasingly powerful voice in cabinet. In 1923 Hoare's roller-coaster career lay ahead of him, and Baldwin's protégé was a rising star – a man with a mission, and soon a man with a vision. He was the first cabinet minister CRF could claim to know reasonably well. Indeed Hoare was the only cabinet minister until the late 1930s with whom he could share a degree of intimacy and common interest. Dick Fairey was not an easy man to impress, but in January 1924 he wrote warmly to Hoare lamenting his departure from Adastral House, unaware that the brief tenure of Ramsay MacDonald's first Labour government would see the minister back in harness before the end of the year.[20]

In fact Hoare served as Secretary of State for Air for all but 11 months between October 1922 and June 1929. Even his harshest critics would concede that this was a period of real achievement, with an energetic Hoare protecting the RAF's procurement budget, promoting his chief of staff's modernisation programme, and pioneering commercial routes to Karachi and the Cape. He forged a powerful partnership with Sir Hugh Trenchard, and shared Fairey's high regard for Sir Sefton Brancker, the eccentric-looking but single-minded Director-General of Civil Aviation.[21]

While the biographer of the Chief of the Air Staff (CAS) reflected postwar antipathy towards Hoare, Trenchard himself had been generous in his praise. Likewise Lord Brabazon was insistent that over time the air minister and the air marshal forged an unlikely but 'irresistible' combination.[22] Moore-Brabazon was a fellow Old Harrovian whose close personal and working relationship with Hoare predated his firm friendship with Dick Fairey. Sir Samuel's record in office rested on the talented team of junior ministers, officials and ad hoc advisers he recruited to the Air Ministry. Most highly regarded was the Air Member for Supply and Research, Sir Geoffrey Salmond, a future chief of staff. His older brother, Ian, would also become CAS, having commanded counterinsurgency operations in Iraq, and from 1926 the defence of British air space.[23] The Salmond siblings were the RAF's rising stars, and the head of Fairey Aviation maintained close contact with both of them. Hoare's parliamentary private secretary was the Cambridge historian Geoffrey Butler, a fellow of Corpus Christi keen that the Air Ministry work closely with the University's Department of Engineering and its subsidiary field of Aeronautics. This collaboration, consolidated by the choice of Cambridge for the RAF's first university air squadron, echoed the formal and informal connections Maurice Wright and Vincent Nicholl maintained with their former tutors.[24]

Fairey Aviation was assiduous in cultivating contacts at every level of government and the armed forces. While civil servants rightly avoided any potentially compromising encounters outside of office hours, RAF officers in Adastral House or at stations across the south of England enjoyed Richard Fairey's hospitality whenever the opportunity arose. The chairman and managing director cleverly cultivated the young men who flew his firm's aircraft as well as the senior staff who evaluated and ordered them.[25] By the mid-1920s CRF was regularly entertaining service guests, Whitehall mandarins and overseas visitors in the Savoy Grill. Famed for its cuisine, decor and quality of service, the Savoy was Fairey's favourite hotel in central London. Here could be found the capital's most fashionable chef, François Latry.[26] Yet for routine business lunches he preferred the Café Royal, a chic meeting place for artists and aristocrats but rarely politicians. Another favourite was the Carlton Grill, where in November 1927 he bumped into H.G. Wells, inviting his hero to motor out to Hayes and 'meet some of your own earnest disciples.' When the invitation was renewed, Wells gently pointed out he was wintering in the south of France, but the kind offer had not been forgotten. No visit ever materialised.[27]

Sam Hoare was not the sort of chap who frequented the Café Royal or the Carlton Grill, but he was inspired by two charismatic characters who certainly did. One was his under-secretary of state for air, the wealthy and flamboyant Sir Philip Sassoon, a pioneer of aerial inspections east of Suez.[28] The other was the evangelising Sefton Brancker, a monocled man of action acclaimed by his minister as 'a superb propagandist.' Eager to promote flying as a popular mode of transport, whether at home or further afield, Hoare, Sassoon and Brancker all saw a need to capture the public's imagination. Inspired by the intrepid Sassoon, the minister and his wife undertook their own high-profile flights, each voyage further than its predecessor, and each covered extensively by the newsreel companies and the national press. Sir Samuel and Lady Maud's furthest imperial expeditions were to India, courtesy of De Havilland, in 1926–7, and to the head waters of the Nile in March 1929, courtesy of Fairey. Here were the two aviation companies most committed to testing the endurance and robustness of their machines, and true to form Hoare spent his final weeks in office checking out Sudanese staging posts for the much anticipated Cape to Cairo air service. Crucial for carrying out the task were the three Fairey IIIF seaplanes assigned to the ministerial party in Alexandria.[29]

The IIIF and its variants (the beefed-up Gordon and Seal, built for the RAF and Fleet Air Arm, respectively) were composite and then all-metal successors of the Type IIID. They flew in a variety of guises from 1926 up until the eve of World War II. The IIIF embodied the Fairey virtue of a template adaptable to various roles on land and at sea (as opposed to CRF's *bête noir* of the overweight hybrid), hence its popularity in South America and the Baltic states. At home, nearly 600 machines entered service, their reliability and performance consolidating Fairey Aviation's popularity with both ground and air crews. Modifications to the IIIF saw it develop the smooth lines and aerodynamic appearance familiar to the sleekest and most attractive biplanes of the late 1920s and early 1930s. No aircraft was subject to greater experimentation between the wars, with flight deck tests of the catapult-adapted variant laying a foundation for standard Fleet Air Arm tactics and techniques once the Royal Navy regained full control of its aerial strike force. If not undertaking reconnaissance and attack duties aboard surviving carriers from the Great War, or deployed on home bases as light bombers, the ever-evolving IIIF was undertaking a variety of duties across the Raj and the Near East. From 1927, RAF squadrons based in Egypt and Jordan had step by step established an 11,000 mile route from Cairo to Cape Town. Meanwhile, at home, a passenger-carrying IIIF had come into service ferrying VIPs such as the Prince of Wales.[30]

The latest version of the IIIF was thus an obvious choice for Sir Samuel and his intrepid companions as they set out to convince General Hertzog's wary administration in Pretoria that cross-continental long-haul flight would see travel to London measured in days not weeks. It clearly never crossed Hoare's mind that recidivist Boers with long memories might not greet a tightening of the imperial bonds in the same enthusiastic fashion that he so clearly did. Yet somehow he secured agreement for Imperial Airways to run a heavily subsidised service the length of Africa. This took the number of air miles on established routes to well over a million – a doubling of distance in five years, and a fulfilment of Hoare's pledge to the 1926 Imperial Conference. As with Norman Macmillan's premature and ill-fated attempt to fly around the world in 1922, and the IIIDs' inaugural flight to the Cape four years later, in March 1929 collaboration between Fairey Aviation and the RAF was vital: civilian and uniformed engineers worked together to provide the Secretary of State's formation flight with specially adapted two-seater seaplanes. Hoare returned to Egypt and then home to Hendon, 'showing the public that it was as practicable to keep to a timetable in Africa as it had been in India.' He had flown 10,000 miles in a fortnight, half of that distance in an open cockpit at the start of the dry season. It is not surprising that Dick Fairey, instinctively suspicious of politicians, held Sam Hoare in high regard.[31]

When Labour returned to office in June 1929, Charles Grey lamented the loss of a fervent 'advocate of aviation' who as a minister had 'done more than any one man to make the British People at home and in the Dominions overseas, airminded.' Privately, Moore-Brabazon congratulated a 'great Minister', who had triumphed over adversity, and to whom a grateful nation owed 'a great debt of gratitude.' Not everyone viewed Hoare's record at the Air Ministry so enthusiastically. The Admiralty resented his readiness to endorse Trenchard's policies, not least the exercise of firm control over the Fleet Air Arm at the expense of the Royal Navy. In 1929, however, such criticism remained largely within Whitehall, and Fairey, hostile initially to someone so insistent on a public subsidy for Imperial Airways, shared his friends' regret when Hoare left Adastral House for the last time.[32]

Hoare's encouragement of efforts by first the RAF and then Imperial Airways to forge commercial routes across Africa was partly his own initiative and partly a response to the commercial ambitions of record breaker, patriot and all round good egg, the newly knighted Alan Cobham. In the winter of 1927–8, Sir Alan's fame was such that he could borrow a flying boat from the RAF, secure sponsorship from the likes of Rolls Royce and Shorts, and set off to survey the area south of Sudan which Sir Samuel would visit a year later. Encouraged by the Secretary of State and Sir Sefton Brancker, Cobham broke off his eventful, much publicised journey southwards to attend a joint Colonial Office and Air Ministry conference in Nairobi. He left Kenya convinced his destiny lay in an African airline criss-crossing the continent free from competition. Cobham's grand plan was never realised, not least because Whitehall quickly concluded that connecting Cairo with the Cape was far beyond the capacity or the capability of one severely underfunded private venture – even if that venture was headed by someone as enterprising as Sir Alan Cobham.[33] In less than four years Cobham had become a national hero, first making his name as the pilot who chauffeured Sefton Brancker the breadth of Europe, and then on as far as Calcutta. Cobham brought the great man safely home, convincing Geoffrey de Havilland that the future lay in cabin-based aeroplanes opening up cross-continental routes and operating as colonial carriers. As chairman of the SBAC Fairey took quiet satisfaction in his role as fundraiser to facilitate the 17,000 miles round flight, the Treasury having refused to sanction a subsidy.[34]

De Havilland built suitably robust passenger-carrying aircraft, most notably the DH50 Cobham used for all but the last of his globe-trotting imperial expeditions. In the winter of 1925–6 air enthusiasts across the Empire followed Cobham and his crew on their barnstorming flight to and around South Africa; all approved that his first appointment upon landing in England was an audience at Buckingham Palace. Only a few months later the whole nation roared with approval when King George V rewarded Cobham with a knighthood for flying to Australia and back in record time. A brilliant self-publicist, Cobham ended his epic journey with a perfect landing on the Thames directly opposite the Houses of Parliament; within minutes he was across the river shaking hands with Hoare, Brancker and de Havilland. One can only speculate, but it would be surprising if Fairey was not among the crowd of well-wishers on the Commons terrace. Oswald Short had masterminded conversion of Cobham's beloved DH50 into a seaplane, with the Medway serving as the much modified machine's test bed and launch point. It was Richard Fairey, however, who provided floats for the return flight, Cobham having transformed 'a sea-bird into a land-bird' upon arrival in Darwin.[35]

Alan Cobham's astonishing achievement in flying to the opposite end of the world, touring Australia, and then returning home with scarcely a crease in his suit – *and* surviving his flight engineer being shot by a random Bedouin bullet in southern Iraq – signalled the extent to which pioneer flights were dependent upon collaboration across the industry. Partnerships with aero-engine manufacturers were of course the norm, but rival plane makers working together was invariably for a specific purpose. Although fiercely competitive, Fairey had a healthy respect for his fellow manufacturers, working closely with them in advancing the cause of British aviation. This first generation of engineer–entrepreneurs shared a keen sense of camaraderie and *esprit de corps*. The 1920s marked the apogee of peacetime collaboration, invariably focused upon prestige projects attractive to Hoare and his advisers inside the Air Ministry. A De Havilland/Short/Fairey combination, intended to ensure a British pilot's success in flying halfway round the globe and back, was one of several such endeavours. The joint project was unusual but not unique in being initiated by an individual and not an organisation, albeit a very forceful, energetic, persuasive and, above all, single-minded personality. However, for all the suppliant's undoubted charm, CRF would never have sanctioned sending floats to the far side of the world had he not been convinced Cobham was more than capable of flying his seaplane all the way home.[36]

Winning the Schneider Trophy in 1927 confirmed the value of British companies working together for the good of the nation. As we shall see, Fairey Aviation is the unsung hero in the story of Supermarine and Rolls Royce combining to build Reginald Mitchell's super-fast monoplanes, the S6 and S6B. Earlier in the decade the prize-winning feats of rival nations, notably the United States, would inspire Richard Fairey to ignore Air Ministry orthodoxy, spurn home-grown power units, and design high-performance combat aircraft, which the RAF simply could not afford to ignore. The end result was the Fairey Fox, of which more later: inspired by Fairey's sharp-eyed vision of the future, Ralli, Lobelle and their design teams at Hayes were challenging the likes of Hawker's Sydney Camm to create similarly clean-lined, high-speed single-engine biplanes.

Meanwhile their boss, Major Barlow, sought to build on the company's prominent role in breaking records and pioneering imperial routes. Before Charles Lindbergh crossed the Atlantic in May 1927, Hawker and the RAF briefly held the record for the longest flight without landing to refuel. The intention had been to reach India, but the expedition ended ignominiously in the Persian Gulf, only hours before the *Spirit of St Louis* landed triumphantly in Paris. Awkward parliamentary questions as to the cost of a venture so demonstrably overshadowed by a transatlantic triumph saw the Secretary of State for Air defending long-distance flight as a vital means of experimentation. Once Hoare had made this claim in the House of Commons then to save face the Air Ministry invited tenders for a long-range prototype capable of conversion into an imperial mail carrier. The latter idea was intended to see off backbench criticism of spending so much money on retrieving the long-distance record from the Americans – soon succeeded by the French and the Italians as the mileage advanced rapidly towards 5,000. Barlow and his boss enthusiastically took up the challenge of building a single-engine, high-wing monoplane that could carry a 1,000 gallons of fuel and fly 5,000 miles non-stop. CRF insisted upon a hands-on role throughout the lengthy period of design and development, spending long periods of time at Cranwell after his aircraft had been handed over to the RAF. Breaking the record became very much a personal mission, and any decisions made by the RAF were subject to final approval by the tall intimidating man in the white trench coat, his Stutz Blackhawk coupé ever present on the edge of the college runway.[37]

Endeavouring to maximise strength and lift, lead designer D.L. Hollis Williams located the fuel tanks within a wide cantilever wing: wooden spars and steel pyramid bracing spread the load, and a sophisticated pumping system fed Napier's 570 horsepower liquid-cooled 12-cylinder engine. An upholstered cockpit contained the latest equipment such as a rate-of-turn indicator and accessible oil filters, as well as a blow-up bed in the rear cabin area. A lengthy take-off necessitated an unusually wide undercarriage, with wheels mounted on roller bearings. In *The Times* Fairey insisted his proto record breaker would 'push science to the limits', demonstrating that in order to extend their range civil aircraft must adopt streamlining and discard wind-resistant radial engines.[38] Geoffrey de Havilland was already thinking along the same lines, but on the whole it was American designers who looked to combine looks and performance: for the next two decades airliners across the Atlantic were a generation or more ahead of their British counterparts.

While Sam Hoare pounded the streets of SW3 urging his Chelsea constituents to keep the Conservatives in office, Richard Fairey rolled out his fabulous-looking aeroplane. The one-time architect of skyward soaring model gliders rightly purred with pride. No wonder that, when the 'Postal' was handed over to the RAF, Norman Macmillan asked Fairey if he might join the RAF's elite flight of endurance aeronauts. Secondment was denied, which with hindsight was regrettable as the team based at Cranwell experienced serious differences over selecting the most suitable air crew. In fact the whole project was dogged with personnel and technical problems. For example, the aerodrome may have boasted a prestigious name, but it was wholly unsuitable: a heavily loaded aircraft was expected to take off up a steeply sloping runway and against the wind. A non-stop flight to India saw poor weather run down vital fuel reserves, and the distance record remained intact. Nevertheless, Fairey was inundated with letters congratulating him on his streamlined machine making Karachi.[39] As we shall see, upon its return to Britain the monoplane proved a star exhibition at that year's Aero Exhibition, at Olympia.[40]

Five months later, in December 1929, a combination of appalling conditions and a faulty panel altimeter caused the Cape-bound monoplane to come down in the Atlas Mountains south of Tunis, with fatal consequences. Antoine de Saint-Exupéry, North Africa's most experienced commercial pilot, and the future survivor of a desert air crash, no doubt applauded the tenacity of Squadron Leader Jones-Williams and Flight Lieutenant Jenkins. At the same time he surely asked why they should risk their lives so late in the year, and without a specialist navigator.⁴¹ Trenchard posed the same question, and a subcommittee of the Air Council instigated an inquiry into why Jenkins was chosen. Fairey was similarly concerned that Hugh Dowding in his role as project director had, with the best of intentions, selected the wrong man for the job. According to Geoffrey Hall his half-brother was deeply upset by the loss of life, having by this time forged a close working and personal relationship with the two dead officers.⁴²

Within 18 months the putative record breakers at Cranwell had in their hands an even more futuristic looking machine; its art deco appearance reinforced by the acquisition of wheel-spats, intended to reduce drag. The second and last of Fairey's long-range monoplanes boasted an elaborate feeder system to maximise fuel economy, more reliable and sophisticated instrumentation, and, most exciting of all, a primitive form of automatic pilot begrudgingly accepted from the Royal Aircraft Establishment at Farnborough. By the autumn of 1931 a Depression-ravaged Britain was experiencing acute financial and political crisis, but in Lincolnshire RAF mechanics quietly went about their business mounting ever more ambitious test flights. Finally, in February 1932, a successful attempt on the non-stop record was mounted, the partnership of O.R. Gayford and G.E. Nicholetts nursing their fragile aircraft as far as Walvis Bay in modern-day Namibia. A flag-waving tour of southern Africa duly followed, and in May 1933 the pair were welcomed home as heroes by Lord Londonderry, the National Government's first secretary of state for air. However, to the deep frustration of the RAF, and of a nation desperate for good news, only three months later an Italian crew flew that bit closer to the magic figure of 6,000 miles.⁴³

Fairey's greatest gamble – finding an engine and funding the Fox

Italian *aviatori* pursued records and prizes with gusto and passion throughout the 1920s and on into the following decade. Rising stars of the *Regia Aeronautica* were aggressive competitors at every staging of the Schneider Trophy, securing victory twice at home and once overseas. Yet in 1923 it was the Americans who claimed all the glory, retaining the trophy on home waters two years later. By this time Richard Fairey had moved from a fringe competitor to a discreet adviser and distinguished guest of the Royal Aero Club. In due course his collaboration with Supermarine and Rolls Royce would render him a major stakeholder in Britain's serial success from 1927 to 1931. Late in September 1923, on the eve of the Schneider Trophy being raced around the Solent, Fairey and Geoffrey Hall found time for a walk on the downs behind Portsmouth.

They were ideally placed to admire the test flight of a low-flying US Navy racer, which in looks and performance would soon render its European rivals obsolete. Having studied photographs posted from America, CRF already knew how advanced the Curtiss machine was in terms of monocoque fuselage and wing structure. He and his half-brother caught the ferry across to the Isle of Wight, where Curtiss mechanics were servicing the seaplane's lightweight D12 engine, its vee-type single block construction housing two aluminium castings of six cylinders: 'much simpler, more compact and hence lighter than the conventional design in which each part of the cylinders was machined separately from a forging.'[44]

Furthermore, the 450 horsepower Curtiss engine was liquid-cooled, its radiators forming part of the wing's leading edges. This gave the R3 seaplane its clean lines, from the engine cowling back down the length of the fuselage. The combination of aerodynamic design, optimum power–weight ratio, rapid lift and high cruising speed saw the American entry sprint to victory, at an average of 177 mph. Dick Fairey had seen the future.[45] He was impatient to jump a generation and build a British aircraft inspired by transatlantic technology. Long hours were spent alone at home drawing preliminary sketches and playing cards – throughout his life repeated rounds of Patience gave CRF mental space for problem-solving. He and Marcel Lobelle, his project manager, saw no problem coming up with the right air frame – the R3 provided suitable inspiration. As so often, the problem lay in finding a high-performance home-grown power unit. It was soon apparent that no such engine existed.[46]

Nor did any company in Britain manufacture an all-metal propeller to a standard which a Richard Fairey, or indeed a Ludwig Wittgenstein, would consider appropriate for piston-engine aircraft capable of flying at nearly 200 mph. The company founded by pioneer aviator Glenn Curtiss, whose wartime work with the Navy Department paralleled the Short brothers' support for the RNAS, had an exclusive contract with the Reed Propeller Company. S. Albert Reed was a research engineer who in retirement devised a process to machine propellers, complete with suitable pitch and correct aerofoil section, out of solid duralumin forgings. Until superseded by more advanced technology later in the decade, Reed propellers were renowned for their durability and efficiency. Facing bankruptcy after the war Curtiss had been rescued by Clement Keys, a Canadian venture capitalist who played a key role in financing the growth of America's fledgling aircraft industry. Although Fairey found a kindred spirit in Glenn Curtiss, all future business would be with Keys and his hard-headed colleagues. Their factory was located in upstate New York: Glenn Curtiss had been born in Buffalo, and it was there he founded the 'Aeroplane and

Motor Company' which bore his name.[47] Clement Keys made a deep and lasting impression on Fairey, as confirmed by a glowing tribute when the Royal Aero Club hosted Curtiss's chief executive in the summer of 1926. The lunchtime bonhomie on that occasion contrasted starkly with the meeting in Buffalo 30 months earlier: with formal pleasantries over, the two men began the hard bargaining.[48]

Fairey could scarcely ignore the painful contrast between the beautiful aircraft America's navy fliers had brought to England and the aesthetically challenged machines his company had created since its inception. Not only did the Fawn – still to enter production when Lieutenant David Rittenhouse USN (US Navy) powered his way to victory in the fall of 1923 – look like a turkey but it performed like one.[49] This was the moment when profit could no longer take priority over idealism – in any case, surely the two were not incompatible. CRF and his A-team of Lobelle, Barlow and Ralli set themselves the challenge of rendering the Fawn obsolete even before it entered service: their next project would be a speedy streamlined two-seater bomber which could out-pace the RAF's fastest fighter. They worked quickly, drawing on advice from operational RAF personnel such as Sir John Higgins, the mastermind in Iraq of a singularly brutal aerial campaign against insurgent tribesmen. In due course Fairey Aviation submitted detailed plans to a cash-strapped Air Ministry, hopeful of capturing ministers' and officials' enthusiasm for a radical new design. In November 1923 Fairey's proposal was turned down, with the company instructed to focus upon development and assembly of the Fawn.[50]

This left an irate Fairey to contemplate building a prototype without government support, and then convincing Trenchard and his political masters that here was an aircraft they could scarcely afford to ignore. A late autumn board meeting decided to accelerate Fawn production, but at the same time set aside funds for design of an immediate successor.[51] Such an aeroplane could only be built if, in Macmillan's words, it was powered by a 'clean engine of small frontal area' – in other words, if one or both of the two largest aero-engine manufacturers, Napier and Rolls Royce, invested heavily in designing and developing a British counterpart to Curtiss's 12-cylinder lightweight vee-type. Neither firm was interested in such a costly exercise.[52] Napier relied on successive versions of the Lion, its much-admired, widely adopted flagship engine. Rolls Royce saw luxury cars as generating a higher profit margin than short-contract aero-engine sales.[53]

In the winter of 1923–4 the cheapest – and most controversial – option was to turn to the United States and shoulder the cost of purchasing, or securing licences to manufacture, state-of-the-art engines, propellers and wing surface radiators. Fairey, working closely with the multi-linguist Maurice Wright, had already scoured Europe for alternatives to Napier's Lion engine, with its huge but unlocked potential. For various reasons neither Hispano-Suiza nor Mercedes power units were deemed technically or politically suitable.[54] Thus, if it required the Americans to facilitate construction of Fairey Aviation's generation-jumping combat aircraft, then so be it. Given that the company was funding the project alone, the Air Ministry could scarcely prevent Fairey from looking abroad for vital components.[55] At some point early in 1924 Fairey and Dawson crossed the Atlantic for the first but by no means last time. Sailing from and to Southampton would in due course become a familiar experience, interrupted only from 1940 to 1945 when wartime expediency required that CRF fly the Atlantic courtesy of a Pan Am Clipper or a US Army Air Force Skymaster. This was the great age of the ocean liner, and sailing the north Atlantic was a rare opportunity to relax for five days. Fairey always travelled courtesy of Cunard. A rare urgent cable might reconnect him with the outside world, but otherwise he could luxuriate in first-class accommodation. Although invariably invited to dine at the captain's table for the duration of the voyage, in later years he and Lady Fairey preferred to take their meals alone, or with friends; their favourite on-board restaurant was the *Queen Mary*'s famous Verandah Grill.[56]

The inaugural trip necessitated more modest expectations and expenditure, but Dick Fairey and 'Wuffy' Dawson could be sure not to stint on creature comforts. Surprisingly, no record survives of their journey, unlike the second visit they made, in September 1925. On that occasion they sailed for America on the *Aquitania*. Duff and Diana Cooper, travelling on the same liner two years earlier, had found it 'of tremendous size, great luxury, and remarkably good taste.'[57] In New York CRF and Dawson checked into the Plaza Hotel, on Broadway. *No, No, Nanette* had opened only a fortnight earlier, and it is hard to imagine such dedicated devotees of the musical not finding time to compare the American production with its British counterpart, already in to the seventh month of a lengthy West End run. As on their first trip to the States the year before, Clem Keys' clients from the old country headed north: checking out of the Plaza, Fairey and Dawson crossed midtown Manhattan to Grand Central Station where they boarded the West Shore Railroad connection to Buffalo. In 1925 the Fairey directors would meet their counterparts for a second time, in Baltimore: on 26 October they found Keys in a generous mood, having just watched Lieutenant James Doolittle USAAC (US Army Air Corps) scream across Chesapeake Bay in his Curtiss V-1400 racer to maintain America's firm hold on the Schneider Trophy.[58]

At the start of 1924 Fairey Aviation's reserve capital had amounted to around £23,000, which, with the dollar floating at less than $4.70, gave CRF considerable leverage when negotiating with the Curtiss company (and therefore with Reed, by now a subsidiary firm). During their first visit to the United States, Fairey and Dawson secured the necessary licences, and pending production at Hayes bought pre-assembled D12 engines for speedy employment back home.[59] At least one of these aero-engines was ready for immediate despatch. Wilf Broadbent claimed to have been present when a large box arrived at the Hayes factory in the early months of 1924. However apocryphal his story, he insisted that the 'old man' had booked a stateroom for the voyage home from New York, giving him sufficient space in which to house a sizeable container marked 'personal luggage.' Inside the box was a D12 engine and a Reed propeller. However improbable the scenario, Broadbent was insistent that the Curtiss board facilitated a smooth transit through US Customs, and that Fairey charmed his way out of the Cunard terminal at Southampton: 'He simply kidded them [HM Customs and Excise staff] along, keeping them on the right side without being subversive about it, and using his personality and his sense of humour until in the end they didn't know what to do either with the box or him. So they just said: take it away – we haven't seen this.'[60]

This still sounds somewhat implausible. There clearly was a box, but as for it travelling to England under Fairey's direct surveillance, this seems most unlikely – how was something so large and so heavy carried in and out of the cabin? Wilf Broadbent's is a great story, but the more prosaic version has a conventional crate shipped as freight. An imminent dock strike meant a rapid turnaround for CRF's ocean liner after its arrival in Southampton. With the ship already steamed up and ready to leave, unloading the crated machinery became a priority. Working with undue haste the dockers dropped their load. Fearing serious damage, Fairey fretted until he got back to the Hayes factory. Once unpacked both engine and propeller were found to be undamaged.[61]

Setting up as an aero-engine manufacturer would clearly take time. Fairey lamented Rolls Royce and Napier falling so far behind the Americans that he should even contemplate such a costly venture: 'as improvement in aircraft is 80% engine and 20% machine I want to get engine development into my own hands so as to pursue the general policy of progressive improvement.' A more immediate priority was tooling up to manufacture Reed's duralumin propeller.[62] If there was heavy demand, as soon proved to be the case, then the propellers could be brought over from America speedily and at minimum extra cost. Milling machines were installed at Hayes to commence propeller production.

At the same time an expensively equipped engine test house was built to analyse the performance of both the D12 and its equally revolutionary propeller.[63] From the outset, any British company which sought to import a propeller soon received a stiff letter from Charles Crisp pointing out that Fairey Aviation remained Reed's sole overseas agent. In the spring of 1926 Fairey took it upon himself to squash Imperial Airways' decision to order Reed propellers directly from Curtiss: the director responsible quickly backed down. Three months later CRF engaged in an increasingly heated private correspondence with the editor of *Automobile Engineer*, the latter refusing to believe Farnborough had infringed Fairey Aviation's exclusive right to wing-based radiators.[64]

Fairey's loathing of the Royal Aircraft Establishment was never deeper than in the mid-1920s. He complained vehemently to the *Observer*'s air correspondent that insidious 'RAE propaganda' coloured the electorate's view of an institution which in reality represented 'a great drag on the progress of aviation as well as a great waste of the public funds.'[65] Farnborough's state-sponsored technocrats lacked imagination, vision and, above all, a readiness to acknowledge superior American expertise.

Fairey Aviation never did build Curtiss engines under licence. The cost of tooling up from scratch was huge, and too risky in the absence of definite orders. For all CRF's lobbying of Trenchard, the Air Ministry was unhappy with CRF's latest enterprise, believing capital investment would generate a better return if concentrated on existing aero-engine manufacturers. Fairey complained that Hoare 'quite lost his nerve following the political pressure and although I have offered to lay down plant for manufacturing British engines of the very latest design he says he can give no support.'[66] Instead, around 50 D12 power units were imported, and duly christened the Fairey Felix. Some engines were sold on to government agencies for testing, but the majority were installed in the original versions of the Fox and its fighter equivalent the Firefly. The director for engine development within Adastral House used test data from the 'Felix' to put pressure on Napier and Rolls Royce to build a comparable 12-cylinder engine. Although he had no authority to do so, the RAF's L.F.R. Fell (as late as the mid-1920s retaining his half-colonel rank) suggested to Henry Royce that without a radical new design, 'an order for Curtiss engines will go to the USA which would be extremely dangerous and a national calamity and a disgrace.'[67]

At the same time a D12 was despatched to Derby for close inspection. Soon after, Rolls Royce adopted single aluminium castings in its design of the Kestrel, forerunner of the 'R' engines which would power Supermarine's seaplanes to Schneider Trophy glory. CRF undoubtedly attracted considerable flak for buying aero-engines from abroad, not least the charge of acting in an unpatriotic fashion, but his dramatic intervention provided a reality check for complacent engine builders at home. Even after Fairey's two trips to Buffalo, Napier carried on modifying its tried and trusted Lion engine. This reluctance to create a fundamentally new model was a consequence of the company's size and international reputation, not least in Weimar Germany. Napier became complacent, refusing to invest in significant R&D or to cut production costs through greater efficiency. Correspondence in the 1920s between CRF and Herbert Vane, his counterpart at Napier, suggests an organisation resistant to change. Vane resented CRF's dealings with Curtiss and his repeated attempts to drive down the cost of Napier engines and spares. Rolls Royce on the other hand rose to the challenge, albeit belatedly. Directors in Derby bore no grudge against Fairey or its founder, and by 1927 their company was the principal engine supplier for the popular IIIF and later variations of the Fox. Basil Johnson, Rolls Royce's managing director, was some years older than CRF, but the two men got on extremely well, whether sharing convivial business lunches in the Café Royal or sizing up the competition on the quay at Cowes. Johnson saw Fairey Aviation as a natural partner, discounting the price of individual engines and not taking offence when CRF chided him for sharing confidential data with rival manufacturers.[68]

Not only was Fairey a partner on the Fox I and its successors, but its state-of-the-art propellers enabled Derby's V-12 'R' engine to secure Supermarine a second and third Schneider Trophy, the company repeating its 1927 success in 1929 and again in 1931. Fairey counted as very much a VIP when the race was once more staged in home waters, but as we have seen he enjoyed similar hospitality across the Atlantic in 1925; the same was true two years later when the Royal Aero Club relocated to the Venice Lido.[69] Especially pleasing for CRF was the Under Secretary of Air's preferred means of winging his way from Westminster to La Serenissima: the Right Honourable Sir Philip Sassoon alighted at Sant'Andrea, having flown to Italy in the passenger seat of the company's custom-built Fairey IIIF, piloted by Maurice Wright. One suspects Dick Fairey chose not to join the minister in sampling the great art and gay life in and around Saint Mark's Square. Wright was no less indifferent to his erstwhile passenger, being preoccupied with shipping replacement propellers out from England. Meanwhile, flying high and fast above the northern waters of the Adriatic, the RAF at last brought the Schneider Trophy back to Britain.[70]

A fortnight later the victorious RAF flight reported to the Savoy, as guests of honour at a luncheon and reception presided over by a delighted Sir Samuel. Ironically, Hoare had declined Mussolini's invitation to visit Venice, fearing the Duce would 'turn an Italian victory into a triumph of dictatorships over democracies.' Relieved by the result the minister ordered his officials to organise what turned out to be a glittering occasion, co-hosted by the Royal Aero Club, the SBAC, the Air League and the Royal Aeronautical Society. Everyone who was anyone in the British aviation industry attended. Surprisingly, CRF sat at a lowly table eleven. He after all had a year earlier launched the Royal Aero Club appeal to fund the Empire's record breakers. Nevertheless, Fairey soon found himself in the spotlight when he joined his counterparts at Napier and Supermarine in accepting the Society's prestigious Silver Medal.[71]

In September 1927 Supermarine's legendary designer, Reginald Mitchell, had relied on a heavily modified Lion engine to secure victory. Four months later, with Hoare and Trenchard having secured Treasury agreement that the RAF establish a fresh High Speed Flight, Supermarine began planning its next campaign.[72] The Southampton firm again looked to Hayes, and in particular P.A. Ralli. Fairey's mathematical wizard of airscrew design duly provided the blueprint for a third-generation high-spec propeller, this time for the S6, the radical new racer necessary to retain the trophy.[73] However, both Mitchell and his boss, James Bird, recognised Napier's reluctance to design and build an aero-engine capable of matching competitors from the continent and North America. Supermarine therefore turned to Rolls Royce for the necessary power plant. Thus was born the forerunner of the legendary Merlin engine (similarly, Mitchell's trophy-winning seaplane shaped his later design of the Spitfire).[74] Rolls Royce was yet to become the dominant force within the British aero-engine industry, but when it did, Richard Fairey – directly via collaboration on the Fox and its variants, and indirectly via the Supermarine S6 and its successor – could boast a key role in rousing the sleeping giant.[75]

Basil Johnson and his board might not have viewed Fairey so benignly had they known of his brief dalliance with the notion of taking over their company. In March 1926 CRF suggested to the company solicitor, Charles Crisp, that he should invest in both Napier and Rolls Royce: once Fairey Aviation became a major shareholder in one or both companies then it could influence or even dictate investment policy. Fairey even contemplated a takeover of Rolls Royce, but lacked the capital or credit-worthiness to buy up the company *and* to maintain the necessary modernisation and expansion of the Hayes plant.[76]

The accusation of being unpatriotic cut deep, at a time when Fairey was subject to considerable stress. Throughout the middle years of the 1920s he responded furiously to hostile comments in the press, with the *Daily Telegraph* and the *Morning Post* his fiercest critics.[77] This may account for the huge row he had with Fred Duncanson, although the latter had ceased to be a trusted member of the design team long before he stormed out to join Blackburn.

In the spring of 1925 Fairey was physically and mentally so low that his doctor ordered him to take an extended holiday in the south of France, recommending months not weeks.[78] He left for the Riviera assured by Norman Macmillan that his Curtiss-inspired – indeed Curtiss-powered – private initiative 'was a winner.' Fairey's chief test pilot flew the Fox I at a then stunning 156 mph, a speed nearly one-third faster than the officially sanctioned light bomber: 'This was not regarded with undisguised pleasure by the officials whose duties were concerned with Air Ministry specifications, and who had dealt with the Fawn but had nothing to do with the Fox. In the circumstances it was no easy matter to sell the Fox to the Air Ministry, despite its performance, which exceeded that of then current single-seat fighters.'[79]

Not surprisingly, Fairey authorised development of a fighter version, the eventual Firefly I. This solitary aircraft first flew 11 months after the Fox, but the design was scrapped when Hoare's officials insisted the original D12 engine make way for the much-anticipated Kestrel: anticipating the new Rolls Royce engine would add weight, Fairey disputed Air Ministry predictions of improved performance. Both the aborted Firefly I and the Fox were built predominantly of wood, albeit with steel tubing to reinforce the wing-bearing central section of the fuselage. The winter of 1924–5 was too early to come up with an all-metal aircraft, and CRF resisted his designers' desire to move on. Vincent Nicholl was more receptive. Taking advantage of Fairey's absence abroad, he authorised an experimental IIID, with sheet aluminium covering the fuselage. When the managing director resumed his duties he was incandescent, but on reflection acknowledged the error of his ways. Reconciled with Nicholl, Fairey recognised the significance of what would become the first of three prototypes, each significantly different from its predecessor but together known as the Ferret. This first-generation machine impressed the Fleet Air Arm in flight trials, but not enough to generate an order. Nevertheless, the Ferret did inspire an all-metal version of the popular and versatile IIIF, and for Fairey Aviation it signalled a new era in aircraft production.[80]

Marcel Lobelle and his colleagues were already contemplating an all-metal Fox II when, to their delight, Fairey's high-risk strategy paid off, and the RAF placed an order for the original model. On 28 July 1925, with CRF back in harness, Fairey Aviation played host at Northolt to a high-powered delegation from the Air Ministry. Heading the party was Sir Hugh Trenchard. The Chief of the Air Staff was accompanied by an eventual successor, Sir Geoffrey Salmond, at the time Air Council Member for Supply and Research. If any service personnel had the power and authority to make arbitrary decisions regarding procurement, then these two could. Fairey's chief test pilot put the Fox I through its paces, and on landing was led away by Trenchard for a quiet chat. The CAS queried whether such a fast aircraft could be safely flown by young and relatively inexperienced pilots. Thirty years later Norman Macmillan recalled his answer: 'I told him frankly that it was one of the easiest and most viceless aeroplanes I had ever flown. We walked back to the hardstand, and Sir Hugh Trenchard looked at Dick Fairey and said in his booming bass: "Mr Fairey, I have decided to order a squadron of Foxes."'[81] The CAS saw development of the Fox as an 'example of enterprise', and insisted the Secretary of State publicly endorse his arbitrary decision. For Trenchard this was a test of Hoare's sincerity and his strength under fire: the minister duly passed, and 'We were partners after that.'[82]

In the Commons, Sir Samuel stoutly defended an initial order for 18 machines. In due course a further ten confirmed Trenchard's faith in the Fox. All aircraft were powered by D12 engines fresh off the Curtiss production line, except for three test models built to put an early version of the Kestrel through its paces. The importance placed on this new home-grown aero-engine was reinforced by the D12's mixed results when tested at Farnborough: the 'Felix' received only qualified approval from the Royal Aircraft Establishment, thereby reinforcing Fairey's prejudice towards government-employed engineers – if these people were any good then they would have switched to the private sector, like Tom Barlow. As for the Air Ministry, 'with the eyes of the trade on them they will not dare to give us a full certificate.' Macmillan shared his employer's surprise that there should be any reservations regarding either engine or airframe. He felt the Fox to be, 'the most stable aeroplane I have ever flown', and, if anything, he experienced more problems when the engine derived from Derby not Buffalo. After demonstrating to the Fleet Air Arm the Kestrel-engined Fox's credentials as a carrier-based aircraft, but with no tangible result, Macmillan understood why Fairey felt so aggrieved.[83]

That keen sense of grievance was compounded two years later after the Air Ministry decided against an all-metal Fox II when selecting the RAF's next two-seat day bomber. Whether by accident or design, Fairey was excluded from the invitation to tender. CRF strongly suspected a decision inside Adastral House to isolate Fairey Aviation from the regular tendering process. In June 1926 he complained to Hoare that he was being penalised for having drawn on American expertise to kick-start technological innovation at home.[84]

Fairey was furious when he discovered that the recommended prototype bomber – Sydney Camm's Hawker Hart – was a metal-framed biplane heavily indebted to the revolution in British aircraft design the Fox had brought about. Fairey believed performance data concerning the Fox, supplied in confidence to the Air Ministry, had shaped the specification officials later circulated to his rivals. Salmond and Trenchard were sufficiently impressed by Fairey's protestations that they agreed to a belated entry into the competition. The speedily constructed Fox IIM was an impressive aircraft, capable of reaching almost 200 mph. However, in design and performance it was not that dissimilar from the Hart, and so the original decision stood. A staggering 989 Harts were built. It is undoubtedly true that the RAF's 1928 contract proved the making of T.O.M. Sopwith's resurrected company.[85] Given how alike the two aircraft were, Tommy Sopwith surely felt uncomfortable at the way Dick Fairey emerged from the whole Fox episode with only a modest return upon his investment. Yes, much of the new technology facilitated a significant updating of the ever popular IIIF, thereby extending its life well into the 1930s, but Fairey Aviation never secured the profits it might have anticipated given the revolutionary changes initiated by CRF's original decision to adopt American technology and to emulate American design.[86]

Fairey's only consolation was that even after the Hawker Hart entered production the Fox I remained a totemic aeroplane for RAF bomber crews. Only 12 Squadron was equipped with the Fox, but the aircraft's deployment was a significant boost to service morale, reinforced when successive exercises from 1926 to 1931 signalled how easily the Fairey day bomber evaded laggardly fighters. In 1928 the pilots of Fighting Area, the precursor of Fighter Command, pooled ideas on how to hunt a Fox. Their demand that Trenchard and his staff purchase the single-seat Firefly fell on deaf ears. A year earlier the commandant of the RAF Staff College, Edgar Ludlow-Hewitt – a future air chief marshal and a great admirer of Richard Fairey – gave 12 Squadron a fox's mask. Draughtsmen at Hayes adopted the mask as the squadron's crest, with the accompanying motto 'Leads the Field.' Keen to generate favourable publicity, Fairey Aviation briefed the press. Photographs of the Fox with an appropriate crest on its tail fin caught the imagination of the general public. Thus, a manufacturer in Birmingham supplied members of the 'Shiny 12' with enamelled fox head badges. Into the 1930s Fairey's advertising made much of 12 Squadron's offensive prowess, with magazine and newspaper adverts pointing out how the Fox had proved so successful in RAF bombing competitions it was now handicapped in order to give fighter pilots a better chance of defending the capital. Fairey himself went so far as to suggest no 'defending scout' boasted the performance required to intercept the latest generation of offensive aircraft. With hindsight such claims, however accurate, sowed the seeds of fatalism, as expressed most famously in Stanley Baldwin's November 1932 assertion that 'the bomber will always get through.'[87]

On a more positive note, the company's involvement in devising 12 Squadron's crest prompted CRF to acquire something similar as a corporate badge – what today we would call Fairey Aviation's logo. Fairey's sister, Margery Anne Needell, boasted an art school training and was an accomplished artist. Her niece recalls Aunt Margery as, 'probably my father's favourite sister although he was deeply fond of all of them. She was very intelligent and had a witty and sometimes rather sharp tongue, but a kind heart.' Here was someone ideally qualified to come up with a suitable motif. In consequence a domestic design process became the basis for global branding, the upright winged ears and inverted triangular head of the ubiquitous Fairey badge enjoying wide recognition across the Empire and beyond.[88]

Fairey hits Olympian heights: the 1929 International Aero Exhibition

Familiarity with Fairey's freshly conceived badge dates from the summer of 1929. On 16 July His Royal Highness The Prince of Wales opened the seventh International Aero Exhibition. 'David' was extraordinarily good at feigning interest in matters military and industrial, but he does appear to have been in awe of aircraft. Perhaps the relatively modest rank he held in the RAF – group captain – signalled the fact that he took seriously his role as regal patron of the nation's newest service. 'Boom' Trenchard treated him with genuine respect, personally approving the adaptation of service aircraft for regular use by the heir to the throne.[89] At Hendon there was no rush to depart the Royal Box when the RAF staged its annual display of aerobatics and combat capability. Nor did HRH display his father's impatience and indifference to new technology when touring the huge exhibition hall at Olympia; he spent time at all the important stands, asking informed and pertinent questions. The Prince of Wales already displayed worrying signs that he was unsuited to become King, but few could fault his fascination with aeronautics. He needed no prompting to remind exhibitors that just 20 years had passed since Blériot crossed the Channel and yet packed into Olympia was the most astonishing array of technically sophisticated aircraft.[90]

The evolution of the piston engine aircraft, accelerated in the hothouse conditions of the Great War, was a truly remarkable phenomenon. Reviewing the prototype-packed display areas, C.G. Grey noted the absence of traditional materials, and the ubiquitous presence of steel and light alloys. The Fairey revolution was complete, as symbolised by the post-Reed propellers mounted on each corner of the stand, and the stark, striking skeleton of a IIIF, stripped down to reveal its welded engine mountings and steel tube airframe plus duralumin structured wings. The further presence of a Fox, dual variants of the Firefly, and another two IIIFs (one a seaplane with folding wings) explains why Fairey Aviation was that year's largest exhibitor. Symbolic of a marked shift in the pecking order was the presence of the long-range monoplane, seemingly floating in mid-air (courtesy of sturdy Victorian roof beams) and directly facing the Short Brothers stand. Beneath the huge cantilevered wings of this slender, eye-catching machine an equally elegant white office housed Fairey's quietly satisfied sales team.[91]

Among Olympia's many thousands of visitors that summer of 1929, one admirer commented upon Fairey Aviation's success in adapting an old design (the Type III) to embrace the new orthodoxy of clean lines. Indeed, 'The engine cowlings of all Fairey machines form a striking feature. They appear to have the engine poured in.' The reality was that by this time cowlings were no longer hinged. Instead they were secured by a simple clip mechanism, patented by Fairey and adopted across the industry. Each clip cost two shillings, and thousands were sold each year. Production costs were minimal and the company derived significant profits from a simple but very effective invention.[92]

Even more profitable was the composite and later all-metal IIIF, with no fewer than 621 machines built in various forms. Its predecessor, the IIID, had sold in similar numbers, compensating for the relatively few Flycatchers bought by the RAF on the back of the Fox's popularity with pilots. Fairey Aviation's commercial success in the second half of the 1920s depended heavily upon the Type III. Thus 127 of the 164 aircraft produced in 1927 were IIIFs, as were all but one of the following year's 251 machines built at Hayes. In 1929, with most rivals other than Tommy Sopwith desperate for contracts, Fairey manufactured 200 aircraft, of which 170 were the ever reliable IIIF. The plane was so versatile that bespoke versions could be built for foreign buyers at prices considerably higher than each machine cost the RAF. At least 25 were sold abroad, Fairey Aviation extending its reputation outside the Empire by sending seaplanes to South America, China and southern Europe. A single sale via Arcos, London's short-lived Soviet trade agency, established a useful link with Stalin's Russia.[93] With the IIIF as its bedrock Fairey could afford a major expansion of the Hayes plant. In old age Wilf Broadbent was insistent that at the end of the decade, 'we had more than fifty per cent of the total ministry vote.' An 'embarrassed' Labour Government put pressure on CRF to subcontract work to rival firms, much to his annoyance.[94]

With numerous overseas visitors attending the International Aero Exhibition, Maurice Wright was ever present in Fairey's gleaming white office. He after all was the company's most senior – and most cosmopolitan – salesman. Four years earlier CRF had lobbied hard for the RAF to release his old friend. As an executive director with recent service experience, the retired Squadron Leader Wright had proved invaluable in selling machines abroad. On his first day as a director Wright had sealed a deal to sell the Dutch Navy six seaplanes; 24 hours later he could be found extolling the IIID's virtues at the Ministry of Marine in Copenhagen.[95] The number of potential customers at Olympia was in reality small, with the bulk of those touring the exhibition excited enthusiasts understandably drawn to a much publicised and undoubtedly impressive display of aviation hardware. There were those from abroad eager to learn, although without employing covert and illegal means only limited intelligence could be gleaned from examining the aircraft on display. Clearly precautions had to be taken, but Richard Fairey and his colleagues were no little Englanders. Wright was as at home in Montparnasse as Mayfair, while CRF and Dawson had fully embraced the work hard, play hard ethos of 'roaring twenties' America. Unsurprisingly, upon their return from New York in November 1925, Fairey fed technical information to the air intelligence officer inside Adastral House.[96]

Any intelligence gathered abroad was usually sent straight to the top, namely Sirs Geoffrey Salmond and John Higgins, successive Air Members for Supply and Research. Fairey wrote frequently to both men, and at considerable length.[97] A further recipient of the information-gathering activities of the chairman and his colleagues was the Air Ministry's Director of Technical Development. The latter would when appropriate reciprocate, witness a 1927 report from Madrid on the Dornier Wal flying boat which conveyed one Colonel Francisco Franco from Spain to Argentina.[98]

Richard Fairey cultivated fellow industrialists in North America and Europe, partly because he enjoyed sharing the high life with men who, like himself, ran large companies they themselves had founded and grown, and partly because these movers and shakers were the ultimate sources of information. Thus, even before CRF took up sailing he had become a close friend of France's Louis Breguet, whose eponymous company had much in common with Fairey Aviation. The two men became acquainted during the course of the Paris Air Show in late 1926, in the first instance establishing a pattern of reciprocal visits to respective factories for senior staff.[99]

While, given the legacy of the Great War, Fairey had no illusions regarding the great Dutch designer Anton Fokker, he held him in high regard. A prickly personality, Fokker had high expectations and a quick temper. Not surprisingly, therefore, in mid-April 1925, CRF was incandescent when poor service at the Savoy infuriated his guest on the occasion of the two men's first meeting.[100]

If fond of the cinema then a fortnight later the jaundiced Fokker may have seen Pathé's newsreel celebrating the sale of six Fairey seaplanes to the Dutch Navy. Perhaps it was the film's title – 'Mr Fokker … Please Note!' – which prompted the old adversary to initiate a mischievous campaign in the Dutch press portraying the IIIDs as poorly built. Maurice Wright was soon back in Holland reassuring reporters that quality control within Fairey Aviation was first class.[101] In later years, and with no illusions, a wary CRF maintained contact with the Dutchman, albeit at a distance once the latter focused his business activities on the United States. Had Anton Fokker arrived unannounced at Olympia in July 1929, Wright would have displayed the hospitality due such a distinguished visitor, but there would have been no warmth in his welcome – with Fairey similarly polite towards a fellow professional, scarcely a friend.

The Great Crash was still three months away and, for all but the most prescient economist, commercial prospects appeared promising. Too many plane makers in Britain were still suffering from retrenchment, but a privileged few had grounds for optimism. That positive outlook translated into forestalling foreign competition or seeking overseas alliances. With the Weimar Republic more stable and the German economy riding on the back of geopolitical rehabilitation and manufacturing revival, Wright persuaded CRF to meet the veteran industrialist Hugo Junkers at his factory in Dessau.

As it transpired, the two men travelled to the banks of the Elbe in the immediate aftermath of Wall Street's 'Black Tuesday.' On arrival in Germany they could scarcely have envisaged to what degree their hosts' fortunes were being shaped by catastrophic events taking place thousands of miles away. Within three years of Fairey's visit, economic upheaval and Nazi control of Dessau council would see closure of the Bauhaus School. Founding father Walter Gropius was already back in Berlin, and one can only speculate on whether CRF found time to admire the School's strikingly modernist construction.[102] One designer he did encounter in Dessau was Junkers' Ernst Zindel. Here presumably was an admirer of Bauhaus's functionalist principles, if not its revolutionary politics. Zindel was a fellow pioneer of employing duralumin in airframe and cantilever wing construction *and* as a robust corrugated alternative to stressed skin. The legendary – for some, notorious – Junkers 52 transport aircraft was yet to fly, but by the time of the Englishmen's visit it was already at the development stage. The JU52 embodied all Zindel's favoured design principles, and a like-minded engineer had the nous to recognise a winner when he saw it: in later years, as he viewed film footage from Spain, Poland, Holland, Greece or Russia, Richard Fairey could gain no satisfaction from knowing he had identified a resurgent Germany's key asset even before it left the drawing board.[103]

Back in London CRF briefed Hoare's successor, Lord Thomson, as in 1924 the Labour Government's Secretary for Air.[104] Fairey warned that in Germany 'he had seen factories which could easily be the foundation of a future aircraft industry.' Yet already the German economy was imploding, and the minister deemed his visitor unduly alarmist. So too did Ramsay MacDonald. The Prime Minister's staff kept Fairey out of Downing Street, but by chance the two men met when dining at the National Liberal Club. Geoffrey Hall recalled his half-brother's belief that 'he was treated in a most scathing manner.' Distaste turned to deep loathing when MacDonald charged CRF with misleading ministers in order to secure more orders.[105]

Responding to the rise of Labour – Richard Fairey's politics

The encounter with MacDonald in late 1929 was purely by chance, and the appointment with Thomson the result of repeated requests. From May 1929 to August 1931 Fairey enjoyed scant access to senior ministers. The same was true of Labour's first, very brief period in office. Whichever party was in government, close acquaintance with cabinet members, other than at official functions or factory visits, came only in the late 1930s. A decade earlier Fairey might have claimed to know Sam Hoare quite well, but almost wholly on a professional basis. As we shall see, acquisition of the Bossington estate allowed Fairey to offer leading politicians a day out with gun or rod. Before that, CRF's contacts at Westminster were junior ministers and loyal backbenchers. When Labour was in office Fairey relied upon a network of senior civil servants and RAF personnel to maintain an informal as well as a formal working relationship between the Air Ministry and its biggest supplier.

Plane makers in 1924 and again in 1929 were reluctant to recognise the readiness of Labour's Secretary of State for Air to promote their industry. They saw only Lord Thomson's support for the public sector and his enthusiasm for airships. Critics regarded the ex-brigadier as responsible for his own and the 47 other deaths resulting from the R101 airship's crash near Beauvais on 4 October 1930.[106] One of those killed was the popular Director of Civil Aviation, Sir Sefton Brancker, who CRF knew well. Branker had confided to CRF his fears as to the R101's airworthiness; his death no doubt compounded Fairey's distaste for perceived traitors to their class, not least socialists from privileged backgrounds with ministerial ambitions.[107] Between the wars the likes of Hugh Dalton and Sir Stafford Cripps lived their lives almost wholly separate from that of Dick Fairey, Tommy Sopwith and their peers. Neither Hampstead progressives, nor proletarian purveyors of power like MacDonald and Arthur Henderson, found themselves in the private dining rooms or hunting lodges of the self-made men shaping the future course of British aviation. For all his shameless socialising, soirées with Lady Londonderry and lengthy tenure at Number Ten, 'Ramsay Mac' was scarcely the sort of chap with whom the country's ills could be sorted out over a select claret, a sound port, a stiff brandy and a seasoned cigar.

Fairey's initial distaste for MacDonald was compounded by the Labour leader's reluctance to condemn mass action by the trade union movement in May 1926. Two days into the General Strike, Charles Crisp visited the Home Office to assure the Chief Civil Commissioner that Fairey Aviation was available to provide all necessary assistance. A follow-up letter from the chairman and managing director was more specific, offering a stricken nation 'the fastest two-seater aeroplane in the world plus chief pilot Captain Norman Macmillan.' According to his daughter, Fairey's determination to see the strike broken saw him volunteer to travel as a bodyguard on London buses, armed with a length of lead piping.[108]

Annoyance turned to anger when industrial action kept Fairey away from a meeting of the Advisory Committee on Aeronautical Education, and an opportunity to promote vacation placements for engineering students: he saw his company as a role model for employing undergraduates on a nominal wage. Here were young men willing to work and eager to better themselves.[109] One such young man was Geoffrey Hall, in the spring of 1926 commencing an apprenticeship at Hayes following his two-year City and Guilds course at Finsbury Technical College. Thanks to his half-brother, Hall would in due course progress to a bespoke course in aerodynamics at Imperial College. For the duration of the General Strike he served as a volunteer bus driver, cheerfully driving through picket lines and giving as good as he got.[110]

Young Geoffrey could have stayed at work if he wanted to cross the picket line. The Fairey Aviation management estimated that around 60 per cent of employees at Hayes wanted to work, and duly arranged transport. Fairey convinced himself that, 'Of the remainder of our people who are out more than half would come in but for fear of the Unions.' On the sixth day of the strike he berated Baldwin's government for being too quick to talk and too slow to 'deal firmly with the trade unions.' As we shall see, by the 1940s CRF believed the BBC to have been taken over by wily left-wing propagandists. Similar sentiments were aired in 1926, with managing director John Reith and his beleaguered staff seen to 'talk a lot of nonsense about being impartial which is the last thing to be in this world if you want to attain your object, and always everybody here who feels a bit weak or doubtful call themselves impartial.'[111]

CRF himself had no problem with partiality. Five months prior to the start of the strike he was insisting to fast-track RAF officers that trade unions were an anachronism, dragging Britain in the direction of subsistence Russia rather than free market America. The weakness of organised labour and the strength of a system based on payment by result rendered the United States an anti-Bolshevik, ultra-capitalist role model: 'There is no dole, no workhouse and very little charity to help a man out of work. There is on the other hand ample opportunity to the man who chooses work to live at any standard that he desires.' Addressing the same audience after the General Strike, Fairey saw recent events as vindicating his view that trade unions were a blinkered and malign force, driving living standards down by thwarting the beneficial consequences of an efficient capitalist system. Britain should not only emulate the Americans but invite them to join the Dominions and the 'mother country' in an extra-imperial economic union.[112]

Fairey's Panglossian view of American society was proclaimed in more public venues than the RAF Staff College. In a lengthy speech at the House Dinner of the Royal Aero Club on 19 January 1927 he refuted the charge of his critics – notably the veteran Labour politician George Lansbury – that he was in hock to American financiers. Rambling and intemperate, Fairey's address treated the Baldwin Government with almost as much contempt as the softheaded exponents of 'the Socialist doctrine' – here was a short-sighted administration that had surrendered control of the Dominions at the Imperial Conference and control of Germany at the Locarno Conference. Treaties, and trade, with Europe were unnecessary if the Empire stood inviolable and the RAF constituted a credible deterrent. Britain's manufacturers, not least its plane makers, should meet urgent needs at home, starting with defence, and then look beyond Europe for fresh markets: 'We do not wish to share the world's production. We want it all. Why should we share our Aviation when we have the power to take most of it for ourselves? Are we ultimately going to share the assets of the British Empire with those comic conglomerations of Latin, teutonic and Slav peoples who form the present League of the Nations of Europe?'[113] Note the labelling of the League – for CRF the absence of the United States simply confirmed the organisation's irrelevancy. Ramsay MacDonald, even prior to his second premiership already seen by admirers as embodying 'the spirit of Geneva', must have read Fairey's speech in *The Times* and been horrified.

By this time a shocked Leader of the Opposition was of scant concern to the head of Fairey Aviation. To adopt the parlance of the nation he so admired, by the middle years of the interwar period Dick Fairey was a player. As if to confirm this, he welcomed the nickname 'Tiny.'[114] The regimental-style sobriquet signified a formidable presence and a forceful personality. Here was the proverbial 'big man', in every sense of the term. For his friends and colleagues he was a 'man's man' in a man's world. This particular world was one of fixed values and firmly-held opinions, invariably of a deeply conservative hue. Yet as we can see, such views were not instinctively Conservative, for all Fairey's courting of ministers and dislike of Labour. Although a member of the Conservative Party he scorned Stanley Baldwin's 'One Nation' Toryism, and by 1930–1 was surely sympathetic to the leadership challenge of press barons Beaverbrook and Rothermere: the 'Empire Crusade' gave voice to Max Aitken's passion for imperial protectionism and Esmond Harmsworth's predilection for making mischief.[115]

Fairey felt keenly that the National Government did not do enough to insist the Dominions – especially Canada – buy British rather than American aircraft.[116] Like Charles Grey and another pillar of the aviation establishment, the Master of Sempill, CRF saw aviation as crucial to a regeneration of the Empire, sending a powerful signal to dissident nationalists – not least Congress – that the promotion of new technology confirmed Britain's fitness to rule its imperial domain. Thus Fairey Aviation's pioneering flights to India became an inspiration for the much grander 'techno-political display' of the 1933 Houston Mount Everest Expedition.[117]

CRF shared Beaverbrook's belief in the 'English-speaking races' as a vehicle for economic revival.[118] It is scarcely surprising they became good friends when finally their paths crossed in June 1940. In the early 1930s Fairey would enthusiastically endorse the *Daily Mail*'s campaign for expansion of the RAF; for nearly a decade he had enthusiastically supported the Air League's call for a huge reservoir of aircraft, available the moment war was declared. Where he differed from Lord Rothermere was in eschewing a parallel call to accommodate Germany's wishes.[119] Unlike Rothermere or Lord Londonderry, the hard-headed factory boss had no illusions regarding Hitler and his acolytes: Reichsmarschall Göering represented a future threat, not a charming if eccentric former foe.[120]

Remarkably, the *Daily Mail*'s James Wentworth Day was even more right-wing than his employer, but he shared Fairey's deep suspicion of German intentions. Proud of his East Anglian roots, Jim Wentworth Day's paens to rural life earned him an invitation to join Dick Fairey for a weekend's shooting. The two men became firm friends, with Wentworth Day excitedly profiling his new countryside companion in the *Daily Mail*: 'a modern Elizabethan ... If you wanted to find an Englishman of the English who might have stepped from the pages of Kipling, walk into the Managing Director's black-and-mahogany office at the Fairey Works...' While Wentworth Day retained a sneaking regard for the Duce, by mid-April 1939 he had drafted with Fairey an extra-governmental programme for covert anti-Nazi propaganda, to be led by 'a man of vision and imagination, of fire and enthusiasm with sufficient coolness of mind to temper that fire with discretion.' Wentworth Day's career up to and after World War II was built on being anything but discreet, and yet – even if the necessary credentials suggested his co-author – there was no doubt who he had in mind for the job.[121]

Three years earlier CRF must surely have disapproved of Murray Sueter's presence at the 1936 Nuremberg Rally.[122] He would have been equally unhappy that his old friend the Master of Sempill was a member of both the Anglo-German Fellowship and the more secretive Link, and as late as 1939 continued to urge an accommodation with Hitler's wishes. In early 1940 MI5 informed the Royal Navy that Sempill, by this time an Admiralty adviser, had for several years been selling secrets to the Japanese. Given his high level of security clearance, CRF must have been privy to this information, no doubt sharing Churchill's keen sense of personal betrayal.[123]

To his credit Fairey was never seduced by 'Tom' Mosley, not least given the latter's past connection with the Labour Party. An industrialist so firmly committed to minimal state intervention and the sanctity of the market was unlikely to flirt with fascism. The same could not be said of Moore-Brabazon, Sueter, Wentworth Day and C.G. Grey, all of whom at different times looked favourably upon Sir Oswald. Enthusiasm for the British Union of Fascists reflected Sueter's admiration for Mussolini, and Charles Grey's visceral anti-semitism.

The Aeroplane's editorials were astonishingly racist, even by the standards of the day, with rearmament celebrated as the means to join Italy and Germany in a grand coalition to defeat both the Russians and the Japanese: '…enemies of the White Race, or human sub-species, as Prof. Julian Huxley would have us say.'[124] Grey's belief in a pan-European Jewish conspiracy, repeated demand for an alliance with Hitler, and insistence that Churchill's faith in the *entente* was tantamount to treason, prompted a spirited exchange of correspondence with Moore-Brabazon in early 1938. The latter, appalled by Grey's public and private diatribes, made clear their 30-year friendship was at an end.[125]

Moore-Brabazon condemned Mosley's weasel words concerning 'the children of Israel.' He admired the man not the movement – like Dick Fairey he never shared the English upper classes' antipathy towards the Jews.[126] Grey's undisguised prejudice warped his political judgement such that he was labelled a threat to national security: open approval of Nazi persecution saw him dismissed as editor of *The Aeroplane* in June 1939.[127] No wonder Fairey distanced himself from his long-time admirer as the 1930s progressed; at the same time he cancelled his membership of the Right Book Club.[128] The summer and autumn of 1937 saw CRF entertain and correspond with the first (and last) Lord Queenborough: the colourful Almeric Paget was a long-serving commodore of the Royal Thames Yacht Club (RTYC), with multiple military and business interests, and despite his advancing years and eccentric views he remained a figure worth cultivating. Fairey cut off all contact with him once the paranoid peer's warnings of world revolution became a passionate plea for Nazi Germany to defend Europe by whatever means the Führer deemed necessary. Queenborough was never wholly ostracised by the Tory leadership, but his friendship was potentially toxic.[129] The stop–start programme of RAF expansion rendered respectability vital in a highly competitive environment. With Baldwin adamant that Mosley was 'a wrong un', demonstrable support for the National Government – both its policies and its personalities – was clearly CRF's wisest course of action.

Fairey did share one character trait with Mosley, in that both men were instinctively hostile to the puritan element within British radicalism, be it Nonconformist sabbatarians or kill-joy Fabians. Like Evelyn Waugh, he was especially scornful of left-leaning intellectuals intent on idealising working-class values and social mores. Richard Fairey senior would have applauded his son's contempt for petit bourgeois small-mindedness. Here was a family which had endured the humiliation of genteel poverty, and felt no guilt in enjoying renewed good fortune. Indifferent to socialist charges of capitalist excess, Fairey was insistent that he had earned his wealth through blood, sweat and toil (and not a few tears), entrepreneurial enterprise, engineering expertise, boundless energy and, above all, a relentless readiness to take risks.

Restructuring the company across the decade

At Olympia in July 1929 *The Aeroplane* had placed two exhibitors in the vanguard of performance and design: Hawker and Fairey Aviation. Charles Grey saw obvious parallels between the two companies, not least their success in securing major contracts at home and abroad. Both firms recognised the value of the university-trained scientist and statistician, but still left the final decision to the 'practical engineer', namely Fred Sigrist and Dick Fairey. Thus the Hayes plant boasted 'an engineering team of which each member is a brilliant specialist, and the lot are overlooked by one man whose only form of argument is to say "That will do" or "Take it away".' The editor of *The Aeroplane* still considered CRF a 'genius', insisting that all the world would feel the same way once Fairey Aviation began building airliners.[130]

Delegation and clear lines of accountability were crucial to the company's future success: Fairey Aviation would soon be too big for the man at the top to exercise a hands-on role regarding every aspect of the manufacturing process. Yet by his own admission the chairman and managing director found it hard not to maintain close scrutiny of even his most senior colleagues. Surviving correspondence from the 1920s shows how, when absent for reasons of business or pleasure, CRF was in constant communication with Nicholl and Wright. By the end of the decade, as fresh interests beyond the factory floor took up more of his time – notably sailing – Fairey weaned himself off total absorption in work. Physically and emotionally, this widening of the hinterland would prove beneficial. Ironically, the operational efficiency of the firm improved. The 'old man' remained of course an ever-present, keeping everyone on their toes. Problems only arose after 1940 when his absence in the United States left a vacuum. Arguably Vincent Nicholl's death 13 years earlier had deprived Fairey of a natural successor.[131] 1927 proved an especially sad year for the company as Arthur Amos also died, albeit with his recently appointed deputy, Archie Hazell, an obvious choice as company secretary. Hazell held the position until his own death only seven years later.

Working closely with Charles Crisp and his solicitous colleagues, both A.A. Amos and A.G. Hazell played a critical role in Fairey Aviation's complex financial dealings across the course of the 1920s. The company created in 1921 gave Fairey and Dawson the capacity to acquire further capital without surrendering tight control of corporate policy, including the size of the annual dividend. That tight grip on the firm remained when in 1925 Fairey Aviation widened its share base and became a more orthodox private joint stock company: Crisp and Amos drafted successive articles of association which not only gave CRF and Dawson the overwhelming majority of both preference and deferred shares, but also allowed them to call and control board meetings whenever they wished. This enabled them to place successive companies in liquidation if such action was deemed strategically necessary. This is what occurred in 1925, 1927 and 1928; albeit in the latter case 'Company No. 5' was purely transitional; this 'legal fiction' was a consequence of Fairey and his closest colleagues opting to float the company in the spring of 1929.[132]

The board's decision prompted the Air Ministry's procurement section to ask why Fairey Aviation was embarking upon such a major restructuring. Adastral House sought confirmation that a public joint stock company would be commercially robust, honour any liabilities to HM Government and – ironically, prompted by CRF himself – inherit existing government contracts without renegotiation. Treasury legal staff provided assurance on all three matters. A withdrawal of £150,000 capital in the transfer of liquid assets from one company to the next in 1927 was noted but not commented upon, given that the Director of Contracts had accepted Hazell's explanation. Indeed, when chronicling Fairey Aviation's history since 1915, Whitehall chose not to explore the motives for serial liquidation and reincorporation.[133] While voluntary liquidation was a vehicle for realising assets, it also served to minimise the amount Fairey paid in Super Tax, a key consideration when the company 'folded' in 1927. In June of that year a barrister advised Crisp and CRF to pre-empt taxation of undistributed profits by accelerating the process of creating a new company from the shell of the old: neither the 1922 Finance Act nor the bill presently passing through Parliament empowered the Inland Revenue to claim fiscal liability had passed to a reborn Fairey Aviation. In other words, the new company had no obligation to pay the 'tax at source', which the old company was required to deduct on behalf of its shareholders, but could not as it was no longer operative.[134]

Furnished with the Treasury counsel's final report, Sir Henry Self, the Deputy Director of Contracts and a future colleague of Fairey in wartime Washington, gave retrospective approval to the flotation.[135] In March 1929 a final transfer of assets from one enterprise to another took place, with Hamble valued at £20,486 and Hayes/Northolt at £595,048; stock and work in progress carried over from the interim company was estimated at £100,000. However, the public was given the global figure of an astonishing £680,000 for orders in hand and contracts under negotiation; such a sizeable sum was partly explained by spares accounting for some 40 per cent – keeping all those Type IIIs in the air had become a highly lucrative business. Fairey with some justification sold itself to the stock market as a prime contractor of the RAF. Fairey Aviation plc announced nominal capital of £500,000, divided into 1 million ordinary shares of ten shillings each. £300,000 of the capital would be held as a first mortgage debenture on the fixed assets. In fact shares had already been allocated before the new firm's prospectus entered the public domain: Fairey, Wright and Dawson dropped Charles Crisp – always a token shareholder – but gained Hazell, Barlow and Broadbent as executive directors. The presence of such loyal lieutenants in the boardroom maintained the founders' unchallenged authority – the Sheppey survivors had secured a substantial infusion of capital without compromising their control of the company.[136]

The chairman and managing director secured a fresh five-year contract: as well as his director's fee and patent royalties, Richard Fairey received a *net* salary of £2,400 as well as £1,200 to defray travelling expenses. CRF's regular income was supplemented by capital released on the occasions when Fairey

Aviation had gone into voluntary liquidation, and by the dividend return upon his substantial stake in successive companies. CRF's 'disbursement' when his firm assumed its penultimate incarnation in 1928 was equivalent to 67 per cent of the stock. Soon after, in preparation for Fairey Aviation's flotation, the company's secretary made a crude calculation of its total pre-tax profit for the period 1923–8. Hazell came up with a figure of £1,179,718 (which, based upon the Retail Price Index, would be around £62,020,000 at today's prices).[137] CRF was insistent that success deserves due reward, and by the end of the 1920s he found himself a very wealthy man. Many wealthy men in 1929 would see their fortunes tumble in succeeding years.[138] The same would not be true of Richard Fairey, his restructured, re-energised enterprise riding the stormy years of the Depression to remain a premier plane maker – at home, in Europe and across the Empire.

CHAPTER 6

Into the Thirties, 'This Age of Crooners and Safety First'[1]

Fairey the man

For Dick Fairey, man of stature literally and metaphorically, initiative and hard work entitled him to enjoy the pleasures of life, albeit within certain parameters. Thus large sums of money might be spent, but only if circumstances warranted such expenditure. CRF was happy to dig into his pocket if confident he was getting the best, whether that be Hardy reels or bespoke Purdeys.[2] For example, when fishing for tunny off Scarborough became highly fashionable, Fairey availed himself of the finest heavy duty tackle: he sought assurance from Filey Bay's best known 'big game' fisherman that he had the right line and rod.[3] Before making any major purchase CRF or his staff undertook extensive investigation into the value and utility of the item in question. Fairey's maxim was simple: when in doubt secure the very best. Thus, when exposed to the stunning beauty of the Highlands in September 1937, his immediate response was to order a new Leica.[4]

Across the 1920s a major cause of tension between Fairey and his first wife was Joan/Queenie's assumption that relative wealth justified serial retail therapy. Faced with outstanding bills from fashionable shops in Bond Street and Belgravia, CRF paid up under protest and promptly closed the account: designer jewellery and haut couture were for special occasions, and scarcely conceivable as everyday acquisitions. In February 1927 a brusque note summoned an especially persuasive art dealer back to Grove Cottage, as Mrs Fairey, 'does not desire to keep the third picture left here by you.' Not that her husband was particularly prompt in settling bills, with the Fairey household regularly receiving final demands. These polite, regretful requests from local tradesmen resulted in the necessary cheque and a suitable excuse elegantly penned by Miss Burns, CRF's loyal and keenly efficient secretary. Golf club subscriptions and school fees were cases where delay would reflect poorly upon Fairey, and any such demands were addressed immediately.[5]

Here was someone careful with his money, and yet CRF was in no way a mean man. He received a steady flow of polite requests for contributions to charitable causes, responding generously almost every time. Friends would on occasion receive gifts out of the blue, usually in gratitude for a good day's shooting or an invitation to dwell on the nineteenth hole.[6] Dick and Joan loved to entertain, and every summer throughout the 1920s they would host weekend parties at Grove Cottage. The guest list reflected new money and a greatly expanded state: self-made men and rising technocrats. It was demonstrably not high society, on a par with say Dickie and Edwina Mountbatten's chic shenanigans at Adsdean. There would be no Coward or Novello at Grove Cottage, the sleek and smart seeking their amusement in less stolidly bourgeois surroundings. Instead, the RAF's two Hughs – Trenchard and Dowding – might be found on court cultivating their forehand, or staring disapprovingly at the high jinks and practical jokes their host found so amusing. Guests were assumed to have packed their tennis whites, with CRF a formidable partner in the men's or mixed doubles – singles was a tad too competitive for such civilised occasions. Picnics were organised on the grand scale, with Fortnum and Mason supplying extravagantly loaded hampers, along with a suitably impressive array of spirits and fine wines: champagne was clearly *de rigueur*.[7]

Children were in no way excluded from the weekend fun and games. Their absence might well be a consequence of boarding, but from July to September they would relish the opportunity to run wild in Grove Cottage's spacious gardens and surrounding fields. When not outdoors young Richard and his chums would be entertained by Chaplin and the Keystone Cops, courtesy of the Kodascope Library's newly launched 16 mm films. CRF owned both projector and camera, with Kodak providing a steady supply of film stock to facilitate home movies. Thus began a cinematic chronicling of the Fairey family, consistently maintained for the next 30 years.[8]

When up in town Fairey regularly led a party of friends on an evening in the West End – in a golden age of cabaret and musicals, he relished going out to a show and then on to the Savoy for supper. Charles Crisp shared CRF's passion for musical comedy and light theatre, with both men as happy in the rear stalls as the royal box.[9] When not indulging in their shared passion for variety, the two friends might be found together on the fairway – Fairey found his solicitor's advice invaluable in driving down his handicap. When CRF began sailing seriously, Crisp soon joined in the adventure, his natural enthusiasm compensating for any previous indifference to matters maritime.

Dick Fairey played golf off a handicap of 18, which suggests he could more than hold his own on any course. Near home he was a member at Denham, Hendon, St George's Hill and later Sunningdale and Sandwich's Prince and Royal St George, but to enjoy links golf he joined Felixstowe Ferry, the Suffolk coastal club famous for its Martello Course. When playing on the dunes Fairey stayed at the neighbouring Felix Hotel, a large establishment run by the LNER. Not that CRF relied upon the railway, preferring to drive or fly to Felixstowe or Burnham-on-Crouch, convenient locations for visiting the Air Ministry's experimental station at Martlesham. From the mid-1920s he increasingly mixed business with pleasure. For example, when Joan's Harley Street specialist recommended convalescence in Cannes at the start of 1927, her husband spent that spring commuting between Hayes and Suffolk. Monday 28 February proved a memorable date in his personal sporting calendar: he teed off at Moor Park in a doubles match play contest against theatre producer Herbert Clayton and matinee idol Jack Buchanan – as president of the Aero Golfing Society, Moore-Brabazon had invited his friend Dick, that year's club captain, to skipper his side against the Stage Golfing Society. That same day *Flight* applauded Fairey's recent state-of-the-industry address to the Royal Aero Club, while 6,000 miles away in San Diego Charles Lindbergh was commissioning construction of the *Spirit of St Louis*. Only four months later Fairey and his fellow golfers would be at the Savoy congratulating the audacious aviator. By then his golfing sabbatical was over: as Captain Lindbergh flew his battered machine from Paris to London, so Joan Fairey followed the same route courtesy of *Le Train Bleu*. That same weekend CRF sailed his newly acquired 12-metre cutter, *Modesty*, for the first time; henceforth cruising and racing would leave little time for extended golf breaks.[10]

Whether at the Savoy's American Bar or Moor Park's nineteenth hole, Dick Fairey was no doubt a deep-pocketed and companionable host. As the head of a large and growing enterprise he was neither mean nor mean-spirited. Seasonal gifts from Fairey Aviation (cigars or scotch for the men and continental chocolates for the ladies) extended well beyond the obvious recipients, while often spontaneous presents for family and friends reflected a natural spirit of generosity. Similarly, cold requests for donations invariably generated a positive response, with CRF's name prominent on a lengthening list of good causes. In 1931, when Claridge's demanded full payment ahead of the Royal Aero Club banquet to celebrate Supermarine's third Schneider Trophy triumph, Dick Fairey was the first to chip in. Even before the Depression began to bite hard, the chairman and managing director received letters every week pleading that he find work for ex-RAF officers, redundant aero-engineers, or bright young men leaving school or college with scant prospects of securing a decent job. These heart-wrenching requests, which multiplied in the early 1930s, were invariably turned down courtesy of carefully drafted replies, always polite and never peremptory. Prized employees at Hayes who fell ill might find themselves visiting Harley Street for treatment paid for by the 'old man.' Here, as in other facets of life, Fairey was prepared to pay a high price if convinced that the service in question was undoubtedly the best.[11]

Over the 1927 Easter weekend David Hollis Williams, the talented young designer Fairey had recruited from Hawker, crashed his self-built aircraft on its maiden flight. The 'Dove' was a write-off but CRF insisted the experimental shop at Hayes rebuild it to a higher specification and with no expense spared. When asked why, he told Williams that, 'I want to help chaps like you.' Rechristened the 'Pup', the aeroplane survives today in the Shuttleworth Collection.[12]

Not surprisingly, property and boats generated extensive investigation ahead of purchase or commission. A precedent lay in the rapid turnover of luxury saloon cars between the wars. As soon as CRF began to make money he indulged his passion for high-end motor vehicles. Ironically, for someone so keen to parade his patriotic credentials, he preferred Detroit's finest to sedate British marques such as Daimler and Wolseley (Bentley and then Rolls Royce or Bentley would come later in life, from the mid-thirties). Fairey believed British car designers were indifferent to the comfort of exceptionally tall men, such as himself. He considered American cars to embody New World inventiveness and advanced engineering, hence their superior performance.[13]

Here was an obvious parallel with his admiration of transatlantic aero-engines in the years following World War I. Maurice Wright shared his friend's passion for stars and stripes technology. However, Curtiss consistently fulfilled its English customers' expectations in a way the likes of Buick, Packard, Chrysler and Stutz demonstrably did not. CRF's preferred supplier of American big-engine coupés always gave advance warning of the latest model to cross the Atlantic. As with all major purchases, Fairey would forensically test the manufacturer's claims. If suitably satisfied then he would trade in his current model for the Motor City's latest limousine. It was then that the problems began, with a perfectionist engineer demanding a level of finish and performance far beyond Michigan's most fastidious keeper of quality control. A constant stream of complaints, suggestions and requests ensured seemingly endless repair, modification and upgrading. With only the flimsiest of warranty, Pass and Joyce Ltd surely made a loss every time they sold Richard Fairey a state-of-the-art automobile. These were cars which were difficult to service and dependent on costly imported spares.[14]

When voicing his demands CRF spoke from direct experience of driving these constant sources of disappointment; rarely did he call on the services of a chauffeur.[15] Only in 1938 did the disillusioned driver at last secure the perfect automobile, built not in Michigan but Middlesex – the 12-cylinder, silkily smooth Lagonda gave its owner 'a renewed pleasure in driving.' Also flying the flag for British motor manufacturers was the second Mrs Fairey: in September 1939 Esther Fairey was running her powerful SS Jaguar roadster on a ration allocation of only nine gallons a month.[16] Fairey hurtled around the Home Counties at the wheel of exceptionally speedy vehicles in an era when well nigh every road remained single carriageway – and he drove very fast. Here was a major industrialist with a diary crammed full of important appointments. Always a man in a hurry, he found himself in frequent altercations with fellow motorists and the local constabulary. Fairey fought a running battle with suitably censorious magistrates. As in his dealings with the tax inspector, he would hold out for as long as possible before pleading guilty and paying the consequent fine.[17] The number of charges for speeding or careless driving would today mean multiple penalty points and long periods off the road. The cars CRF chose to drive were fast and fashionable, but frankly they were too big for the back roads of pre-freeway America let alone the narrow lanes of the Thames Valley.

The fines incurred for serial motor offences were a minor irritant for a man of Fairey's wealth.[18] Their significance lay in the presumed abuse of the state's entitlement to limit the actions of the individual. Chief villain was Leslie Hore Belisha, the high profile Minister of Transport who, in 1934, introduced the first enforceable speed limits.[19] Moore-Brabazon, who loathed the minister, had been the wide-ranging Road Traffic Bill's most vocal opponent as it passed through Parliament; over 7,000 were dying each year in motor accidents, but the argument for non-intervention rested on the fact that most dogs now avoided cars: 'There is education even in the lower animals. These things will right themselves.'[20] CRF bristled at regulation and intrusion, and held the local constabulary in low esteem: he confided to Moore-Brabazon early in 1936 that the police were 'getting away with altogether too much', not least in disguising the 'thousands' of accidents for which their vehicles were responsible.[21]

A growing sense of persecution saw Fairey rail against the ostensibly heavy-handed treatment of otherwise law-abiding and respectable citizens.[22]

As a property owner he encountered similar supposed examples of excessive state power, insisting upon a fundamental right to maintain his estate without external interference. As we shall see, defence of the status quo, not least the free flow of the River Test, became crucial to Fairey's stewardship of his Bossington estate following the family's move to mid-Hampshire.

CRF could never be described as an easy-going guy, but Geoffrey Hall noted a hardening of the shell in the years following the debâcle over the Fairey Fox, 'a turning point in his whole life': '…at that stage he became *very* cynical … He was quite blunt about what he felt about the Civil Service … it was then he started looking to see, first of all how he could build his own fortune, and also making up his mind that he was going to enjoy life instead of working himself to death and letting bureaucracy treat him the way it did … I don't mean it unkindly, *but he was a disgruntled pioneer*, feeling it very deeply indeed.' Hall believed his half-brother entered his fortieth year physically and emotionally exhausted. Harley Street's finest insisted Fairey cut down his workload and, with the company once again profitable and secure, he took the doctors' advice. The tendency to micro-manage declined sharply, but staff remained keenly aware who was boss.[23]

CRF spent even more time on the golf course, but below par scores now offered only a passing thrill. More excitement could be gained from shooting, which Fairey adopted with alacrity from the late 1920s: within a few years he was renting a shoot at Flixton, and hosting parties on not one but two newly acquired estates. While never losing sight of the firm, he sought fresh challenges. A work-related initiative was to back Maurice Wright's expansion plans in Belgium, of which more in due course. At home he made ruminative plans for life after Grove Cottage, envisaging the eventual purchase of a large country house and an accompanying estate. This ambition came to fruition in 1930 when Fairey moved the other side of Iver Heath to Woodlands Park, an Edwardian residence with 97 acres of land owned originally by Lord Curzon. Over the next few years the house was subject to extensive refurbishment, with a large annexe added to house a heated swimming pool. In 1939 as the prospect of war loomed ever greater, CRF finalised plans for his family's move to Bossington, and the house was prepared for a permanent residency. As we shall see, that move duly occurred in September 1939, but not as planned. In March 1936 Fairey acquired a second estate, purchasing Oakley Hall in north Hampshire: although the heath and woodland was put to good use, the large Regency-era house was never lived in.[24] Similarly, he rarely visited his holiday home outside Salcombe at North Sands Bay, overlooking the coastal ruins of Fort Charles. Esther Fairey stayed there in the autumn of 1939, during which time her husband seriously contemplated selling the eight-bedroom Edwardian villa for £6,500. With no sale imminent CRF handed the house over to his eldest sister, Effie Hulme, who spent the war years in south Devon with her daughter.[25]

It is worth noting that once he became wealthy Fairey was generous in supporting his relatives.[26] He paid for his younger sister Phyllis's house in north London, and subsidised accommodation for other members of the family. Such generosity was affordable given the sizeable revenue generated by his lucrative property dealing. CRF's portfolio extended well beyond his three homes, and when selling on houses – plus tennis courts in Potters Bar – he provided the buyers with mortgages. Thus, as well as profiting from the original transaction, he earned interest up to as much as five per cent on the financing of the purchase. When in the late 1930s Fairey bought the Hind's Head Hotel in Maidenhead his solicitor devised a scheme which generated over £50 in debenture interest twice a year: for someone as wealthy as CRF this was a tiny sum, but satisfaction lay in maximising the return on his investment.[27]

Finally, in mapping out his grand scheme for the future, Fairey 'cast his eyes on to yachting. This came out of the blue as far as I [Hall] can see. He certainly hadn't done any intensive sailing with anyone ... it *suddenly happened*. He just decided to go sailing.'[28] However surprising the family found Fairey's purchase of a yacht, a clear signal was sent out the previous summer when the factory's annual closure coincided with Cowes Week: while his clubs and plus fours were wending their way to the Felix Hotel, CRF could be found on the Solent admiring Tommy Sopwith's helmsmanship.[29]

Extravagantly celebrating his fortieth birthday in 1927, Fairey was still too young to worry unduly over who might succeed him. His one concern was that an enterprise bearing his name would always retain a close connection with the Fairey family. The size of his shareholding, even after the company's flotation in 1929, ensured close relatives' continued involvement in the unlikely event that CRF suffered cruel misfortune. Yet the chairman and managing director anticipated more than his extended family's grateful receipt of the annual dividend. He relished the notion of working alongside his nearest and dearest, secure in the knowledge that they would act in his and the firm's best interest once he surrendered executive control. Geoffrey Hall gave his half-brother that assurance and security. Having completed his apprenticeship at Hayes, Hall complemented academic study at Imperial College with experience of working on a major engineering project: the test programme for the Rolls Royce's 'R' engine, intended to power Supermarine's trophy-seeking S6 seaplane. After four years in Derby, Hall returned to Fairey Aviation in 1932, initially living with

CRF's second family at Woodlands. Esther Fairey was fond of Geoffrey – solid and reliable, but terribly dull – but she found his presence increasingly intrusive; secretly she longed for him to leave.[30]

As a young man Geoffrey Hall was the ideal role model for Joan and Dick's little boy, directly and indirectly cultivating an inherited fascination with technology. CRF naturally anticipated the day when his son could stand beside him on the shop floor, generating the same recognition and respect that he himself enjoyed. More immediately, Fairey looked to give young Richard a stability and continuity he saw as absent from his own schooling. Many would consider CRF to have gained from a privileged education, but clearly the hasty departure from Merchant Taylors' had hurt – why else would he airbrush St Saviour's School out of his life story? When educating his only child, Richard Fairey wanted nothing but the best.[31]

As a little boy Richard junior was educated close to home, but from the age of seven he boarded at Twyford, which claims to be the oldest prep school in England. Surprisingly, given the Fairey family's instinctive antipathy towards the established church, Twyford was and is an Anglican foundation. Presumably Richard's parents were swayed by the school's success in preparing its pupils for the common entrance examination, ensuring their progression to England's premier public schools (the so-called 'Clarendon Schools').[32] Twyford is located close to Winchester College, but the entrance exam would have been too testing for someone far happier on the playing field than in the library. Richard was a bright lad – and, like his father, a big lad – but always his own man: stripping down a motorbike made sense, whereas studying Classics clearly did not. The less cerebral, more rumbustious atmosphere of Harrow was better suited to his character, and in September 1930 he was duly accepted into the school on the hill by Cyril Norwood, the most distinguished headmaster of his day and by coincidence an old boy of Merchant Taylors.' However attractive the pavilion or the workshop, Richard applied himself with enough energy to secure a scholarship to Christ's College, Cambridge, in 1935. That autumn he secured his pilot's licence after only 15 hours training and joined the varsity flying club, of which his father was a vice president. Not long after, a serious back injury forced young Richard to interrupt his studies, and, after a long period of recuperation, recommence his degree in October 1936.[33]

As President of the Royal Aeronautical Society from 1930 to 1934, Fairey lectured on 'The Evolution of the Aeroplane' at Harrow and at Merchant Taylors', as well as Westminster, Marlborough, Charterhouse and Stowe. He considered these talks a 'striking success', insisting that, 'If we are to instil airmindedness into the population we can hardly start with better material than the public schoolboy...' Members learnt that over 15,000 pupils (all boys of course) had benefited from CRF's illustrated talk, and, by donating his collection of lantern slides and photographs, he could establish the Society as a 'custodian of aviation history': today the National Aerospace Library boasts a collection of 'over 100,000 books, articles, reports, and images.'[34]

Fairey regularly employed lantern slides, for example, when briefing RUSI on future aircraft design in February 1931. This proved a prescient – yet at the same time flawed – technical discussion of the problems entailed in transforming the S6 trophy-winner into a high-performance fighter aircraft. If present that night, Supermarine's Reginald Mitchell would have endorsed the speaker's plea for a powerful and reliable engine: the ubiquitous Merlin was still five years away from entering production. Unlike Mitchell, Fairey failed to make the conceptual leap from seaplane to monoplane fighter, arguing that an improvement in overall performance would reduce a prototype's top speed to only marginally more than current biplanes could achieve using half the S6's horsepower.[35] To be fair this was a widely shared view at the start of the decade, yet Mitchell and Sydney Camm were already convinced of a monoplane fighter's capacity to match speed with manoeuvrability.

Needless to say, that autumn the President of the Royal Aeronautical Society was a VIP guest when a third victory for Supermarine meant the Schneider Trophy would stay permanently in Britain. Fairey Aviation shared in the triumph, once again supplying the seaplane's propeller, and advising on redesign of the floats. Only a fortnight after retaining the Trophy, the RAF's High Speed Flight again took to the skies above the Solent, flying the S6B at a record-breaking 407.5 mph.[36]

Illustrated talks formed the core of factory visits from the Imperial Defence College and its naval counterpart at Greenwich: the message for both parties a simple 'Build more aeroplanes!' Lantern slides were equally in evidence for the President's extraordinarily long welcoming address – in English – to a delegation of the *Société de Navigation Aerienne*. One can only hope the visitors' enthusiasm for reviving the *entente cordiale* kept them awake. Soporific conditions were surely absent when an audience at the Royal Society of Arts heard CRF vehemently denounce the Disarmament Conference's dual aim of regulating civil aviation and promoting arms limitation: 'Our late defeated enemies will not pay their debts and have permission to arm. We, on the contrary, must both pay and disarm ... Disarmament as a principle is right but disarmament by this country in the case of a world that is increasing its armament is very certainly not right.' Such sentiments echoed an earlier warning that ministers' naive idealism, so evident in Geneva, presaged disaster for the nation's air defences.

Fairey's tone was intemperate but his fears were genuine – and only 13 days after his RSA speech Adolf Hitler entered the Reich Chancellery.[37]

Yet, serving as the Royal Aeronautical Society's high-profile ambassador in an era of severe economic recession and growing international tension was not all gloom and doom. Yes, Fairey performed his Cassandra-like role with particular passion and conviction, but he could ooze charm and goodwill when required. At Society functions he dutifully flattered the Marquess of Londonderry, despite privately believing him a sorry successor to Sam Hoare. He was generous in his praise of Amy Johnson (always 'Mrs Mollison') when a joint dinner with the SBAC and Royal Aero Club celebrated her solo flight to South Africa in July 1932: '…in the trail of her achievements follows pride in human achievement and good will between the peoples she meets and this country.' Three months earlier CRF experienced the greatest triumph of his tenure as President. When he met Amelia Earhart only days after her solo flight across the Atlantic, Fairey invited her to the following week's Wilbur Wright Memorial Lecture. To his surprise and delight the world's most famous aviatrix said yes. Even better, Miss Earhart's acceptance of the Society's invitation meant she would miss the Congress of Trans-Atlantic Flyers, thereby snubbing her Italian hosts. What would have been a dry and dusty affair turned into a high-profile occasion, with Fairey acclaiming his guest of honour's remarkable achievement in eclipsing Lindbergh. A group photograph of the Society's white-tied, be-medalled members depicts a solitary woman in the middle of the front row, with a suitably smug President in close attention beside her. Fairey and his guest gelled so well that on 3 June she crossed the Channel courtesy of his motor yacht, *Evadne*. From Cherbourg, where she found husband George Putnam waiting to congratulate her, Earhart travelled to Paris for an official reception.[38]

The presence of both Amelia Earhart and Amy Johnson at events organised by Richard Fairey cemented his reputation as one of the key players within the aviation establishment. Throughout his tenure as President he not only raised the profile of the Royal Aeronautical Society, but of the aircraft industry as a whole. It was Fairey who covered the cost of the RAeSoc's gold and silver medals, awards established to recognise innovation in advanced aeronautics. At the same time he cemented his position as a key figure on the executive committee of the Society of British Aircraft Constructors (SBAC). Fairey was a high-profile company boss, but acknowledgement by his peers of CRF's achievements as an engineer came with the award of the prestigious Wakefield Gold Medal in 1936.[39]

Yet throughout the years when Fairey's persona was shifting from tough company boss to industry spokesman, his private life was proving deeply stressful. Exercising and enjoying status, standing, power and influence in public makes the absence of authority and control at home that much harder to handle. Like most fathers, Fairey's patience was tested by the truculence of his adolescent son: 'an overwhelmingly strong character ... he was a man of such powerful emotions.' CRF's son by his second marriage, John Fairey, late in life recalled how, 'when my father was in a temper I doubt there was a man in England who could stand up to him. I did not go out of my way to annoy him.'[40] Richard was less acquiescent and emollient, although ironically his half-brother was subject to a far stricter upbringing. 'Light hearted and disinclined to be serious about anything much if he could avoid it', Richard frequently incurred his father's wrath, and to avoid yet another confrontation he became increasingly evasive.[41] The clash of two such strong personalities was not rendered any easier by Richard senior remarrying so soon after his first marriage ended in divorce: his only child was still too young to view this dramatic change of events in a suitably mature and rational fashion.

However, at the start of the 1930s any serious deterioration in relations between the father and his son was still in the future. Dick Fairey was coming to terms with the disintegration of his marriage to Joan. His private papers from the late 1920s contain odd items of correspondence, which, with the benefit of hindsight, suggest all was not well. To take just one example, on 26 July 1927 the ever discreet Miss Burns informed the secretary of one Captain Deane that Mr Fairey refused to meet and discuss what Miss M.M. Howell considered 'a rather important but absolutely personal matter.'[42] CRF would have known perfectly well what this particular matter related to, viewing an invitation to meet as at best the intervention of a prurient busy-body and at worst the first step to blackmail.

The detailed evidence Fairey provided in his May 1931 petition to the Probate Divorce and Admiralty Division of the High Courts of Justice suggests extensive use of private detectives, and possibly over an extended period of time. Without going into detail, Joan – in the official documentation again Queenie – maintained an intimate relationship with a wing commander for several years before he was posted to Egypt. The affair resumed when Joan/Queenie found herself in Cairo en route to Kenya, from where she flew on to Cape Town. However, while in Africa there were subsequent sexual liaisons, with three other RAF officers, all of whom were summoned home to give evidence should the divorce petition be contested. It was not, and the marriage was duly dissolved as a consequence of proven adultery. Richard Fairey was now a single parent, the court awarding him custody of his son.[43] In the aftermath of the divorce he maintained minimal contact with Joan.[44] She eventually remarried, settling down on a 'lovely farm' in north Hampshire. Now Mrs Buxton, she rarely saw her son, although a wartime letter to Moore-Brabazon confirmed an unfaltering maternal affection and concern.[45]

Strangely, young Richard still saw his celebrity uncle: CRF remained friendly with George Clarke, home from Hollywood in the hope of reviving his comic career.[46] Fairey did not remain on such good terms with his erstwhile mother-in-law, whose failing eyesight forced her to enter a nursing home in the autumn of 1932. Mrs Markey insisted her daughter's ex-husband held bonds on her behalf; these were supposedly of sufficient value as to cover her care costs. An increasingly testy exchange of letters saw CRF continue to insist no such bonds existed, despite 'mater' and Joan repeatedly demanding their return. Faced with the prospect of exorbitant legal costs should she press her case, an embittered Mrs Markey conceded defeat.[47]

Fairey the manager

Joan Fairey and her mother-in-law attracted the same depth of suspicion and antipathy as speed-sensitive constables, impatient tax inspectors and cost-conscious civil servants. True to his libertarian instincts, Dick Fairey bristled at almost any form of government intervention. His readiness to roll back the state was over half a century ahead of his time. Like a later generation of free marketeers he believed firmly in the efficacy of trickle-down economics, even if the term had as yet to be invented. Thus wealth-creating entrepreneurs, once they had fulfilled their tax obligation, were entitled to spend their money any way they liked. Nor should officialdom dictate how they treated their estate staff, or their factory workers.

An aura of authority and acuity surrounded the founder of Fairey Aviation; yet in the boardroom, the hangar and the design office he eschewed Olympian aloofness, maintaining a hands-on approach to all aspects of the production process. There was a lifetime attention to detail, not least in the day-to-day operation of Hayes, Hamble and their satellite works – woe betide a presumed slacker who came face to face with the omnipresent and omnipotent managing director.[48] Notwithstanding a 'ferocious temper' when annoyed, Richard Fairey's style of management might best be described as tough paternalism.[49] Fairey Aviation's founder displayed great generosity of spirit to those he valued and appreciated, and cold indifference towards those employees with whom he found fault. Such indifference could translate into dismissal for an offence deemed especially grievous, occasionally with wider consequences, such as the five-day unofficial strike in March 1937 over alleged wrongful dismissal of a fitter and a charge-hand. CRF was convalescing from influenza courtesy of a six-week cruise when more than 2,000 employees walked out. The chairman's return coincided with settlement of the dispute. By this time he suspected most if not all shop stewards to be Communists, refusing to accept them as 'the sole accredited representatives of the workpeople', by being 'in a sense in opposition to the genuine Trade Union leaders.'[50]

A position scarcely credible within the home factory at Hayes was even harder to maintain after rearmament saw Fairey Aviation establish a 'shadow factory' in the north-west of England. Fairey suspected organised infiltration of his Stockport works by 'agitators, usually Communists, whom nobody wants and who, once in, are very hard to get rid of.' His suspicion was confirmed when members of the Amalgamated Engineering Union (AEU) across the north-west, including the Fairey plant, struck for better pay and conditions in September 1937.[51] With no shared history or sense of corporate identity and tradition, the Lancashire workforce viewed Fairey Aviation, not as a paternalistic business, but just another profit-driven employer. That view was increasingly prevalent down south: freshly recruited employees in Hayes may have lacked the hard edge of their northern counterparts, but they viewed management with a suspicion largely lacking among workers who had been with Fairey since the 1920s or even the war. In May and November 1938 the company narrowly averted further strikes over employment of non-unionised labour. A year earlier, at Hamble, sheet-metal workers downed tools when the company started to recruit non-unionised metal skinners. The dispute prompted an SBAC survey, the outcome of which was every employer endorsing Fairey's refusal to recognise a closed shop.[52]

Up in Stockport AEU members stopped work when the fledgling plant gratefully accepted young men from a local training centre dedicated to precision light engineering. Management had spurned skilled labour desperate for a job but drawn from heavy industry. Given the cost of schooling older workers and their resentment at having to retrain, the company chairman considered them unemployable – unlike the jobless cotton operatives he felt confident could be trained up in a fortnight if 'obstructive unions' allowed. Employing unskilled, non-unionised workers avoided demarcation disputes, seen by CRF as the biggest obstacle to improved productivity.[53]

The Air Ministry was always sceptical as to how well management dealt with organised labour once rearmament began to drive a dramatic expansion in aircraft production. Fairey accused officials of displaying prejudice towards his company, insisting that industrial unrest in Stockport was unavoidable given such an ambitious initiative. T.M. Barlow, formerly chief engineer and from late 1935 managing director at the Heaton Chapel site, went so far as to threaten legal action against Henry Disney, head of the Ministry's Directorate of Aeronautical Production (DAP). Senior RAF officer turned successful businessman, the DAP director was heavily critical of Fairey's aggressive approach to the dilution of labour. CRF and his board deemed Disney to be shamelessly ambitious and openly malignant in his dealings with their company. Colonel Disney's power lay in his direct line to the secretary of state and his industrial adviser, Lords Swinton and Weir, neither of whom were as receptive to informal lobbying from the likes of Dick Fairey as in the days of Sir Samuel Hoare and Lord Londonderry. Fairey and his fellow members of the SBAC were reluctant to complain too much, for fear of stricter government control or even nationalisation. In reality Swinton (until November 1935 Sir Philip Cunliffe-Lister) preferred his officials to maintain an advisory role, albeit down to shop floor level. Thus he was deeply reluctant for the state to assume a direct role in management, let alone take the draconian step of bringing the industry under public ownership. Weir was similarly hostile to the notion that nationalisation was conducive to greater efficiency and superior design. Between 1935 and 1938 Swinton distanced himself on a day-to-day basis from company patriarchs like Dick Fairey, leaving Lord Weir to keep open informal lines of communication.[54]

In September 1937 CRF sent Weir a lengthy memorandum on the problems at Heaton Chapel, issuing a personal invitation to visit Lancashire and see for himself the poor calibre of so many workers.[55] Seven months later Fairey found himself briefing Whitehall officials on why industrial relations in Stockport were so poor. In Parliament, Communist MP Willie Gallacher had questioned the local police's handling of alleged factory sabotage by party members: Barlow had insisted that as many as six Battle medium bombers had been wilfully damaged, and he named those shop stewards supposedly responsible. This incident did Fairey Aviation no favours, reinforcing ministerial suspicion that the company was too heavy-handed in its response to shop floor grievances.[56]

After Swinton was sacked in May 1938, Fairey launched a charm offensive on the incoming secretary for air, Sir Kingsley Wood. By now he had a close contact inside the Cabinet, maintaining regular correspondence with Ernest Brown, Minister of Labour and prominent National Liberal. Holder of the Military Cross (MC) and born into a fisherman's family, Brown boasted a stentorian voice, whether addressing the House of Commons or instructing the crew of his 12-metre yacht. United in a love of racing and a dislike of trade unions, Ernest Brown and Dick Fairey were a marriage made in heaven. Where they differed politically was over Germany, with Brown's appeasement credentials securing his place on an impressive array of high-society guest lists.[57]

Although Sir John Simon, Neville Chamberlain's successor at the Treasury, was never anything but chilly and remote, Brown came across as friendly and approachable. Succeeding Simon as leader of the National Liberals in May 1940 meant Brown remained in office throughout Churchill's tenure as prime minister: after the coalition with Labour broke up in late May 1945, he became the last – and certainly the least known – Minister of Aircraft Production. Ironically, it was probably Ernest Brown who arranged for Churchill's most Machiavellian enemy inside the Conservative Party, Sir Joseph Ball, to fish on the Test: in June 1937 Chamberlain's chief fixer and dirty tricks operative was a fellow guest when CRF invited the Prime Minister to spend a quiet weekend at Bossington. It is likely that the two men were previously unacquainted, even if Richard Fairey would have been a familiar name inside Number Ten. Three years would pass before Bossington again played host to the Prime Minister. With the mining of Norwegian waters imminent, Chamberlain spent Easter Sunday 1940 quietly casting his fly, for, in the words of Lord Lloyd, 'he feels this may be his only chance for some time.' It was George Lloyd – veteran pro-consul and enthusiastic aviator – who arranged the premier's visit (the sexagenarian pilot and helmsman had already cemented his brief but breezy friendship with CRF by engineering an appointment with the First Lord of the Admiralty – one can only imagine a ruminative and tearful Churchill sharing old memories of Prince Louis's fearless young men confidently reaching for the sky). Lloyd, Brown and Fairey all shared a simple view of leadership: whether ruling Egypt, leading your chaps over the top, or reefing a spinnaker in a force seven, you needed to be certain your orders would be carried out, and to the letter.[58]

Fairey Aviation's founder presided over a management hierarchy made up of tried and tested lieutenants, almost all of whom were fiercely loyal to 'the big man' and the company he had created.[59] Like so many industrial organisations before and after World War II there were few graduates, with directors and senior managers qualified by dint of part-time study and a quick brain. In the case of the aviation industry these were the first and second generations of plane makers, with pioneer engineers like Fairey and Fred Handley Page displaying impressive longevity. Their enterprises embraced a quasi-regimental ethos. This was scarcely surprising as several managers had wartime experience or were closely acquainted with the military. Thus senior staff would take their meals in the mess, with the workforce (in peacetime overwhelmingly male) perceived as a respectful body of men for whom one bore the same sense of responsibility as a platoon or company commander. Within this subaltern scenario, deferential charge hands and foremen assumed the persona of trusted NCOs, while disrespectful shop stewards stood condemned as the worst form of barrack-room lawyer. 'Tiny' Fairey stood at the head of this industrial army, his shop-floor battalions multiplying as fresh factories came on stream.

Russia

Fairey maintained close contact with captains of industry in North America and continental Europe, but the most intriguing connection was with key figures in the Soviet aircraft industry. As President of the Royal Aeronautical Society CRF fostered close relations with Andrei Tupolev, regarded in the west as the USSR's premier designer and accorded the same respect as a William E. Boeing or an Anthony Fokker. Tupolev's reputation among British designers grew in the years following World War I, but he became more widely known following a brief visit in 1929. That summer his prototype airliner, the ANT-9 'Wing of the Soviets', embarked on a high-profile promotional tour. Ten trusted correspondents and party officials visited European capitals, including London. Having by now met Tupolev and no doubt sensing a business opportunity, CRF invited the Russian delegation down to Hayes. The party included Soviet aviation's most senior technocrat, P.I. Baranov, a figure of power and influence, open-minded, dynamic and highly efficient. Fairey kept in contact with Baranov until the latter's death in a flying accident four years later. Each man professed to hold the other in high esteem, with Baranov orchestrating CRF's visit to Moscow in the summer of 1931. The Wing of the Soviets visit also allowed Fairey Aviation to establish close links with the Bolsheviks' key organisation for aviation design and development, the Ventral State Aero-Hydrodynamic Institute (better known by its Russian acronym, TsAGI).[60]

CRF thus found himself in contact with Sergei Ilyushin, an unquestioning servant of party and state who would soon eclipse Tupolev in terms of power and influence, if not innovative design. The ultra-loyal Ilyushin enjoyed Stalin's favour and survived the purges, his organisational abilities rendering him vital to national defence. There are obvious parallels between Tupolev and CRF, in that the two men demonstrated a capacity to design and build aircraft on the grand scale, and yet it was S.V. Ilyushin who shared Fairey's mental toughness, ability to adapt, and surefootedness in the corridors of power: both men masterminded the construction of complex, multi-functional manufacturing plants before dire national crisis saw them assume key procurement roles within their respective state apparatus.[61]

Although acclaimed designers of astonishing longevity, both Tupolev and Ilyushin gained notoriety abroad for reverse-engineering. Most famously, at the onset of the Cold War Tupolev based his four-engined TU-4 on Boeing B-29 bombers stranded in Vladivostock. Yet it was the younger man who first pirated western technology: Ilyushin became known to the Bolshevik leadership in the autumn of 1919 when his team took apart an Avro 504K biplane, swiftly reincarnated as the mass production U-1 trainer.[62] The end of the Civil War saw the 'Red Workers-Peasants Air Fleet' flying five Fairey IIIC reconnaissance planes abandoned by the RAF. The Russians' fledgling air force and civilian airlines flew a mixture of combat aircraft captured from pro-White forces and multi-purpose machines acquired from British, French and Dutch manufacturers. In the 1920s nearly 90 per cent of Soviet aircraft were imported or based on foreign design, the latter increasingly German as a consequence of close collaboration with the Weimar state's surviving plane makers and uniformed aviators. Bolshevik victory in the civil war saw Russia and Germany launch a joint passenger service, with the *Deruluft* airline operating until as late as 1937. While the two countries' manufacturing plant near Moscow was an open secret, both governments denied knowledge of a joint training and experimental base at Lipetsk, a remote airfield over 200 miles south of the capital. The key German collaborators were Junkers for production and Fokker for experimentation, their partnership programmes peaking in the late 1920s.[63]

German engineers and aircrew operated inside Russia until 1933, but after 1928 Stalin's real or convenient perception of a military threat from the imperialist west ensured a greater emphasis upon self-sufficiency in the design, development and manufacture of Soviet aircraft. The aviation programme was a key element within the first Five-Year Plan, with a huge rise in the training and employment of engineers and technicians. Aircraft output quadrupled between 1928 and 1933, with over 50 prototypes translated into 11 mass production fighters and bombers (some with a civil equivalent). Later Plans presumed near exponential growth in order to equip *Aeroflot* and establish the newly created VVS as the world's largest air force: 'Under the Soviets, the airplane occupied an exalted place at the very centre of public life, as a metaphor for progress and an enduring measure of national achievement ... the Russian approach to aviation acquired a peculiar style of its own, a blend of Western technology and indigenous engineering.'[64]

The Russians developed their own design philosophy, with the emphasis upon building aeroplanes which were uniform, rugged, reliable and simple to maintain. Tupolev's transports and bombers exemplified the over-arching principles of home-grown aircraft construction. Yet his travels across Europe in 1929 and 1930 reflected Russian designers' readiness to risk the Kremlin's disapproval by remaining open to advice from the West. In 1931 Ilyushin travelled to Britain, where he visited the Hayes factory. By that time several Russians had inspected the plant, although it is unclear whether the initial guest list had included Tupolev.[65]

Stalin never hesitated when forcefully reminding high-profile personnel who bore ultimate responsibility for Soviet aviation's expansionist programme. Yet the premium placed on the work of Russia's top designers was evident when in 1930 Nikolai Polikarpov's draughtsmen joined him in Butyrka prison to design the country's first state-of-the-art fighter aircraft. Frequent crashes were the price of pursuing a revolution in aircraft design, and Polikarpov was falsely imprisoned on suspicion of sabotage. Ultimate rehabilitation came in the form of Soviet 'volunteer' pilots' air superiority in defence of the Republic at the start of the Spanish Civil War. Soviet surrender of the skies above Madrid coincided with the imprisonment of Tupolev and his wife, on a trumped-up charge of selling secrets to the Germans. Bizarrely, Tupolev was later accused of being a French spy, and in total he spent four years building aircraft behind bars. By endeavouring to maintain contact with their Western counterparts, veteran aircraft designers were obvious victims of Stalin's purges. Similarly, the officer corps of the VVS was decimated in the late 1930s.[66]

A.N. Tupolev paid the price for being courted by his admirers in Britain, not least Richard Fairey. In October 1932 the President invited Tupolev to address the Royal Aeronautical Society on his stewardship of the TsAGI. A diplomatic illness prevented the great man from taking up the Society's expenses-paid invitation; unsurprisingly, Tupolev's readiness to visit London at a later date was never realised.[67] His standing with the authorities contrasted sharply with that of Sergei Ilyushin. Throughout the 1930s Ilyushin's advisory role within the Kremlin left him untroubled by the Soviet secret police: working closely with the aviation section of the NKVD he remained head of an enormous, eponymous design and production plant outside Moscow. Ilyushin was always keen to pick up ideas from close examination of Western technology, but he did so discreetly, not least when scrutinising machines made by Fairey Aviation.[68]

Although fiercely anti-communist, CRF took full advantage of the thaw in Anglo-Soviet relations which followed Labour's return to office in May 1929. For the next three years Fairey cultivated members of the Soviet Embassy, the Soviet Trade Delegation and Arcos, the trade delegates' autonomous enterprise responsible for overseas procurement. He enjoyed the hospitality of Grigori Sokolnikov, the Soviet ambassador, and reciprocated by inviting embassy officials to lunch at the RTYC. Fairey Aviation became an enthusiastic member of the Russo-British Chamber of Commerce, appointing the demonstrably well-connected Alexis Ignatiev as its London-based trading agent. In 1932 Ignatiev took the precaution of acquiring British citizenship, yet his advice to CRF revealed someone remarkably well-informed regarding events back home.[69]

Both Ignatiev and Fairey were in close contact with Aleksander Ozersky, who for six years after 1931 oversaw his country's trade mission from within the Soviet Embassy. In late 1937 Stalin initiated a drastic purge of the Commissariat of Foreign Affairs. Ozersky was summoned back to Moscow where under torture he made a false confession, and was promptly shot. Almost certainly it was Ozersky who, before leaving for home, asked CRF to look after his son. The boy was boarding at a prep school outside London, and on news of his father's death Fairey phoned the headmaster to ensure he was safe. He learnt that it was too late as two officials from the Soviet Embassy had already taken the boy away.[70]

Ignatiev and Arcos negotiated the Soviet air force's purchase and delivery of a Fairey IIIF in October 1930. Had he not pulled out at short notice, no doubt Norman Macmillan would have provided *Daily Mail* readers with a lively account of travelling across Europe to facilitate the machine's assembly and testing. Instead, his deputy, C.B. Baker, waited over a decade before providing a no-frills account of a three-week trip, which, 'though grim in parts, was a unique experience and one that I was very glad to have had.' Macmillan's replacement was royally entertained by senior air force staff, suitably impressed by their purchase. Also striking was the level of support he received from the British Embassy at a time when Ramsay MacDonald's government was keen to encourage trade with the Soviet Union. While MacDonald stayed on in Downing Street after a Tory-led coalition replaced Labour in August 1931, the National Government's cooling of Anglo-Soviet relations saw diplomats offer more discreet support when a second machine was delivered in the autumn of 1932.[71]

Only one man went to Moscow in October 1930, whereas a large party accompanied the Fairey Firefly sold to the VVS two years later. In between these visits Richard Fairey decided to see 'the world's first socialist state' for himself. He reached Moscow in July 1931, and was regally entertained by Sir Esmond Ovey, the British ambassador: MacDonald's minority Labour government was clinging to office, but still adamant that restoring diplomatic relations with Russia would open up a fertile market. Spurred on by Ignatiev and the comrades at 13 Kensington Palace Gardens, Fairey saw a business opportunity. He sought to make money out of the Bolsheviks whether HM Government's policy towards the regime was benign or hostile. Yet at the same time he kept the

Air Ministry fully aware of his dealings with the Soviet state apparatus, whether in London or Moscow.[72]

Fairey, Maurice Wright and senior salesman L.A. van de Velde crossed Europe by train, arriving in Moscow dirty, sweaty and anticipating a hot bath. Fairey's travelling companions were keen to improve their Russian, but they left their hosts vulnerable to indiscretion by remaining silent and relying upon a female interpreter. A venerable open-top Rolls Royce whisked the three distinguished guests to a suite of rooms in the Savoy Hotel on Rozhdestvenka Street, north of the Kremlin. On arrival CRF complained that there was no hot water with which to wash, but this surely seemed a minor hardship once he looked out the window to find that across the square stood the forbidding walls of the Lubyanka. Descending for dinner he found the restaurant curtains drawn for fear of protests from starving Muscovites staring into the hotel. Fairey was acutely conscious of food shortages across the city, surreptitiously photographing bread queues from the back of the limousine. At one point his camera was seized, but then quickly returned, its reel of film unspoilt. P.I. Baranov, the Fairey delegation's host, was a powerful technocrat within the party bureaucracy, and concerned that nothing should spoil a crowded itinerary of factory inspections, theatre visits and intense discussion as to the potential for future collaboration. CRF met fresh and familiar faces from the highest echelons of Soviet aviation, with no less a figure than Molotov welcoming him on behalf of the Politburo. Expensive yet malodorous fur coats were presented to Fairey and his acolytes; irritatingly for their benefactors, the three men's Saville Row suits attracted admirers each time they ventured out on to *Ulitsa Rozhdestvenka*. His hosts laboured hard to convey a favourable impression of the Soviet Union at the height of collectivisation and the onset of the first Five-Year Plan; yet a cynical CRF was constantly appalled by the dirt, squalor and malnutrition that elaborate propaganda and stage management had failed to disguise.[73]

Others may have returned home their illusions shattered, but for CRF it was a case of existing prejudices ferociously reinforced. A year later he recalled how, 'On my return from Moscow I was amazed to meet people who had seen the same places and institutions as I had, and had formed diametrically opposite views on everything.'[74] For Dick Fairey visiting the Soviet Union was a revelation, in that conditions were even worse than he had anticipated, thereby reinforcing a visceral anti-communism. At the same time, business was business. In the early months of 1932, Fairey and Ignatiev entered high-level negotiations with Baranov, via Ozersky and other senior embassy officials. The Russians sought technical assistance for the construction of a large, state-of-the-art factory, which would operate as a supply centre for subsidiary production plants. The probable location of this joint enterprise was Balashikha, conveniently located on the southern edge of the Moscow metropolis. The Soviet negotiators anticipated hostility in Whitehall, but the Home Office proved surprisingly helpful in providing visas for Baranov's three-man commission to visit London in June 1932. An initial agreement was duly signed, with Lord Londonderry's tacit approval.[75] Presumably the Air Ministry and the Foreign Office saw the provisional deal as an ideal opportunity for British aviation to establish a physical presence inside the Soviet Union, and thus a rare opportunity to gather vital information. Ministers and officials must have anticipated their Soviet counterparts arriving at the same conclusion, but both sides saw value in taking talks to a higher level.

In the autumn of 1932 van de Velde spent a fortnight seeking to finalise arrangements with two of Baranov's senior *apparatchiks*, M.V. Dushkin and S.I. Makarovsky. Conveniently, their boss and his deputy were both on leave, enabling *Glavaviaprom* (the powerful Aviation Industry Main Directorate) to continue tapping into Fairey Aviation's technological expertise without having to sign a final agreement.[76] Much was made of rigorous tests to be carried out on the Fairey Firefly that van de Velde's technical team had brought with them to Moscow. Bad weather, multiple visits to the ballet, opera and cinema, and extended tours of TsAGI facilities delayed the Firefly's test flight. When at last the reassembled aircraft took to the skies it left a favourable impression upon the watching array of commanders and commissars, most of whom would within a few years fall victim to the purges. This august body included the then mastermind of Stalin's defence strategy M.N. Tukhachevsky, accompanied by other key members of the Politburo and the War Council. Among a similarly impressive – and similarly impressed – audience of technocrats was the ever reliable Sergei Ilyushin, clearly not fated to share Tukhachevsky's experience of a bullet in the back of the head.[77]

Dinner that night saw a gushing Ilyushin compliment Fairey Aviation on the quality of its aircraft, and the firm's founder for his enlightened relations with leading lights of the Soviet aircraft industry. In reply, van de Velde was equally flattering of his hosts. As the vodka flowed and the toasts multiplied, CRF's man in Moscow remained sober enough to take on board a coded message from the mysterious Comrade Jadroff – the still absent Baranov's 'right hand man' – that any final deal depended upon the 'head of the British aircraft world' returning to Russia.[78] In his account of the visit van de Velde seriously underestimated how advanced the Russians were in design, development, output and quality control. Clearly the notion that he and his colleagues were being used scarcely entered his head. Van de Velde advised Fairey that a contract for the Firefly, and confirmation of a joint factory outside the capital, necessitated direct negotiations with Baranov in Moscow. The Dutchman seriously under-estimated the extent to which the Russians could produce high-quality aircraft in large numbers, and clearly his hosts did nothing to disabuse him. Having bought the Firefly they took their time inspecting it, and duly concluded that there was little to learn from an ageing design: in the spring of 1933 the aircraft was tested against the VVS's principal fighter, the Polikarpov I-5, and found wanting. Soviet collaboration with Junkers in the 1920s had proved an unhappy experience, so Fairey Aviation had to offer something truly exceptional for Stalin to sanction an exception to the prevailing orthodoxy of socialist self-sufficiency. Negotiations withered on the vine, with no obvious ill will on the British side given CRF's reluctance to revisit a country and a system he demonstrably loathed. That loathing was reinforced by the final remark in van de Velde's report: 'There is evidence at every step that control exercised by the authorities over the lives of the entire population is more complete than ever before and becomes daily more firmly established with the growth of new generations.'[79]

Fairey loathed the ideology, the dictatorship and the miserable conditions they had combined to produce, but this in no way diminished his admiration for the aeronautical pioneers who had made the long journey to Hayes a year or two earlier. He continued to court them, and in early 1936 the Royal Aeronautical Society again invited Tupolev to address a large and prestigious audience on a topic of his choice.[80] Nor did the demise of Fairey's grand but unrealisable plan to work with the Russians stifle the pleasure of selling them aircraft and making money on the deal: in late 1935 Ignatiev and Arcos orchestrated the sale of two Fantôme biplanes, both of which were supposedly sent to Spain and flown by Russian fighter pilots defending the Republic.[81]

Fairey Aviation had a mechanism for evading official sanctions should the overwhelmingly Conservative National Government veto sales to the Soviet Union. A Russian order could, if necessary, be serviced courtesy of the company's Belgian subsidiary, Avions Fairey – as was the case with the two Fantômes built at Hayes for the Russians in 1936. The link with Belgium dated back to CRF establishing his company in 1915. Among the refugee workforce had been a young aeronautical engineer who for the following four decades would dedicate his working life to Fairey Aviation. Ernest Oscar Tips was hugely admired by Dick Fairey, and a close family friend. The two men corresponded frequently after Tips moved back to Belgium, with CRF repeatedly urged to cross the Channel and be royally entertained. In London Fairey was a friend of the Belgian ambassador, Baron de Cartier de Marchienne, regularly attending embassy receptions. The baron joined CRF's shooting parties at Bossington and

Oakley, and both men teed off in the annual Anglo-Belgian golf tournaments.[82] In late 1937 Fairey was made a Commander of the Order of Leopold, his delight in the award tempered by the strict protocol as to when the decoration could be worn.[83]

Having served for many years in England as Fairey's 'go to man', early in 1931 Tips was despatched to French-speaking Belgium's industrial heartland, south of Brussels. There he set up a factory in the then mining town of Gosselies (today a northern suburb of Charleroi). Avions Fairey came into existence in September 1931 as a consequence of its parent company selling 25 Firefly IIs to the Belgian air force; the sales contract stipulated that any additional machines would be built on home territory. In the end, over 60 Fireflys were assembled at Gosselies, many of which flew against the *Luftwaffe* in May 1940. It was a similar story for the Fox IIM. Successive marques were built in large numbers, and from 1932 the Fox served in a variety of roles as Belgium's principal combat aircraft. E.O. Tips, who in his spare time designed and built the Tipsy light monoplane, saw his factory destroyed by German bombers on 10 May 1940, after which he fled first to France and then to England. It would be another six years before he saw a replacement plant re-open, and a further four years before Avions Fairey resumed volume production.[84]

The success of Avions Fairey helps explain why Baranov and his foot soldiers at *Glavaviaprom* saw its parent company as well worth talking to. They mistakenly assumed Fairey Aviation could raise enough capital to fund a joint enterprise capable of producing as many as 5,000 aircraft a year. Yet so long as Labour remained in office the Foreign Office was opposed to a commercial proposition which flew in the face of its disarmament initiative. When briefed on CRF's visit to Moscow, Arthur Henderson – Foreign Secretary, and future chair of the Geneva Disarmament Conference – saw state support for a plan to build combat aircraft in Russia as fatally undermining the MacDonald administration's efforts to promote peace. In July 1931 Henderson ordered his officials to keep Fairey and his company at a distance, but within six weeks Labour was out of office. With MacDonald remaining in Downing Street, the National Government still adhered to a disarmament agenda, and at the same time a Tory-dominated administration reversed Labour's rapprochement with the Soviet Union. In practice, senior civil servants at home and in Moscow discreetly encouraged British manufacturers' courting of entrepreneurial apparatchiks approved by the Kremlin.[85]

In his report on the autumn 1932 visit, van de Velde listed a remarkable number of senior civilian or military personnel who had previously visited Hayes and/or Gosselies. These names ranged from the VVS chief test pilot, Aousan, to the military's second-most senior procurement official, Gouschin. The latter refused to confirm an order for the Firefly, but 'emphasised that no aircraft firm enjoyed the prestige of The Fairey Aviation Company in Russia.' The arrest and trial for 'sabotage' of six Metropolitan-Vickers engineers two months later gave a strong signal that it scarcely mattered how high Soviet officials supposedly valued a foreign company: its overseas staff were still fair game should Stalin choose to ignore the advice of Litvinov, his commissar for foreign affairs, and translate isolation into provocation.[86]

Maxim Litvinov, exiled in Edwardian London and married to an English woman, encouraged rapprochement with the West long before the Soviet Union joined the League of Nations, the Comintern promoted an anti-fascist 'popular front' with social democratic parties, and Stalin anticipated a renewed grand alliance with the French and British. The Munich conference of September 1938, when Czechoslovakia was forced to accept its Sudeten borderlands' secession to Nazi Germany, was a body blow to Litvinov's western-oriented strategy. In May 1939 Molotov succeeded him as Russia's foreign minister, although Litvinov returned to favour after Hitler invaded in June 1941, serving for two years as ambassador to the United States.

Fairey must have met Litvinov many times in wartime Washington, but in the 1930s it was Ivan Maisky, the long-serving ambassador to the Court of St James, whose hospitality he enjoyed on a regular basis. Maisky – arguably Litvinov's closest confidant – served in London for an astonishing 11 years, from 1932 to 1943.[87] His arrival preceded Stalin's shift in 1933–4 from pursuing an isolationist 'class against class' strategy to an accommodation of the western capitalist powers in a system of collective security designed to contain Nazi Germany. In consequence, Russia's envoy in London could pursue a policy of rekindling Anglo-Soviet relations (starting with the Metro-Vickers engineers' early release), identifying mutual security interests and projecting a less suspicious and more outward-looking USSR to sceptical politicians, journalists, bankers and industrialists. Maisky enjoyed the bourgeois pleasures of life to the full, and his crowded social calendar included an annual visit to the Royal Aeronautical Society's garden party.[88]

While neither Maisky's published nor unpublished diaries make mention of Richard Fairey, photographs and correspondence confirm the latter's readiness to attend embassy functions and accept regular gifts of prime caviar.[89] Nor did Fairey restrict himself to invitations from the ambassador, for example, making space in a crowded diary to pass an evening 'At Home' with the military/air attaché and his wife. As his personal papers confirm, CRF's normal response when invited to a function was a polite decline, but he usually found time to attend events held under the umbrella of the Soviet Embassy. For example, on 5 March 1936 Fairey subjected himself to an evening watching film footage of Red Army manoeuvres in the Ukraine.[90] He regularly attended receptions and parties, although ironically on 1 March 1939 he found himself unable to attend Maisky's grandest prewar gathering of ministers, MPs and magnates: over 500 'pillars of society' were in attendance, all of whom 'stopped in the middle of their sentences and rushed childishly to have a look at Chamberlain in the interior of the Soviet Embassy.'[91]

If a grim-faced Prime Minister could venture into the Bolshevik outpost then one can see why the Chief of the Imperial General Staff was happy to join CRF for a long lunch with Monsieur Maisky as late as 11 August 1939, only 12 days before the signing of the Molotov–Ribbentrop Pact. Lord Gort clearly had no idea that Anglo-French efforts to forge a tripartite alliance were foredoomed, and that a previously inconceivable realignment of military power in Europe was imminent – but then neither did Maisky.[92]

The obvious question is why someone so hostile to the Soviet Union maintained such close relations with its representatives in London, particularly when any opportunity to sell the Russians more aircraft had clearly passed. For a start, the quality of cuisine at 13 Kensington Palace Gardens rarely matched Fairey's high expectations: in the spring of 1936 the equally demanding Harold Nicolson found lunch with the Maiskys an especially grim experience.[93] This was ironic given the ambassador's reputation as someone genuinely concerned for the well-being of his guests: like Fairey he loved to put on a good party. However deep the ideological divide, and however contrasting their characters, 'Jean' Maisky and Dick Fairey found much on which to agree. Both men admired Lord Beaverbrook and H.G. Wells, with a healthy respect for Winston Churchill and mutual disdain for the Labour Party. Both men loved aeroplanes, but rarely enjoyed travelling in them.[94] Most important of all, they agreed on the need for rapid rearmament in the face of undisguised German expansionism. One can imagine Fairey echoing the head of the Foreign Office: 'I rather like Maisky, although – or perhaps because – he's such a crook.' Similarly, the only time CRF might agree with 'Rab' Butler was when the Foreign Office's keenest appeaser labelled Maisky 'an agreeable scoundrel.'[95]

With rapprochement between Hitler and Stalin seemingly impossible, it made sense for Fairey to play the long game: however odious its regime the Soviet Union constituted a future ally, which in time of war would require more aircraft than its domestic industry could produce. Why burn bridges? The irony is that between June 1941 and May 1945 Fairey Aviation sold no aircraft to the VVS, as none fitted a combat role on the Eastern Front. Just as he had a penchant for socialising with minor royalty, Fairey enjoyed the company of diplomats with access to the corridors of power, whether back home or in Whitehall. These encounters probably prompted a quiet word with the right chap in MI5 or MI6, but no available file makes reference to any such conversations.[96]

Maisky may have been *persona non grata* in Downing Street, but his diaries confirm how frequently he met ministers and senior officials at the Foreign Office. He had an astonishing array of friends and acquaintances in London, as well as back home. By 1940 the same could be said of Joe Kennedy, assumed by all to have a direct line to the Oval Office. Soon after he and his high-profile family arrived in Grosvenor Square, the American ambassador became acquainted with Richard Fairey. The latter found the straight-talking Bostonian of like mind other than in the need to tame German ambitions, and in due course he invited Kennedy to cruise on his racing yacht *Evaine*.[97] As we have seen, the Belgian ambassador was similarly welcome to spend a weekend out of town. Any such invitation to Maisky was of course inconceivable, especially after 'Hitler's latest achievement in rereleasing the floodgates of Bolshevism.' Like so many of his peers, Fairey believed 'Communism and Nazi-ism are exactly the same thing', and the *démarche* of August 1939 allowed Moscow and Berlin 'to impress these doctrines on the world by force.'[98] Yet even as he railed against Soviet duplicity he implicitly excluded Maisky, blaming Stalin's unholy alliance with the Führer on a lifelong failure to engage with an enlightened, cosmopolitan West.[99] Had Fairey found himself in the same room as the Soviet ambassador between the autumn of 1939 and the spring of 1941 it is hard to imagine him snubbing the most anglophile, the most 'airminded' and the most bourgeois, advocate of proletarian revolution.

Dealing with disarmament, 1930–6

Minimal overseas sales, witness the miserable complement of Fairey aircraft despatched to Moscow, compounded the problem of short production runs. Across a struggling industry, few manufacturers achieved economies of scale. Fairey Aviation was an exception in the 1920s, and Hawker in the early 1930s. Only four of almost 40 different aircraft types ordered by the Air Ministry between 1919 and 1935 generated bulk orders. Officials sought to avoid any one firm attaining a size larger than what was necessary to meet the Government's future needs: 'The object of this policy was to maintain capacity … particularly design capacity, which would form the nucleus of an expanded aircraft industry in time of war. Control was exercised through market power…' As opposed to the late 1930s and into the war, the Air Ministry deliberately did not use its purchasing power to encourage rationalisation and centralisation, as 'this would have strengthened the market power of the sellers.'[100] Thus the 'ring' of recognised contractors remained intact, and yet between 1923 and 1930 the industry was building an average of only 646 aircraft per year. Nevertheless, most of the major manufacturers secured over £1.8 million in military sales across the same period. A chronic failure to standardise meant the RAF maintained no less than 44 different models at the start of the 1930s, with almost as many engines. Naval procurement totalled a mere 18 machines between 1929 and 1932, sowing the seeds for a severe downturn in Fairey's fortunes.[101]

Predominant in Britain's postwar aircraft industry, Fairey Aviation's market share had risen from £457,000 in 1923 to £915,000 seven years later, far outstripping its nearest rival, Vickers. Yet the Fox losing out to the Hart on such a large military contract signalled the end of Fairey's pre-eminence: in just one year – 1930 – Fairey saw its net profit switch from double that of Hawker to half. While in the short term net profits continued to rise – totalling almost £200,000 in 1932 – thereafter there was a steep decline, bottoming out in the mid-thirties at just over £39,000. Determined to retain control of his company, CRF was insistent Fairey should remain an independent contractor. This was in contrast to Vickers and Hawker, each of which would by 1935 form the core of the industry's two dominant conglomerations; between them they shared nearly 80 per cent of the Air Ministry's military contracts.[102]

Prior to the mergers that created Hawker Siddeley, the Hart's prime contractor was unable to meet the number of aircraft on order. With most manufacturers so hard hit by the economic downturn, the Air Ministry ignored the protests of Sopwith and the SBAC, and sought subcontractors via competitive tendering. At a conference in March 1931 Richard Fairey urged an alternative approach, whereby Whitehall sought two types from each operational class, thereby keeping the most inventive design teams intact. With a besieged Labour government focused upon consolidation and driving down costs, CRF's suggestion was ignored.[103] Solidarity within the SBAC crumbled as members competed fiercely to secure work building the RAF's latest bomber, the Hart. Hawker's board noted the irony of a disproportionally large contract resulting in lower profit margins: struggling rivals were taking full advantage of Kingston's finite manufacturing capacity. Here was a powerful argument for acquiring companies with a proven track record and/or potential for growth. A further incentive was the changing fortunes of Fairey Aviation, demonstrating the vulnerability of a market leader so dependent upon Air Ministry patronage.[104]

From early in the decade Fairey Aviation found itself suffering in the same way as every rival except Hawker: short-run military contracts were not conducive to generating a return on development costs *and* a healthy profit. The problem was compounded by the firm's failure to diversify. With Napier clearly a fading force, CRF revived his desire for Fairey to challenge the established aero-engine manufacturers. Without shareholders' approval he formed an engine division, headed by the vastly experienced aero-engineer Captain A. Graham Forsyth. Pilot, technocrat and inventor, Forsyth served as chief engine designer until World War II, after which he headed Fairey's vertical take-off programme. Here was a magician of piston engine technology whose 'Prince' project produced the PV12, a credible alternative to Rolls Royce's signature power unit, the Merlin. The Air Ministry's heavy investment in the Merlin meant CRF failed to secure £30,000 of state funding. Fairey urgently required a cash injection in September 1935 so Forsyth could conclude his alarmingly expensive test programme. The PV12's succeeding design project was stunningly original, second-guessing what future carrier aircraft would require rather than responding to any present need of the Royal Navy: the P24 was a revolutionary design which comprised two supercharged, liquid-cooled, vertically opposed 12-cylinder engines; their opposite-rotating crankshafts, housed within a common crankcase, drove two three-blade co-axial propellers. Design, development and testing of the P24 continued on both sides of the Atlantic until as late as 1942, and it cost Fairey Aviation around £100,000. Such a substantial outlay undermined the profitability of the airscrew division, which continued into the war mass-producing and servicing the Fairey-Reed fixed-pitch propeller.[105]

The engine division was testimony to CRF's lifelong determination to promote cutting-edge technology, but in this instance it was a financial black hole. Institutional and individual shareholders not represented on the board could applaud the chairman's visionary instincts, while criticising a demonstrable lack of transparency. Richard Fairey could initially justify his hiding heavy capital expenditure and significant ongoing costs within the company's overall accounts: there was a perceived need to hide from Rolls Royce the threat of competition. However, secrecy became harder to defend later in the decade, and a non-production engine division became a major drain upon Fairey Aviation's finances.

Fairey's principal asset was its property portfolio, a major part of which was the firm's 150-acre airfield at Harmondsworth. Initially labelled the 'Great West Aerodrome', the Middlesex development was bought for £15,000 in 1929, with the intention of building a companion to the Hayes factory. Extensive assembly facilities never materialised, but Harmondsworth quickly played an important role in the testing of Fairey's future front-line aircraft, starting with the twin-engined Hendon heavy bomber. This all-metal monoplane was tested at the newly acquired site in a number of guises and for a number of years: half a decade of experimentation saw a once revolutionary design become obsolete, and only 14 aircraft went into service. Equally expensive was the clunky, chunky and chronically underpowered G.4/31 light bomber, of which only one multi-purpose model was built – the four-year project must have cost Fairey a fortune. Less visually striking but more lucrative machines flying out of the newly acquired aerodrome were the Gordon and Seal advanced variants of the ever reliable IIIF: sales of these aircraft at home and abroad ensured a steady if unspectacular income throughout Fairey Aviation's fallow years.[106]

As always, Norman Macmillan took a keen interest, but the man who for the next 11 years made Harmondsworth his second home enjoyed an even closer working and personal relationship with the company chairman: in 1931 the speedster Chris Staniland, fresh out of the RAF and a familiar face at Brooklands, became chief test pilot following his immediate predecessor's fatal plane crash.[107]

By the mid-1930s the now rechristened 'Heath Row Aerodrome' was a vital element in its owner's efforts to maintain a high profile, witness the annual gathering of the great and the good when Fairey hosted the Royal Aeronautical Society's spring garden party – by 1937 *The Tatler* considered the event a highlight of the social calendar. To mark George V's silver jubilee, in early May 1935, CRF and John Moore-Brabazon together masterminded a spectacular flying display, a thrilling demonstration of fire-fighting, and a visually striking history of the Society. The garden party's hero was the thrills-inducing C.S. Staniland. Its guest of honour was the Marquess of Londonderry, whose wife had earlier that week hosted a grand reception for everyone who counted within the British aircraft industry. Lady Londonderry's social skills clearly failed to impress Stanley Baldwin, as five weeks later her husband found himself unceremoniously ejected from the Air Ministry.[108]

As President of the Royal Aeronautical Society, a senior member of the SBAC, and the head of a company faced with a severe downturn in sales, Richard Fairey was unsurprisingly vocal in arguing the case for a rapid expansion of the RAF. Hitler's coming to power in Germany, and Lord Rothermere's high-profile campaign for a far larger air force, saw CRF restate the case for creating a reserve of combat aircraft so big that Britain could sustain huge losses at the start of a war and still strike back. In January 1934 Fairey urged *Daily Mail* readers to heed the paper's call for an air force of 5,000 machines. He lamented the inertia of a pacifist-inclined National Government. To avoid inevitable defeat in any future conflict, Ramsay MacDonald and his ministers should revise their strategic – and thus their spending – priorities: 'To build up an adequate reserve would cost about £9 million which is only about £2 million more than the cost of a capital ship like the Rodney. Unless something is done about this very vital matter we might just as well give up our Air Force – fine and efficient as it is – and rely on the League of Nations.'[109] That same month an expansionist-minded CRF set out his store in a briefing paper for the National Press Agency. Meanwhile, students at both the Imperial Defence and the RAF Staff College were being lectured on the procurement failures of what was an overwhelmingly Conservative administration. The prospect of a resurgent Germany confirmed Fairey's longstanding conviction that disarmament was noxious, infectious and deeply debilitating. MacDonald's surrender of the nation's military advantage was to be expected, but Baldwin's support for the Disarmament Conference had been unpardonable.[110]

Urged on by Rothermere, Fairey was on occasion incautious in his comments concerning the National Government. In April 1934 he expressed greater alarm over ministers' reluctance to secure 'in this troubled world … command of the air' than the personal invective of 'disarmament cranks' like Fenner Brockway. In his recently published polemic, *The Bloody Traffic*, Brockway had included Fairey when profiling industrialists grown rich on manufacturing weapons of war. This was the first in a succession of ostensible exposés detailing the malign influence of named arms manufacturers, including of course the mastermind behind Fairey Aviation. Unaware of Brockway's affiliation to the decidedly non-Stalinist Independent Labour Party, CRF insisted the attack on him had been written by 'one of our communist leaders, and communist followers are not unheard of in such august establishments as the Air Ministry or Farnborough. Some of them have had the advantage of a trip to Moscow, as have some of us, the great point of which has been to prove that that visit has never yet altered anybody's political opinion.' One can imagine the sharp inhalation of breath among those attending the Aeronautical Inspection Directorate's annual dinner when the guest speaker made his singularly injudicious remarks.[111]

Yet CRF carried on delivering similar speeches until the onset of Chamberlain's premiership, lambasting government agents of disarmament and defeat, whether closet Communist or cautious Conservative. Baldwin, along with Lord Londonderry, was an easy target, having reassured MPs in November 1934 of the RAF's overwhelming air superiority, yet a mere six months later conceded Hitler's claim of German parity. The *Daily Mail* found both men's promise to restore air parity 'still inadequate. It should have been in hand years ago.' Isolated inside Whitehall and lambasted by Churchill in Parliament, the Secretary for Air proved easy prey for Fleet Street's largest mass-circulation newspaper.[112]

Fairey scarcely lamented Londonderry's demise in the late spring of 1935. Less commendable was Rothermere's endorsement of Mosley, let alone his courting of the Führer. Similarly, CRF distanced himself from the press baron's aggressive campaigning ahead of the 1935 general election. Although Norman Macmillan and Murray Sueter surely sought Fairey's support, he never joined Rothermere's National League of Airmen. When it came to promoting 'airmindedness' he remained loyal to the long established – and eminently respectable – Air League of the British Empire. The Air League enjoyed royal approval and, when required, CRF could be suitably obsequious in complimenting the Prince of Wales ('the trained mind', boasting the 'long experience of a genuine aviator' – after the Abdication a high-flying Duke of Kent was similarly flattered).[113]

In 1933–4 Fairey's profits fell by an alarming 42 per cent. Accordingly, the value of ordinary shares in the company dropped, bottoming out at ten shillings. The five per cent tax-free dividend was half that of the previous year, in both proportion and sum.[114] Both government and shareholders acknowledged the impact on profits of having to rebuild the Hamble plant after a disastrous fire left most buildings gutted, with the Hayes factory having to absorb all assembly work. Equally debilitating by 1934 was the loss of lucrative overseas contracts after disarmament-minded ministers chose to embargo sales to conflict zones in South America and the Far East. This drop in exports was compounded by Canadian and Australian governments, emboldened by the Statute of Westminster, placing orders with American rather than British suppliers. Fairey saw the National Government's slavish adherence to a disarmament agenda as the reason why Downing Street never insisted the Dominions purchase similar machines to the RAF.[115]

To the relief of CRF and his fellow directors, City anticipation of rearmament spurred a recovery in share value across the next 18 months: by 1935 ordinary shares were selling at 35 shillings. Hawker's share price was consistently lower, but the company's share capital of £787,000 was over a third higher than that of its rival: building the Hart in such large numbers gave an ambitious and aggressive board the financial leverage it required. Ironically, a need to expand in response to rearmament saw Hawker Siddeley's share value fall between 1936 and 1938 as the conglomerate sought further capital in order to finance fresh plant. Faced with its own bulging order book Fairey Aviation would experience a similar need to raise capital via heavily subscribed share issues, scaling back only when the Treasury belatedly acknowledged a need for government intervention.[116]

Fairey's 1934 issue of preference shares, underwritten by the chairman on a two per cent commission, set a pattern for securing fresh capital without the six-man board surrendering its tight control over the company. Fresh investment boosted a considerable cash reserve, which even in the lean years continued to grow.[117] The mid-decade drop in dividend payments was offset by directors' substantial salaries and expense allowances, plus for CRF significant patent fees. Thus the firm's founding fathers, not least its chairman and managing director, still benefited from sizeable net incomes. Not only were board members re-investing in an enterprise ideally placed to benefit from delayed expansion of the RAF, but each of them could plough money into other 'sunrise industries' that had weathered the worst effects of prolonged economic recession. They were all by now wealthy men, but their leader was in a league of his own, supporting a lifestyle and standard of living unmatched by his acolytes.

By mid-1936 CRF boasted at least three lucrative share portfolios, spread across two continents. All shares in traditional heavy industry had been sold off other than a stake in the rising steel company William Cory & Son. Having bought cheaply in the aftermath of the Great Crash, Fairey rode the wave of economic recovery. Across the Atlantic he shrewdly ran down his American portfolio ahead of the 1937 slowdown in recovery, before investing heavily north of the border in anticipation of Canadian industry expanding dramatically to service Britain's wartime needs.[118]

Fairey Aviation's accountants and auditors were adroit in minimising investors' tax liabilities in that the company absorbed the tax burden rather than the preferential and ordinary shareholders. By 1937 the tax-free dividend had recovered to a lucrative 12.5 per cent on a net profit of £248,678. The *Sunday Times* profiled CRF and urged readers to continue investing in Fairey Aviation. Those ordinary shareholders who heeded the newspaper's advice picked up a modest tax-free cash bonus in 1938. Directors were clearly sensitive to any suspicion that they were benefiting disproportionately from the company's impressive recovery.[119]

Two years earlier, in letters to *The Times* and the *Morning Post*, CRF vehemently contested the claim of veteran pacifist MP Frederick Pethick-Lawrence that thanks to rearmament Fairey Aviation had paid shareholders an excessively generous dividend. Wisely deleted from Fairey's published letter was the question, 'I wonder who is the greatest rogue, the inventor who provides the finance for a pioneer industry, or the politician who, under cover of privilege accorded by the House of Commons, resorts to falsehood to further his ends?'[120]

Pethick-Lawrence's criticism came in the wake of CRF's robust defence of the aircraft industry when called to give evidence before the Royal Commission on the Private Manufacture of and Trading in Arms, announced by Sir John Simon in a Commons statement on 22 November 1934. The Commission took evidence throughout 1935 and into 1936, its creation a calculated move by MacDonald to calm the growing clamour for a government clamp down on the global trade in arms, and stricter control of weapons production in Britain and across the Empire. Enthusiasts for disarmament from both left and right saw a moral imperative in public ownership of all arms manufacturers, viewing the aircraft industry as a prime candidate. Ironically, there were advocates of rapid rearmament who, in the interest of efficiency, also saw a case for the Government assuming direct responsibility for aircraft production. The arguments for and against nationalisation were fought out in the Commission's public hearings, the protagonists attracting considerable attention in the national press and on the BBC. Throughout the summer and autumn of 1935 an impressive array of critics, led by David Lloyd George, arraigned the 'merchants of death.' Fairey's belief that Farnborough harboured communists almost certainly rested on its employment of prominent leftist engineer, Ronald McKinnon. Insistent that free enterprise hindered rather than enhanced development and design, the RAE technocrat's call for a state-run Corps of Aircraft Constructors made him a star witness. He clearly impressed Commission members instinctively suspicious of large, profitable aircraft manufacturers. Yet McKinnon and like-minded witnesses were out-gunned by powerful voices defending the status quo: citing exponential wartime expansion as evidence, Whitehall and the RAF defended private-sector competition as conducive to maximum efficiency, with Sir Hugh Dowding, Lord Weir and Cabinet Secretary Sir Maurice Hankey especially persuasive.[121]

At last the chance came for the industrialists to defend themselves, with public and press eager to hear from Britain's best-known plane makers, summoned to appear at Middlesex Guildhall, in the heart of Westminster, on 7 February 1936. Fairey took the lead in ensuring the SBAC's seven council members were properly briefed concerning each member of the Commission, whether sympathetic or hostile, were familiar with the Society's 36-page memorandum which he and director Charles Allen had repeatedly redrafted prior to its January submission, and were cognisant of every reference to aviation at every sitting since the previous May. Between them Fairey and Allen, supported by respective secretarial staff, made certain the SBAC delegation was meticulously prepared. On the day, CRF and Frederick Handley-Page, as figureheads of their respective companies, fielded the majority of questions, almost all of which were deeply hostile. Fairey, working hard to control his temper, offered a spirited defence of the free market and its 'intensive competition', and repeatedly clashed with veteran war correspondent and man-of-letters Sir Philip Gibbs. He dismissed suggestions that Fairey Aviation had secretly sold machines to Russia, Germany or extra-European combatants, had secured contracts in Peru through bribery, or allied with the *Daily Mail* simply to sell more aircraft. Gibbs' knowledge of domestic aeronautics in August 1914 was ridiculed via a brief lecture on Sueter, Sopwith and the Short brothers, and his suggestion that military aircraft should be built within the public sector attracted withering contempt. Courtesy of a prepared statement, CRF clarified the status of Avions Fairey, insisting that the 'Belgium branch operates under the Belgian government.'[122]

Fairey felt that he and his fellow council members 'had done very well', although after reading the evidence, 'My general impression … is that I talked too much.' The Commission's minutes strongly suggest that, in what was clearly a formidable, even an intimidating performance, Fairey came across as truculent and irascible, scarcely bothering to disguise his contempt for a veteran radical like Gibbs. CRF's irritation was no doubt compounded by the knowledge that he had been criticised by earlier witnesses without challenge. Generally speaking, Fleet Street's coverage of the SBAC delegation's individual and collective performance was favourable. Needless to say, Charles Grey in *The Aeroplane* labelled Dick Fairey's performance a triumph, devoting no less than six pages to how the aircraft industrialists effortlessly saw off 'the charges of Pacificists, Disarmamists, Communists, Socialists and Unrealists generally.' The American press was more sensationalist in its reporting: London correspondents hinted that there might be mileage in the story of 'Peru War Plane Graft' or the suspicion that British firms had secretly helped Germany rearm. The following week Fairey thanked naval architect Sir Charles Craven, present on the day, for his letter of congratulation, expressing satisfaction that the SBAC's careful preparation had borne fruit. On reflection he may have qualified his satisfaction, acknowledging potential damage to Fairey Aviation given its newly acquired negative image on the far side of the Atlantic.[123]

Not surprisingly, the Government saw no reason to act upon the Commission's recommendation that – while not as yet enjoying full control over aircraft manufacture – the Air Ministry should establish a design and training centre, with the potential to pioneer fresh methods of mass production. In Parliament the published report was brusquely dismissed by Sir Thomas Inskip, newly appointed as Minister for the Co-ordination of Defence, and by the time a formal response appeared in the spring of 1937, the Commission had been all but forgotten. Only 48 hours after a brief Commons debate on a largely redundant white paper, a decidedly 'airminded' Neville Chamberlain moved next door to become Prime Minister. The SBAC's evidence had been important in killing any prospect of radical change, but the contribution of Whitehall mandarins and senior air force staff had been vital. Yet it was realism not rhetoric which dealt the Royal Commission's disarmament agenda a killer blow – on 7 March 1936 Hitler had initiated Germany's remilitarisation of the Rhineland, and such a stark defiance of the Treaty of Versailles concentrated minds wonderfully within the Foreign Office, the service ministries and, above all, the Treasury.[124]

CHAPTER 7

Through the Thirties, That 'Low Dishonest Decade'[1]

Cometh the hour, cometh the man – and cometh his legacy: the Swordfish

> The Swordfish fly over the ocean,
> The Swordfish fly over the sea;
> If it were not for King George's Swordfish
> Where the 'ell would the Fleet Air Arm be?[2]

Yes, the delta-winged, supersonic FD2 was Fairey Aviation's crowning achievement, still seen 60 years after seizing the world speed record as one of the most beautiful aircraft ever built. Yet, for all its spin-off technology, the firm's last fixed-wing prototype is scarcely remembered today. Not so the Swordfish: the Fleet Air Arm's most enduring marque ensures the continued presence of the Fairey name within the nation's collective memory. The Fairey Swordfish is to the Royal Navy what the Supermarine Spitfire is to the Royal Air Force. Both aircraft are the crown jewels of respective services' historic flights, and both played a pivotal role in 1940 when the British people 'stood alone': the Spitfire in the Battle of Britain and the Swordfish in the destruction of the Italian fleet at Taranto. Although both aircraft flew throughout World War II, the Spitfire saw serial modification and the Swordfish scarcely any: whether aeroplane or seaplane the fundamental design hardly altered.

As a canvas-covered, open-cockpit, radial-engine biplane, the three-man 'Stringbag' was already obsolete at the onset of hostilities. The Swordfish seemed an antique survivor of a bygone age, but its strength, versatility, manoeuvrability, range and low stall speed rendered it a potent instrument of naval air power in the early years of the war.[3] The day of reckoning came only in

February 1942 when a whole squadron was annihilated attacking the cruisers *Scharnhorst*, *Gneisenau* and *Prinz Eugen* as they and their escorts passed up the English Channel. Before then, the Royal Navy's principal strike aircraft had hit the *Kriegsmarine* hard in Norway, struck the Vichy French fleet at Mers-el-Kebir and off the Syrian coast, sunk or disabled four capital ships of the *Regia Marina* at Taranto and Cape Matapan, and crippled the *Bismarck*.[4]

Best known as a torpedo-bomber, the Swordfish fulfilled multiple roles, including radar reconnaissance, hunting submarines and flying with floats off cruisers equipped with catapults (a function shared with the highly functional but long forgotten Sea Fox, a seaplane celebrated only for its crucial role in the Battle of the River Plate – its presence at the sinking of the *Graf Spee* meant Fairey aircraft helped destroy Germany's three best known battleships). When no longer flown off the Royal Navy's fleet carriers, the Swordfish continued to serve on escort carriers and – courtesy of rocket-assisted take-off – merchant aircraft carriers. If no longer vital to the Mediterranean Fleet, in the latter years of World War II the Swordfish remained uniquely equipped to protect the North Atlantic and Arctic convoys in all but the harshest weather.[5] While two machine guns provided pitiful protection, the aircraft's range of ordnance was formidable, not least its highly destructive 1,620 lb, 18-inch aerial torpedo. The Swordfish could cover formidable distances, remaining on patrol at low speed for long periods of time.[6] Its pilots were trained to dive at nearly 200 knots, level out at less than 500 feet, throttle back to 90 knots, and then drop to 60 feet for a wave-skimming run in to the target. Relying on an ingenious system of calibrated light bulbs to pinpoint the enemy ship's speed and direction, the pilot waited as long as possible before finally releasing his torpedo. Night-time attacks were especially deadly as a scarcely discernible 'spidery silhouette' glided down to the target at half throttle: flying almost at sea-level and on the shortest of torpedo runs the Swordfish silently delivered its deadly load.[7]

Making few concessions to modernity, this venerable aircraft's unique qualities ensured its front-line role and continued manufacture long after a more sophisticated version entered service. As many as 800 of the Swordfish's enclosed cockpit successor, the Albacore, were built, but pilots cursed its poor handling, and mechanics bemoaned its unreliability.[8] In contrast, a staggering 2,399 Swordfish rolled off the assembly line between 1936 and 1944, generating universal affection among air crew, artificers, armourers and riggers. Ironically, Fairey Aviation manufactured fewer than 700 of the aircraft synonymous with the firm and its founder: building the RAF and Royal Navy's first generation of monoplanes, notably the Battle and the Fulmar, necessitated subcontracting

Swordfish assembly to Blackburn, a company with spare capacity once the Admiralty recognised the operational failings of its rival designs.[9]

Note these numbers – 13 of the Fleet Air Arm's 20 first-line squadrons in 1939 flew Swordfish, soon to be joined by a further 12. Across the war the total number of squadrons more than doubled, to which could be added RAF units and single ship-based seaplanes.[10] The size and substance of the Swordfish's contribution to the Fleet Air Arm highlights how vital Fairey Aviation was to the war at sea: as well as the Albacore and Sea Fox, by September 1945 the Royal Navy's 59 carriers and 69 first-line squadrons flew or had flown several thousand Fulmars, Fireflys and Barracudas – all bearing the Fairey badge. This was a huge contribution to the British war effort, complemented by the Fairey Battle's front-line service with the RAF in 1939–40. Given that the fall of France cruelly exposed the Battle's shortcomings as a medium bomber, the credibility of Fairey Aviation – and of its chairman and managing director – rested upon the achievements of a single aircraft as the Royal Navy wrestled for control of the Mediterranean and the Western Approaches.

In late 1933 Richard Fairey responded positively to a request from the newly appointed Director of the Naval Air Division (DNAD) that his company initiate the sort of project he loathed, namely the development and design of a multi-purpose aircraft. The irony was that for once the Air Ministry was wholly unsympathetic to the idea. Captain H.C. Rawlings considered himself a friend of CRF, although it is hard to see how the two men could have been close. The DNAD told Fairey he was keen to take up a suggestion from officers on the carrier *Courageous* that, in light of the Fleet Air Arm's straitened circumstances, its reconnaissance aircraft should carry a bomb or a torpedo.[11]

For all his reservations regarding Rawlings's deeply flawed specification, Fairey told Marcel Lobelle and Wilf Broadbent to build a TSR (torpedo/spotter/reconnaissance) prototype. Blackburn's preferred design for this multi-purpose carrier aircraft, the Shark, was already being tested, so Fairey set a seemingly impossible deadline – he gave Lobelle 14 weeks, from drawing board to first flight.[12] Not that the design team was starting from scratch. On the instructions of the managing director they reworked a project initiated by the Greek Navy but never followed through. CRF realised his company could recoup considerable development costs by upgrading the TSR prototype intended for the Greeks. A radically reworked aeroplane could meet British requirements for folding wings, an arrestor hook, a cowled Bristol Pegasus radial engine, and a unique load-carrying capacity (in weight and volume, hence the nickname 'Stringbag'). The TSRII was without doubt a big biplane, its duralumin-ribbed upper wing swept back by four degrees to compensate for an unduly long fuselage. Across the winter of 1933–4 Fairey was the driving force, monitoring progress day by day and leading from the front. This was the last project micromanaged by the founder of the company, and the final outcome can be seen as very much a personal achievement. His deadline for design and development was – just about – met. Renamed the Swordfish, the Fleet Air Arm's primary strike aircraft first flew in April 1934 and entered service on board HMS *Glorious* just over two years later. The Shark, arguably a more advanced aircraft, had failed to enthuse pilots already impressed by the Swordfish's handling and reliability. By late 1938 speedy re-equipment of serving carrier squadrons left the Swordfish as the Fleet Air Arm's sole strike aircraft.[13]

Two years later the Norwegian campaign demonstrated to what extent 'the stately Swordfish … a kind of one-aircraft air force' was an exception so far as multi-purpose machines were concerned. Yet, for all its success in Norway, the sedate biplane could never match the performance of its more specialist adversaries, or survive attack by *Luftwaffe* fighters such as the Bf109. The same basic flaw would undermine the effectiveness of the first-generation Firefly, developed as a reconnaissance aircraft in the late 1930s but only operational from 1943, and the two-man Fulmar, which served as the Fleet Air Arm's first choice fighter until succeeded by high-performance single-seater machines such as the British Seafire and the Lend-Lease Corsair. At the start of the war traditionalists in the Fleet Air Arm believed fast aircraft could not operate off carrier decks, and – unlike their Japanese and American counterparts – assumed fighter pilots needed navigational assistance from an observer. Ironically, the solid, sluggish Fulmar cost £2,000 more than the naval version of the Spitfire; but on the plus side it was more robust and easier to repair. Like the Swordfish, both the Firefly and the Fulmar ran counter to Fairey's fundamental belief in designing aircraft tailored to a particular purpose; yet clearly his company had to accommodate the demands of the Air Ministry and the Admiralty. By meeting the Fleet Air Arm's demanding yet superannuated specifications Fairey Aviation generated large profits on the back of lengthy production runs, whether in-house or via a subcontractor: nearly 900 of the original Firefly and precisely 600 of the Fulmar were built at Hayes and Heaton Chapel, respectively.[14]

Fairey Aviation manufactured a number of other aircraft across the 1930s, mostly variants on existing marques and in modest numbers. The Swordfish, the Albacore, the Fulmar, the Firefly and the Battle together ensured the company's dramatic revival in fortunes once the Treasury and the Air Ministry agreed on an agenda for rearmament. At least four of the five development projects resulted in production on an unprecedented scale, invariably in partnership.

Only one of these designs could be deemed a success – but what a success. Fairey Aviation's wartime reputation would rest on the remarkable if unlikely achievement of a demonstrably obsolescent aircraft. The Swordfish reflected the poor state of the Fleet Air Arm in 1938 when full autonomy was restored. Nevertheless, on operational duties from the Arctic to the Aegean, again and again it triumphed over adversity. For all the overblown propaganda this was an outmoded design living on borrowed time, and yet the Swordfish literally flew in the face of defeat.[15] Between 1939 and 1942 it maintained a valuable morale-boosting role, and in so doing consolidated Richard Fairey's reputation as one of the nation's foremost plane makers.

Aircraft and an agenda for rearmament

Back in 1936 those same plane makers felt emboldened by ministerial indifference to the findings of the Royal Commission on the Private Manufacture of and Trading in Arms. Unlike several senior officials, Lord Swinton resolutely ignored the final report's recommendation that the Air Ministry adopt a more interventionist policy. With a long winter's day at Middlesex Guildhall a fast-fading memory, the Society of British Aircraft Manufacturers set up yet another committee. CRF was once more in charge, overseeing the appointment in late 1936 of a no-nonsense high-profile accountant to act as the SBAC's negotiator, lobbyist and spokesman. Sir William McClintock had clout and was widely respected in Whitehall. Within a month of his appointment he had already met the minister.[16] Meanwhile, Fairey Aviation's chairman and managing director complained that the profit margin sanctioned by the Air Ministry was intolerable given the industry's exceptional development costs and narrower customer base. Three years later, looking back on the last year of peacetime production, Fairey remained indignant that the state should exercise such a tight control over the level of profit. At the same time he continued to rail at the high level of corporate taxation. Yet Fairey Aviation's net profit for 1939 was £248,122, and shareholders found the rigours of the 'phoney war' eased by news of yet another 12.5 per cent dividend.[17]

Throughout the 1930s Richard Fairey's twin-track vision of volume production in both airframes and aero-engines ignored the Air Ministry's insistence on consolidation and rationalisation. Rearmament was driven by the Government concentrating airframe and aero-engine manufacturers in two distinct sectors, each with subsidiary specialist groups. If individual companies embraced fresh areas of R&D this meant an even greater delay between initiating a project and delivering it. Streamlining the process was vital, as under Scheme A for expansion of the RAF the Treasury would sign off 30 per cent more contracts than in the previous year. In July 1934 the Chancellor of the Exchequer had challenged the cautious forward projections of Lord Londonderry and the then CAS, Sir Edward Ellington. Neville Chamberlain secured cabinet agreement to rebalance tri-service procurement heavily in favour of the RAF, with the 52 squadrons recommended by the CID's Defence Requirements Sub-Committee more than doubled.[18]

November 1934 saw Sir Hugh Dowding – in his role as Air Member for Supply and Research considerably more ambitious than the CAS – brief Fairey and his fellow managing directors on an unflattering comparison with their overseas rivals' record of project management: the British were painfully slow in developing a prototype and tooling up for volume production. Dowding, a frequent house guest at Woodlands Park, accepted a need for the Air Ministry to simplify its technical specification, and urged that the industry pool its expertise. Fairey felt vindicated, and throughout the winter of 1934–5 the SBAC lobbied for officials to issue simple briefs, apply the light touch throughout the development stage, and scrap the Ministry's year-long testing at Martlesham Heath prior to agreement that a prototype was cleared for manufacture. Not that this signalled a dramatic transformation, as in 1935 fewer than 900 aircraft entered front-line service, only a third more than the previous year.[19]

Scheme C, intended to cover the period 1935–7, still set relatively modest targets, and yet the gap between planned and actual production continued to grow. A projected 50 per cent increase in annual output provided firms with certainty and stability, yet it largely ignored the lack of skilled labour and, with shadow factories as yet just an idea, the absence of an adequate infrastructure. High manpower costs forced greater investment in sophisticated machinery, in itself expensive and slow to install. Retooling and rejigging for a wide variety of aircraft meant the absence of uniform plant across the industry, and thus little opportunity to secure economies of scale.[20] Thus, ambitious targets were based on miscalculation of the man-hours and the sophistication of equipment necessary to build larger, more technologically challenging aircraft. This applied to both the medium bombers and the fast monoplane fighters ordered in March 1936 under the RAF's Scheme F. With hindsight, Bomber and Fighter Commands – each established in 1936 – benefited from the delay in commencing and then accelerating their re-equipment: just as the economy escaped premature disruption from the concentration of finite resources upon aircraft manufacture, so the RAF avoided dependence upon a stockpile of prematurely obsolescent aircraft in an era of unusually rapid aeronautical advance.[21]

Naturally there were exceptions, and, as we shall see, Fairey Aviation was one manufacturer whose front-line aircraft were seen by some as unfit for purpose at the advent of hostilities. Scheme F would prove profitable yet problematic for Fairey, the company's founder delighted that at last the Treasury saw a need for plentiful aircraft in reserve, totalling over twice the size of the RAF's front-line complement.[22]

Furthermore, the planned strength of the Fleet Air Arm was raised to around 500 aircraft; still miserably small, but the potential for further growth cheered board members at both Fairey and Blackburn. Overall, the group of manufacturers dedicated to naval air power had little to celebrate given the absence of projects to develop and build the next generation of carrier aircraft. Given that the Fleet Air Arm only regained full autonomy in 1938, the potential for Admiralty-driven R&D ahead of hostilities contrasted starkly with the hothouse conditions of 1912–14. Not that contractors such as Westland, Boulton & Paul, or even Blackburn, could necessarily rise to the challenge, witness the poor performance of those aircraft they did develop and produce. Their management shared the stubborn independence of a Dick Fairey, and the same instinctive suspicion of pragmatic solutions, but where they differed was in efficiency. Other than Fairey Aviation, the Fleet Air Arm's contractors were viewed within the Air Ministry as being fundamentally incompetent, hence a demand for changes at board level and below before contracts could be awarded.[23] In wartime these companies would prove invaluable as subcontractors, but they were seen as minor players when the Air Ministry launched its Shadow Scheme in 1936.[24]

Whitehall saw the Shadow Scheme as a means of ensuring that spare manufacturing capacity was readily available should international tension and the threat of war necessitate a rapid increase in the rate of rearmament. The scheme was publicly funded, and initially intended to boost aero-engine output. However, it soon embraced airframes, and, although shadow factories were state-owned and state-funded, car makers and plane makers ran them as if they were their own. The Treasury agonised over how assets might be split if a final triumph of deterrence and appeasement enabled plant closure, but the onset of war rendered any such discussion academic.[25]

Persistent production problems and a growing conviction that war with Germany was unavoidable prompted the Air Ministry in early 1939 to consider an extension of state control, and where necessary the establishment of one or more government-run factories. This was never a serious proposal, and in any case major motor manufacturers were by now key partners in a painful if ultimately productive relationship with the aircraft industry, witness

Austin Motors' troubled dealings with its impatient partner when tooling up to build the Fairey Battle. In the Commons, Moore-Brabazon – Richard Fairey's voice in Parliament – had already launched a pre-emptive strike against any further role for the 'Civil Service technician' in the design and manufacture of combat aircraft.[26]

Yet with war less than a year away, a variation on the Shadow Scheme was already firmly established: protected financially by the McClintock Agreement between the Air Ministry and the SBAC, the major aircraft manufacturers established fresh factories, often at considerable distance from their parent plants. Latecomers enjoyed a state-funded safety net, but Fairey Aviation had been in the vanguard of expansion, hence the need for a rights issue. As early as 1935 CRF and his fellow directors convinced institutional shareholders that the size of both current and anticipated contracts justified establishing a factory in north-west England. Output across the entire organisation was expected to treble once the new venture was fully operative. That autumn Fairey purchased a former vehicle manufacturing plant at Heaton Chapel, Stockport, and, under Barlow's supervision, commenced a major programme of redevelopment. In 1938 the new factory acquired a satellite plant, Errwood Park, with final assembly and flight-testing taking place at the grandly named Ringway Airport, near Wilmslow. The unveiling of Fairey's assembly plant in June 1937 was a much-publicised municipal event, followed soon after by a royal visit. The company shared Ringway with the RAF during World War II, but succeeding decades saw the site transformed into today's Manchester Airport.[27]

Accelerated rearmament generated an obvious tension. In the second half of the 1930s the RAF evolved a doctrine of effective air defence and strategic bombing which a Chamberlain-dominated administration had little option but to hope constituted a credible means of deterrence; and yet peacetime conditions made nonsense of any claim that Britain boasted the immediate ability to wage 'industrial war.' Scheme F set a target of April 1939 for the RAF to achieve its planned front-line strength, yet it remained the case that 'the creation of industrial capacity or wartime potential was a secondary consideration.' The process of mechanisation initiated by Scheme C had, however, reached critical mass, facilitating a significant increase in output and productivity as and when required. Japanese expansionism in the Far East, continued conflict in Spain and Abyssinia, and Italy signing the Anti-Comintern Pact, all signalled a heightening of international tension across the course of 1937. At the end of the year officials inside Adastral House calculated that total domestic output in the first year of any imminent conflict would be 7,080 aircraft, less than half the number demanded by the RAF; with a deficit of nearly 5 million square metres there was an urgent need to double manufacturing space in order to achieve inflated targets, and here the major car makers were crucial in realising the shadow factories' productive capacity. The Air Ministry's calculations were in support of Scheme J, an ambitious but ultimately abortive programme predicated on the Cabinet placing aircraft production on a quasi-wartime footing: front-line fighter strength would grow by an astonishing 90 per cent, with quality (Hurricanes and Spitfires) matching quantity.[28]

When Fairey met Swinton a month before he left office, he and the other manufacturers called for 'nothing short of a last whole-hearted effort to overtake the German production': at long last the moment had come to give the Reich 'something to think about.' *Anschluss* and the Sudeten crisis forced a shift in the Chamberlain Government's strategic priority from attaining a pre-determined combat capacity to providing the RAF with a 'War Potential' programme of steeply rising manufacture: 17,000 aircraft in the first year of a war commencing in October 1939, and a monthly output of 2,000 machines by December 1941.[29]

In May 1938, with war over the Sudetenland a very real possibility, Moore-Brabazon sat on the Government backbenches certain that the Cabinet seriously under-estimated 'the seriousness of the German technical achievement': listening to an upbeat Prime Minister in the Commons, he feared Chamberlain had been misled as to the superiority of the RAF over the *Luftwaffe*. Fairey agreed to provide a detailed but anonymous corrective which Moore-Brabazon could pass on to ministers.[30] By this time the two men and their wives had become extremely close, holidaying together on Fairey's motor yacht *Evadne* and pursuing mutual passions (namely golf, fishing, shooting, model railways and the search for the perfect electric razor).[31] Where the ultra-loyal Tory and the fiercely independent industrialist differed profoundly was in their view of Britain's future fighting potential, with Moore-Brabazon's support for appeasement rooted in deep pessimism and fealty to his party and its leader. Fairey believed that a concerted effort, with government and industry equal partners, could secure eventual air supremacy: while 'the Socialists' bore a heavy responsibility for delaying rearmament, it was the party of Baldwin and Chamberlain which in the 1920s had 'disbanded the greatest Air Force in the world and crippled the Industry without which it could not exist', and thereafter failed to plan strategically for volume production of high-performance combat aircraft. As late as May 1940 the staunchly Chamberlainite magazine *Truth* was blaming manufacturers, not ministers, for a supposedly under-strength RAF. This left Fairey incandescent over the 'libel' propagated by Joseph Ball, who controlled a weekly which before the war had been consistently pro-German.[32]

Moore-Brabazon may not on this occasion have agreed with *Truth*, but even after Chamberlain left Downing Street he refrained from any direct criticism of the National Government's prewar record. Such loyalty may explain a fierce argument with CRF at Bossington during the Battle of Britain. Next morning, when Moore-Brabazon failed to appear for breakfast, Fairey swiftly scribbled a message of reconciliation, 'because I have enjoyed and valued your friendship more than I can tell, and on no account would I have intentionally offended you.'[33]

The Prime Minister's return from the Munich conference in September 1938 signalled implementation of the suitably monickered Scheme M, an uninterrupted realisation of manufacturing potential, of which Fairey Aviation was a major beneficiary. Across the succeeding 12 months, investment, both private and public, in new and existing factories at last bore fruit. A unique concentration of machinery and manpower enabled British aircraft production to overtake that of Germany and continue to grow: throughout 1940, output – if not productivity – would significantly exceed the Reich's ambitious targets for re-equipping a *Luftwaffe* hit hard by offensive operations in Poland, France and Britain.

After Munich, manufacturers benefited from Whitehall planners' long-term projections, with war now seen as probable not possible. The Air Ministry's design, development and production divisions became more integrated, with heavy recruitment of suitably qualified staff to the newly designated Department of Development and Production; this fresh system for the coordination of procurement would provide the Ministry of Aircraft Production (MAP) with an obvious model in the late spring of 1940. From mid-1938, a vital decision-making body was the Air Council's subcommittee for supply, with SBAC chairman Sir Charles Bruce-Gardner a permanent member, and senior executives such as CRF regularly invited to attend. Instead of informal consultation, manufacturers now had a guaranteed voice. Fairey Aviation's ad hoc member on the Supply Committee took satisfaction in the key policymakers being military, not civilian personnel: CRF's correspondence with Dowding's successor, Sir Wilfred Freeman, confirms his healthy respect for an air marshal qualified to discuss experimental aeronautics while at the same time articulate a service perspective on problems of production.[34] After Lord Beaverbrook established the Ministry of Aircraft Production (MAP) in May 1940, Freeman sought a return to regular duties, but Sir Stafford Cripps made him MAP's chief executive in mid-1942, thereby restoring a close if uneasy working relationship with Richard Fairey, by then two years into his Washington posting at the British Air Commission.[35]

Freeman endorsed CRF's view that each firm's designers were still forced to interpret unduly complex and time-consuming technical specifications, and in 1938 he took action to simplify the Air Staff's requirements. The Air Marshal listened sympathetically when Fairey complained to the Supply Committee on 8 September 1938 that the slow progress of design and development across the industry was attributable to a shortage of draughtsmen and drawing office section leaders. On the cusp of rapid expansion, in 1935, Fairey Aviation had employed more than 200 draughtsmen, but over the next three years its principal drawing office at Hayes was unable to sustain the same density of technical expertise. The industry's expansion meant a serious lack of suitable staff, with the technical colleges unable to provide an adequate supply of trainee designers. This acute manpower shortage was compounded by British firms eschewing the Americans' project team system, and assuming their chief designers could supervise the work of the entire drawing office(s).[36]

Dowding believed the manufacturers should be left to devise their own solution to systemic problems of design, which helps explain his close acquaintance with Dick Fairey. Freeman, however, was not averse to pooling design expertise under a government umbrella, and perhaps even establishing a factory to build prototypes. These ideas were never realised, but as the de facto head of procurement his words carried weight when he warned the SBAC in September 1938 that scarce manpower should not be wasted working on unsolicited designs. Fairey doubtless appreciated the irony of this advice as the TSRII/Swordfish had been a project initiated without Air Ministry sanction.[37]

Fairey had a grudging respect for Freeman, but mixed feelings towards Sir Charles Bruce-Gardner. SBAC's spokesman was respected for his knowledge of business reorganisation, but he claimed no executive experience inside the industry. At a tense meeting with Air Ministry officials in December 1937, CRF and his fellow SBAC members had been told to accept an independent chairman, or face stricter government controls. Bruce-Gardner was naturally viewed with suspicion by Fairey, and the two men clashed over CRF's refusal, first, to fund expansion of the Hayes plant without state subsidy (added assembly lines came on stream following the declaration of war, but at the cost of much stricter government control), second, to redirect scarce managers and engineers to a second shadow factory on the far side of Stockport, third, to accept without protest government-imposed subcontracting, and, finally, to endorse the Harrogate programme of September 1939, which saw the SBAC cease to coordinate procurement negotiations, leaving firms to deal directly with Whitehall (believing the agreement favoured large companies like Hawker Siddeley, and that official production targets were unrealistic, Fairey felt vindicated when supply shortages slowed output in the winter of 1939–40).[38]

For the Royal Navy the glory of the Swordfish, for the RAF the misery of the Battle

The winter of 1939–40 was anything but a *drôle de guerre* for the ten squadrons flying Fairey Battles from their forward bases in north-eastern France. The RAF's hapless and short-lived Advanced Air Striking Force (AASF) saw well over 200 medium bombers shot down between deployment to France at the start of the war and withdrawal in June 1940. Many of the aircraft lost were Battles built at Heaton Chapel, or originating from Longbridge courtesy of Fairey Aviation's painful partnership with Austin. Poor productivity and low output plagued the west Midlands shadow factory, with Richard Fairey and his management blaming Herbert Austin and his management for assuming vehicle production methods could be applied to aircraft construction.[39] The car makers mistakenly believed they could bring the Battle into production faster than jig and tool specialists in Stockport. CRF saw the problem as systemic, rubbishing Austin's claim that on-site advice from Major Barlow would address the issue of consistently low productivity.[40] Lord Austin attended the opening of Ringway aerodrome, and inspected the Battle's initial assembly line, yet he clearly underestimated the complexity of the operation. *Flight* generously observed of the RAF's belated acquisition that 'The real cause for surprise is not that it should have taken so long, but that it has gone into production so quickly ... the amount of tooling necessary before there were any visible signs of progress was prodigious.'[41]

Over a year earlier CRF had realised Fairey Aviation was committed to an aircraft design fundamentally unsuited to volume production. Fairey had no grounds for complaint when the Air Ministry turned down the suggestion that Lobelle and his team design a stripped-down Battle, thereby facilitating 'a repetition system for really large quantities.' Here was a deeply challenging technical and organisational project, embracing as many as five assembly plants (three Fairey and two Austin, if including each company's Ringway and Elmdon aerodromes). It is scarcely surprising that productivity was adversely affected by a shortage of suitably skilled workers and a prickly relationship between management and unions, especially in Stockport. By the summer of 1937 there was a serious storage problem at Heaton Chapel, as the supply of components exceeded the speed of assembly down on the shop floor; yet, only a year later production at Fairey's northern outpost had improved markedly, with the workforce meeting and then exceeding its Air Ministry target.[42]

Initial problems with production were a consequence of the monocoque, monoplane Battle being over-engineered by the standards of 1935, and yet such was the speed of prewar aeronautical development that when finally the single-engined, three-man bomber entered service its modest speed, poor cockpit protection, haphazard fuel system, and sparse armament together rendered the aircraft highly vulnerable.[43] The Battle's Merlin I engine was not only slow to come on stream, with the Hurricane and then the Spitfire claiming priority, but it was demonstrably inadequate for such a heavy payload. When Rolls Royce supplied later versions of the Merlin the improvement in overall performance was marginal, with the Messerschmitt Bf109 still almost 100 mph faster at operational height. The high attrition rate experienced during the 'phoney war' compared favourably with the losses sustained by bomber squadrons heroically carrying out suicidal low-level daylight attacks on German columns in May and June 1940.[44]

The *Luftwaffe*'s fighter pilots decimated the AASF and, following the fall of France, British and Polish squadrons flew only a handful of sorties before the Battle was withdrawn from front-line duties. In the Horn of Africa and the eastern Mediterranean, British, South African and Greek squadrons flew less costly missions between August 1940 and April 1941, after which the Battle found itself in a variety of non-combat roles, from training to tug-towing. The problem for Allied air forces, especially the RAF, lay in finding a use for so many Battles, notwithstanding the numbers lost trying to stem the German invasion. Between them Fairey and Austin built an astonishing 2,200, of which around half made their way to Canada and Australia as part of the Commonwealth Air Training Plan (CAT).[45]

Given the relatively small numbers sold to friendly foreign powers the RAF was responsible for maintaining in large numbers a machine which, however much it impressed senior staff in the mid-1930s, was demonstrably obsolescent by the end of the decade. There is clear evidence that the Battle quickly became a 'stop gap' aircraft, with the Air Ministry acceding to CRF's request that production continue well into the war so as not to lose suitably skilled workers ahead of replacement programmes, notably subcontracted work building heavy bombers at Stockport. At Hayes, additional contracts for the manufacture of Fulmars and Albacores were anticipated (with plant conversion from late 1940 to production of the Fleet Air Arm's main midwar fighter and strike aircraft, the Firefly and Barracuda). Evidence to support the unnecessary continuation of Battle production is the Air Ministry in 1938 reversing its previous decision to cut back production, CRF's bargaining with procurement officials inside the Air Ministry, and the CAT's heavy dependence upon an aircraft wholly unsuited to serve as a first-choice trainer – twin-cockpit dual-control versions only appeared on the eve of war, largely built at Longbridge.[46]

The history of the Fairey Battle is a sorry story, contrasting starkly with that of the Swordfish. Both machines were out-dated by the time they commenced front-line duties, but only one overcame its unsuitability to prevailing conditions. At a moment in aeronautical development when piston engine aircraft were advancing rapidly towards maximising their performance potential, the Swordfish ran wholly against the grain. It enjoyed an extended period of grace, terminated abruptly with 825 Squadron's destruction over the Channel on 12 February 1942. The Battle enjoyed no such good fortune, its baptism of fire above the Ardennes in practice a serial cremation at the hands of the same fast flying Jagdgeschwader that two years later would expose the Swordfish's operational shortcomings. No mythology survives the Fairey Battle. It holds no special place in Britain's collective memory, let alone – like the Swordfish – our imaginary pantheon of genuinely iconic aircraft. Unsurprisingly, the ratio of surviving Swordfish to the Battle is more than three to one.

Yet the Battle was as much Dick Fairey's creation as the Swordfish – it benefited from that same drive, energy, determination and quiet ruthlessness, and it shared the same input from Lobelle, Broadbent, Barlow, Staniland and the myriad other pillars of Fairey Aviation who day after day realised their leader's wishes. The irony is that, for all its obvious failings – evident even as it entered service – in commercial terms the Battle was a striking success. What appeared at the outset to be financially a rare miscalculation proved in the long term a major revenue earner, confirming just how much profit the company could enjoy if long production runs drove down unit costs. Nor was the Battle unique by being measured in thousands not hundreds, as proved to be the case for all Fairey's wartime aircraft, notably the Barracuda.

Taking on the Americans – war kills off Fairey's state-of-the-art airliner

Manufacturing combat aircraft in large numbers, with the expectation of future orders on a similar scale, enabled cross-subsidy. Thus, one consequence of the Battle remaining in production for so long was the freeing up of funds to finance Fairey Aviation's first non-military R&D programme. CRF was a man with strong views on the provision of medium and long-haul passenger/mail services, objecting strongly to the near monopoly and the government subsidy enjoyed by Imperial Airways, which later became BOAC.[47] He clearly felt this position was unsustainable in the medium to long term, and Fairey was not alone in anticipating that privatised airlines would require multi-engined monoplanes. This view chimed with the Directorate of Civil Aviation's concern that Britain lagged far behind the United States in the development and production of land-based passenger aircraft: urgently required was an ambitious programme comparable to Shorts' supply of 42 C-Class Empire flying boats to Imperial Airways and its associate airlines.[48]

Even if CRF and Whitehall officials anticipated war, planning for peacetime provision could not remain in stasis – the alternative scenarios had to be that of conflict in Europe being avoided, or in the event of war the British Empire ultimately emerging victorious. In either scenario the long-term demand for military aircraft would decrease dramatically, as, of course, would be the case should Britain experience defeat at the hands of the Axis. Given the Air Ministry's desire that a short- to medium-haul airliner enter production no later than December 1940, and the Treasury's readiness to support the building of a prototype, Fairey Aviation had a commercial incentive to tender for the contract. Hard bargaining throughout the second half of 1938 saw the company confirmed three months later as the preferred contractor. Fairey was given a guaranteed profit of 10 per cent on two prototypes costed at £78,000 each, and the assurance that at least 12 airliners would enter service with BOAC. This initial order was seen as a precursor to volume sales both sides of the Atlantic, with CRF looking to build over 100 airliners. For all the rumours in 1945 of resurrecting the project, the FC1 (Fairey Commercial Number One) became the great what if – on 17 October 1939 the Air Ministry, for obvious reasons, cancelled the contract.[49]

Anyone visiting the experimental shop at Hayes in the summer of 1939 would have been surprised to discover a well advanced and remarkably complex mock-up of Fairey Aviation's first civil airliner. Especially striking was the size of the aircraft given the FC1's modest passenger load. Even more impressive was the level of technological sophistication. Here was a hi-tech project intended to match the latest advances in speed and comfort achieved by American manufacturers such as Douglas and Lockheed. The flight-deck simulator confirmed that this was a safety-conscious design with autopilot, pulsating de-icers, beam-based navigation and instrument landing, back-up power systems, a fail-safe pressurised cabin facility, strict temperature and noise controls, fuel jettisoning provision, tailwheel availability for runways unsuited to a retractable tricycle undercarriage, and a planned four-piece power unit of 24-cylinder sleeve-valve pressure air-cooled engines, which together would generate an impressive 4,800 horsepower. For over a year prior to cancellation of the project, Fairey Aviation's operations manager, A.C. Campbell Orde, liaised closely with his counterpart at Imperial Airways/BOAC. This was a genuine collaboration, predicated on building an aircraft that embodied the very best in British engineering, and which would project BOAC as a global competitor to the likes of Pan Am.[50]

There was a veto on components suppliers from overseas, although the cantilever low-wing, stressed-skin structure clearly drew on the pioneering work of American designers earlier in the decade. At first sight the same seemed true of the fuel-efficient flap system, but the revolutionary Fairey-Youngman flap – quickly resurrected for use on the Barracuda and the Firefly – was a simple device that owed much to CRF's obsession with wing design early in his career. In the FC1's wing design could be seen the direct involvement of the managing director, Fairey gaining quiet satisfaction from the favourable data generated inside the Hayes wind tunnel and its counterpart at the National Physical Laboratory: as his personal papers confirm, despite his wide range of commitments he still received all test results as soon as they came back from analysis. On 20 February 1941, only five days after the fall of Singapore, a remarkably optimistic report in *Flight* magazine observed that the airliner project 'was unavoidably shelved by the outbreak of war, and one may reasonably expect that the Fairey company will keep it up to date, so that when we are at peace once more the production of a modernised F.C.I will be forthcoming with a minimum of delay.'[51] Sadly, this was not to be, and Sir Richard Fairey could never relish the satisfaction of cruising at 15,000 feet – and at well over 200 mph – in an aircraft of his own invention.

Life beyond the boardroom – out on the water

> There were those who scoffed at the new-fangled ideas of these 'flying millionaires', saying that Fairey and Sopwith between them were making the sport too expensive. That was nonsense, for as Dick once said to me: 'If a fellow starts asking how much it costs to go yachting, the answer is that he can't afford it!'[52]

By February 1941 Fairey was familiar with the luxury of an Empire or Clipper flying boat, if not the super-streamlined, inter-continental airliners rolling off the Douglas, Boeing and Lockheed assembly lines. A Douglas Skymaster would take him to New York in June 1943, but otherwise flying long haul on

American leviathans proved a postwar luxury; back home there is no evidence that the late 1940s saw CRF fly on board the only completed Brabazon, Bristol's eight-engined failure to beat the Americans at their own game. With the return of peace, Sir Richard and Lady Fairey regularly crossed the Atlantic courtesy of the great ocean liners. After all, from the mid-1930s, luxurious sea journeys had become a familiar experience for a young family with high expectations – and an enviable lifestyle made possible by a resurgent Fairey Aviation's success in riding depression and disarmament. Tommy Sopwith's challenge for the America's Cup in September 1934 saw CRF sail out on the *Olympic* and home on the *Aquitania*, his preferred choice of ships in the spring of 1937 when accompanying Richard to America for the first time.[53]

Fairey travelled to Newport in the late summer of 1934 so that every evening he could test Moore-Brabazon's golfing prowess on the links of Rhode Island, and every morning he could cheer on the crew of the giant J-class *Endeavour*.[54] When racing the New York Yacht Club (NYYC)'s *Rainbow* for the prize of sailing's most prestigious trophy, Sopwith flew the pennant of the Royal Yacht Squadron (RYS). He had been elected an RYS member in 1930, having dominated 12-metre racing throughout the second half of the 1920s. As with his super-luxurious motor yachts, Sopwith had commissioned the speeding sloop *Mouette* from the Solent's premier boat builder, Camper and Nicholsons; the yacht was built to the complex rating rule for 12-metre yachts which operated from 1920 to 1933, and which Fairey found such a challenge once he became serious about sailing.[55]

The driving force behind Camper and Nicholsons was the veteran Charles Nicholson, an engineer who both Sopwith and Fairey considered every inch their equal. Remarkably, *Endeavour* was the third J-class built in his shipyard. In creating a genuine contender for the America's Cup, Nicholson's design team worked closely with Hawker's Frank Murdoch, born into a Belgian boatbuilding family and as keen a competitor as his employer. Knowledge transfer was especially evident in the use of wind tunnel testing to determine the most suitable sail settings, and here Murdoch's racing expertise reinforced his engineering credentials. Nicholson eschewed a radical design for *Endeavour*'s steel hull and superstructure, but Sopwith and Murdoch developed a fresh, American-style combination of mast, rigging and winches. Murdoch invented several new devices, notably an electronic wind direction indicator. Fairey enthused over these inventions, but sceptics later maintained that Sopwith relied too much on new technology, at the expense of his natural flair and instinctive seamanship.[56]

Like his fellow plane makers, Tommy Sopwith, Oswald Short and, across the Channel, Louis Breguet, CRF recognised how much hydrodynamic testing could draw on the theory and practice of aeronautics. If one held sufficient power and influence then the opportunity might arise to gain advantage by rewriting the rule book, and remarkably early in his sailing career Fairey endeavoured to do this. Normally, however, 'The designer's legitimate object is to defeat the Rule and ... he can only be successful by applying scientific means, the resources of artistic variation having been long since absorbed.'[57] These were the words of the now veteran racer, 12 years after he first took to the water. By that time CRF and Sopwith had long since been seen as uniquely qualified to take the helm of the world's fastest yachts. In November 1933 *Yachting World* deemed both men ideally suited to contest the America's Cup: 'They know by instinct the feel of the helm to the hand, they readily realise the effect of the wind on the sails, and their minds react to the sensitive observation of the trim of the sheets. They are able to descry and foresee the effects of tacks and manoeuvres. Airmen, it seems, make good seamen in racing, and particularly in the finer arts of racing ... These aeronautical experts have taught us an immense amount about sailing, and if we have any sense we shall go on learning.'[58]

Back in 1927 Fairey's enthusiasm for golf and field sports remained as great as ever, but sailing offered a fresh physical – and intellectual – challenge. That spring he bought his first 12-metre boat, the slow, sturdy and suitably named *Modesty*. Brice Slater was instructed to assume responsibility for all matters maritime, and he continued in this increasingly demanding role until the onset of war. For the sake of convenience Slater secured a permanent mooring on the Hamble. Equally convenient was the presence of the Royal London Yacht Club across the Solent in Cowes, and the residential Royal Thames Yacht Club in Knightsbridge. Fairey joined both august institutions. At the RTYC's sumptuous clubhouse overlooking Hyde Park he found the Coffee Room ideal for lunchtime appointments, and the Britannia Bar perfectly suited to more private business meetings. Meanwhile, on the Isle of Wight, the RLYC's newest member soon found himself serving on the general committee; remarkably, by 1935 he was commodore. Back in the summer of 1927 Fairey confined himself to racing *Modesty* in regattas from Deauville to Devon, along the way staying in the finest coastal hotels and for the first time becoming fully acquainted with friends in high places. Elite sailing was hugely expensive and thus the preserve of the few. Highly fashionable in the interwar years, racing the world's largest yachts secured the approval of high society either side of the Atlantic.[59]

Fairey was invariably present at the most prestigious races, and their accompanying social events: he sailed every weekend in May and June, and dedicated the summer months to regattas in the Channel ports and Devon, proceeding westwards via Cowes Week at the start of August. With the Royal London responsible for the first two days racing in the Solent, it was necessary to rent, as a family base, Pier House, the insurance magnate Sir Edward Mountain's summer residence on the Isle of Wight.[60] He used his motor yacht, *Evadne*, commissioned from Camper and Nicholsons in 1931, as a floating guest house, a support vessel for his yachts, and a second office. Whenever at sea or in port, CRF maintained regular contact with Miss Burns, via cable, telegram, ship-to-shore telephone or ciphered correspondence (all directors were conversant with Fairey Aviation's elaborate coding system); his secretary sent him a daily business report, and senior staff often travelled a considerable distance to brief the chairman on company business. This arrangement worked so well that year on year *Evadne* undertook ever longer tours. For example, in the late summer and early autumn of 1937, Fairey and his second family spent seven weeks cruising around the Hebrides and across the Pentland Firth to Orkney; Oban served as a home port, and the west Scotland postal service was kept busy handling a torrent of telegrams as CRF liaised daily with fellow directors and domestic staff. Home or abroad, *Evadne* was umbilically tied to Hayes, Hamble and Heaton Chapel.[61]

The twin-engined *Evadne* was, by the standards of the day, a big boat, measuring 193 feet in length and 26.5 feet in beam. Inspired by *Vita I* and *Vita II*, his earlier creations for Sopwith, Charles Nicholson designed a suitably large and luxurious cruiser. It carried a dozen guests, serviced and supported by at least as many crew members. In the bowels of the ship lay an impressive array of wines, selected with great care by Barry Neame, who ran the Hinds Head Hotel in Bray, near Maidenhead. Neame had acquired the hotel from Kitty Henry, a former nightclub owner in London. In 1932 he persuaded Richard Fairey to take a major shareholding in the hotel, enabling him to buy a neighbouring cottage and expand. Eight years later Neame's love of the high life, and a distinctly casual approach to settling bills, saw his principal creditor take over the business. Retained as manager, Neame was soon replaced by his secretary, who stayed in charge until 1968. Several owners later, the hotel was acquired in October 2004 by Bray's celebrity chef Heston Blumenthal, owner of the neighbouring Michelin three-star star restaurant, The Fat Duck. Throughout the 1930s and late 1940s the Hinds Head Hotel served as an ideal location for Fairey family functions not held in London or aboard *Evadne*. Fairey kept his luxury yacht at sea for much of the year, paying her captain, first officer and chief steward permanent retainers. The owner wanted value for money, and, although every refit embraced extensive modification, he avoided the extravagant Sopwith's habit of ordering bigger, ever more salubrious boats.[62]

The mid-thirties saw *Evadne* play a major role in her owner's hectic lifestyle. 1936 was an unusually busy year for the motor yacht's crew. Spring saw a

Mediterranean cruise to hasten Richard junior's recovery from a serious spinal injury, and a Whitsun weekend in Deauville where the Fairey family were guests of fellow plane maker and yachtsman Louis Breguet, by this time a close friend.[63] In late June *Evadne* made her third trip to the Baltic, for the annual Kiel Week, a prelude to the Olympic regatta five weeks later. As two years earlier, this entailed a two-way journey through the Kiel Canal, or, as it was still known then, the Kaiser-Wilhelm-Kanal. Fairey was sensitive to the fact that, unlike Sopwith, his racing success had failed to secure his election as a member of the RYS – in January 1936 the German Embassy forwarded to Schleswig-Holstein his request that the canal authorities accord yachts of the Royal London and the Royal Thames the same special status as those sailing under the colours of the RYS.[64]

Unsolicited recognition came with Fairey's royal invitation to attend George VI's opening of the National Maritime Museum on 27 April 1937 – *Evadne*'s unsuitability saw Miss Burns hire the ill-fated river launch, S/L *Marchioness*, to carry guests from Kew to Greenwich. Off Spithead a month later, both *Evadne*, fresh from a fishing holiday off the Irish coast, and the 12-metre yacht *Evaine* formed part of the Coronation Naval Review, their owner again enjoying special status. Fairey's principal guest that day was record breaker Sir Malcolm Campbell, who reciprocated by inviting his host to a large 'Anglo American Party' for the then US ambassador, Robert Bingham. By 18 June *Evadne* was in Torbay, where every evening for the duration of the Coronation Regatta – much of which he had masterminded – Fairey invited a succession of VIPs to dinner; guests ranged from the head of ICI to the heir to the Norwegian throne, with an ever present Ernest Brown providing a direct line to Downing Street. CRF basked in celebrity status, having broadcast live from Torquay a personal view of the Coronation Regatta and the state of competitive yachting at the start of a new reign. It was only ten years since he had acquired his first boat, and yet in 1937 Fairey was the BBC's first choice when the Home Service sought a speaker of substance to address the nation on behalf of the sailing establishment. One wonders what CRF made of the young man who oversaw the rehearsals for his talk at Hayes and Broadcasting House – Guy Burgess was already well practised in disguising beneath a veneer of conviviality his complete contempt for capitalist plutocrats.[65]

Arriving at international regattas by motor yacht, enjoying royal approval, and being toasted as club commodore and championship contender, was a scenario Fairey could scarcely have imagined back in 1927. Well qualified to address issues of air flow and stress, he established test facilities at Hayes to calculate how rerigging and a reshaped hull would allow *Modesty* to lose weight

1. The Fairey family, late 1890s.

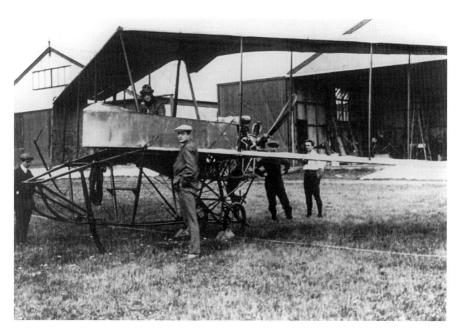

2. CRF with J.W. Dunne's D8 prototype.

3. Dick Fairey with the model aircraft he sold to Gamage's for production, 1911.

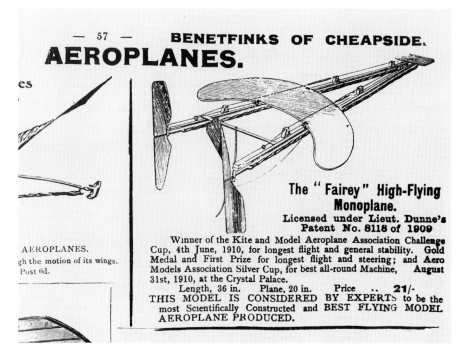

4. Glider model advert, 1911.

5. Lord Brabazon of Tara, 1908.

6. Murray Sueter, founding father of the Royal Naval Air Service.

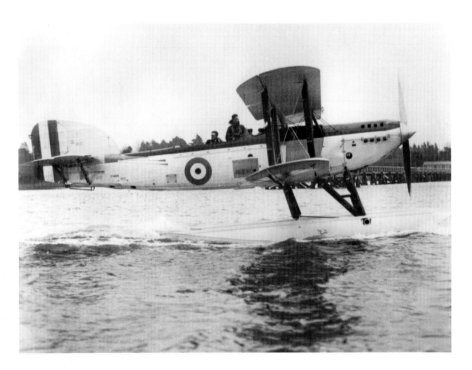

7. Fairey IIIF, mainstay of interwar sales.

8. CRF in the 1920s, company founder, chairman, and managing director.

9. CRF and fellow yachtsman and aviation industrialist, Louis Charles Breguet.

10. Richard and Esther Fairey, July 1935.

11. Fairey Swordfish, Fleet Air Arm icon.

12. Lord Halifax chairs a meeting of British and French mission heads in wartime Washington.

13. Postwar shooting party (Richard Fairey behind his father and Lord Brabazon to CRF's left).

14. Sir Richard and US Ambassador Lew Douglas at Bossington, 1947 or 1948.

15. Fairey Delta 2, world air speed record breaker, 1956.

16. CRF inspection of FAA Flight703X, April 1954.

and gain speed. Observing Sopwith at the helm, a single-minded Dick Fairey sought to match his rival's formidably high standard of skill and seamanship. Having drafted in the technical staff required to rebuild and maintain *Modesty*, the fledgling helmsman acquired a crew of canny old salts and fit young yachtsmen eager to gain experience sailing a big boat. Tutored by his skipper, the redoubtable Herbert 'Dutch' Diaper, the first-time owner learnt fast, and with growing confidence commissioned Charles Nicholson to build a seriously fast Bermuda cutter. *Flica* was a state-of-the-art yacht, which, after a quietly impressive start in 1929 and mixed fortunes in 1930, dominated British and European 12-metre racing from 1931 until the rule rating was revised at the end of the 1933 season. In the winter of 1932–3 *Flica* underwent a radical rebuild, and the following summer won or was placed in all but six of the 39 races she entered.[66]

Yet still *Flica*'s master wasn't satisfied. From March to May 1933, daily wind tunnel tests produced eight bulky reports, which together recommended a radical revision of *Flica*'s sail settings; close attention was given to how far Slater's riggers could exploit loopholes in the regulations without risking disqualification. Acutely aware of his position – he was now a senior committee member at both of Britain's oldest and most distinguished sailing clubs – Fairey could not risk any suspicion, however slight, of having bent the rules. Thus no fundamental changes were made, although the use of duraluminium in Fairey aircraft inspired the adoption of lighter yet sturdier spars. A further innovation was the quadrilateral jib, soon used to good effect on the J-class yachts *Shamrock V* and *Endeavour*. Fairey and Diaper saw the sail as their invention, a claim challenged by Charles Nicholson's son, John.[67] Such stringent testing meant Fairey began the 1933 season with contingency plans, rooted in hard empirical evidence, for every conceivable combination of wind and tide; whatever the attraction of racing under sail, it clearly was not the romance of the sea.

Fairey spent July 1933 competing in the Baltic, at Hanko winning the Norwegian royal family's Jubileums Regatta; throughout the Scandinavian cruise *Evadne* served as *Flica*'s support boat, performing a similar role at home. Back in England, Fairey, Diaper and their hand-picked crew again secured the championship pennant, winning or coming second in nearly 90 per cent of her races: *Yachting World* declared *Flica*'s owner 'in the front rank of British helmsmen.'[68] Almost certainly it was in Norway that CRF first met Crown Prince Olaf, a keen yachtsman who became a regular visitor and guest crewman across the following six years.

Meanwhile, encouraged by the test results generated in the Hayes wind tunnel, Fairey initiated the first of two attempts to contest the America's Cup in a boat smaller than the NYYC's preferred J-class. Throughout the first half of 1933 Fairey lobbied the International Yacht Racing Union (IYRC) to put pressure on the Americans to accept a British challenge to race 12-metre yachts. CRF was convinced *Flica* could triumph in any such contest, and, according to *Yachting World*'s New York correspondent that spring, the Americans thought so too. Nor would the revision of the 12-metre formula from 1934 have stopped Fairey, as he ordered Charles Nicholson to modify *Flica*'s lead keel to conform to the new regulation. This instruction was never implemented, and neither was Fairey's call in late 1933 for Camper and Nicholsons to build him a 'third rule' successor to *Flica*. Fairey's ambitious plan to force the NYYC to compete on his terms was thwarted in October 1933 when the IYRU's international conference agreed on immediate adoption of the 'third rule.' The following day the same body recognised that the NYYC was bound by a deed of gift for the America's Cup to accept the challenge mounted by the RYS and its J-class yacht, *Endeavour*. In December 1933, Fairey bounced back from Tommy Sopwith's success at his expense by challenging the North American Yacht Racing Union to contest the Twelve Metre International Cup. The trophy dated back to 1930 but had proved a victim of the Depression: year on year, competition for the cup had been deemed too costly. A lack of enthusiasm for Fairey's challenge in sailing circles both sides of the Atlantic prompted the sale of *Flica* only a few weeks later.[69] One wonders if longstanding, experienced yachtsmen in New York saw Fairey as an *arriviste* with a lot of money but scant respect for tradition. No doubt there were older members in the clubhouses of Cowes and Knightsbridge who shared similar feelings.

With the advent of the 'third rule' in 1934, Scottish shipbuilder Arthur Connell bought *Westra*, a bespoke 12-metre racer designed by Nicholson to the new specification. *Westra* proved a winner, but CRF was certain he could make her go faster. To Fairey's intense frustration her owner refused to sell, and, to add insult to injury, after a third successful season he had the boat mothballed.[70] Having sold *Flica*, Fairey again looked to Camper and Nicholsons. The resourceful and ever reliable Charles Nicholson sold him the half-built but unwanted *Evaine*, ensuring her completion in time to chase *Westra* hard across the 1936 season.[71]

The following two seasons are regarded by the sailing *cognoscenti* as a high point of 12-metre racing, with Vernon MacAndrew's *Trivia* in both years edging out *Evaine* as overall title winner. Fairey's poor health meant a passing of the helm to Bob Garnham, born into a distinguished sailing family and a future vice commodore of the Royal Thames. With Hayes wholly focused upon rearmament contracts, wind tunnel and test tank experiments were now a serious distraction.

The task of making *Evaine* sail faster therefore fell to a specialist team in Lymington. Not that the yacht fell out of the limelight, and *Picture Post* featured it in an impressive spread on 28 June 1939. With the prospect of war rendering a regular crew hard to maintain, *Evaine* managed only fifth in the 1939 championship ratings. At the start of the year her owner had deemed imminent war an adequate reason not to race, but in early May he changed his mind after Ernest Brown passed on Lord Halifax's fear of public alarm should newspapers report the absence of both Fairey and Sopwith from the new season's roster: the Foreign Secretary was eager for the millionaire Harold Vanderbilt to ship his 12-metre yacht across the Atlantic, and he would only do so if the best boats were in competition. In the interest of maintaining calm and an impression of normality the two aviators agreed to race as normal.[72] Although permanently out of the water once war was declared, the mighty *Evaine* remained in the Fairey family until Sir Richard's death in 1956, after which she enjoyed an Indian summer serving as a 'trial horse' for the next futile effort at wresting the America's Cup away from the firm grip of the NYYC.[73]

While *Evaine*'s contribution to the war effort was psychological – a tangible reminder that better days would one day return – her watchful companion returned to Gosport for adaptation to ASW (anti-submarine warfare) duties. Fairey suggested better use could be made of luxury cruisers the size of *Evadne* by selling them for dollars in the United States and then compensating the owners in sterling. Both Churchill, by now back at the Admiralty, and the Chancellor, Sir John Simon, took this suggestion seriously, but Treasury officials calculated that requisition of comparable vessels would cost more than any profit made from selling off the yachts.[74]

Meanwhile, CRF complained that the *Evadne*'s long-serving captain, George Courtman, had been replaced by her first officer. Unlike his number one, Courtman lacked the master's certificate to be commissioned in the Royal Naval Reserve at an appropriate level of command. Fairey took a lot of convincing, but finally he backed down. Consequently, a mollified owner maintained regular correspondence with the new commanding officer: Lieutenant H.N. Taylor RNR continued, in a suitably deferential and reassuring manner, to brief his former employer on how HMS *Evadne* and her new crew were being made ready for war, courtesy of machine guns, depth charges and a four-pounder gun.[75] On 13 December 1939, with *Evadne* moored on the Mersey, a relaxed CRF congratulated Taylor on how well he had handled a difficult situation: 'I gather she is a very happy ship, doing good service, with officers held in high esteem. I understand you will shortly be going into dry dock but at the first opportunity when you are in commission again I hope to be able to look you up.'[76]

This intention was overtaken by events, with Fairey absent in America while *Evadne* escorted convoys from Liverpool to assembly areas in the Western Approaches. However, later in the war Sir Richard could periodically catch up with his beloved boat.[77] It is hard to believe that an Admiralty decision to station *Evadne* in Bermuda was coincidental: throughout 1943 and 1944 her daily task was to patrol a reef line beyond the main channel into the dockyard. CRF increasingly saw Bermuda as a second home, with its naval base a suitable location for HMS *Evadne*'s decommissioning at the end of the war. Three months prior to the German surrender *Evadne* was on escort duty the far side of the south Atlantic when it identified an enemy submarine: its depth charge attack left the U-300 crippled and led to her sinking three days later.[78]

Evadne was returned to a man whose reputation now rested on his readiness in 1940 to serve his country in the best way he could, by hard bargaining with the Americans to secure desperately needed aeroplanes. This contrasted sharply with Fairey's image ten years earlier, when more and more he was seen as a self-made man intent on fulfilling fresh ambitions. In Fleet Street the *Daily Express* was not alone in profiling the boss of Fairey Aviation as 'yachtsman first and air designer second': readers were reliably informed that for the 'builder of the most beautiful fighting airplanes in Britain … yachting is his passion.' Considered 'a probable British challenger for the America's Cup', by the mid-1930s Fairey had long since emerged from the shadow of his fellow plane maker.[79]

While *Endeavour* was under construction in 1933–4 Sopwith had purchased the previous challenger for the America's Cup, *Shamrock V*. He and his full-time crew spent weeks out on the Solent learning how to maximise the potential of a J-class yacht, for which the main sail matched a mast over 150 feet high; meanwhile, Frank Murdoch was able to work out what warranted replication on *Endeavour* and what clearly did not. With Camper and Nicholsons' sleek, shapely but phenomenally heavy creation freshly fitted out and ready to race, CRF asked Sopwith if he could buy *Shamrock V* from him. Having bought the boat he set a team to work, identifying how modification of its giant rigging could generate greater speed and manoeuvrability. Changes undertaken ahead of Cowes Week saw the renamed *Shamrock* dominate the second half the 1934 season, and Fairey ended the year having helmed his big beast to victory in over 20 races.[80]

The 1934 America's Cup proved a controversial contest, and, although Sopwith lost 13 of his most experienced sailors in a walk-out over wages, *Endeavour*'s scratch crew came remarkably close to bringing the trophy home. Eight days before leaving for Newport the professional yachtsmen had sought remuneration for working abroad and a share of any prize money. A latent hostility to paid hands within the sailing establishment saw magazines such as the *Yachtsman* label the strike a shameful mutiny. Fairey and Charles Nicholson both urged Sopwith to sack the strikers and create a wholly amateur crew, drawn from *Shamrock* and the Royal Corinthian Club. CRF claimed that the new recruits had more recent experience of racing a J-class yacht than the old salts initially signed up to man *Endeavour*. This was certainly the case for the amateur apprentices being trained up on *Shamrock*, but the ten other crewmen were on a steep learning curve. Young, fit, and enthusiastic, these dinghy sailors were proven winners, as demonstrated by the presence of an Olympic gold medallist. They learnt fast, and shared Frank Murdoch's enthusiasm for ceaseless invention, in this case devising a new spinnaker drill. In a close if contentious competition *Endeavour* failed in its challenge, and yet responsibility lay not with the amateurs but their senior shipmates, notably the navigator; also, Tommy Sopwith was a brilliant helmsman, but he displayed poor judgement in inviting his wife and Charles Nicholson to form part of the crew.[81]

Having come so near in 1934, Sopwith returned to Newport three years later; but at 160 tons *Endeavour II* was a heavier build than Nicholson's original design, and this time the technology was demonstrably inferior to that of the Americans' all-conquering boat.[82] This was the last occasion on which J-class yachts competed for the America's Cup, but CRF had already decided to sell *Shamrock*. Funding a 22-man crew for the season was a serious financial drain, even for someone as well off as Richard Fairey – this after all was someone maintaining not one but two large yachts, and a hugely expensive 'mother ship.' In any case, driving a boat as big as *Shamrock* proved far less fun than pushing a 12-metre to the limit. For Fairey the future – including the America's Cup – still lay in less demanding, more responsive yachts, with the likes of *Westra* and *Evaine* ideally suited to international competition at the highest level.[83]

Sopwith was able to mount a second challenge for the America's Cup in 1937 because his rival had once again failed to convince the NYYC that the great days of the J-class were over. An end-of-season celebration in 1935 saw Fairey inform his fellow diners that 'the Js were too large, too expensive, and not fast enough, and that they were a relic of the past when owners gave their guests hospitality which was beyond the means of any modern yachtsman.'[84] This was ironic given the experience of those invited to sail with CRF. Over 40 years later, Royal Corinthian yachtsman Beecher Moore – the only American on board *Endeavour* in 1934 – recalled his apprenticeship racing *Shamrock*, a boat 'very much like a well-run country house in that the gentleman does not go into the kitchen and on a well-run J-Class the owner does not go forward of the mast.'[85] Beecher Moore, in later life a successful boatbuilder and sailing grandee, was highly critical of his patron's seamanship. A trophy-laden dinghy racer, the ex-pat Ivy Leaguer saw himself as a professional, and the likes of Dick Fairey a gentleman amateur. Needless to say, Dick Fairey saw himself, and his commitment to competitive sailing, in a very different light.

In 1935, partly to circumvent any problem with the deed of gift, he urged the American sailing establishment and the IYRU to accept his proposal, supported by the Royal Yacht Association, to race in the smaller size 'K class.' At around 75 feet in waterline length, the few K-class boats in existence weighed 50 tons lighter than the J-class and carried two-thirds the sail area. The difference between the established and embryonic classes was not that great. This meant reformers within the yachting fraternity were less inclined to support CRF when he faced fierce opposition from vocal traditionalists in high places. These fierce critics of Fairey urged Tommy Sopwith to build a new J-class boat and to mount a fresh challenge for the 'Auld Mug.' Nevertheless, in his formal capacity as commodore of the RLYC, Fairey did issue a challenge to the NYYC. He called on the Americans to mount a wholly fresh competition, with each side racing a brand new, internationally recognised K-class boat. At face-to-face meetings in New York he was again politely rebuffed. Had he issued a challenge with the intent of sailing *Shamrock* or another J-class boat then the NYYC would presumably have accepted, given its record of never declining a contest. Instead, the NYCC insisted that any future race must be between 'yachts of the largest and fastest class racing at the time.' The Club argued that the global political and economic situation was not conducive to investment in a fledgling class, and in any case it declined to host a challenge ahead of a presidential election. This left the field free for Sopwith's freshly commissioned *Endeavour II* to cross the Atlantic in the late spring of 1937.[86]

Fairey's many critics were unaware that Tommy Sopwith had previously entered a gentlemen's agreement with his rival. Sopwith accepted that, having mounted a challenge in 1934, his rival should be next to take on the Americans: he would only compete for a second time in RYS colours if the commodore of the RLYC failed to secure a contest on his own terms, namely racing the still unproven K-class yachts. This had proved to be the case, hence the over-engineered *Endeavour II*'s ill-fated enterprise off Rhode Island in the summer of 1937.[87] What is striking is the spirit in which dealings between Fairey and Sopwith, and between Fairey and representatives of the NYYC, were conducted. It seems to have been a refreshingly polite and mutually respectful set of negotiations, and Fairey went to great pains in newspaper interviews and private correspondence to emphasise the absence of ill will at every stage of a process which, for him personally, ended in great disappointment – he never did get to helm a challenger in what remains the world's most famous and most glamorous sailing competition.[88]

Life beyond the boardroom – back on land

With Fairey and Sopwith such keen rivals in the boardroom and out at sea, the obvious temptation is to portray their personal relationship in a largely negative light. Yet, while in no way bosom pals (always surnames, never 'Dick' and 'Tommy'), they clearly respected each other's strengths and achievements, and got on well. The two men worked closely within SBAC, although CRF saw Frederick Handley Page – President of the SBAC on the eve of war, and of the Royal Aeronautical Society in the first years of peace – as a more natural ally. Fairey would be a frequent guest at Sopwith's shooting parties, and reciprocated with invitations to Oakley and later to Bossington. They would become close neighbours when Sopwith moved to King's Somborne in 1945, having discovered the delights of the Test Valley six years earlier when invited down for a day's shooting.[89] This was December 1939, and Dick Fairey's personal circumstances as the decade ended were very different from his troubled life ten years before.

The 1930s was the most eventful and the most enjoyable decade of Charles Richard Fairey's 69 years. The company he had built from scratch was flourishing, his investment portfolio was exceptionally profitable, and his lifestyle was enviable and exciting.[90] A continuing accumulation of wealth enabled a standard of living and quality of life few in Britain could comprehend, let alone enjoy. Yet that enjoyment would count for far less if personal happiness failed to complement material well-being. Fortunately for Fairey his second marriage, in 1934, provided the security and satisfaction absent first time around.

Esther Sarah Whitney was the lively, attractive and stylish daughter of a Home Counties bank manager. Fairey found in Esther every quality he felt his first wife lacked: honesty, loyalty, discretion, insight and, above all, common sense. He trusted her completely, and in return she proved the ideal companion, both at home and in public. In the words of her daughter: 'She deeply admired my father and was in love with him all her life, certainly she never got over his death.'[91] Wherever she found herself – at a Belgravia soirée, a Chevy Chase cocktail party or a Hamilton dinner dance – the second Mrs Fairey was always cool, calm and in control. Woodlands or Washington, Bossington or Bermuda, Esther Fairey was the perfect hostess, with a laudable attention to detail. The weekend shoot or the bank holiday cruise could proceed without hiccup or embarrassment, Dick Fairey confident that catering arrangements were in hand. House staff or cabin crew similarly appreciated the exercise of quiet authority, not least in Mrs Fairey's tactful handling of difficult and demanding guests. Unlike Joan, Esther would never contemplate a shopping spree in Bond Street. No longer irritated by the need to return jewellery, her husband could take pleasure in the purchase of expensive gifts. He celebrated their fifth wedding anniversary by buying Esther an emerald and diamond clip worth over £2,600 at today's prices.[92]

Here was Dick Fairey the family man, relishing a secure domestic base and a supportive wife, for whom his physical and emotional well-being were paramount. The birth of their son John in April 1935 demonstrated how Fairey fostered popular perception of himself as the all-action patriot endeavouring to defend the skies above Britain and bring home the America's Cup *and* the devoted husband caught on camera with consort, child and canine companion.[93]

Fairey lived an unusually crowded and carefully organised life but never at the expense of his family, whether immediate or extended. Thus CRF retained a keen sense of obligation towards his sisters, and he felt equally protective towards Geoffrey Hall, who for much of the 1930s tested Esther's patience with his reluctance to leave Woodlands Park.[94] Perhaps it was Hall's continuing presence at the dinner table which saw Mrs Fairey take her maid and chauffeur to Westgate-on-Sea for the penultimate month of her second pregnancy. She was back at Woodlands in plenty of time to give birth to a daughter on 17 July 1937. CRF adored Jane, his third child, but – as with John – sought not to spoil her in the way he felt he had Richard junior. If anything, Fairey over-reacted with John who, as his sister recalls, 'was very strictly brought up.'[95]

If the second son was kept firmly under control there was no lingering resentment, and in the early 1960s John recalled his father with great affection: 'He had the ability to stimulate curiosity and a healthy scepticism. He taught us never to take anything at its free value ... one of the most valuable gifts that he could hand on to children.' Jane endorsed her brother's praise of someone who in public appeared so distant and domineering. She saw CRF as an unusually empathetic parent, with an instinctive understanding of how children acquire knowledge: 'He was a wonderful teacher ... No matter what it was you *wanted* to learn.'[96] Both John and Jane relished Fairey's accomplishments as a practical joker *and* as a teller of tall tales: on occasion he put pen to paper, most notably a ghost story written on request in 1951.[97]

Fairey now had the joy of a loving wife and two young children. Yet he made sure Richard did not in any way feel excluded, witness the scale of the 21st birthday party Fairey put on for his oldest son: not one but two floors of the Dorchester Hotel booked for the night, a guest list running into thousands, and a budget of over £21,000.[98] CRF recognised Richard's resentment regarding his remarriage, and in the view of an earlier biographer sought to buy his son's approval. Throughout the second half of the 1930s Richard clearly enjoyed a privileged lifestyle, if too often taking his good fortune for granted.[99] Random examples of 'Mr Richard' testing his father's patience include running out of road in the Rolls Royce, with expensive consequences; abruptly quitting a family cruise in the Caribbean and catching a flight to California; organising an alcohol-fuelled party for gate-crashing guests on the back of CRF's Air Council invitation to the RAF display at Hendon; preferring Paris to Hayes when due back at work; and booking the decidedly racy Max Miller to amuse guests at a family function in the Dorchester.[100]

The jokes may have been a tad risqué, but young Richard was invariably good company, and 'Life around him was always fun.' He was witty, charming, selfless, attractive and generous to the point of extravagance. 'Light hearted and disinclined to be serious about anything much if he could avoid it', Richard clearly lacked gravitas.[101] CRF reserved judgement, albeit infuriated by the haste with which his spendthrift son married a fledgling film star in 1938: Aino Bergö was a Swedish opera singer who found herself on the set of *Thistledown* at Teddington studios, playing the scorned Austrian spouse of a feeble laird. This hastily shot musical failed spectacularly, and so did the marriage. It was over long before the couple divorced in 1943. Despite his fierce disapproval, as a wedding present Fairey gave the couple Sutherland House, a large house with substantial grounds located at Oakley Green in Windsor. He subsidised his son and daughter-in-law's extravagant lifestyle well into the war, and cleared their debts whenever the bailiffs threatened.[102]

For all Richard's failings his father envisaged him playing a key role within Fairey Aviation, and by the late 1930s he fulfilled various functions within the firm, albeit under the eagle eye of Barlow, Broadbent and their fellow directors. A lengthy spell working for Marcel Lobelle, followed by several months in the jig and tool office, provided the heir apparent with invaluable practical experience. Later he graduated to a junior management role on the shop floor, fostering good relations from an initially sceptical workforce. However, Richard junior's greatest asset was his competence as a pilot, put to good use by the company as the war approached. His skill as an aviator, in even the most adverse conditions, suggested hidden depths of courage and fortitude, as confirmed in the course of the war: denied priority entry by an overstretched RAF, Richard would in due course join the Air Transport Auxiliary.[103]

The late 1930s were the years when Richard senior gained access to a variety of quietly influential organisations, manned – literally – by the great and the good. For example, in the spring of 1937 Fairey was invited to dine at the Savoy Hotel with Lord Derby and fellow members of his eponymous club, and to become a liveryman in the Worshipful Company of Shipwrights. Dinner with the Derby Club proved impossible as CRF was again in Manhattan, lecturing the NYYC and later the Institute of Aeronautical Sciences on the need for aerial deterrence.[104] 1937 also saw Fairey join the Savage Club. This was at the time located in Carlton House Terrace, the club's address belying its declared aim of being 'bohemian.' The British Sportsman's Club (BSC) made no such claims, priding itself on its royal and aristocratic connections. A champion sailor and second category golfer, Fairey found himself elected to the BSC, regularly lunching at the Savoy in the company of visiting sports personalities and touring teams. CRF may have skipped lunches for the New Zealand and West Indies sides in 1937 and 1939, but it is hard to believe he missed Don Bradman addressing a packed dining room in the spring of 1938.[105]

By the time Australia's finest cricketers arrived in England, Fairey had been appointed a governor of Westminster Hospital, and, less predictably, commissioned as an honorary colonel in the Territorial Army (TA): he succeeded fellow industrialist Rookes Crompton, the original CO of the 27th (London) A.A. Battalion (L.E.E.). Before assuming nominal command the veteran colonel had led his men in both the Boer War and the Great War. CRF could boast no similarly impressive service record, but he did share Crompton's technical expertise when inspecting the equipment operated by the London Electrical Engineers, a part-time searchlight unit.[106]

The TA commitment was minimal, and ended abruptly after only five months.[107] Nevertheless, it is easy to see CRF's life in the late 1930s as focused largely upon family, sailing, golf, field sports, fast cars, club life and light entertainment. This would be far from the truth, as Fairey remained a man whose *raison d'être* was to build aircraft, and in so doing make money. Thus a typical weekday – in this case Wednesday, 7 June 1939 – comprised a ten o'clock appointment with Sir Kingsley Wood, reporting back to fellow SBAC executives after lunch, a dental appointment, and a teatime business meeting at the RTYC.[108] If Fairey was on holiday – as he had been earlier that spring, for a fortnight in Cannes – he maintained the same system of encrypted communication with Hayes as when cruising on *Evadne*. His *modus operandi*, at home and abroad, only broke down if the ultra-efficient Miss Burns was indisposed. When the now Mrs Haydon became seriously ill, Fairey realised just how dependent he had become on his secretary, trusting her to oversee his day-to-day financial affairs in the same way that she planned his working week.[109]

As his diary confirmed, Charles Richard Fairey was by now one of Britain's most prominent industrialists. Yet, given that his MBE had been awarded as long ago as 1920, the absence of further honours is noticeable. The cynic would judge this especially strange when it seems the Conservative Party saw Neville Chamberlain's elevation to prime minister as a fresh opportunity to secure CRF's support. The presence of Stanley Baldwin in Downing Street, and Leslie Hore-Belisha at the Ministry of Transport, had been seen by Fairey as sufficient reason not to support the Conservatives. Chamberlain's fresh administration, with Hore-Belisha now at the War Office, saw Fairey being courted by a senior fundraiser. Assigned to the task was Hampshire land owner George Penny, a low-profile City banker who had made his money in the Far East before returning home to a safe seat and an astonishing 11 years as a whip. Elevated to a peerage and appointed party treasurer, in 1938 Penny began to see Fairey on a regular basis. When later that year Fairey organised shoots at Flixton Hall, the newly ennobled Lord Marchwood was in attendance, staying at the same Norfolk hotel as his host.[110] Another four years would pass before CRF was knighted, in recognition of his wartime work in Washington. This might suggest Marchwood enjoyed only modest success in persuading his new friend to part with his money.

If the golf course or the yacht club were convenient locations for informal business conversations, the organised shoot was ideally suited to meeting politicians, albeit in the lower ranks of government or on the backbenches.[111] Ernest Brown was an exception to the rule, being both a keen yachtsman and a cabinet minister. Brown was so keen a yachtsman that on the weekend Britain went to war he was far from Westminster, spending the final hours of peace out in the Channel with his good friend Dick Fairey.[112] Looking to the far end of the cabinet table, CRF lacked the aristocratic and moneyed connections to enjoy a day's shooting with the likes of Lord Halifax, with an invitation to Oakley surely declined. Chamberlain preferred rod to gun when escaping the pressures of office, and dry fly fishing upstream on the Test was a powerful incentive to enjoy Fairey's hospitality – so powerful an incentive that the Prime Minister deemed the river a welcome relief from the test of war.[113]

The Bossington estate, south of Stockbridge, boasts six beats on the River Test and a two-mile stretch of the tributary Wallop Brook, complete with wooden foot bridge, thatched fishing hut and hatch pool. Beside the main drive up to the house can be found a small Early English-style church, St James, consecrated in 1839. The front facade of this early Victorian greystone house seems stern, even austere, when compared with the lounge and dining room at the rear: the view across the lawns from the patio down to the river – the House Beat – is always a pleasure on the eye, especially in summer, not grand and panoramic like Chartwell, but crowded and vivid, with a clear, fast flowing stream, and the intense, sun-lit greenery of reeds, rushes, oak, ash and willow. For Churchill the easel, but for Chamberlain such an alluring scene meant only one thing. His host must have responded similarly the first time he set eyes on Bossington, his determination to buy the estate tempered only by an initial failure to secure full fishing rights. Fairey's agent urged patience, 'as I know of no other sporting estate of this size within easy reach of Woodlands that one can get hold of.' The sale went ahead, and by July 1937 Bossington belonged to Richard Fairey. Purchase of the estate prompted the sale of Oakley Lodge to Lady Peel for £67,000, but the deal fell through because Fairey feared the estate would be broken up and plots of land sold off for building.[114]

With refurbishment of the house likely to take years, Bossington's value before the war was as a fledgling shoot, and a prime site for privileged fly fishermen. The 40-mile River Test is arguably England's finest, most eminent chalk stream: clear, alkaline water is slowly released from an underlying chalk aquifer, which, when combined with the temperate climate of central southern England, creates a fertile environment for salmon, grayling and, above all, trout, the latter feeding off multiple species of mayfly, their hatches crucial to the sustainability of the river. Hampshire man of letters John Waller Hills lauded the Test, 'with its crystal streams and its portly trout, with its lovely valley, its bridges, its trees, its chalk cliffs and its broad water meadow ... she is still the greatest trout river in the world.' For CRF, *A Summer on the Test* was the fly fisherman's bible. Echoing Hills, Fairey celebrated casting his line in 'the loveliest of all valleys, on the best of all trout streams that ever brought peace to the soul of man.'[115] The Test is home to Britain's oldest and most exclusive fishing club, the Houghton, its historic beat neighbouring Bossington. With his own stunning stretch of river, Richard Fairey had no incentive to join the Houghton Club, in the unlikely event of his being invited to do so, but he had friends who were, including the American ambassador Lewis Douglas, granted honorary membership in the late 1940s.[116]

With six servants in residence, the Bossington estate was well suited to the country house weekend, but in the final years of peace Fairey saw his new acquisition as primarily a rural retreat: 'The Test meadows in summer with the trout on the feed, the cuckoo flowers, the snipe drumming – an hour or two of that sets me up for the week.'[117] Geoffrey Hall felt his half-brother 'came naturally to Bossington with the hand and eye of the sportsman and the patience and the skill of the fisherman ... he felt that this was made for him and he was meant to be there.' The first time Hall saw the planned successor to Woodlands was on a midnight picnic in the grounds, with the headlamps of CRF's limousine lighting up the house: 'He just sat there behind the wheel, looking at this lovely floodlight effect on this beautiful frontage and simply said "Now, doesn't that house look wonderful!" And I [Hall] certainly had to agree with him. I was surprised too, because to me it looked enormous ... And then he said again: "I think it's beautiful, don't you?" And then, thoroughly practical, he added: "I'm hungry. Let's have our food." ...It's worth telling because it illustrates his *feeling* about the place ... It was a deep sentimental feeling which showed itself. And it was quite spontaneous.'[118]

Buying Bossington would prove the most significant property purchase Fairey ever made, but acquiring the estate did not stop him seeking to expand his portfolio. In the summer of 1939, on the recommendation of Tommy Sopwith's right-hand man, Fred Sigrist, CRF decided to buy land in the Caribbean.[119] The coming of war saw him put the plan on hold. Meanwhile he made arrangements to visit Canada that autumn, with a view to purchasing property – that summer he changed his mind, as 'I have received a big hint from the powers that be that my presence is required at home for the time being.'[120] At the time of writing he was staying in none of his homes, using the Hyde Park Hotel in Knightsbridge as his London base and Torquay's Imperial Hotel as a safe location for the family in the event of war. A rented flat in Sloane Street was an alternative *pied à terre* in town, until a cautious Fairey ordered its furniture be put in store. In late summer 1939 Esther and the children moved along the coast, taking up residence at Fort Charles. In October she became ill and travelled up to London to convalesce at Claridge's, before arriving at Bossington in time for Christmas.[121] Woodlands Park was no longer their home. As we shall see, from the start of September it was no longer fit to be anybody's home.

By early 1938 Fairey's exaggerated fears that the *Luftwaffe* could secure aerial supremacy in the skies above southern England convinced him that government planning for defence of the capital was fundamentally flawed. He laid out for Lord Beaverbrook ambitious, expensive and frankly unrealisable long-term plans for placing all London's arterial roads and principal railway lines underground. Not surprisingly, he received no reply after suggesting Express Newspapers urge the Government to spend £378 million on implementing his grand plan.[122]

If CRF was delusional in anticipating a radical restructuring of London's infrastructure, he was listened to more sympathetically when briefing Sir John Anderson on an alternative approach to the evacuation of women and children from big cities. Senior staff from Fairey Aviation had recently toured German aircraft factories, and been duped into believing that monthly overall production exceeded a thousand machines. They duly reported back to a receptive managing director. Convinced that the *Luftwaffe* could tolerate a high attrition rate in order to maintain near continuous bombardment of London, Fairey recommended contingency planning for a wartime skeleton city, with massive camps housing the permanently evacuated. His alternative model assumed local government was incapable of implementing a voluntary based scheme for evacuation.[123]

Surprisingly, well into the autumn of 1939 Fairey remained adamant that Germany was significantly out-producing Britain in aircraft production. Through Moore-Brabazon he secured Churchill's support for the manufacture of cheap, easily assembled army support aircraft – an idea adopted in cabinet committee but then lost in Whitehall inertia.[124] Fairey was equally convinced the German people were psychologically better equipped to withstand the exigencies of total war than their British counterparts. His fellow civilians' negative response to wartime regulation, and the Government's feeble efforts to rally the nation, reinforced his belief in a rapid collapse of morale once bombing began: 'The German nation disciplined and rationed to a degree our people would not stand will more than likely outlast us having less to lose, and our delicate economic system is tumbling.'[125] Sensitive to the charge of defeatism Fairey sensibly kept to himself his fear of imminent defeat. By this time Anderson and his colleagues had initiated the first wave of evacuation, with municipal authorities in Southampton and Portsmouth despatching children and teachers across Hampshire and Wiltshire.

At the start of the year Romsey and Stockbridge Urban District Council had enquired whether the owner of Bossington could house a number of evacuated children in time of war. The council chairman was politely informed that the Fairey family would take up residence in the event of hostilities, and that in any case children were best billeted in smaller houses, 'where the rental is more in proportion to the cost of their upkeep and any possible damage they might do.' CRF's main point, however, was the same as he was making to ministers, namely the need to build camps and plan evacuation on a grand scale – anything else was mere improvisation, and failed to address the scale of the problem.[126]

Meanwhile, for all Fairey's protestations, Woodlands Park was designated as a home for 16 evacuees. When war was declared he forestalled imminent action by the local billeting officer. On 1 September 1939 the family home was stripped bare, its contents moved to Heal's furniture store in Aylesbury. Over the next six weeks a substantial amount of foreign currency held in the Woodlands safe was deposited at Barclay's Bank. More remarkably, CRF kept gold bullion in his home, and this was duly transported to the vault of the Bank of England, presumably ending up in Canada. In the first week of September Woodlands Park was officially registered as an ancillary to the drawing offices at Hayes. Eton Rural District Council was consequently informed that no children could be balloted in a property requisitioned by Fairey Aviation for necessary war work.[127] Fairey and his half-brother – in Torquay on *Evadne* when war was declared – continued to use Woodlands as their base, even if Geoffrey Hall was the more permanent resident. CRF stayed more and more in London, or was away visiting production plant up and down the country. He relocated his personal office – headed as always by the formidable Miss Burns – to Buckinghamshire, where he rented a large house in Taplow.[128]

Relieved that he retained a firm grip on his two principal properties, and that his family appeared safe from aerial attack in the west of England, Fairey focused on how well Britain was waging war. Deeply pessimistic, and convinced Communists both sides of the Channel were intent on undermining the Allied war effort, he speculated on the potential for splitting the Axis given the presumed distaste of Fascist grandees such as Air Marshal Balbo and Count Grandi towards 'Hitler's latest achievement in the rereleasing of the floodgates of Bolshevism.' On 12 October 1939 Fairey wrote to Ernest Brown suggesting Lord Lloyd, 'diehard' Tory and one-time High Commissioner in Egypt, secretly visit Italo Balbo, the governor of Libya. Lloyd would convince Balbo that he should openly challenge the pro-German position of Count Ciano, Mussolini's son-in-law and foreign minister.[129]

This was not an idea that came out of nowhere. Fairey would have encountered the egregious Grandi when he was ambassador in London, and he must surely have known Italo Balbo, Italy's most famous aviator and a friend of Philip Sassoon. Both Grandi and Balbo had been *ras*, of Bologna and Ferrara, respectively, and as Fascist founding fathers resented an *arriviste* like Ciano enjoying the ear of the Duce. For six years Balbo had languished in Libya, and informed circles knew of his opposition to the costly campaigns in Abyssinia and Spain, not least because prolonged war had left Italy dependent on Nazi Germany.[130] Fairey also knew that Lord Lloyd, like Murray Sueter, had a more than grudging respect for the Fascist regime. He admired Mussolini as a man who got things done, and regretted that the Stresa Front had not translated into a bilateral treaty with Italy. Both the Baldwin and Chamberlain administrations used Lloyd as a back channel for communicating with members of the Fascist Grand Council. Lloyd's appointment as a surprisingly proactive chairman of the British Council facilitated discreet meetings with senior Fascist Party members. In January 1939 Lloyd accompanied Chamberlain and Halifax on their much publicised but singularly unprofitable visit to Rome.[131] Writing to Grandi in September 1939, Lloyd interpreted the Molotov–Ribbentrop pact as a sign that 'what the Nazis have lost in ideology, we have a thousand times gained, for the war is now a clear issue of Christianity against that which opposes it.' This was the same message as Fairey conveyed to Ernest Brown, suggesting that he had already spoken to Lord Lloyd.[132]

Dick Fairey considered George Lloyd his sort of chap, not least because he hated the Nazis and was well to the right of the present Conservative leadership. Like CRF, Lloyd had taken up sailing in middle age, and he had also qualified as a pilot.[133] Both men shared a similar hinterland, and by February 1940 Lloyd had convinced Churchill, a long-time friend and ally, that inviting Fairey to the Admiralty would be in both men's interest.[134] In May 1940 the new Prime Minister made Lord Lloyd colonial secretary in the coalition government, but eight months earlier he could not convince his predecessor that the only member of the House of Lords to have flown a Spitfire was better suited to running the Fleet Air Arm than the British Council, a view Richard Fairey doubtless endorsed.[135]

Back in October of the previous year Brown wrote to Fairey from the Ministry of Labour thanking him for his suggestion, reassuring him that the Italian situation was being closely monitored, and acknowledging that there was a need to exploit 'the recoil from the Russo-German pact.' By this time Fairey had already sought a meeting with Lloyd to suggest he secure Whitehall support for a meeting with discontented Fascists when visiting British Council offices in the Balkans and eastern Mediterranean.[136] Arguably CRF's initiative did reach fruition: in due course Lloyd travelled to the Balkans and then on to Rome, albeit for a meeting with Ciano and not his critics.[137] Fairey's dream scenario did come about, but not until July 1943 when *all* members of the Fascist Grand Council united to depose Mussolini, support the King, and seek an armistice with the Allies: Balbo was already dead, in a suspicious plane crash three years earlier, but Grandi was the principal rebel and lived to tell the tale. However, Ciano found himself in front of a firing squad, on the orders of a vengeful father-in-law.[138]

Ciano's final testament revealed a man reconciled to his fate, but in October 1939 the same could scarcely be said of Richard Fairey. For the first time in his life he found himself gravely ill. The ethos of work hard, play hard had, give or take the odd bout of bronchitis, stood him in good stead as he advanced into middle age. His regular doctor mistakenly assumed CRF's heart was in sufficiently stout condition that it could tolerate a regular diet of good food and fine wine, heavy smoking and frequent use of a Benzedrine inhaler.[139]

Increasing heart palpitations and consequent lack of sleep saw Fairey undergo a cardiograph test and consult a Harley Street specialist. He was told in no uncertain terms that, irrespective of the war, an extended break was crucial. At this point a strangulated hernia arising out of an old appendix wound complicated matters. Fairey found himself in acute pain, and in consequence could not travel to Torquay where Esther was similarly incapacitated. His consultant 'rescued' him from an alarmingly relaxed GP, and carried out life-saving surgery. After his operation Fairey's condition rapidly improved. Again, he was told that rest was paramount, or at some point sooner rather than later his heart would give out. This was not the news the head of a company as vital as Fairey Aviation wished to hear when his country was only six weeks into an unprecedented struggle for survival.[140] He retreated to Bossington, and long spells out on the river, interrupted only by equally long telephone calls to Miss Burns and the production teams in Hayes and Stockport – there were fish to be caught, and there was a war to be won.[141]

CHAPTER 8

Wartime in Washington – the British Air Commission

'Lord Beaverbrook requests…'

Just before noon on Saturday, 24 August 1940, W.P. Crozier, the extraordinarily well-connected editor of the *Manchester Guardian*, found himself in the lobby of Imperial Chemicals House, the splendid neo-classical building opposite Lambeth Bridge requisitioned to house the Ministry of Aircraft Production. Less than four months had passed since Winston Churchill, in one of his first acts on succeeding Neville Chamberlain as prime minister, had relieved the Air Ministry of its procurement responsibilities: MAP was a creation of wartime emergency, only briefly outliving the breakup of the coalition in May 1945. The man Churchill appointed as Minister of Aircraft Production was Lord Beaverbrook, Britain's best-known press baron. Aggrieved Chamberlain loyalists viewed the owner of Express Newspapers' appointment as an act of supreme folly, while others – such as Richard Fairey – saw it as confirmation that henceforth the war would be conducted in a far more decisive and effective manner.[1]

Seen by his critics as an unscrupulous crony of the equally maverick Churchill, Max Aitken's prewar credentials as an anti-appeaser were near non-existent.[2] This was in sharp contrast to Liberal leader Sir Archibald Sinclair, whose friendship with the Prime Minister through good times and bad was duly rewarded when he succeeded Sam Hoare as Secretary of State for Air. Sir Samuel's second spell at his favourite ministry had lasted little more than a month, but was long enough for him to attract ferocious criticism from both sides of the House over the RAF's feeble contribution to the Norway campaign. Surprisingly, the *Manchester Guardian*'s editor enjoyed Hoare's confidence, and yet Crozier was clearly happier taking tea with the like-minded Sinclair – as on the day before his meeting with 'The Beaver.' Sir Archibald made plain his dislike of Beaverbrook, and of Dick Fairey's favourite pro-consul, Lord Lloyd, sent by Churchill to stiffen sinews at the Colonial Office.[3]

As a veteran correspondent and leader writer, and a shrewd judge of character, Crozier could see for himself if MAP's early impact depended upon the man with 'the fixed and penetrating eye' busy dictating events from the top floor of his Millbank citadel. As a master of self-promotion Beaverbrook was eager to impress, orchestrating an hour-long display of executive action. His observant guest noted the harshness of the Canadian's strong New Brunswick accent when barking orders at high volume, in a 'manner always incisive and sometimes abrupt.' Part of the show was a pre-arranged call to the Ministry's key player in New York, who was duly instructed by Beaverbrook to take the first flight to DC and finalise an agreement on planned output for the coming year. Senior officials and RAF personnel were then called in to join the visiting journalist, and all present learnt from the secretary of state that once in Washington his man would seal the deal. That man – the security-conscious Crozier's 'Mr. X' – was almost certainly Charles Richard Fairey.[4]

MAP was established only four days after Churchill came to power, its staff a combination of senior officials from the Treasury and the Air Ministry, movers and shakers from Express Newspapers, and a handful of industrialists notorious for their unorthodox methods. The transfer of Sir Charles Craven and Sir Wilfred Freeman, joint heads of the former Department of Development and Production, meant CRF retained his direct line to the top. Crozier noted how both men had been summoned to Beaverbrook's office before the call to New York, with a clear signal in the subsequent conversation as to who was in charge.[5]

Beaverbrook was in awe of no-one, insisting that the motor manufacturers bow to superior expertise re aircraft production methods, and, where appropriate, as with the Civilian Repair Organisation, play to their strengths. The peremptory treatment of Lords Austin and Nuffield surely earned Fairey's approval, as no doubt did the messages boldly posted in Beaverbrook's office: 'Committees take the punch out of war' and 'Organisation is the enemy of improvisation.'[6]

Well, up to a point, Lord Copper – at a moment of dire national emergency a top-down system of rapid decision-making with minimal consultation could indeed energise tried, tested, and possibly tired methods, but at the expense of accountability and collective agreement. Too much depended upon Beaverbrook's personal priorities and prejudices, and in any case Sinclair and his staff were adamant the Air Ministry had already laid the foundations for accelerated aircraft production. Dowding, whose respect for Beaverbrook was loudly reciprocated, insisted that Fighter Command would for ever be in his debt, a claim endorsed by MAP's permanent secretary Sir Archibald Rowlands. Sir Maurice Dean, in 1940 fast-tracking his way to head the Air Ministry, viewed Beaverbrook's single-mindedness and narrow focus as a short-term solution and a long-term catastrophe, a view endorsed by another high-flier, Air Vice-Marshal Tedder, after he left MAP later that same year.[7] In his memoirs Moore-Brabazon applauded Beaverbrook's stunning achievements, before recording how, having succeeded him as Minister of Aircraft Production, he immediately announced the restoration of formal planning procedures and an end to crisis management.[8]

Reassured by the continued presence of Craven and Freeman, CRF revelled in an environment where his telephone was likely to ring at any time of the day or night. The caller would bellow out his staccato instructions like a St John longshoreman. Fairey had always admired Beaverbrook, their common view of empire over-riding past differences re the German threat. A son of the manse, young Max Aitken had grown up on the banks of the Miramichi, the same timber river into which Fairey's father had invested so much time, money and emotion. CRF must at some point have mentioned the New Brunswick connection, as for Beaverbrook it explained why he was, 'a fine man ... [with] a good character, excellent reputation, splendid presence, charming manner, an agreeable personality.' The admiration was mutual, as Fairey greatly admired Beaverbrook's 'dynamic work' and 'ferocious energy.'[9]

The minister's frequent calls were flattering, but the chairman of Fairey Aviation was well aware every other plane maker in Britain had been phoned from Stornoway House, the Canadian magnate's home and headquarters until it was badly damaged when the *Luftwaffe* began bombing London. However, only a select few had been asked to serve on MAP's Emergency Committee, a discreet body set up to cover any interruption in executive decision-making. The letter of invitation made specific reference to contingency planning should Beaverbrook be indisposed, thus covering the consequences of aerial attack or even invasion.[10] In the early summer of 1940 CRF served as the secretary of state's go-to man, visiting regular production plants and shadow factories in an advisory capacity, and reporting back to Beaverbrook. Being available 24 hours a day across a two-month period made heavy demands of someone who, although still in his early fifties, no longer enjoyed the best of health.[11]

This may have been Britain's 'finest hour' but CRF recognised that there was no way he could match the 24/7 commitment of Beaverbrook's younger acolytes. Rest and recuperation came via a few hours fly fishing at Bossington. Moore-Brabazon was a frequent guest, Churchill having left him on the backbenches as a sign of disapproval: joining Mosley in overtures to Berlin on the eve of war had displayed a familiar lack of political nous.[12] Back in London 'Brab' lobbied energetically on behalf of Briggs Manufacturing Co., a Michigan enterprise with the grand ambition of building British dive bombers south of the 49th Parallel. Ideally in partnership with Fairey Aviation, this Anglo-American operation would supply HM Government directly and outside of any delivery agreements negotiated with the Federal Government. Unsurprisingly, nothing came of this proposal, if only because the Air Ministry and the RAF insisted MAP recognise it to be little more than a speculative venture an under-employed Moore-Brabazon had come up with having enjoyed one glass too many of Bossington's best brandy.[13] On 21 September 1940 Beaverbrook in effect told Moore-Brabazon to stop wasting his and everyone else's time, and make a more valuable contribution to the war effort: no aircraft of British design would be built in the United States across the course of World War II. Yet one must assume 'The Beaver' had a quiet word with Churchill, as only a fortnight later Fairey's fishing companion was appointed Minister of Transport and made a member of the Privy Council – with ennoblement looming, Lord Brabazon's war had belatedly begun.[14]

Earlier in the summer Moore-Brabazon had unknowingly sowed the seeds of an idea in Lord Beaverbrook's febrile mind. Eager to visit the Briggs plant and discuss plans for future aircraft production 'Brab' sought permission for CRF to accompany him. Fairey would brief the Michigan engineers on state-of-the-art manufacturing methods; while in New York he could energise a Purchasing Commission indifferent to the nation's parlous condition. Beaverbrook poured cold water on Moore-Brabazon's intention to visit the United States, and refused to release Fairey from his present duties on behalf of MAP.[15]

Less than a month later CRF was spending a rare night at Woodlands when the telephone rang; picking up the receiver he heard a familiar voice order him to make ready for an important mission – within a few days he would be flying to New York, where he would add weight to MAP's American outpost, the Air Section of the British Purchasing Mission. He would work alongside Morris Wilson, an old banking associate of Beaverbrook who in June 1940 agreed to serve as the minister's personal representative in his native Canada and in the United States. Wilson performed this task admirably, but Fairey's arrival would facilitate an eventual return to running the Royal Bank of Canada.[16]

The Air Section of the British (briefly Allied) Purchasing Mission soon assumed a fresh identity, as the British Air Commission (BAC). Sir Henry Self, the BAC's first Director-General, had played a key role in the Air Ministry's exploratory mission to Canada and the United States in April–May 1938, visiting several potential contractors. The immediate outcome had been the purchase of 400 Harvard and Hudson trainers, and a largely critical view of American combat aircraft. Ironically in light of later developments, the SBAC's highly vocal opposition to the American contracts was a significant factor in Lord Swinton's decision to quit the Air Ministry. Self's recommendation to buy trainers from Lockheed and North American Aviation facilitated transatlantic agreement on reassembly, interchangeability of parts, and pre-service inspection. This set an agenda, and provided a model, for Anglo-American collaboration across the course of the next six years. It also gave Lord Halifax leverage in persuading a prime minister instinctively suspicious of Americans to accept a wide-ranging trade agreement, and to respond more positively to White House gestures of friendship. While acknowledging the superiority of American civil aircraft and their military variants, with regard to aerial combat the British presumed a qualitative advantage well into the war, and there were tangible benefits where this really was the case – most notably in advanced development of the P51 Mustang. The downside was that, to safeguard its status as a priority client, Britain covered the cost of all technical improvements to the original American designs. 1939 saw a growing number of contracts placed with US manufacturers: in the first seven months of the war Britain ordered 1,320 aircraft and 1,200 engines, and in April 1940 contracts were signed for a further 2,440 machines. At the start of the Norwegian campaign orders placed in the United States amounted to almost 15 per cent of the Air Ministry's total procurement.[17]

Meanwhile, prewar failings in the French aircraft industry saw the Daladier government negotiate even larger contracts, buying fighters from Curtiss on a scale far beyond the company's capacity to produce. The RAF considered the Curtiss P36 and P40 fighters unreliable and obsolete, as became evident when aircraft originally meant for the French were deployed by Fighter Command in North Africa. Frustratingly, the machine tools necessary to service French contracts were held back from export to British manufacturers impatient to build state-of-the-art fighters, most obviously the Spitfire and the Hurricane. However, for Churchill in June 1940 the performance of the P40 Warhawk was irrelevant – like Dick Fairey a quarter of a century earlier, the new prime minister saw the Curtiss plant in Buffalo as symbolic of a new and potent transatlantic partnership. In this instance a partnership forged at a unique moment in history when together 'the English speaking peoples' faced their fiercest test.[18]

Six months before Churchill reached out to Roosevelt for help, production plans for meeting British and French needs were so ambitious that Henry 'Hap' Arnold, Air Corps service chief and the chair of the Air Board, had voiced his concerns to the President. Nor was this the first time Arnold had complained. Roosevelt appeared remarkably well informed about the Allied missions' activities, with Arnold soon discovering that the White House relied heavily on wire-tapping. While in his published memoirs the Chief of the United States Army Air Corps (from July 1941, US Army Air Forces – USAAF) lauded the RAF's achievements, he consistently questioned the Administration's prioritising of British needs. This scepticism persisted long after America entered the war in December 1941. Ironically, although 'Hap' Arnold's initial complaint was ignored, American factories were demonstrably ill-equipped to meet demand. Only 2,076 machines were supplied to the RAF across the whole of 1939 and 1940. To put this in perspective, the RAF accepted into service well over 11,000 aircraft from British suppliers in the first ten months of the war. American capital investment paled by comparison with that of the *entente* prior to the French defeat, witness the £65 million invested both sides of the Channel in aero-engine production between January and June 1940.[19]

The French shared their ally's frustration over the modest American output: well over 3,000 machines were still to be delivered when German forces reached the Channel. The British and French had shared a purchasing board since January 1940. With the fall of France the British Government assumed responsibility for all contracts previously agreed between Paris and Washington. Sanctioned by the Anglo-French London Committee, an agreement was signed in New York before dawn on Monday, 17 June 1940. This was only hours before the US Treasury froze French assets. Most members of the French Air Mission chose home over exile, but key figures like Jean Monnet stayed on in Washington. Twelve days after Marshal Petain assumed power in Vichy, and with Fighter Command already in action over the Kent coastline, Lord Lothian addressed the American people. In his broadcast the British ambassador made a stark and simple appeal for help: 'In the long run we need airplanes.'[20]

Lothian made his plea comforted by the knowledge that Roosevelt had secured the Democratic Party's nomination for an unprecedented third term as president. Furthermore, he was about to authorise a '3,000-per month' scheme from an initial target of around 750 aircraft. The aim was to address a huge backlog on current orders while at the same time satisfying overseas contracts, which by now stretched as far ahead as the summer of 1942. Britain in July 1940 was committed to spending in Canada and the United States the vast sum of $1.1 billion, with that sum sure to rise exponentially as a result of continuation orders. Spending on this scale would soon exhaust foreign exchange reserves, and at this point any settlement akin to what became Lend-Lease remained a distant prospect, lying well beyond the presidential election in November.[21]

While the US Treasury talked tough to its Whitehall counterpart, in public the White House conveyed an image of munificence, tempered only by acknowledgement of still solid support for American neutrality. Intense radio and newspaper coverage of events in Europe meant FDR could scarcely ignore Beaverbrook's exhortations to build more aircraft, the press baron's publicity machine operating at full tilt both sides of the Atlantic. Lord Beaverbrook made his case as only he could, but he was astute enough to ensure MAP's men in Washington addressed Arnold's fears for an Air Corps left stripped of front-line aircraft. Beaverbrook was reassured that neither the USAAC's chief of staff nor the Treasury Secretary would torpedo his ministry's intention to place large orders with all available manufacturers. He therefore set about convincing Churchill that a desperate Britain should gamble on Roosevelt's third administration facilitating the means to support the '3,000-per month' procurement programme. The Prime Minister shared Beaverbrook's belief that following the presidential election a deal could be done, and not surprisingly in late August the Cabinet agreed. A fortnight later MAP restated its ambitious medium- and long-term targets for aircraft supply across the Atlantic, securing Cabinet approval. By this time the minister felt confident he had the organisation and the personnel in situ to support a truly huge programme of collaboration with American manufacturers and the relevant federal agencies.[22]

Furthermore, in Philip Kerr – Lord Lothian – Beaverbrook had a valuable ally, unafraid to mount an aggressive lobbying campaign the moment Roosevelt was re-elected: an unusually public display of adroit diplomacy culminated in FDR seeking congressional approval for a 'Lend-Lease' agreement with Britain and its allies. Lothian cultivated the press, and Churchill cultivated the President, his cable of 9 December insistent that the very future of civilisation depended upon America keeping Britain solvent. Lend-Lease secured a largesse of material and financial support, its strict conditions only becoming obvious once the war in Europe was won. Aid on such a grand scale was Lothian's legacy as the ambassador died a fortnight before FDR's broadcast pronouncement that America must become the 'arsenal of democracy.' The next envoy in Washington would be Lord Halifax. CRF was no great admirer of the former foreign secretary, but with both men in Washington between January 1941 and April 1945 he had ample opportunity to change his mind.[23]

By late 1940 the United States' Neutrality Act had been repealed for over a year, but back in September 1939 strict enforcement of the Act had rendered delivery of American aircraft highly problematic, with an embarrassed White House keen to assuage its isolationist critics courtesy of a formal – and transparent – mechanism for negotiation with French and British purchasing agencies. This was a view shared with Whitehall, where officials had welcomed Lord Riverdale's eve-of-war report recommending a permanent mission on the far side of the Atlantic. Thus the British Purchasing Mission began its life in Ottawa only weeks after war was declared. By March 1940 the now Allied Purchasing Mission had relocated to New York, with Self seconded to head the Air Section. What by now was a large, multi-purpose agency for both Britain and France operated out of 15 Broad Street, a vast late twenties skyscraper famous then and now for being wrapped around its diminutive neighbour, 23 Wall Street. A Manhattan location, far from Washington DC, deflected congressional suspicion that the Roosevelt Administration was hand-in-glove with the British. Yet the Mission maintained a significant presence off Pennsylvania Avenue, in accordance with Riverdale's belief that non-specialist embassy staff could not properly brief senior Democrats anxious to support the Allied cause. Not surprisingly, therefore, in November 1940, very much at the instigation of Richard Fairey, an autonomous BAC separated from the Purchasing Mission and relocated to the federal capital.[24]

Sir Henry Self embodied the senior civil service's finest values, as an admiring Richard Fairey – no lover of technocrats – readily acknowledged.[25] Yet, for all his integrity and intellect, Self lacked the expertise, the authority and, above all, the transatlantic contacts unique to the founder of Fairey Aviation.[26] Here was a plane maker who more than any other had actively sought Anglo-American collaboration. Dick Fairey was the ideal envoy to despatch westwards at a moment of supreme crisis, with the Battle of Britain being fought out in the skies over southern England. At the BAC he served as Self's Deputy Director-General, but CRF's initial appointment was as the Air Section's Controller of Production: he would liaise directly with North American contractors, individually and collectively, and with the federal administrations in both Washington and Ottawa. Given Canada's status as the Empire and Commonwealth's principal weapons supplier, the dominion's bilateral working relationship with London was relatively straightforward: the Ottawa-based British Supply Board oversaw weapons delivery until June 1940, after which responsibility lay with the federal government's munitions and supply department.[27]

This allowed Fairey to focus on securing the maximum number of combat, reconnaissance and transport aircraft, *and* engines, components and machine tools, which American manufacturers were able to supply a client still 18 months off becoming an ally.[28] The number of machines available to the British was determined by volume of output, namely production capacity over and above that required to meet the ever-increasing demands of the USAAC and the USN. Quality was as important as quantity, with complementarity both sides of the Atlantic a further vital element. While an immediate priority was the supply of existing aircraft, Anglo-American collaboration required strategic decision-making regarding future types of aircraft, both piston-engine and jet-powered. This necessitated someone who could speak with authority on the development and design of machines still on the drawing board, or as yet no more than a proposal.

Whoever negotiated at national level on Britain's behalf not only had to have the necessary technical expertise, but he also had to know what was taking place in factories the length and breadth of the United States. That meant being welcomed in boardrooms and hangars by industrial patriarchs who, like their British counterparts, still ran the enterprises they had founded before or during the Great War. No plane maker was better known in the United States than the man who gave the SBAC and Royal Aeronautical Society a global voice, the entrepreneur who brought to Britain both the Curtiss engine and the Reed propeller, the yachtsman who fiercely but fairly fought the eastern seaboard sailing establishment, and the generous socialite who hosted every homesick Yankee flier from Amelia Earhart to Glenn L. Martin.

One federal fast-tracker who worked closely with Fairey found him 'much more American than English in his approach ... He liked to live, and he liked to live well; he lived thoroughly and to the utmost.' In 1940 Britain's governing and properties classes were still viewed with considerable suspicion. Yet Fairey came across to Americans as an imposing figure who, unlike most of his countrymen, chose not to equivocate or indulge in tiresome under-statement. Above all, he was trustworthy: 'You could do anything. You could fight. You could still remain the best of friends, and get along very well.'[29] Although otherwise they had little in common, there is an obvious parallel with another Englishman displaying distinctly transatlantic characteristics: Churchill despatched Mountbatten to Washington in the summer of 1941, and again in 1942, because Lord Louis loved life in the States and because he understood how to charm Americans and get what he wanted. CRF could not work a room like Dickie Mountbatten, but he could match his hosts when it came to straight talking and doing a deal. Also, Fairey's longstanding love of vaudeville left him attuned to the American sense of humour: at the movies he was in his element watching slapstick cartoon series such as *Looney Tunes* or *Merrie Melodies*.[30]

From a personal perspective, the BAC appointment offered an ideal opportunity for Fairey to be reunited with his family: Esther had previously been persuaded that she should take the children to Canada in case of invasion. Any invidious suggestion that CRF agreed to go solely so he could join his family in a safe location could easily be dismissed: it was an order from 'The Beaver', not a polite request. Having settled in Quebec, the future Lady Fairey had no idea her husband was being sent to the States: she rushed to meet him only after he had phoned to say that he was waiting for her in downtown Montreal.[31]

However attractive the prospect of a family reunion, a lengthy stay in the United States raised awkward questions regarding the future health of the company. Fairey Aviation was a firm with a tight hierarchical organisation built around the founder and chief executive: to remove such a powerful figure, albeit temporarily, was to create an obvious power vacuum. With Tom Barlow overseeing Fairey's northern operation, and Wilf Broadbent coordinating construction across an ever-increasing number of home plants and subcontractors, Maurice Wright was the obvious successor. Wright was CRF's most trusted colleague, but his biggest asset was as a salesman. Untested as a mastermind of manufacturing industry, could Wright run what was now a huge operation at Hayes and its satellite sites, and at the same time fulfil a strategic role heading the whole company? Time would tell, but in the meantime CRF was forced to suppress serious doubts about the future direction of Fairey Aviation, accept the absence of contingency plans to cover his absence, and bow to Beaverbrook's dictat.[32]

Twenty-five years after he accepted Murray Sueter's challenge to build his own aircraft, Dick Fairey took another decisive step. He informed his fellow directors that for the foreseeable future they were on their own. The chairman and managing director was handing over the reins of power, with immediate effect. The next time the board heard from him, he would be in New York. Once in America Fairey would exercise only nominal control over his company; a very different state of affairs from cruising in the *Evadne*, maintaining near continuous contact with Miss Burns and Charles Vinson. In August 1940 all CRF could do was to convince himself that Fairey Aviation's present management structure was sufficiently resilient as to accommodate any perceived weaknesses on the part of his most senior colleagues.

Fairey took off from Poole harbour on the afternoon of Saturday 3 August as one of only three VIP passengers on the maiden trans-Atlantic flight of the S30 C-class flying boat *Clare*. Appropriately, he crossed the north Atlantic courtesy of a mid-air refuelling system pioneered by his old friend Sir Alan Cobham, and in an aircraft the construction of which he would have overseen had he stayed with Short Brothers. Ironically, he was travelling courtesy of BOAC, the state-owned airline he considered an insult to the spirit of free enterprise. The Empire flying boat G-AFCZ had been rebuilt to facilitate a ten-hour, low-level crossing from Dorset to the Maritimes, after which it would fly on to Montreal and then down the north-east coast to New York. The *Clare*'s wartime functionality reinforced CRF's belief that only private enterprise could provide a necessary level of comfort and service – vital sustenance came courtesy of the sandwiches and champagne presciently packed in the hand luggage. More frugal repast was on offer in Newfoundland before the *Clare* proceeded to the St Lawrence and a welcome reunion with Mrs Fairey. There was palpable relief on Esther's part as, apart from feeling terribly homesick, she was dependent on 'Wuffy' Dawson – wartime currency controls meant CRF's Canadian assets had been frozen. Dick and Esther spent two days together in Montreal, and then took the train to New York City.[33]

By then, G-AFCZ had long since reached its final destination – Flushing Bay – the float plane annex to New York's North Beach Airport (today, LaGuardia). With scant regard for security, east coast newspapers heralded the inauguration of a pioneering route to England. In actual fact the *Clare* made only three further crossings before a north Atlantic service was deemed too costly and dangerous to continue. The *Baltimore Sun* named the inaugural flight's three passengers, noting that the only VIP left to alight at Flushing Bay was one Colonel William J. Donovan, ostensibly visiting England on behalf of Frank Knox, the Secretary of the Navy. Decorated veteran, Republican state attorney, and appalled witness to the worst excesses of European fascism, 'Wild Bill' Donovan's antipathy towards isolationism had seen him switch allegiance to FDR. Using Knox as a cover, from the spring of 1940 Roosevelt employed Donovan as an emissary to Churchill, thereby excluding his defeatist ambassador, Joe Kennedy. Bill Donovan was flying home to convince a reluctant Air Corps that Churchill should have his symbolic P40 fighters, thankfully unaware that the RAF would in due course turn the offer down. Given Donovan's presence on the flight it is no wonder the RAF gave G-AFCZ a Spitfire escort across the Irish Sea. The man who over the next two years would revolutionise the intelligence services of the United States, and whose whole life read like a novel, was surely the ideal travelling companion for Dick Fairey. The champagne-fuelled, anecdote-filled exchanges between CRF and 'Wild Bill' helped the hours pass as their aerial leviathan flew westwards a mere 1,000 feet above the turbulent waves.[34]

Fairey's work base on his arrival in the United States was 15 Broad Street, but his family took a suite at the Wardman Park Hotel, its 'founder's gift of an in-city resort to Washington's elite and influential.' New York in high summer was hot and humid, and in any case neither he nor Esther wanted to set up home in Manhattan. Washington offered mildly more efficient air conditioning, more congenial society, a well-connected British community which was growing by the day, and speedy travel connections to all points of the Union. The climate was hot and humid at the height of summer, and the Foreign Office listed the Washington mission as a tropical posting. Yet the federal capital was an ideal transit point for flights or railroad trips to factories building airframes, aero-engines, parts and machine tools for despatch across the Atlantic. Sir Edwin Lutyens' magnificent embassy on Massachusetts Avenue offered invaluable diplomatic support, with direct access to senior members of chancery, including the ambassador himself. The other Allied purchasing agencies, including an embryonic Free French mission, chose to operate out of Washington, not New York. Critically, being based in the capital enabled easy formal and informal contact with US service personnel, senators, congressmen and members of the Administration. The ultra-fashionable Wardman Park Hotel was located high up Connecticut Avenue, in a semi-rural suburban setting that belied the cosmopolitan nature of its clientele. From August to November Fairey commuted between his new home and the Big Apple, but in the autumn of 1940 the now autonomous BAC relocated to Washington.[35]

A high-level US delegation had visited London in September, and the BAC's move was partly prompted by the arrival of a reciprocal mission: the presence in Washington of future CAS John Slessor – in November 1940 still an air commodore, albeit Air Ministry director of planning – provided invaluable support for Dick Fairey, a fellow realist. With the White House projecting 4,000 aircraft for Britain every month by late 1942, a worldly-wise Slessor urged caution. After three months becoming acquainted with assembly lines the length and breadth of the country, a similarly sceptical Fairey must have welcomed an RAF counterweight to the overly optimistic colleagues left behind in New York. Beaverbrook endorsed Fairey's dim view of the British Purchasing Mission in his correspondence with Churchill. By now both CRF and Sir Henry Self shared Slessor's view that, however shambolic the USAAC's administrative processes, its senior staff were gaining ground in their resistance to Britain securing combat aircraft at the expense of home squadrons. In the winter of 1940–1, as Lend-Lease proceeded through Congress, and Slessor's extended stay enabled the first proper staff conference between the British, the Canadians and the Americans, the Air Corps chief was seen by the BAC as a stumbling block to future collaboration.[36]

Whatever warm feelings Fairey might later feel towards 'Hap' Arnold ('a topper and quite outstanding among his contemporaries'), they were clearly qualified in early 1941.[37] Yet, by comparison with many of the generals, admirals and industrialists with whom CRF came into contact during his first year as Controller of Production, Arnold was a model of politeness. Old rivalries and the return of mutual suspicion after Versailles had fostered considerable antipathy towards Britain in many quarters across the interwar period, only partially countered by a highly successful royal visit in the spring of 1939.[38] Notwithstanding Roosevelt's re-election, isolationism remained a potent political force, fuelled by many commentators' long-standing prejudice towards Britain. Anglophile journalists like Ed Murrow and Quentin Reynolds endeavoured to counter isolationism in their sober yet highly partisan reports on Londoners' under-stated response to the Blitz. Aircraft manufacturers were especially hostile, seeing British firms as credible commercial rivals. In reality Fairey was guaranteed a warm welcome in only a minority of boardrooms. Elsewhere, he had to work hard in encouraging his hosts to view transatlantic collaboration as a force for good, and not simply an opportunity to maximise profit by exploiting the present plight of a long discredited British Empire. These early visits brought him into contact with suspicious plane makers, and with building contractors in no great rush to build fresh factories on green field sites – the United States acquired 50 new aircraft and engine plants between 1940 and 1945, with the speed of construction accelerating dramatically after Pearl Harbor. A key reason for the failure to supply even half the aircraft Roosevelt promised would reach Britain in 1941 and 1942 was the lengthy lead-in time for manufacturing plant: in these respective years deliveries totalled just over 5,000 and just under 8,000. The directors of companies such as Curtiss, Lockheed and Bell would have been even harder to deal with had they known how poorly the RAF rated their fighters, a number of which were shipped to the Soviet Union after June 1941.[39]

Working in Washington

> One day that *was* happy and exciting was the day the *Bismarck* was sunk, and Sir Richard was so proud of the Swordfish.[40]

The BAC was housed at 1785 Massachusetts Avenue, a Beaux Arts apartment block better known as the Andrew Mellon Building. On 'Embassy Row', and only seven blocks north of the White House, the Dupont Circle location was ideal for accessing officials, politicians, and visiting industrialists. Contract negotiations remained in New York, but the seven other departments and sections established in Washington a large bureaucratic organisation, with middle-ranking service personnel (including US liaison officers) out-numbered by civilian managers. By 1945 the BAC comprised over a thousand officers, officials and secretarial or support staff. Although the Fleet Air Arm came to depend heavily on American carrier aircraft, the full complement of BAC staff included only one Royal Navy officer. Beyond the federal capital, 'Inspection Offices' were established in Chicago, Detroit, Pittsburgh, New York and Toronto, with 'Resident Inspection

Units' embedded in all major US and Canadian aviation plants. In 1940–1, 70 per cent of deliveries to Britain originated on the West Coast: Fred Sigrist, who Fairey 'found in New York in search of medical aid and conscripted', set up a Californian outpost of the Planning and Production Department. This was the department headed by CRF until his promotion to Director-General at the start of 1942.[41]

As well as dealing with sensitive company bosses and obstreperous construction engineers, Fairey had to adopt a fresh approach when managing his desk staff, many of whom were American or Canadian. For security reasons CRF's two personal secretaries, who stayed with him throughout the war, were Canadian. Mary Bell, his chief secretary, was Ontario's answer to the formidable Miss Burns, holding the fort back home. Looking back on her time in Washington, she judged her old boss 'a warm and certainly not a difficult person to work for.' Fairey displayed great patience when training her for the job, but 'he was *not* patient with incompetence, and, I would say, had definite likes and dislikes about people.' When Mary Bell's cousin was killed flying a Royal Canadian Air Force (RCAF) bomber, Fairey moved heaven and earth to discover in what circumstances he had died, and where he was buried. Not surprisingly, she considered him at heart 'a kind man … He was devoted to his family, Lady Fairey, John, and Jane…'[42]

The BAC depended heavily upon North American personnel as they boasted a level of local knowledge all British officials – including Fairey – demonstrably lacked. The Planning Department, for example, required specialist expertise which only American-trained project managers could provide. The same was true of the numerous aircraft inspectors scattered around the country – tension usually arose when British or Canadian were *in situ* and clashing with their US military counterparts. The BAC's native-born senior staff were highly valued and suitably remunerated. Neither they, nor the largely local-born clerical and support staff, were expendable. This meant department heads had to handle all colleagues with due sensitivity. CRF found himself adopting management methods which constituted the norm on the eastern seaboard and the west coast, but were as yet unknown back home. He recognised the importance of formal recognition for outstanding performance, particularly when the person concerned was many thousands of miles from home. He wrote on a regular basis to Sir Archibald Rowlands and Lord Halifax, recommending that they reward the hard work of junior colleagues.[43]

Fairey's role was very much hands-on, and he was operating at a distance from the bilateral negotiations conducted between the US Treasury Secretary Henry Morgenthau and Arthur Purvis, head of the British Purchasing Commission until his death in August 1941. For all his intimacy with Beaverbrook and long hours spent in the company of Bill Donovan, CRF was even more remote from the critical channel of communication maintained between Churchill and FDR. Shuttling between the White House and Downing Street, veteran New Dealer Harry Hopkins proved a powerful and trusted intermediary.[44] Fairey must have met Hopkins, but would have been on better terms with future secretary of state Edward Stettinius, who from the summer of 1941 ran a Lend-Lease programme totalling $7 billion, of which $2 billion was allocated to aircraft procurement. Stettinius's background was in industry, but Fairey must invariably have felt uncomfortable with the New Deal administrators who would become so crucial to the federal war effort, and who saw Lend-Lease as wholly consistent with their commitment to deficit funding and aggressive state intervention. This was anathema to someone whose political beliefs and economic thinking were rooted in a minimalist state and largely unfettered competition. Yet Fairey always acknowledged the incompatibility of total war and the free market, enabling him to deal effectively and harmoniously with progressive Democrats who in peacetime he would instinctively condemn.

The commencement of Lend-Lease in mid-1941 meant that the BAC no longer negotiated cash contracts directly with American manufacturers. A centralised mechanism for procurement identified the number and type of aircraft for inclusion in USAAF or USN contracts, and what proportion would be allocated to the British in accordance with the appropriate BAC requisition order. As is obvious, appropriations to the British Account were generous, and, excluding a freeze on overseas deliveries post-Pearl Harbor, grew at an ever faster rate until the second half of 1944. In some cases the selection of British-bound machines might occur surprisingly late in production, but key to the process was an inter-service planning unit at the Wright-Patterson airfield near Dayton, and a strategic decision-making body based in Washington, the Joint Aircraft Committee (JAC). By the summer of 1942 over 500 British personnel had been posted to Dayton, liaising with the USAAF's Material Division headquarters.[45] Fairey was a frequent guest at the Ohio air force base, his high-profile visit in July 1942 attracting extensive coverage in the *Dayton News*: in an interview and a reported speech the newly knighted Sir Richard was fulsome in his praise of Britain's newest and most generous ally.[46]

CRF's most crucial role in 1940–1 was as Britain's representative on the Administration's small but highly influential JAC. Established a month after CRF arrived in the United States, the JAC was technically a subcommittee of the President's Liaison Committee, a larger body for the coordination of foreign and domestic coordination. Its shape, size and significance shifted after the transition to Lend-Lease, but in essence the remit of the JAC was to coordinate all aviation requirements of America, Britain and its allies, to standardise equipment where possible, and to exchange operational experience. It grew from a small but vital decision-making body to a larger, more representative supervisory body – where once Fairey sat alone, by early 1942 the British contingent numbered four.[47]

US Treasury representative and de facto secretary at JAC meetings was Philip Young, once of Harvard Business School but since 1938 a government adviser on American arms exports. His father, Owen, was a farm boy from upstate New York who ended up president of General Electric. In 1929 Owen Young had formulated a successor to the Dawes Plan for restructuring Weimar Germany's reparations payments. In due course the Young Plan was derailed by Nazi denunciation and the onset of the Depression. A decade later, and with war looming, Owen's son found himself working inside the Treasury: Philip Young worked for the Office of Emergency Management and as Secretary Morgenthau's special adviser, in which capacity he serviced the JAC. From March 1941 he helped Harry Hopkins set up the Lend-Lease Administration, and by the end of the year he was Edward Stettinius's chief trouble shooter. Presumably motivated by a sense of duty, Young ended the war in uniform. Far from Washington he served as a humble lieutenant commander in naval logistics. A distinguished postwar career would see the one-time New Dealer switch allegiance to the Republican Party, working in the White House as Eisenhower's chief of staff, and then as US ambassador in The Hague. Philip Young, the youthful scion of a patrician New England family, and Richard Fairey, the middle-aged self-made man, came from very different backgrounds. Yet they established a close working and personal relationship, with Young always given a warm welcome at Bossington. His wife Faith became Esther's closest friend in Washington. After Faith died in 1963 Philip Young married his old friend's widow – Esther Fairey, now Esther Young, lived quietly in upstate New York and Virginia until her death in 1978, nearly ten years before her second husband died of a heart attack in Arlington Hospital.[48] Long after his death Philip Young would remain a powerful presence within the collective memory of the Fairey family.

Fairey formed one-third of a triumvirate, with the JAC's other two members being General Arnold and Rear Admiral Jack Towers, chief of the Naval Bureau of Aeronautics. The BAC representative doubtless had a healthy respect for Arnold, but Towers was surely one of the few men in whom Fairey stood in awe.[49] A protégé of Glen Curtiss and an American counterpart to Charles Sansom in pioneering naval air power, Towers was as competent, courageous and comfortable in the cockpit or on the bridge as behind a desk at the Navy Department. Between 1939 and 1942, when he assumed command of the

Pacific Fleet's principal carrier force, Towers secured an increase in naval aircraft from around 2,000 to a staggering 39,000.[50] Fighting on the far side of the Pacific in the final months of the war, Jack Towers was reminded of his old sparring partner in Washington whenever he saw Barracuda squadrons flying off the fleet carriers London insisted must form part of the final assault on Japan.

With the Fleet Air Arm largely relying on British – often Fairey – aircraft in the early years of the war, CRF was more likely to clash with 'Hap' Arnold. A sea change in relations followed the Air Corps chief's two-week visit to Britain with Harry Hopkins in April 1941. The Air Ministry, MAP and the RAF provided VIP treatment and showed Arnold everything, from the Whitehall war room to the control centre at Bentley Priory. He visited several companies, but not Fairey Aviation: Hawker and De Havilland made the deepest impression, courtesy of the Tempest and the Mosquito – the Battle, Fulmar or Barracuda could not compete with combat aircraft of this calibre, hence the Hayes factory's absence from an astonishingly packed schedule. Arnold's hosts for the visit were Slessor, by now an air vice marshal, and the Chief of Air Staff himself, Sir Charles Portal. MAP pulled out all the stops, with Beaverbrook – despite having to deal with the loss of Stornoway House – seemingly available at all hours of the day. Arnold was courted by Churchill over a long weekend, and regally entertained at Buckingham Palace. He flew home impressed by Fighter Command, underwhelmed by Bomber Command, and annoyed by Coastal Command – for not immediately deploying newly delivered B24 Liberators better suited to bombing Berlin. The delay in modifying these aircraft for anti-submarine warfare remained a sore point across the following 12 months. While returning home in a better frame of mind towards the British, Arnold remained unconvinced that the RAF was making the best use of aircraft, which otherwise would be equipping front-line squadrons from Maine to Manila. Back in Washington he provided Roosevelt, key cabinet members and the ever-present Harry Hopkins with detailed briefings on his trip. In the weeks that followed collaboration stepped up several gears, with initiatives ranging from an RAF training programme on American soil to a transatlantic ferry service.[51]

A photograph of Arnold at the controls of a Wright brothers aeroplane *c.*1912, signed with a warm dedication, graces the mantelpiece of the dining room at Bossington. In late 1945, reflecting on their time together bargaining hard over which aircraft would go where and in what numbers, the USAAF Chief of Staff would write to Fairey in glowing terms: 'The great victories that we have won could not have been achieved without the aid of such men as you who gave so unselfishly of their time and their knowledge and experience to blueprint the vast production of airplanes which won us mastery of the skies.'⁵² In reality the two men may have considered themselves friends, but they were never close – as implied in Philip Young's account of how the JAC operated. Given the detail of 'Hap' Arnold's diary, the absence of any reference to Dick Fairey is striking. To be fair, Arnold's diaries are episodic, so everyday business in Washington rarely warrants a mention. What they do reveal is that Arnold's goodwill towards Britain following his first visit was short-lived, even if he never became as openly Anglophobe as say Admiral Ernie King or 'Vinegar Joe' Stilwell.⁵³

Having witnessed the worst effects of the Blitz, Arnold's dealings with Fairey were now rooted more firmly in a common purpose. Yet their partnership on the JAC was shortlived. In August 1941, by now Chief of the Army Air Forces, Arnold attended the first Anglo-American summit at Placentia Bay, off Newfoundland. During the conference he agreed with Portal, head of the RAF, that from now on key decisions regarding aircraft procurement should be the preserve of the relevant service chiefs and their staffs. The conference confirmed that this was already happening, with Sir Wilfred Freeman, now Vice Chief of the Air Staff, arguing for the RAF to receive nearly 2,000 heavy bombers by the end of 1942. Arnold agreed a final figure of less than 600. Endorsement of the new inter-service arrangement came from a familiar figure: although he had left MAP to become Minister of Supply, Beaverbrook still wielded considerable influence. He recognised that joint staff procurement decisions could only be influenced by the intervention of Hopkins and Averell Harriman, FDR's personal representatives: the RAF Delegation in Washington and the BAC together claimed a 'moral right' of appeal to the President. Yet if the Commission was to have any indirect influence then, in Beaverbrook's words, it had to have 'toughs' lobbying the Administration. Within six months Self had moved on and CRF was in charge.⁵⁴

Seeking to influence decision-making is very different from being an integral part of the process. Fairey was no longer party to a bilateral process of aircraft allocation, let alone strategic decisions concerning future priorities; neither in the first instance was the Admiralty. The absence on the British side of a civilian or naval voice was confirmed at the 'Arcadia' conference in Washington following the United States' entry into the war: in early January 1942 Arnold, Portal and Towers agreed on the types and quantities of Lend-Lease aircraft to cross the Atlantic over the next 12 months. That May in London the three men thrashed out with Churchill a ten-point programme of collaboration for Roosevelt to sign off on. A month later the service chiefs met again in Washington, with overseas deliveries significantly cut given the urgent need to re-equip American squadrons decimated by the all-conquering Japanese. If anything, divisions over the supply of US heavy bombers had deepened, with the Americans still sensitive to what they saw as the RAF's inappropriate deployment of Liberators and B17s already in Britain. Meanwhile, the needs of the Royal Navy were all but ignored: a staggeringly high attrition rate for American carrier aircraft in the Pacific left the Fleet Air Arm heavily dependent on Fairey Aviation and Supermarine for its fighters, and Fairey for its strike aircraft – the loss of all six Swordfish sent to sink the three capital ships 'dashing' up the Channel on 12 February 1942 signalled an obvious need for high-performance torpedo-bombers like the Grumman Avenger.[55]

Succeeding Sir Henry Self as Director-General of the BAC, Fairey's job was to influence, and not actually make, the big decisions, and then to see those same decisions implemented in the best interest of his country.[56] By the start of 1942 he was already focused upon the latter task, and here his executive skills and manufacturing experience were at a premium. Clearly he still had to court public opinion, with the BAC developing a slick PR operation, but his access to senior members of the Administration was increasingly via the British Supply Council (BSC). This latter body, supported by a large secretariat, coordinated the work of all British missions. Self served as its chief executive for the final two years of the war. His successor as Director-General sat on the Supply Council, which met weekly at the Willard Hotel on Pennsylvania Avenue, and from late 1942 was chaired by a resident minister accountable directly to Downing Street. Every fortnight he attended a liaison meeting of all British and Commonwealth heads of mission, 'when each would give a brief account of anything of interest that was happening, and everybody was so kept generally in touch.' Halifax was in the chair, his main purpose being to ensure that in their dealings with the Americans representatives of the United Kingdom and the Dominions were all singing from the same hymn sheet. The same was true for reporting back home.[57]

Fairey kept the Embassy fully briefed, while at the same time reserving certain business for his own direct line to London. This was code-named 'Fairland', with messages enciphered separate from the BAC's regular signals, and cabled home via the Embassy's high-security system. CRF communicated almost daily with Archie Rowlands, MAP's permanent secretary, and he liaised directly with the ministry's Director of Canadian and American Purchases, a post the title of which changed three times between 1941 and 1943. CRF's third key link with home followed Sir Wilfred Freeman's return to MAP at the end of 1942. Freeman's appointment as chief executive confirmed how much the ministry now saw the need for realistic planning, as opposed to Beaverbrook's exhortatory and teleological approach. Fairey could always communicate directly with Rowland or Freeman on his 'scrambler' telephone, but only on rare occasions did he speak with Beaverbrook and his successors. To Moore-Brabazon's credit, during his brief tenure at MAP he resisted an obvious temptation to call his old friend across the water.[58]

Moore-Brabazon left Transport in April 1941 to become the second Minister of Aircraft Production. CRF was of course delighted with the appointment: 'So you have become beaver!' Unknown to Self, Fairey sent his old friend a 'personal insight' into the present state of the BAC. He painted a generally positive picture of procurement arrangements in Washington and across the country. Here was no Iagoesque betrayal, with Fairey unstinting in his praise of Sir Henry. Surprisingly, he displayed similar goodwill towards both Towers and Arnold, believing them sincere in wanting to make the JAC work well for both the United States and the United Kingdom.[59] Moore-Brabazon was minister for less than a year, but in that time he maintained an unofficial line of communication with Washington. In September 1941 he insisted CRF ignore rumours of an imminent return to an ailing company as MAP believed his work in Washington to be 'of outstanding importance and could not be done by anybody else in the world.' A grateful Fairey pointed out how much he valued his present job: success outside of industry would prove conclusively that past achievements were based on more than mere luck. The price of staying in the States was not seeing Moore-Brabazon, as 'one cannot replace friends. Americans even the best of them are dull drops, and quite blind to the virtues of such matters as the Einstein theory or model railways as subjects of conversation.'[60]

Within months Moore-Brabazon had plenty of time on his hands to study relativity or scan the Hornby Dublo catalogue. A leaked indiscretion at a private lunch in Manchester cost him his job: the late summer of 1941 was not an ideal time to suggest Britain benefited from the bitter fighting between the Soviet Union and Nazi Germany, even if Moore-Brabazon claimed his comments were merely an appeal for industry to exploit the absence of imminent attack. Any attempt to clarify these remarks made matters worse, with critics ignoring a beleaguered minister's claim that supplying front-line aircraft to the Russians was always his priority. On 21 February 1942 Churchill accepted the harsh political reality. He sacked Moore-Brabazon, sending the soldier turned politician John Llewellin to MAP for nine months, after which Sir Stafford Cripps headed the ministry until May 1945. In his memoirs Lord Brabazon maintained that he left office with no regrets, the award of a peerage having eased the pain of dismissal: 'Anything more miserable, or harder work with less thanks, than being a Minister in war-time I don't know.'[61] A shocked Fairey felt 'the loss to our war effort is a real one ... I shall miss the personal touch that pertained before.'[62]

Even without Moore-Brabazon, when Fairey took tough decisions out in the field he was confident that back home MAP would support him, as proved the case in May 1943 when the BAC cut its losses on contracts for Brewster's unwanted scout-bomber, the Bermuda. The BAC Director-General was in the unusual position where he was free to amend or terminate contracts, but could not authorise capital expenditure without London's agreement. He must have found such an arrangement intensely frustrating. Yet CRF was ready to risk congressional criticism when he cancelled delivery of the Bermuda, an obsolete aircraft which, for diplomatic reasons, had been forced on the RAF and the Fleet Air Arm. Parallels with the Battle's later life were obvious, although it is unlikely Fairey appreciated the irony. The Bermuda, and its sister aircraft the Buffalo, demonstrated how, among the remarkable array of American marques sent to Britain, there were some truly terrible designs: supply had to be controlled or curtailed without causing offence inside DC, and beyond.[63]

Here Fairey became acutely conscious of the need to prevent potentially damaging stories being taken up by the American press. Back in Britain those machines the RAF deemed unfit for operational use were discreetly stored or despatched to Delhi and the Delta. MAP and the BAC were reluctant to cancel contracts if this endangered delivery of those aircraft the RAF really did want, notably the Catalina flying boat. Less problematic was the machine transformed by modification, of which the Merlin-engined P51 Mustang is the most famous example. A key factor in the RAF's critical evaluation of American front-line fighters was the poor performance of their Allison engine at high altitude: with the Castle Bromwich factory building Spitfires in such large numbers, why accept inferior aircraft? Fighter Command's criticism was dismissed in the United States, fuelling charges of British ingratitude and complicating already tough negotiations with 'Hap' Arnold. The immediate success of a Rolls Royce-powered P51, as proclaimed in newspapers on both sides of the Atlantic, was a major boost for the Allies. Yet it raised awkward questions regarding the overall performance of other USAAF fighters, notably the Lightning and Airacobra. This further complicated inter-staff wrangling over future Lend-Lease allocations, such that in mid-1942 Britain initiated a reciprocal agreement in order to appease Arnold and his staff. Thus, while the British might claim some or all of the credit for improving upon the original Mustang, avoiding an image of Limey superiority was paramount. The same was true when the Army Air Forces' squadrons were seen to be flying types built in Britain. Fairey needed to be sensitive to American feelings, and yet in the early years of the war he could take quiet satisfaction from the fact that qualitatively Britain's best combat aircraft had the edge over their American counterparts. Ironically, the obvious exception was carrier aircraft, with Grumman, Douglas and Chance Vought generating designs superior to those already realised or currently under development at Blackburn and Fairey Aviation.[64]

For four years from 1938, American car manufacturers matched their British counterparts in mistakenly assuming vehicle production methods were easily transferable to airframe assembly. Similar to Britain, their productivity levels were low, and as with the shadow factories there was a steep learning curve: until costly lessons were learnt heavy capital investment in new plant was no guarantee of volume production.[65] Nevertheless, production of American aircraft increased at a truly stunning rate in the latter part of World War II, peaking in 1944 with an output of 96,318 – 70,000 more than were built in Britain that same year.[66] The unprecedented size of American manufacturing plants, and the length of production runs, facilitated economies of scale unattainable in Britain. This was despite expanded workforces significantly increasing their productivity from a low prewar base.[67] With Towers relaxed about the demands of the Royal Navy, it was the Fleet Air Arm which benefited most from a marked improvement in the quantity and quality of American-built aircraft.[68]

The RAF had less reason to cheer as future deliveries were largely to replace losses sustained by those squadrons flying previously supplied aircraft. This was a consequence of the modified agreement Arnold and Towers signed off with Portal and his staff in June 1942, and it acted as a critical brake on the rapid re-equipment of RAF squadrons in the Far East.[69] By the end of 1942 the scale of the challenge facing Fairey and his colleagues remained formidable: Britain had received 22.2 per cent of total US output, constituting only 42 per cent of the target agreed with Slessor during his stay in Washington nearly two years before. In the course of the year MAP received fewer than 2,000 aircraft from the United States, and yet British firms shipped overseas nearly half the 16,329 machines that rolled off UK assembly lines.[70]

Eighteen months after Hitler's invasion of Russia, Britain, not the United States, was bearing the brunt of Lend-Lease supplies to the Soviet air force; albeit at the same time offloading unwanted US fighters. Once the Administration realised US aircraft were being passed on, officials counted them within the United States' Lend–Lease commitment to the Russians, at the expense of Britain's overall allocation. Here was another problem Fairey had to confront, and one that was never really resolved. Meanwhile, Arnold became ever more suspicious of British intentions. The goodwill shown in May 1942, during his second visit to Britain, soon ebbed away. Tensions emerged at the Casablanca conference in January 1943, with Portal stunned to learn just how big the USAAF intended to be by the end of 1944: a complement of well over two million, with around 52,000 combat aircraft. Later in the year Arnold visited Burtonwood, the Merseyside reception depot for US deliveries, unfavourably comparing the turn-around time for engine repairs of British mechanics with that achieved by Americans brought in to replace them. While in England Arnold came to the firm conclusion his Eighth Air Force was out-performing Bomber Command in their day–night joint offensive over Germany. 'Hap' Arnold's criticisms and his instinctive suspicion of British intentions fuelled resentment among senior Air Force staff in London, Cairo and Washington – CRF had to keep the Americans happy *and* appease his uniformed colleagues in the RAF Delegation.[71]

Fairey's gruelling work load would have challenged a younger, healthier man. He divided his time between administrative and committee work in Washington (and in Ottawa), and visiting factories and air force bases the length and breadth of North America. As CRF explained to his old friend 'Mary' Coningham, the RAF's freshly appointed C-in-C in the Middle East, and a singularly poor choice, 'It is one long struggle getting you aircraft, and many of them are going to you. American production methods are very different to ours and take much longer to get going and it often takes much deft handling to pry them loose when they are made.'[72]

Sundays in Washington were spent playing golf at the ultra-fashionable Chevy Chase Club north of the capital; in the club house or out on the green Fairey took every opportunity to cultivate valuable contacts and fly the flag. In reality, therefore, across the whole week he rarely took a day off. CRF's secretary recalled the long hours her boss spent at his desk, often returning to the office after dinner. Saturdays constituted a normal working day, and time off for public holidays was kept to a minimum. Working lunches were the norm, whether with senior colleagues at the Metropolitan Club, or with federal and military contacts at the Army-Navy Club or the Pentagon dining room. A formidable amount of time was spent entertaining visitors, not all of whom had a direct interest in the work of the BAC. While an hour or two with Jean Monnet could be considered time well spent, afternoon tea with the veteran thespian Cedric Hardwicke offered little more than a trip down memory lane.[73]

As we shall see, rare visits to England in the course of the war entailed long and tiring flights across the Atlantic. In later life Lady Fairey lamented the physical cost of CRF spending so long at high altitude in poorly heated cabins, especially when doctors in Washington warned of the consequences for her husband's heart: 'they had some terrible trips. But it simply had to be done, even though he never liked flying…'[74] Then, when he was back home, Fairey found scant opportunity to quit Whitehall briefings and seek sanctuary in his beloved Test Valley. Maintaining such an arduous lifestyle was not unusual in time of war, but Fairey sacrificed both his health and a normal family life, at a time when John and Jane were still very young, and eager to spend quality time with their father.⁷⁵

Living in a 'bourgeois town' – Washington DC

> Home of the brave, land of the free
> I don't wanna be mistreated by no bourgeoisie
> Lord, in a bourgeois town⁷⁶

Wartime Washington was a crowded city, busy and alive both day and night. It was unique in being subject to congressional administration, and the southern 'Dixiecrats' remained a powerful and unreconstructed element within FDR's broad Democratic coalition. This meant the District of Columbia was very much a city of the South, with institutional segregation. African Americans were heavily discriminated against, and within Washington they constituted a literally servile class. Visitors from Europe encountered a wholly new experience, with almost all service and support staff black and poorly paid. On arrival at Wardman Park the Fairey family found that everyone not in a supervisory role was black. African Americans constituted around 35 per cent of what was a deeply stratified residential population. They needed easy access to their places of work, hence the close proximity of de facto ghettoes to the most affluent and the most public parts of a capital city only a century old.⁷⁷ While visitors from South Africa might take for granted an urban and racial landscape marked by such a stark contrast between wealth and poverty, most Europeans found themselves living in an environment very different from anything they may have experienced back home.

The federal capital hosted representatives from every member of the grand alliance, with by far the largest number drawn from Great Britain and the dominions. Thus the British constituted a major presence within Washington society. Lord Halifax – an aristocrat serving at the heart of the Republic – was the public face of a seemingly ubiquitous ex-pat community. The British ambassador's seniority and status was matched only by his military counterpart, the universally liked and well regarded Field Marshal Sir John Dill.⁷⁸ FDR once remarked that Washington society boasted a trio of diversions: 'the saloon, the salon, and the Salome.'⁷⁹ While conscripted intellectuals like Isaiah Berlin might attend the salon, puritan and punctilious members of the British missions avoided all three distractions. Led by Lord and Lady Halifax, the pursuit of pleasure was for a purpose: high-profile appearances and whole-hearted immersion in the social life of Washington's political elite were seen as vital to oiling the wheels of a still embryonic 'special relationship.'

Military and civilian personnel, and crucially their wives, needed to suppress the British upper classes' inherent sense of superiority: for too long privileged and powerful Americans had been judged brash and materialistic. Such impolitic prejudice placed further strain on a traditionally difficult diplomatic relationship between the United States and the United Kingdom. Few temporary residents in DC, other than the diplomats, would have read Henry James, or explored the complex and remarkably hierarchical network of the eastern sea board's most prominent families, from the Cabot Lodges to the Rockefellers. Family ties cut across party loyalties, witness the Roosevelts' historic importance in both major parties. Notwithstanding fierce divisions fostered by the New Deal, the Washington establishment was surprisingly free of partisan bitterness, and long before the United States declared war, Republican non-isolationists in Congress were privately, and sometimes even publicly, assisting the Administration. British movers and shakers, and their spouses, had to comprehend the subtleties and complexities of Washington power politics, and at the same time overcome most Americans' instinctive suspicion of a nation which since 1776 had more often been an enemy than an ally. After the war Halifax recalled how FDR 'used to delight in reminding me that there were more English in Washington in 1942 than there had been when they had burnt the White House in 1814.'[80]

The Director-General of the BAC was unusual in knowing the United States so well. He liked Americans, and he could claim a history of having promoted his host country's commercial interests in Britain between the wars. Wealthy in his own right and always a generous host, Fairey could entertain in style, complementing the meagre allowance allowed by HM Government for wining and dining Washington's version of the great and the good. These duties were shared with a wife, partner and companion who shone in the full glamour of society life on the banks of the Potomac. In early 1942 the *Washington Post* profiled Esther Fairey as an accomplished mother, a stylish and attractive representative of the best in British *haut couture*, and an intelligent and accomplished woman of many parts. Here was someone urban and urbane, but at the same time an archetypal English lady, with her roots firmly in the countryside: 'She loves … riding, but is too fond of animals to hunt [like CRF], is fond of reading, and now and then enjoys a good fishing trip.'[81] A carefully lit, well-composed portrait confirmed for readers that here was someone who in looks and appearance represented all that was best about the British: newly acquired allies who every American patriot would feel privileged to fight alongside. The piece was a fitting and timely tribute to the future Lady Fairey, and – by appearing in no less a newspaper than the *Washington Post* – it constituted a triumph of public relations.

Fairey received his knighthood five months later, with the announcement triggering a deluge of letters and telegrams. Predictably, the fullest messages of congratulation came from his peers within the aviation industry. Everyone from Geoffrey de Havilland to Tommy Sopwith paid tribute to Dick's multiple achievements. However, the only correspondents from within the Cabinet were Kingsley Wood, an unfortunate choice as Chancellor of the Exchequer with Keynes restored to the Treasury, and the ever reliable Ernest Brown, an unfortunate choice as Minister of Health with Beveridge's welfare proposals due that autumn. Among senior Tory ministers Churchill clearly had neither the time nor the inclination to put pen to paper, and Eden had probably never met Sir Richard. The absence of letters from Labour MPs was scarcely surprising, although Bevin may have uttered a begrudging 'Well done' if he encountered Fairey in a Whitehall corridor. Of the other letter-writers, Lords Kemsley and Iliffe conveyed the congratulations of Fleet Street, but the most famous press baron was noticeable by his absence. More likely, Beaverbrook conveyed his good wishes in person.[82]

Notwithstanding his work with the BAC, Fairey's services to the aircraft industry were of sufficient merit as to warrant a knighthood. Nevertheless, unpaid service for King and Country surely deserved an honour appropriate to its recipient's status and reputation. Sir Richard's services were pro bono, but the Treasury maintained a miserly view of his expense claims. Similarly strict control was exercised over Fairey's personal bank accounts, on both sides of the Atlantic. Whitehall officials insisted on a punitive rate of currency exchange, and full disclosure of the Fairey family's financial interests in Canada and the United States. Fairey, eager to repay Dawson the generous sum lent to Esther in Montreal, was left fuming at the inflexibility of wartime bureaucracy – which of course he deemed no different from that of peacetime bureaucracy. Throughout the war Fairey conducted a running battle over travel expenses with the Treasury's senior representative inside the British Embassy. No payment was allowed to cover the cost of Lady Fairey's presence on any joint tours away from Washington. Access to a car and a regular chauffeur was given grudgingly after CRF's promotion to Director-General.[83]

News of Treasury obduracy came at the end of a torrid few months, relieved only by America's declaration of war on 8 December 1941. A few weeks earlier CRF took his first proper break since arriving in the United States, confiding to his nephew just how much he needed a rest: 'I am beginning to experience something of the feelings of a heavyweight who, in the middle of a long fight, can not remember how many rounds have gone or know how many there are yet to come, and the intervals in which he just sits down and feels tired must get briefer and briefer.' Flying the red-eye express to LA was exhausting, while day-to-day life in DC was a drain on both spirit and physique. Whenever possible Fairey travelled by train, but, for all the creature comforts of couchette and dining car, the criss-crossing of such a vast continent was both time-consuming and tedious. In early December 1941 Mr and Mrs Fairey departed Union Station, their Baltimore and Ohio locomotive heading for St Louis, and the Union Pacific streamliner service to California. This was a railroad route familiar to CRF from earlier visits to West Coast plants. On this occasion Esther accompanied him, the couple keenly anticipating a visit to Hollywood.[84]

The excitement of meeting Walt Disney and touring his studios was marred by news of the attack on Pearl Harbor. That Sunday evening the Faireys were confined to their hotel, but they soon slipped out, and, with Charlie Chaplin and King Vidor at the next table, dined in a suitably solemn eating house. Plans for an ambitious tour of the north-west and Canada were swiftly cancelled. This was doubly disappointing in that visits to Vancouver usually extended to a few days salmon fishing in the Rockies. With all airlines in lockdown, the military flew them to Chicago and the B&O service south. Twenty years later Lady Fairey recalled their unique yet solitary flight: 'An enormous plane and there was nobody else in it at all.' The couple had been lunching in Beverly Hills with the ferociously Anglophobe Donald Douglas when news reached the restaurant that Japanese carrier aircraft had decimated the Pacific Fleet. CRF must have wondered to what extent the success of his Swordfish at Taranto had inspired such an audacious operation. Later it emerged that the Japanese naval staff took a keen interest in how much damage Admiral Cunningham's squadrons had inflicted on the Italian fleet. Furthermore, the US Navy had ignored Downing Street's presidential briefing on the wider implications of the Fleet Air Arm's triumph at Taranto. Despite his dislike of the British, Douglas got on well with Dick Fairey, their two companies sharing an expertise in the design of carrier-borne aircraft. Yet all his prejudices would have been confirmed had he been party to CRF's private thoughts regarding Pearl Harbor: 'Apart from the loss involved, the attack on Hawaii did much more good than harm.'[85]

The news from home was equally bleak, with Woodlands damaged and Fort Charles destroyed. Bossington would soon be partially requisitioned by the RAF – Moore-Brabazon ensured that this was no more than a mildly disruptive occupation. For a melancholy CRF the house, the estate and above all the Test meant 'all that there is of a future for the family and myself, and presumably we shall be lucky if we ever get back a normal life.'[86] In the early months of 1942 peacetime normality must have seemed a long way off: 'The time passes and yet lags. It is difficult at times to imagine whether one has been out here ten years or ten days. I long for the company of my friends and the feel of rod or gun again.'[87] Appointment as Director-General and the knighthood that followed naturally spurred him on, but Fairey had to learn to pace himself. It is not obvious that he ever did.

The dense volume of unofficial correspondence emanating out of Fairey's office signals evenings and early mornings dedicated to his personal affairs, above all the upkeep and maintenance of Bossington and its prized beats. Hand-written notes would introduce American visitors to friends and contacts at home, not least his agent at Bossington, where a day on the Test was always available for any fisherman arriving from the States having met CRF's stringent standards. A regular stream of telegrams conveyed birthday wishes or seasonal greetings to both relatives and estate staff, while every effort was taken to secure state-of-the-art fishing tackle and dispatch it across the Atlantic as surprise gifts for close friends. Fairey was especially generous when thanking those he had seen and/or stayed with when back in Britain, most notably in the spring of 1943.[88]

When writing home a frequent – if exaggerated – complaint was that no-one in America seemed able to talk 'about such important matters as yacht racing, shooting, wild-fowling or fishing.'[89] While the family holidayed on Long Island to escape the spirit-sapping summer heat of Washington, Sir Richard headed north to test his mettle in the trout streams and salmon rivers of Quebec and the Maritimes. Here was a chance to talk fishing with fellow enthusiasts, and to test the newest rods and reels. In October 1942 CRF impressed his New Brunswick guides with 'my up-stream dry fly methods ... I could catch six fish to one against their method of towing a large sunk fly down stream.' In correspondence with his local vicar back home Fairey considered at length why North American salmon were so much larger than their British cousins. The Reverend Daubeney – a man of the cloth more at home in waders than gaiters – was keen to compare notes on how the deep and subtle art of fly fishing differed across two continents. Daubeney was in no way intimidated by Fairey's rigorous application of scientific method, endorsing his parishioner's commitment to preserving the ecological balance of the Test Valley.[90]

Well over a year after arriving in America, Jane Fairey was still demanding 'that she be taken back to Bossington.' Her older brother was 'rapidly absorbing Americanisms' in the playground, thereby confirming his father's jaundiced view of the local schools system. While personally averse to raising his hand in anger, Fairey disliked the absence in American schools of corporal punishment. He believed this prejudiced his son's view of schools back home run by strict, sound-minded disciplinarians. By the spring of 1943 CRF was sufficiently concerned by the attainment gap between John and his class mates that a place was secured at Ridley College, in Canada. Ontario's premier boarding school was deemed to 'follow our standards', not least when it came to enforcing discipline. Pending enrolment in an English public school, life at Ridley would be ideal for someone who 'is a grand fellow and going to be good; he is keen on simply everything.' By this time Sir Richard and Lady Fairey were renting a downtown apartment, but still their daughter yearned to go home: 'Jane does not acclimatize; she demands to be taken back to England and is passionately anxious to live in the country.'[91]

A parent's fears and worries: Sir Richard's eldest son

In January 1942 Moore-Brabazon, mired in controversy over his inopportune comments concerning the Russians, enjoyed a welcome distraction. He secured Richard Fairey passage across the Atlantic so he could visit his father. Fairey joined the Norwegian merchant ship, DS *Ringstadt*, sailing in convoy from Cardiff to Canada. The trip was later justified on the grounds that in the States an experienced pilot with the Air Transport Auxiliary awaiting call-up to the RAF would be of value to the newly established Atlantic Ferry Organisation (AFO): Richard junior could complement the AFO crew ferrying the first Liberator to Britain. In 1960, following Richard Fairey's death, a legal dispute required Lord Brabazon to state the exact circumstances in which he had authorised the visit. Claiming ministerial authority to arrange such a journey, Brabazon maintained that agreement was against his better judgement, and that Fairey's father was not in any way party to the decision: it was vital to the war effort that MAP act on Richard's claim to be more detached than any of Fairey Aviation's directors when briefing the chairman on his company's present poor performance. According to his testimony, Brabazon had agreed that young Richard was uniquely qualified to put Fairey in the picture, after which his father could take appropriate action. In this version of events Richard Fairey was portrayed as the Minister of Aircraft production's personal envoy: when a flight to the States proved hard to secure, Brabazon instructed his officials to arrange a sea passage.[92] The first Lord Brabazon placed his version of events on record, but almost 60 years later his justification for arranging Richard's transit reads like a retrospective rationale. Was he simply doing a favour for his best friend's eldest, in the hope that far from home father and son could be reconciled?⁹³ One thing is certain – within weeks of sanctioning Richard's journey, the minister was bitterly regretting his decision.

The *Ringstadt* was operated by Nortraship, the umbrella body for Norway's substantial mercantile contribution to the Allied cause.⁹⁴ Captain Jakob Knudstad's 37-man crew was still predominantly Norwegian, and their two years in exile had seen them cross the Atlantic around a dozen times. On her final voyage the *Ringstadt* was carrying a cargo of china clay – and, with cruel irony, her destination was St John, New Brunswick. On 13 January 1942 the *Ringstadt* left Belfast Lough as part of convoy O56, but out in the north Atlantic she became separated from the other ships. The ferocity of the winter weather rendered radio communication weak and intermittent, but on 23 January a message did reach the captain that another Norwegian vessel heading for Newfoundland had been sunk. Knudstad changed course to evade the enemy submarine, but a day later Kapitänleutnant Peter-Erich Cremer's U-333 fired its fatal torpedo. Twenty minutes later DS *Ringstadt* disappeared beneath the waves, having launched three of its four lifeboats with no loss of life – Richard Fairey was one of six British-born passengers, two of whom joined him in the captain's motor boat. The U-boat briefly surfaced, its crew asking if there were casualties and advising their victims set a course north-west. With the weather worsening the three lifeboats were separated, with two of them never seen again (the three lost passengers from Britain were a specialist Royal Navy ordnance team en route to Ottawa).

For a horrendous five days Knudstad's storm-tossed boat drifted north-west, the 13 men on board taking turns in sub-zero temperatures to chip away ever-thickening sheets of ice and to bail out a deadly flood of freezing, finger-numbing sea water. Emergency rations and protective clothing, including – crucially – seaboots, were missing from the lifeboat lockers, the assumption being that they had been looted by dockers. A solitary tin of cocoa allowed Fairey and his shipmates to share a drink of questionable nutritional value once a day.⁹⁵ On 29 January, 60 miles south-east of Cape Race they saw the smoke of a convoy, and soon after an escort aircraft flying out of Newfoundland spotted them. Some hours later the USS *Swanson*, an American destroyer, arrived to rescue the sole survivors of the *Ringstadt*. Although suffering from severe frostbite and in a near terminal condition, the ten sailors and three passengers, plus the ship's dog, were all still alive. Remarkably, Richard climbed up the destroyer's scrambling nets. Family folklore has Fairey collapsing on the deck and demanding 'For Christ's sake, get me a drink!' On being told the *Swanson* was a dry ship, he asked to be thrown back in.⁹⁶

In Reykjavik on 5 February 1942 the 13 men were transferred to the hospital ship *Avon Glen*, but only after Knudstad handed his alsation over to the *Swanson*'s skipper. The survivors were incredibly lucky; but also incredibly unlucky in that they had been just one day's sail from a home port. Cremer's crew – *Kriegsmarine* professionals praised by the *Ringstadt*'s chief oiler for their honorable behaviour – enjoyed similar good fortune: unusually for U-boat veterans, most of them survived the war. Indeed, with his impressive war record and subsequent business success, the highly decorated Peter-Erich Cremer seems the sort of chap Sir Richard and his son would happily have hosted for a weekend at Bossington – had he not been a German.[97]

It should also be noted that Richard Fairey's recollection of the submariners' behaviour contrasted sharply with the Norwegian sailors' version of events, as provided to the official inquiry in Reykjavik. He recalled the German crew laughing at the plight of the *Ringstadt*'s survivors, even claiming that Cremer caused lifeboats to capsize by ramming them. He was almost as critical of Jakob Knudstad, insisting the skipper had ignored a favourable wind, and in so doing wasted fuel which should have been held in reserve for when a rescue vessel was sighted.[98]

Informed that his son was missing, Fairey moved heaven and earth to ensure the Canadians mounted a full and continuous search for survivors. Mary Bell recalled how 'despondent' her boss became as it seemed more and more likely that Richard was dead. In later years Esther described her husband as having been 'utterly distraught.'[99] The RCAF search was eventually called off, so it was only by chance that an eagle-eyed bomber crew picked out Captain Knudstad's tender. Advised that a high fever and severely frost-bitten legs were life-threatening, the Director-General secured Richard an emergency flight from Reykjavik to London.[100] Admitted to the London Clinic an impressively stoic Fairey accepted the consultant's advice that immediate amputation of both legs was the only means of saving his life. Despite the greater discomfort, he opted for both legs to be amputated below the knee, in order to maximise mobility. As soon as he could, Richard had himself transferred to Queen Mary's Hospital at Roehampton, where with great courage he learned how to walk with prosthetic limbs.[101]

Crutches were discarded at the earliest opportunity, with Richard discharging himself well ahead of the doctors' timetable for rehabilitation. Fairey refused to accept the label 'disabled' or to feel in any way dependent on others. He set out to lead as normal a life as possible: always an impressive golfer, within a year he was once more playing to a six handicap.[102] Crucially, he resumed flying. Only eight months after losing his legs Richard Fairey joined the Aircraft Armaments

Experimental Establishment as a flight observer. He flew out of Boscombe Down for nearly three years, deferring a return to Fairey Aviation until the war was over. In other words, he eschewed the easy option of an undemanding desk job at Hayes, opting for the arduous task of in-flight data-recording on a succession of high-altitude trial programmes. When possible he would take the controls, demonstrating the same confidence and competence as a pilot he displayed prior to his double amputation. Fairey later extended his licence to include four-engined aircraft. In the late 1940s he trained to fly jets, and the following decade he flew solo in a helicopter after less than two hours of dual instruction.[103]

Before taking up his post at Boscombe Down, Richard Fairey – still dependent on walking sticks and still to shave off the beard that began when suffering in Captain Knudstad's lifeboat – travelled to Washington. This time he flew both ways, the authorities agreeing on compassionate grounds. His stay with the recently invested Sir Richard and with Lady Fairey was by all accounts a great success. Father found son a changed man: worldly-wise, mature and now in his late twenties, Richard remained eager to enjoy life to the full, but with an exuberance tempered by sobriety and reflection. The loss of both legs had proved a deeply painful (in *every* sense), sobering, cathartic and possibly even redemptive experience. It is scarcely surprising that Richard Fairey had belatedly tempered his playboy image, albeit without sacrificing his 'fantastic zest for living.' Some sceptics wondered to what degree he had put his darker days behind him, but none questioned his courage.[104] Physical and personal rehabilitation was rendered that much easier by Richard meeting his future – second – wife Diana Craig. After his remarkably brief stay in Roehampton he lived in Romsey down-river from Bossington, moving to Maidenhead after he remarried. It would be in Berkshire a year after the war ended that Diana Fairey gave birth to a son, Charles.

Although he had separated from Aino Bergö some time previously, when in hospital Fairey was concerned enough to ask how well she took the news that her estranged husband had lost his legs. Richard Fairey and his first wife would divorce in May 1943, but there was an unfortunate postscript. They were both sued the following November for non-payment of the balance owed on clothes which had been supplied three years earlier by a West End couturier. A censorious presiding judge voiced his disapproval that a total of £1,600 had been spent on dresses in 'the most critical year' of the war. He found the former Mrs Fairey wholly liable for the amount owed, accepting her ex-husband's claim that he had never sought credit from the company in question. The reputation of the Fairey family remained intact, and yet Mr Justice Lewis berated CRF for disapproving of Richard's marriage but continuing to 'over-indulge' him: here was someone who had given the extravagant couple a splendid first home in Windsor, repeatedly cleared their debts, and sanctioned a generous annual allowance to supplement his son's not insubstantial salary. Sir Richard must have been incandescent when he read newspaper reports of the court case. It is unlikely that he grieved for long when news reached Washington in July 1944 that his one-time daughter-in-law was the victim of a V1 flying bomb crashing into her Kensington residence. His son, however, was devastated when he heard the news.[105]

Travails at Fairey Aviation

Richard Fairey's visit to Washington in the autumn of 1942 gave him every opportunity to discuss with this father how successfully a Fairey Aviation bereft of its master had adjusted to the demands of total war. The simple answer was, not very well. Sir Richard was, of course, fully aware that at home huge changes were taking place across a vastly expanded aircraft industry. By late 1943 nearly two million workers were employed directly by airframe or aero-engine manufacturers, or indirectly via shadow factories and subcontractors. Capital investment since 1935 totalled £350 million, and production costs were running at over £800 million per annum. On the eve of D-Day MAP calculated that no fewer than 102,609 aircraft and 208,701 aero-engines had been built since September 1939.[106]

After the war Whitehall was keen to claim that, in a running battle between MAP and the SBAC, the will of the state had prevailed; witness the level of output achieved once a powerful and popular Sir Stafford Cripps sanctioned wide-ranging ministerial intervention.[107] Fairey's first instinct was tribal loyalty to the SBAC. He endorsed the view of directors and civil servants who visited the United States in September 1942 and found American manufacturing processes unsuited to the more modest assembly lines back home.[108] Yet, for the duration Sir Richard was a servant of the crown, duly beholden to the taxpayer – if lessons could be learnt from the Americans then he was duty-bound to say so. The same was now true of Roy Fedden, knighted alongside CRF for transforming Bristol's aero-engine division, but soon after sacked in a boardroom row. In early 1943 MAP sent Fedden on a fact-finding tour of the United States, his final report concluding that British manufacturers could learn a lot from their American counterparts. In this instance the minister and his senior staff sided with their major suppliers, fending off Sir Roy's call for fundamental changes in research, development and output as a consequence of a more interventionist promotion of cutting-edge R&D, a more dynamic working relationship between those responsible for design and assembly, and a more systematic and data-based approach to production planning.[109]

Fedden may have been fobbed off by Sir Wilfred Freeman – and even for a while a sympathetic Stafford Cripps – but his vision of manufacturing constituted the future, and CRF was in the unique position of witnessing at first hand the transformation in American manufacturing methods.[110] Fairey was adamant that this time around Britain's plane makers could survive the onset of peace: 'the UK can hold its own if we stick to the principles of competitive private enterprise … "Planning" might beat us, but nothing else will.' Yet he clearly could not ignore what was taking place in the United States, and even to some extent in Canada.[111] Neither could the heads of other British companies once they saw for themselves transatlantic techniques generating a level of performance rarely matched back home. Forty years later the British aircraft industry's harshest critic, Corelli Barnett, claimed a lamentable level of productivity in comparison with the United States. Other historians of aircraft manufacture on the Home Front, such as David Edgerton and Richard Overy, have acknowledged variable levels of productivity; but at the same time questioned the methodology behind Barnett's unflattering comparison of Britain's overall performance relative to that of the United States (and Germany).[112]

Barnett saw several structural and specific factors as contributing to low productivity in the aircraft industry, prominent among which was poor industrial relations. He identified a toxic combination of inept management and shop floor militancy as the trigger for strike action, citing Fairey Aviation as a firm adversely affected by a tired and truculent workforce.[113] With the prospect of invasion and defeat increasingly remote, workers by the middle of the war no longer felt the same sense of urgency as in 1940 and 1941. As we have seen, unofficial action had been a feature of the late 1930s, particularly at Heaton Chapel, and Hayes had outgrown the early plant's strong sense of a 'family' identity rooted in personal loyalty to a paternalistic executive. Notwithstanding close ties to subcontractors such as Blackburn, Fairey Aviation was a huge undertaking when its founder surrendered near-absolute control in the summer of 1940. Ensuing problems were not attributable to dissident shop stewards or obsolete restrictive practices, but the consequence of a power vacuum. Here was an organisation already incapable of meeting over-ambitious production targets. In the wake of CRF's departure, Maurice Wright and his colleagues spent the next two years fire-fighting. For all the proven attributes of Barlow, Broadbent et al., it quickly became clear – not least in Whitehall – that at the highest level Fairey Aviation lacked collective acknowledgement of the need for realistic planning, a clear sense of direction, a healthy respect for strategic thinking and, above all, a quality of leadership comparable to that of its most successful rivals. Production logjams, at both Stockport and Hayes, saw output stall, with replacement marques held back until hangars could be cleared of older models.[114]

Most notoriously, a huge backlog of the ever problematic Albacore prevented Hayes from initiating production of the Firefly, a two-seater reconnaissance-fighter, which in late 1941 claimed the life of Fairey's esteemed test pilot, Chris Staniland.[115] The much modified Firefly took an astonishing ten years to develop, with all involved soon losing sight of its original purpose. The production model only entered service in July 1944, by which time the aircraft's multiple variants rendered it all things to all men. Astonishingly, Firefly strike squadrons saw action in Korea, and the trainer version was flown by the Fleet Air Arm for a further five years. In contrast, the Barracuda was unequivocally a carrier-borne torpedo/dive-bomber, and, despite a switch of power plant, it took less than three years from the drawing board to the prototype. Yet it too was affected by a much-delayed transition, in this case replacing Fulmar production at Heaton Chapel and its partner subcontractors. Thus the unloved, aerodynamically challenged Barracuda II only entered service in January 1943. Nevertheless, 1,700 machines were built, and in the last two years of the war the aircraft belied its poor reputation by crippling the *Tirpitz* and decimating Japanese naval bases. The problem for the Fleet Air Arm was that, like the Fulmar and the Firefly, the Barracuda became available far too late (notwithstanding the fact that all three models were demonstrably inferior to equivalent American aircraft entering service with the Royal Navy after 1941 – as Fairey tacitly acknowledged when design of the Barracuda's successor, the Spearfish, drew heavily on the Grumman Avenger).[116]

Chronic delays at Fairey Aviation's primary plants provoked parallel problems with subcontractors, notably Blackburn, Westland and Boulton Paul. These companies had their own issues with subcontractors and suppliers, thereby compounding the overall problem. Although Fairey oversaw an established system of inter-factory liaison, there was an acute shortage of experienced staff, making the maintenance of quality control that much harder. MAP felt Fairey Aviation was failing to monitor performance at its 'daughter firms', which ostensibly were under the direct control of the Hayes parent plant.[117] To make matters worse, Heaton Chapel and its satellite plant at Errwood Park were not dedicated to constructing their own company's aircraft, rendering it necessary to set up assembly lines for the Barracuda at Fairey's three principal subcontractors.

Bizarrely, much of the firm's northern division was occupied building hundreds of Beaufighters and Halifax bombers, under contract to Bristol and Handley Page.[118] These were urgently needed aircraft, vital for day–night home defence or the accelerated air assault on Germany. A failure to deliver on time prompted Cripps' increasingly interventionist MAP to take radical action. Soon after his appointment the new minister flexed his muscles, insistent that coercion, not persuasion, would be necessary where companies were impervious to change. He soon had Fairey Aviation in his sights.

While it would be an exaggeration to say that, following his return from Moscow, Sir Stafford Cripps was a law unto himself – Churchill always had his measure, and out-maneuvered him at the height of his popularity in 1942 – he exercised ministerial power unilaterally, with scant concern for personal or party political considerations. Here after all was someone still seen by many beyond Westminster as a future prime minister. At the same time Cripps preferred to remain outside the Labour Party following his prewar expulsion for insisting an anti-fascist alliance must include Communists; if anything, he cultivated allies on the centre-right such as Anthony Eden. Crucially, Labour's most powerful minister on the Home Front, Ernie Bevin, tolerated his old ideological sparring partner's readiness to intervene. This was on the grounds that no Minister of Aircraft Production could ever be as bad as Beaverbrook – if a few underperforming industrialists' feathers were ruffled, then so be it.[119]

Cripps's personality and mode of behaviour were disarming for anyone primed to expect a revolutionary firebrand. There were those, on both sides of the House, who, for all his courteousness, simply loathed him. The diaries of 'Chips' Channon and Cuthbert Headlam reflect how so many Tory backbenchers, especially those loyal to Chamberlain, despised yet another maverick elevated to high office as a consequence of Churchill's premiership. Yet less partisan observers chronicled Cripps' remarkable capacity to charm pillars of the established order, not least the Royal Family.[120] For all the high principles and austere lifestyle, Cripps could, when required, turn on the charm. Many of Sir Stafford's fiercest critics acknowledged his intelligence and integrity. Friend and foe alike found it hard to reconcile a familiar charge of political naivety with a proven record of running large organisations. At social gatherings Sir Stafford could always rely upon his charming and popular wife, Lady Isobel, even when the atmosphere in the room became especially tense. Despite themselves, instinctive enemies such as Richard Fairey encountered someone whose executive credentials and readiness to listen, even when he did not agree with what was being said, belied his status as a Bolshevik barrister instinctively hostile to industrial capitalism. CRF's personal correspondence with Cripps suggests a relationship rooted in mutual respect, with a surprising absence of discord.[121] Beaverbrook he held in high regard and Brabazon was his closest friend, but Sir Richard maintained a grudging respect for Cripps while at the same time despising everything he stood for. That deep loathing boiled over when CRF sent the vegetarian minister a salmon, and a grateful Cripps guilelessly admitted that he intended his guests at a Savoy dinner party to eat the fish.[122]

Oswald Short's loathing of Cripps was deep and untempered: his enforced resignation as chairman of Short Brothers in early 1943 prefaced MAP taking control of the company under Defence Regulation 78. Cripps rode a storm of protest, highlighting Shorts' refusal to switch the manufacture of heavy bombers from its own miserable twin-engined Stirling to Avro's mighty four-engined Lancaster. Although the Prime Minister dismissed talk of the aircraft industry coming under public ownership, the SBAC feared postwar nationalisation.[123] Fairey's reaction to a state take-over of Shorts was predictable, yet a year earlier his response had been surprisingly measured when MAP set about restructuring his own company. Perhaps it was because, for all his loyalty to longstanding board members, he was sufficiently distant from events at home as to acknowledge his firm was in desperate need of new blood – only outside intervention could put Fairey Aviation back on track.

Cripps adopted a very hands-on approach to the production problems at Hayes, addressing the work force in the factory canteen at dinner time on Saturday, 19 December 1942 (directors saw his arrival in an Austin 10 as a stunt – they had clearly never met Attlee). He congratulated the 'comrades' on their hard work, and paid tribute to Sir Richard's hard work in America, using his regrettable absence as the key reason for restructuring the management and parachuting in a factory controller, G.E. Marden, a Great War veteran and chartered accountant whose business experience was in the Far East far from manufacturing industry.[124]

This undoubtedly odd choice was a ministerial appointment, but Marden duly joined the board as deputy chairman and, crucially, as managing director. He cultivated the shop stewards, and his positive impact on shop floor morale was recognised even by his fiercest opponents. Yet Marden was a divisive force, boosting the power of the Hayes plant's joint production company, and creating factions across the company's middle and senior management. Those managers who in 1943–4 openly supported MAP's factory controller were later seen by Sir Richard as having betrayed his trust. Personality clashes and power politics saw Marden seek to oust the old guard, while they moved heaven and earth to see him removed. Neither side was successful, with Sir Richard as the major shareholder, and the man with a direct line to the most senior figures inside

MAP, defiant in defence of his loyal acolytes. CRF invited Maurice Wright to Washington in August 1943 for a face-to-face briefing on Marden's performance eight months into the job. By this time Fairey was in regular correspondence with MAP's most senior officials, notably Freeman, to complain how much his present and future position within his company was being undermined. Rowlands, the permanent secretary, was also in Washington at this time, and no doubt subject to the full force of the Director-General's anger. The outcome of Wright's visit to the States was a lengthy policy paper, and a formal letter from the chairman to Sir Archibald outlining his concerns and proposing suitable candidates to succeed Marden. The permanent under-secretary deflected any criticism in the direction of the minister, who was insistent his man remained in situ.[125]

Meanwhile, G.E. Marden MC was still seeking to speed up production and development, with only qualified success. Marden alienated designers and supervisors, a number of whom secured better paid and less stressful jobs elsewhere in the industry – it was a sellers' market. Ironically, this haemorrhaging of technical staff came at a time when Hollis Williams, who had rejoined the firm as chief engineer, was orchestrating a fundamental shift in the way Fairey Aviation built aircraft, with concrete results. Hayes built six times as many Fireflies in 1944 as in 1943, but the rise in output occurred largely after Marden left the company.[126]

The company pioneered in Britain the concept of a 'project team', imported from America. This was the use of a designated group of designers, aerodynamicists and structural engineers who would steer a project through from inception to production. This was a system where specialists were responsible for constituent projects within the overall project of conceiving and constructing a wholly new aircraft; the same principle applied to later variants. Responsibility and accountability were crucial to a holistic approach, which, while not unknown in the aero-engine industry, was anathema to airframe manufacturers moulded by the individualistic 'inspirational' techniques that had been pioneered before and during the Great War. Few firms conceded that 'the growing complexity of design *and* production required team work and a broader distribution of skills than the industry often possessed.'[127] Fairey Aviation was an exception in that Marden articulated an outsider's enthusiasm for emulating the Americans, as did Hollis Williams, who had drawn up a radical agenda for reform. Both men had the organisational clout and – thanks to the backing of MAP – the political will to initiate reform. In different circumstances the chairman would have vehemently resisted change, but first, he was several thousand miles away, and second, he had seen the American model in operation and knew that it worked.

The more politicised members of the workforce at Hayes knew Marden was unpopular with his fellow directors, as indeed was Cripps, who met the Joint Planning Committee but snubbed the board when visiting the plant in December 1942. For the trade unions, MAP's controller was deemed to be Sir Stafford's man and therefore symbolic of a socialist alternative to the status quo. Along with his counterparts at Shorts, did Marden represent the future face of British aviation, as a fully nationalised industry? Did he at the very least symbolise a social democratic model where those companies still in the private sector would be subject to a heavily interventionist system of planning and direction, as in wartime? Marden's departure triggered widespread discontent, translating into strike action on the Hayes site.[128] He quit because Fairey proposed a substitute candidate not even the minister could veto – Sir Richard at last convinced a key player in Washington to resign his post and go back to Britain.[129]

The new deputy chairman was the Australian Sir Clive Baillieu, varsity oarsman, barrister, Anzac war hero, RAF veteran, mining and banking executive, dominion diplomat and a proven businessman whose post-1945 career would embrace running Dunlop and serving as chairman of the Federation of British Industries. Crucially, the future Lord Baillieu had served since 1941 as Director-General of the British Purchasing Commission and then Head of the Raw Materials Mission. While serving in the United States, Bailleu was seen as 'A skilled, patient negotiator, who listened with courtesy and could sum up discussion with a persuasively phrased proposal, he proved as effective in Allied councils as he was in boardrooms…' Here was a man Richard Fairey saw in action at least every fortnight when in Washington, and he was duly impressed. For Fairey and Cripps, Sir Clive Baillieu was the ideal compromise candidate when it came to replacing Marden. The board would welcome someone who enjoyed Sir Richard's respect and friendship, while MAP was certain Bailleu would not go native and abandon an agenda for change. Sir Clive soon stamped his personality on the company, not least in removing many of the privileges enjoyed by senior management, and in trimming unnecessary expenditure. He initiated a more Spartan regime, and in so doing gained the grudging respect of each plant's joint production committee. Like many of his countrymen Baillieu was very good at people management – here was an Oxford Blue wholly relaxed when touring the shop floor.[130]

One suspects Sir Clive introduced changes which Fairey endorsed, but had been reluctant to initiate as they would adversely affect friends and close colleagues. Sir Richard considered Baillieu a great success, and, with concerns about his health growing, agreed in January 1944 to the establishment of a joint chairmanship. One reason Fairey gave later that year for accelerating his departure from the BAC was a genuine fear that Baillieu might leave Hayes before the war was over.[131]

'From this distance ... the socialists appear keener on destroying England than Hitler'[132]

Mr Marden's stormy 18 months at the helm of Fairey Aviation was seen by Sir Richard as evidence of a wider malaise: the Churchill coalition was a necessity, but Labour's disproportionate influence on the Home Front was both punishing and pernicious. Punishing, in that sacrifices willingly made in wartime by more privileged members of society had been ignored by egalitarians pursuing what a later generation of Conservatives would label 'the politics of envy.' Pernicious, in that ideologues like Cripps and 'Herr Doktor Professor Laski' were pursuing a divisive socialist agenda which in the long term would damage social cohesion and in the short term would erode American support for the grand alliance.[133]

A five-minute conversation with Clem Attlee would have silenced Fairey's demonising of Harold Laski, but his real complaint concerned less prominent yet similarly left-wing politicians and public servants. However powerful the politicians, they could be challenged inside Cabinet or on the floor of the House of Commons. Whitehall bureaucrats, whether individually or collectively, constituted a formidable enemy, and they were armed with unprecedented powers of intervention. Yet in the final analysis even the highest mandarin was accountable to his minister, who in turn was accountable to Parliament and the wider public; the biggest problem here was wartime secrecy, and the sacrifice of transparency on grounds of security. Arguably Fairey was more concerned with petty officials and appointees, many of whom in his mind were enjoying and exploiting an unprecedented degree of devolved power as a consequence of wartime conditions. Clear evidence of this was the Rivers Avon and Stour Catchment Board, which from 1941 looked to extend its involvement in the maintenance of the Test, or to create a parallel body responsible for drainage and river management.

CRF saw the Catchment Board flexing its muscles as evidence of 'misguided or even malevolent' town dwellers setting out to create unnecessary jobs, and to attack the landed classes for no good reason other than pure prejudice. Intent on lowering the water table in order to drain the highly fertile water meadows, Hampshire and Dorset's river authority was deemed by Fairey wholly indifferent to wartime rural concerns: complacent Catchment Board members were content to bring about 'the end of the Test as the greatest fishing river in the world and the ruination of England's finest beauty spot.' Any such threat was based not on reasoned argument but 'the appeal to prejudice by which the socialists live ... jobbery rather than patriotism is the motive of our opponents.'[134] Fairey's caricature of board members intent on turning a 'beautiful valley into a concrete banked canal' ignored the fact that they were largely fellow landowners, appointed by overwhelmingly Conservative county and borough councils. His distrust and dislike of the Catchment Board was a consequence of its origins in the 1930 Drainage Act, conceived by the second Labour Government to implement the recommendations of a commission headed by Lord Bledisloe, a close ally of Stanley Baldwin. The board's successor would be similarly damned, after the Attlee Government saw fresh legislation for the statutory supervision of rivers pass through Parliament with minimal opposition in 1948.[135]

Organising resistance at a distance of several thousand miles, Fairey mobilised his agent and solicitor to fight 'with might and main' any attempt by the Catchment Board to exert its authority: 'We have known for a long time that we had to defend the Test with the rest of England against Hitler; I was not expecting an attack from the rear, with the added menace that a bureaucratic victory would last forever, and Hitler would soon have learnt to his cost what invasion meant.' Far from home CRF was insistent that 'the preservation of the Test is one of my principal war aims.' This particular war aim was seen as synonymous with that of the nation as a whole: resisting an over-powerful state in the interest of individual liberty, not least the right to maintain a stretch of river free from damaging interference. Fairey denied any selfish motive: an owner's claim to have the clearest insight into how a river bed should be conserved and maintained was rooted in sound local advice and a healthy respect for the countryside (note, 'environment' was not a word commonly used in the 1940s).[136]

Across the winter of 1941–2, and into the ensuing spring and summer, Fairey lobbied aggressively to prevent the Catchment Board implementing its proposals for dredging the Test and partially draining the water meadows in order to grow wheat. The 'present craze for planning' had gone too far, threatening a truly dreadful future, 'in which every aspect is planned, controlled and rationed, as it perforce must be under socialism.'[137] Within Whitehall and Westminster Brabazon worked enthusiastically on Fairey's behalf, both men ultimately securing direct intervention by Robert Hudson, the quietly efficient Minister of Agriculture throughout Churchill's wartime premiership. Officials found on the Bossington estate were summarily ejected, especially after the Catchment Board resurrected drainage plans which would 'render my water meadows useless', damage pasture land by lowering water levels, and potentially silt up the river bed. Inspectors and planners, with their free cars and generous petrol allowance, were 'advising further vandalism', and the estate staff at Bossington had to do all in their power to stop them. By July 1944 Fairey complained that measures enacted upstream had drained off too much water, lowering the level and speed of the chalk streams, and allowing a mud bank to build up below Rusholme bridge.[138] Here was a further incentive to get home as soon as possible. However, deteriorating health delayed an imminent return. CRF was rendered *hors de combat*, leaving Richard Fairey and the agent for Bossington and Oakley, Algy Maudsley, to defend the integrity of the Test as it flowed down and past the imposing lawns at the rear of the family home.

Troubles on the Test seemed indicative of a wider malaise. Significant, even seismic, social, cultural and political changes were taking place in Britain across the course of World War II. This was a shift in satisfaction, expectation and aspiration which Fairey was not party to; as much a consequence of social status as geographical location – arguably a corporal beside the Burma Road had a clearer idea of what was occurring at home than Sir Richard was sensitive to up on Massachusetts Avenue. Not surprisingly, he could not properly comprehend the nature of such changes, and from his perspective they seemed adversarial and threatening, scarcely conceivable as a force for good. Crucially, judging by the content of his letters home, he failed to differentiate between widely-supported reformist measures to extend social welfare, improve the standard of living, and thereby enhance the quality of life for the least affluent sections of society, and red-blooded state socialism. Paradoxically he lambasted left-leaning advocates of Beveridge-style full employment and welfare provision, while at the same time applauding the Russians for their courage and resistance: 'I certainly do all that I can for them as they are putting up a fine show.' Similar to Churchill linking Labour with the Gestapo and not the NKVD in the opening salvo of the 1945 election campaign, in a letter to his son Fairey compared Britain's traitorous socialists with the Norwegian fascist Quisling and not 'Uncle Joe' Stalin.[139]

Unsurprisingly, such crude labelling and misplaced fears were never supported by proper analysis of the postwar agenda taking shape from as early as 1941. How could Sir Richard offer a convincing explanation of societal change, when the only certainty in his mind was that whatever was taking place could never be a force for good? In such circumstances the intemperate language became ever more clamorous as he sought individuals and/or institutions to blame. One institution that repeatedly felt the full force of his vituperation, yet remained immune from his transatlantic efforts to challenge its content and tone, was the British Broadcasting Corporation.

The BBC was Richard Fairey's *bête noire* before, during and after World War II. CRF's hostile view of Broadcasting House, maintained throughout the war, was best summed up in a letter to Jim Wentworth Day, as deeply reactionary in October 1941 as in the days when he flirted with Mosley. Fairey urged Day to grab the microphone from 'the snivelling socialists so dear to the heart of the B.B.C. who are doing their damnedest, however unwittingly, to wreck American cooperation by pushing anti-British propaganda of the most mistaken sort and who appear to be under the impression that the best way to solicit American aid is to keep on attacking what they describe as the privileged classes, old school ties, and all that sort of thing, to the bewilderment of our friends here who are firmly of the opinion that we were fighting for the same individualistic doctrine which is the background of their entire outlook. Only such people could possibly imagine that their parlour pink brand of communism would appeal to anyone but the isolationists, but it suits them fine.'[140]

Similar sentiments were expressed in a succession of letters sent to friends and relatives at home before and after the United States entered the war: Britain was succumbing to the allure of anti-individualistic 'socialistic parties', with disastrous consequences for how Americans viewed their closest ally: 'it is maddening to note the rotten patriotism (always augmented by insistence on sacrifice for others) of the socialists who after all were responsible for our not being prepared. Perhaps they are trying to make people forget that they voted against rearmament ... From this distance, however, the socialists appear keener on destroying England than Hitler.' Who was responsible for spreading and popularising this mentality, with calamitous consequences across the Atlantic? With the likes of Priestley, Laski and Orwell firmly in his sights, Fairey once again tore into the BBC. Only Lord Halifax's 'excellent speeches' silenced Washington's widely shared belief that Britain was going to the dogs.[141]

Increasingly, Fairey complained that the BBC was fuelling ingratitude towards the Americans by fostering a popular obsession with the Soviet Union. In this respect he would have been better off complaining about Lord Beaverbrook given Express Newspaper's enthusiastic promotion of the 'Second Front Now!' campaign. Especially repugnant were those socialists in parliament, the press, and of course the BBC, who posed for photographers in Home Guard uniforms. Of this 'disgusting crew' one wonders if Fairey had Tom Wintringham in mind given *Picture Post*'s extensive coverage of the Spanish Civil War veteran's efforts at Osterley Park to make the Home Guard a genuine people's militia, and his midwar role in masterminding the Common Wealth Party's by-election success.[142]

In February 1943 the boss of the BAC drafted a lengthy appeal to Harry Crookshank, the Cabinet's freshly appointed Postmaster General, asking him to do something about left-wing bias in BBC broadcasts at home and overseas: earlier complaints to the Ministry of Information had been dismissed out of hand, but the Government could no longer afford to ignore the false impression shared by so many Americans that Britain would soon be a nation 'divided entirely into workers and officials, in which the workers will be left with the choice between the collective farm or the state-owned factory.' Wisely, Fairey seems not to have sent this letter, or, if he did, then Crookshank sensibly ignored it. Within a month or two CRF's complaints concerning the BBC became less alarmist, dismissing the corporation as a mere irritant at a time when the United States' 'respect and regard for England was never higher.' At the same time Fairey's own enthusiasm for the Russians had become more measured, his view of Soviet military success jaundiced by the extent to which – encouraged by the BBC – too many at home 'go dizzy about communism.' In succeeding months the Communists were singled out as the party benefiting most from the war, which – if increased membership is measured in relative terms – arguably it was.[143]

From mid-1943 to late 1944 Sir Richard's preoccupations when writing home were his continued concern over the Hampshire and Dorset Catchment Board's threat to the integrity of the Test, and its survival as *the* prime location for fly fishing in the south of England, and his fury over the compulsory purchase of Fairey Aviation's west London aerodrome. His apparent obsession with the BBC faded away, but it had by no means disappeared. Throughout the six years Labour was in office after the war, and then well into Churchill's peacetime administration, CRF continued to protest over what he saw as institutional bias within an instinctively left-wing BBC.[144] Programmes heard on the radiogram at Bossington after 1945, and on those rare occasions during the war when he could retreat to the country, served only to reinforce Fairey's firm belief that the Reithian ethos of public service broadcasting had been subverted, turning the BBC an alarming shade of 'parlour pink.'

Transatlantic crossings, and a long goodbye

Fairey always had mixed feelings about his trips home. England meant the opportunity to spend a day or two at Bossington, ideally fishing from dusk to dawn. But such unadulterated pleasure had a price, namely the danger, discomfort, tedium and fatigue experienced on the journeys to and from New York ('…it will be a long time before I forget our takeoff in Foynes Bay when I did the pioneer passage on the Claire').[145] Fairey's first flight home came in

WARTIME IN WASHINGTON – THE BRITISH AIR COMMISSION

late May 1942, for briefings and discussions at MAP's Milbank headquarters. The seriousness of the situation in the Far East following the fall of Singapore demanded a reassessment of aircraft procurement, not least the final destination of the far fewer aircraft presently arriving from America. Before returning to the States he made time to visit Bossington, this being the height of the mayfly season. Whether blessed by good fortune or by the Reverend Daubeney, CRF found the fishing exceptionally bountiful: 'I killed seven salmon and lost four, took twenty-eight trout, mostly two pounds or over, and put back at least a hundred.' Here was a catch worthy of St Peter, and a suitably biblical storm forced Fairey's flying boat to turn back when already well into its westward passage. Sir Richard admitted to his nephew just how frightening he had found the experience. To make matters worse, the second time out the Pan Am Clipper was shrouded in fog for almost the entire journey.[146]

Internal flights were far less stressful, and American airliners boasted an impressive array of creature comforts. A tour of the West Coast in the spring of 1943 highlighted the stark contrast between Britain and the United States in the development of short and long haul civil aircraft. Almost a year earlier Fairey had told Richard junior how much 'I wish that we had been able to go on building our own four-engine transports, and what boon their existence will be now.'[147] To add insult to injury, Fairey flew to Scotland in May 1943, and back to New York a month later, courtesy of the four-engine Douglas Skymaster; his carrier was the USAAF, but a newly created Transport Command awaited delivery of 23 C-54s, ordered via the BAC. Fairey's second visit home was prompted partially by Cripps and Sir Archibald Rowlands seeing a need to stiffen the sinews. Back in Washington after his trip to California, the Director-General clearly felt the production and procurement process to be working so well that an early return to Hayes was now a priority. London informed him in no uncertain terms that he was going nowhere.[148]

During his stay Millbank and Whitehall made heavy demands upon Fairey's time, but Bossington offered a welcome retreat. Charles Crisp orchestrated proceedings when it came to his old friend meeting solicitors, agents, bankers and so on. Fairey found time to attend the Handley-Pages' silver wedding anniversary, in the course of which he bent Sir Frederick's ear on the need to revive a discredited SBAC. No doubt Lady Handley-Page welcomed Sir Richard's sound advice on such a special day.[149]

Sir Frederick and Sir Richard attended what may well have been the aviation industry's most glittering event across the whole of the war: the 1943 council dinner of the Royal Aeronautical Society. This was a high profile celebration of the Anglo-American alliance, with Ivan Maisky and the Soviet air attaché noticeable by their absence. Perhaps the Russians boycotted the event when they saw from the invitation that Franco's ambassador, the Duke of Alba, had been placed on the top table between Lord Brabazon and Sir Archibald Sinclair, a former and a serving cabinet minister. With 'volunteers' from Nationalist Spain fighting for the Axis on the eastern front, Alba's presence, let alone his prominent placing, was at best impolitic and at worst insulting. Fairey, sitting opposite, may have felt relaxed about the Spanish ambassador's presence, but Sir Stafford Cripps and the radical-minded American envoy, John Winant, must surely have questioned his presence. CRF was present on behalf of the BAC, not Fairey Aviation, and the predominance of American guests indicated how important the occasion was in terms of consolidating a still uneasy and imperfect working relationship.[150]

Unknown to the organisers when setting a date, the Royal Aeronautical Society chose to cement good relations with visiting American aviators in the immediate aftermath of the second Washington conference, where Churchill, Roosevelt and their service chiefs agreed a timetable for invasion across both the Mediterranean and the Channel. By coincidence, the day of the dinner – Thursday, 27 May – marked a key moment in America's emergence as the arsenal of the Grand Alliance: Roosevelt directed James Byrnes to set up the US Office of War Mobilization, intended to coordinate all government work on the home front. For the BAC, by now heavily dependent upon federal agencies, this was an initiative long overdue.[151]

1943 was a year of relentless travel, as Fairey acknowledged in a self-penned celebration of Anglo-American relations for the *Philadelphia Inquirer*, the high-circulation, pro-Republican morning newspaper in 'The City of Brotherly Love.' This was a remarkable piece of propaganda, not so much for the content, as for its size and prominence: a full page, dominated by a colour portrait of CRF at his desk, and a further half page, including a box biography of the author, and behind the text on both broadsheet pages was the profile of an Avro Lancaster in flight.[152] Features such as this raised Sir Richard's profile, but later in the year came a highly prized acknowledgement of his present standing and importance. In November 1943 Fairey had again flown home, anticipating a speedy return to the States courtesy of the Pan Am Clipper. Bad weather forced him to spend a week at Bossington, where he organised an impromptu shoot, and then reluctantly repacked his bags to cross the Atlantic by convoy. Arriving in Halifax, CRF headed east on yet another arduous tour of Canadian factories. When finally back in Washington, Fairey learnt from the prestigious Institute of Aeronautical Sciences (IAS) that he had been elected an honorary fellow. With directors that included Charles Lindbergh, Howard Hughes, Great War air ace

Eddie Rickenbacker, and the architect of the 1942 carrier-launched attack on Tokyo, Jimmy Doolittle, the IAS's award ceremony at the Waldorf Astoria in early 1944 proved a glittering occasion.[153]

For Sir Richard, recognition by his peers was deeply gratifying, and it was clear that London still felt he was doing a good job. Yet the cost to his health was huge. While focusing on the task in hand, the absent chairman fretted over the condition of Fairey Aviation – he felt helpless when his firm learnt that the 'Great West Aerodrome' would be subject to compulsory purchase, ostensibly to host B29 bombers but in reality as the planned site for postwar London's principal airport. A calculating Department of Civil Aviation gave the company access to Heston aerodrome after the transfer of ownership in 1944, with a later move to White Waltham. The fight for adequate compensation would last 20 years.[154]

At the same time Fairey fumed from afar as Cripps visited Hayes to placate shop stewards certain Sir Richard had orchestrated G.E. Marden's departure.[155] By this time CRF had already suffered a serious thrombosis, while on a summer vacation with Esther and the children in Canada.[156] A year later his heart condition proved so severe that he was forced to spend a month in hospital. Ironically, when Cripps urged him to return home for a month's leave ('a joyous prospect') Fairey had to say no, explaining that his consultant feared for his life if he travelled any earlier than late August. Come the autumn and Fairey was still convalescing in Washington, albeit keen for Cripps and Sir Archibald Reynolds to clarify how soon he could resume the running of his company. The priority now was a return to Hayes, and reshaping Fairey Aviation to meet the postwar challenge of building aircraft superior to those of rival American manufacturers.[157] First, however, Sir Richard had to refocus on the BAC.

In late October he went back to work, despite his doctor fearing another heart attack. Mary Bell found her boss physically much weaker: 'Sir Richard aged during those years. I guess we all did, but he *definitely* did. But he never lost that rolling walk, nor his dedication.'[158] Within weeks of his return Fairey was laid low with severe bronchitis, recovery from which was painfully slow. On 14 December 1944 he informed Rowlands's successor at the MAP that, to avoid the worst of a Washington winter, he was taking his family to Bermuda after Christmas: after a month's convalescence on the island he would fly back to the States fully fit for 'the job that will confront me at home' when finally relieved of his duties in Washington. MAP's permanent secretary duly arranged a flight south for Sir Richard and his family, and at the same time he urged the RAF to vacate Bossington.[159]

While the Director-General had become a rare sight in the building, staff at 1785 Massachusetts Avenue assumed that soon he would be back, fit and ready to steer the Commission through the last months of the war in Europe. In reality Fairey's tenure at the BAC was over, a severe heart attack forcing him to resign in the early spring of 1945. Bermuda's main hospital was located close to the Faireys' house in the mid-island parish of Paget, and it was there that Lady Fairey learnt her husband had only six months to live. Luckily a neighbour, Kenneth Trimmingham, gave Esther the name of America's most distinguished cardiologist, Paul White. The result was an emergency flight to Boston, with the children left at Lyndham in the care of their Canadian nanny, Florence Wakeling. In April 1945 CRF underwent a sympathectomy, a major invasive operation to reroute cardiac nerves and relieve high blood pressure, which had been pioneered at White's home hospital, Massachusetts General. The immediate and longer-term side-effects were harsh, and Fairey suffered acute discomfort for the rest of his life – Paul White rightly estimated that this would be ten years. After lengthy recuperation in hospital, during which Esther spent every day by his bedside, Fairey left the hospital to stay in a neighbouring hotel. Almost immediately he suffered a massive coronary, and was back in Massachusetts General. There followed a further lengthy stay in hospital, after which a return to Bermuda was sanctioned only if Sir Richard maintained a gentle exercise regime in and around his adopted home, and eschewed work for the remainder of the year. White was a pioneer of preventive cardiology, and CRF knew he was being treated by the pre-eminent expert in the field. The plan was for a further consultation the following year, with yet more surgery should his recuperation have faltered.[160]

It would be the spring of 1946 before Fairey and his family at last arrived home. A still grey, drab *Queen Mary* carried them across the Atlantic. Sir Richard's daughter recalls Cunard's frugal fare, served 'in a small mess room. We landed at Southampton and were driven to Bossington. The centre of Southampton was bombed flat and I remember winding through tidily swept streets with piles of rubble on either side. When we got to Bossington the house was more or less in wartime mode, no carpets, etc. It rained every day all summer and the harvest was a wash out ... I was in seventh heaven to be home. I loved animals and spent most of my time on the Home Farm.'[161]

A year earlier Halifax had learnt from London that his most experienced mission head would not be returning. He immediately wrote to Fairey, under intensive care in Boston: 'Apart from my regret at your departure for personal reasons, I know what a splendid job you have done here and how much you will be missed.'¹⁶² The British ambassador had no illusion as to how deeply someone of such strong opinions as Dick Fairey had loathed him before the war, but for the past four years the two men had worked well together as plenipotentiaries of an imperial power demonstrably on the wane. Henceforth their paths would rarely, if ever, cross. With VE Day still a month away, CRF's secondment to Washington had at last come to an end.

CHAPTER 9

Charging to an End, at Supersonic Speed, 1946–56

Given a second chance

For Sir Richard, Boston surgery and Bermuda sunshine proved a life-giving – and a life-enhancing – combination. Together they gave him another ten years as a husband and father, and as chairman of an aerospace company keen to diversify and follow fresh paths of research and development. Note the order of priority: CRF was eager to embrace the jet age, yet at the same time all too well aware that he had been given a second chance. It would be an exaggeration to claim he was living on borrowed time, or that these were his twilight years: Fairey did not perceive himself to be in visible decline, or consider his demise imminent. Yes, there were changes to his lifestyle, not least a permanent farewell to competition under sail, but CRF never wholly abandoned his overarching principle of work hard, play hard. He considered himself mentally as sharp as ever. The big change was that from now on the family would always come first, no matter how urgent the business in hand. Thus, wife and children saw far more of Sir Richard than had been the case in the course of his first marriage. During and after World War I, hands-on control of Fairey Aviation over-rode all other considerations, and any spare time seemed taken up with golf and field sports. Even in the early years of his marriage to Esther, sailing had proved an all-consuming passion, pursued with drive, diligence and single-minded determination: C.R. Fairey esq., commodore and champion yachtsman – helmsman extraordinaire – properly acknowledged his duties as a father, and yet racing had invariably triumphed over domesticity. Post-convalescence, Fairey realised that if he was to see John and Jane reach their adult years then he clearly had to pace himself, not least in his dealings with the firm he had founded 30 years earlier.

Fairey spent every winter in Bermuda, at 'Lyndham', his waterside home in Paget parish, adjacent to Hamilton Harbour. Between 1948 and 1955 the family would never stay later than April, but while in the late 1940s the date of departure was always October, from 1950 onwards it was pushed back to after Christmas. Cruising from Southampton to New York courtesy of Cunard, a brief excursion to Boston would preface a 40-hour trip to Hamilton on board one of the Furness Bermuda Line's luxury mail boats.[1] The return journey was also via New York, but minus the consultation with Massachusetts heart specialist Paul White. 1946 entailed a much longer stay in Bermuda as at the start of the year Fairey had undergone a second heart operation in Boston, from which he took several weeks to recover; pneumonia set in, followed by another heart attack. Remarkably, he bounced back, returning to Lyndham for an extended convalescence. Eventually CRF was deemed fit enough to make the long journey back to Bossington, and again immerse himself in company business. Before the end of the year the Faireys were back in Bermuda. In the spring of 1947 a relapse led to Sir Richard and his family enjoying life among the well-to-do of Long Island – they stayed in a large house on the outskirts of Great Neck, the commuter town reinvented as 'West Egg' in F. Scott Fitzgerald's *The Great Gatsby*. As well as CRF's cardiac condition, a further reason for not returning home was the appalling weather. Britain experienced its worst winter of the twentieth century: with no evidence of the temperature starting to rise, a house the size of Bossington must have been unbelievably cold. Even when the family did finally find themselves back in the Test Valley the great thaw had scarcely begun.[2]

Whenever he was home from Bermuda, Sir Richard chaired the monthly meetings of Fairey Aviation's board. These were always held at Hayes, although the January gathering usually took place at Bossington, presumably in conjunction with a social event hosted by Richard in his father's absence. This arrangement continued until the summer of 1955, after which CRF chaired *in absentia* even when in England.[3] Thus for several months of the yeat Sir Richard Fairey oversaw company business from the far side of the Atlantic. Four years in Washington had left him adept at maintaining regular contact with colleagues in Britain, but at the BAC he had enjoyed full access to official systems of transatlantic exchange. Now he was reliant upon the same means of telecommunications as any other private individual, albeit one with the means to support speedy and – by the standards of the day – reliable phone and telegram links to London and Hayes. Nevertheless, in an analogue era, transmitting or transferring a large body of information from one continent to another was too often a slow and costly process. Courtesy of Cable and Wireless, the Royal Mail and two of the three state-owned airlines, Fairey was heavily dependent on a sophisticated system of imperial communication, by this time mature and tested in war. On the whole that system worked well, not least as senior management and secretarial staff – together with the Fairey Aviation switchboard and mail room – maintained a regular flow of news and background information regarding day-to-day company affairs. For obvious reasons every effort was made to keep the chairman fully aware of what was going on, and when big decisions had to be made Richard junior or C.C. Vinson would fly 5,000 miles for a face-to-face briefing. Fairey Aviation covered the travel expenses of its envoys, as it did those of the chairman when visiting North America on company business and to see his Boston consultant.[4]

Vinson acted as a courier for CRF's solicitor and Charles Crisp's partner, Roland Outen. Crisp had long since passed on responsibility for Fairey's complicated personal finances, which overlapped considerably with those of the company. Outen was both an adviser and a friend, such that he and his wife were house guests of the Faireys early in 1948. He sent the absent Sir Richard a lengthy report almost every week, when necessary recommending suitable action. Thus, in 1949 CRF was sufficiently alarmed by Outen's fear of heavy interest payments causing serious cash flow problems that he ordered Vinson to speed up the sale of loss-making subsidiaries.[5] In 1949 Outen became deputy chairman, acting as CRF's eyes and ears whenever he was out of the country.[6] By 1955 he was bringing the board minutes to Bossington so Sir Richard could sign them off *in absentia*, and he was an obvious choice to succeed as chairman. In his final years Fairey relied heavily on Richard for regular updates regarding company affairs, but before that he communicated almost daily with Vinson, although he clearly did not consider his assistant managing director indispensable. Plagued by kidney problems, Vinson was given time off to convalesce, encouraged to contest Hayes and Harlington as the Conservative candidate at the 1950 general election, and released without regret when he announced his resignation in 1952.[7]

Someone who before the war CRF had seen as indispensable was his ever reliable personal secretary. Although she remained close to her old boss and his family, Miss Burns now shared her responsibilities with Kay Alford, whose office was at Bossington not Hayes. Miss Alford's residential position reflected the extent to which Fairey worked from home when in England: however seriously he still took the role of company chairman, the maintenance and protection of his estate invariably came first.[8] Prioritising Bossington meant frequent visits from Vinson and Chichester-Smith to discuss company business. From the summer of 1946 both men were eager to arrange a 'welcome home' party at the

Dorchester, but were disconcerted to discover how reluctant CRF was to travel up from Hampshire. The function, which boasted an impressive guest list, did finally take place, with ministers, senior officials and military personnel out in force. This was a sign of the times, as perhaps was the absence of industry patriarchs such as Sopwith and de Havilland.[9]

'Big men' running large firms as personal fiefdoms remained commonplace in mid-century British industry. If anything, an autocratic business structure had been strengthened by wartime Whitehall's need to deal directly with company bosses capable of initiating immediate action; lines of accountability were simple if crude. Fairey was unusual in that he had been absent for most of the war, and had then been forced by ill health to take a step back. In different circumstances his extended absences might have prompted institutional investors or less reverential board members to question his continued insistence on having the final say. Continuity not change was a consequence of the chairman retaining, firstly, his controlling interest in Fairey Aviation, and secondly, the loyalty of his directors. For all G.E. Marden's attempts to replace the board, and Sir Clive Baillieu's efforts to reform it, Fairey's acolytes remained in situ. Indeed, CRF's control of the board strengthened throughout the 1940s. Boasting an impressive service and private sector CV, C.H. Chichester-Smith quit MAP in 1942 at Fairey's behest, and in due course became assistant managing director. Geoffrey Hall returned to the fold in 1949 as director of engineering, and, starting as general manager, Richard Fairey began his rise through the corporate ranks. Dick Fairey found Charles – better known as Colin – Chichester-Smith a man after his own heart. The most influential figure on the Fairey board after 1945 was a proven engineer and entrepreneur who had raced both small yachts and light aircraft between the wars. Before that Major Chichester-Smith had flown with the Royal Naval Air Service, in 1916 earning himself the DFC for a Channel patrol during which he chased a Zeppelin and strafed two U-boats. In the aftermath of World War I he had joined a delegation which travelled to Tokyo, survived an earthquake, and advised the Imperial Navy on the creation of a carrier-based air force. Memories of this visit no doubt came flooding back in December 1941 when news came through of the attack on Pearl Harbor.[10]

Before moving to Fairey, Chichester-Smith had spent two years as controller of the Burtonwood Repair Depot, where Ernest Tips worked having fled France in June 1940. A month earlier the Avions Fairey factory had been destroyed by German bombing; senior management, and whatever records and machine tools they could salvage, left Charleroi in requisitioned trucks and headed south. Tips and most of his staff were eventually evacuated, but at a terrible cost: eight colleagues were drowned in the catastrophic sinking of the liner *Lancastria* at the mouth of the Loire on 17 June. It was Tips who recommended Chichester-Smith to CRF, and who followed him to Hayes after the USAAF assumed control of Burtonwood in 1943. E.O. Tips served as chief experimental and research engineer until he returned home in 1945 to oversee reconstruction of the AF factory. Over the following 30 years the Belgian plant became a large and highly profitable operation for the servicing and assembly of cutting-edge combat aircraft. Renamed, and initially with Geoffrey Hall as its chairman, Fairey SA continued to operate until it entered receivership in 1977, at which point the Walloon and Belgian governments nationalised the company in order to safeguard production of F16 fighters.[11]

Not surprisingly, with the same faces in the boardroom, and Sir Richard retaining a high profile when in the country, no-one suggested the chairman spend more time talking to the workforce.[12] Yet a refusal to countenance change at the top did not mean complacency and inertia. The same men who had made Fairey Aviation such a key player in the domestic aviation industry between the wars, in the early years of the Cold War aggressively pursued a global strategy, creating subsidiary companies in Canada and Australia and reviving the Belgian operation. The board was ultra-loyal, if only because – however distant he might be from the action – CRF endorsed Chichester Smith's and Massey Hilton's efforts to diversity in the face of shrinking aircraft contracts.[13] As in the aftermath of World War I, every effort was made to utilise surplus production capacity, with Heaton Chapel the focal point for major engineering projects, starting with Bailey bridges for the Army.

In 1946 Chichester-Smith's enthusiasm for dinghy racing saw Fairey Marine set up in the empty assembly sheds at Hamble Point. The new company's first chairman, in discussion with Alan Vines, another keen sailor and Hayes's deputy chief engineer, spotted an obvious synergy between plane making and boat building. Chichester-Smith saw how mass production of monocoque hulls could be achieved if boatyards adopted the aircraft industry's process for hot-moulding plywood into stand-alone components. This had been most famously employed in construction of De Havilland's 'wooden wonder', the Mosquito. The use of one or more autoclaves for hot moulding meant every hull within a designated class was guaranteed to be the same, and of a uniform quality. Five uniform 12-foot boats were successfully built and tested in the early summer of 1946. A design and sales team of the inventive Uffa Fox and the ultra-competitive Charles Currey exploited a gap in the market. Small boat specialist Currey, a former motor torpedo boat skipper and a future Olympic medal-winner in the Finn class, anticipated a growing demand, both at home and overseas, for a modest range of reliable, simple-to-sail dinghies. With bespoke fittings and aluminium masts, racing or family boats would be readily available and easily transportable, whether singly or in bulk. The scale of that demand, not least in North America and Australasia, and the level of output – each week around 30 boats built – ensured enthusiastic support for Fairey Marine from both Sir Richard and his eldest son, the latter serving as the new enterprise's deputy chairman.[14]

In the first ten years of its existence Fairey Marine produced a remarkable range of yachts of all types and sizes, of which Fox's initial creation – the Firefly – proved the most famous and the most enduring. At least three Fairey designs became Olympic classes, and medals in the Finn have been fought for at every Games since Currey raced at Helsinki in 1952. Designs for large cruisers were on the drawing board by the time of Fairey's death in 1956, but it was another year before Richard junior and an old school chum, Bruce Campbell, convinced Vines and Chichester-Smith that the company's future lay in power boats.[15] Sixty years later, with so many of its craft still in the water, the name of Fairey Marine is better known than that of Fairey Aviation. There's a cruel irony here, not least as the Hamble operation was tiny when compared with a company which in the postwar era still considered itself a global, hi-tech, multi-million-pound enterprise.

Factory space at Hamble was available, because by the end of 1947 final assembly and production, along with experimental flight testing, had relocated to the wartime depot of the Air Transport Auxiliary at White Waltham, near Maidenhead. White Waltham was made available as partial recompense for the Air Ministry requisitioning the Harmondsworth or 'Great West' aerodrome in 1944. Ostensibly the USAAF would fly B29 bombers from Harmondsworth on missions over Germany, but the Superfortress squadrons never materialised, fuelling the Fairey board's suspicion that compulsory purchase of its estate was a prelude to developing London's principal airport once the war was over. This proved to be the case, and confirmation of the Heathrow project in early 1945 saw Fairey offered space at the soon to be obsolete Heston airport. This compensatory site was quickly deemed unsuitable, hence the firm's focus upon White Waltham as a vital complement to the key plants at Hayes and Heaton Chapel. The Harmondsworth 'land grab' was a running sore for Fairey's directors, not least the chairman and managing director: Sir Richard attributed the absence of adequate compensation to Whitehall officials flagrantly abusing their statutory powers under the Defence of the Realm Act. In the courts lawyers acting on the board's behalf argued that action pertinent to peacetime circumstance had been taken under the veil of wartime emergency. Fairey Aviation's legal bill grew and grew, with no resolution of the dispute until as late as 1964, by which time the company no longer existed.[16]

Loss of the Harmondsworth site remained a major grievance when the board organised a dinner at the Connaught Rooms on 22 December 1949 to thank 282 directors, managers and workers who, for a quarter of a century or more, had been employed by Fairey Aviation. Thirty-two guests that night, including most of those at the top table, had worked with CRF since the Great War. These were the loyalest of the loyal, all duly awarded with watches (of different quality, depending upon length of service). Entertainment would be on the grand scale, with the works choir and the acclaimed works band joined on stage by an array of well-known acts – although it is hard to imagine what Sir Richard made of the surrealist routine that sealed Michael Bentine's credentials as a founding Goon. The chairman and managing director took a keen personal interest in arrangements for the great night, drawing up a short list of suitable guests. With Murray Sueter seriously ill, the most esteemed VIP would be Air Marshal Lord Trenchard: 'We should be greatly honoured if you could join us. It will naturally be an evening of reminiscences and would give us an opportunity to express our gratitude to you for your great friendliness and support in our pioneering days.' It was indeed an 'evening of reminiscences', not least those of the firm's founder, and Trenchard's address struck just the right tone – complimentary, but not gushing. Brabazon was uncharacteristically quiet, but Brice Slater and Wilfred Broadbent were given every opportunity to eulogise their boss, who duly reciprocated. All the old campaigners were given due mention, but Sir Richard paid tribute to a workforce which collectively shared a common bond, having made Fairey Aviation a major presence within the British aircraft industry: 'We have been together for what is virtually a lifetime. We have had a common object, a common interest, and a common purpose throughout our lives and we become thereby a tribe – a family – held together by bonds and feelings which the passage can only tighten.' This was no exercise in smug self-celebration. Indeed, the speaker tempered his enthusiasm for the opportunities which a new decade offered, adding a sombre note of foreboding: 'Cheerful as the occasion is I find myself unable to treat it lightly, being concerned, indeed overwhelmed, with the emotional forces it presents.'[17]

By the beginning of the 1950s, in both its home factories and its overseas subsidiaries, Fairey Aviation had transformed itself into an aerospace company, with increasing resource dedicated to the design and development of guided missiles and drones. Sir Richard approved and encouraged these radical changes, just as he supported Fairey's postwar initiatives in rotary as well as fixed-wing aircraft. In other words, he rarely if ever acted as a brake upon fresh projects and acquisitions, and in consequence his detachment from day-to-day affairs was not a serious cause of tension. At key moments he was still proactive, for example, in cultivating old friends across the Atlantic to facilitate an information exchange agreement between Fairey Aviation and the Martin Aircraft Corporation. In 1950 CRF wisely skewered an Admiralty suggestion that Fairey buy a near bankrupt Blackburn.[18] Of course, given Fairey Aviation's unhappy experience in the course of the 1950s, especially at the end of the decade, Sir Richard might have displayed a healthy scepticism and questioned more forensically his colleagues' enthusiastic if ultimately disastrous embrace of rotating-wing technology had he retained his old hands-on, interventionist approach.

One key element of the company's affairs where its chairman could have intervened in the interest of Fairey Aviation's long-term strategy was the size of the annual dividend in the early postwar years. Profits dipped sharply between 1946 and 1952, totalling £2,113,290. Yet the aggregate dividend payment across the first five financial years after the end of the war averaged 64 per cent of the firm's earnings. In the short term this high dividend allocation did not appear damaging, and pre-tax profits peaked at £2,159,568 when Sir Richard presented the 1954–5 accounts in a valedictory address to shareholders: 'This is our record profit and we have no apologies to make on that account, least of all to those political theorists who declaim against all profits as such. On the contrary, I here and now declare our intention of using our utmost efforts to continue to make profits.' However, with relatively little capital in reserve, Fairey Aviation experienced serious difficulties in funding multiple R&D programmes. Delays in initiating a project could have serious repercussions, and nowhere was this more obvious than in the design and testing of the supersonic Delta Two.[19]

The firm of Fairey enters the jet age

Reinvention of the Firefly saw the aircraft flown in multiple guises by the Fleet Air Arm and several other naval air services. The advent of the Korean War prompted substantial orders for variants of the Firefly 5 (FR5), with three carrier-based squadrons deployed in a ground-support role between the autumn of 1950 and the summer of 1953. The initial outlay on R&D had long since been recouped, while redesign, production and conversion costs remained relatively low. This ensured a clear profit on every Firefly built postwar, while at the same time absorbing the development cost of power-folding wings, a feature of the FR5s flown in Korea. Money made on extending the lifespan of the Firefly only partially compensated for the losses incurred in building prototype successors to the Barracuda: with the defeat of Japan and the advent of jet-powered aircraft, the huge and lumbering Spearfish was demonstrably too late, too big and too slow. A lasting legacy of the Spearfish was the creation of Fairey Hydraulics, as the redundant aircraft proved an ideal test bed for pioneering power control systems. Like Fairey Engineering, within which it operated, Fairey Hydraulics proved highly profitable: widespread civil and military use of its cutting-edge technology ensured the firm's survival under a succession of British and American owners for almost 70 years.[20]

Ever more sophisticated avionics and weapons systems, and highly engineered jet or turbine engines, rendered the design and development of one or more prototype aircraft a far more expensive process than before World War II. Lead times grew longer in proportion to the complexity of the development programme, and cost escalation rose disproportionately in the 15 years prior to the effective demise of Fairey Aviation in 1960. Aero-engine makers passed on a substantial proportion of their R&D outlay to the aircraft manufacturers, thereby increasing the overall cost of carrying a project from drawing board to test flight. As aircraft gained in complexity and performance across the course of World War II the state absorbed most or all of rival manufacturers' costs when faced with ever more demanding technical specifications. After 1945, companies like Fairey were forced to absorb a large proportion or, on occasion, all of the cost entailed in competing for a contract which might ultimately be awarded to another tender. The challenge was rendered that much greater by the size of the aircraft industry as a consequence of wartime demand, and the limited rationalisation which took place in the immediate postwar years – Britain had too many companies chasing too little work, and the onset of the Korean War delayed a top-down culling of companies which lacked the substantial capital reserves required to fund ever more expensive R&D. One-third of the SBAC – ten members – had assets of less than £1 million. Korea threw a lifeline to under-capitalised, under-performing companies, with cost-plus contracts and wage-driven inflation reinforcing poor productivity and complacent management. East–West tension fuelled an artificial boom within the British aircraft industry as annual sales rose from £78 million in 1950 to £240 million six years later. Yet 80 per cent of the joint aero-engine and airframe industry's productive capacity was controlled by just six companies, of which Hawker Siddeley Group was by far the biggest conglomerate. Outside the big six Fairey Aviation could consider itself the largest and most viable independent manufacturer.[21]

For both civil and military aircraft the ideal scenario was a succession of substantial contracts for essentially the same design: factories enjoyed optimum production and increased productivity, with the consequent economies of scale driving down unit costs. In the early years of the Cold War a post-prototype contract to supply the RAF and/or the Fleet Air Arm was a commercial lifeline for companies which had expanded rapidly from 1935 to 1945; the numbers were such that most or all of the R&D costs not chargeable to the Ministry of Supply could be recouped. Conversely, an unsuccessful tender signalled significant losses, unless – against the odds – sales were secured abroad. For long- or short-haul civil aircraft, sales at home *and* overseas were paramount in order to maintain a healthy profit margin, hence De Havilland and Rolls Royce in the early 1950s promoting the Comet 3's unique status as a transcontinental airliner powered by axial-flow jet engines.[22]

Military contracts at the start of the 1950s may still have been quite sizeable, but by the end of the decade technological complexity and consequent cost escalation meant that RAF and Royal Navy procurement budgets funded much more modest tranches of new aircraft. Thus, defence contractors in 1960 were supplying only a quarter of the 2,000 or so aircraft that annually rolled off the assembly line a decade earlier. One beneficial consequence of this shrinkage was that domestic plane makers built far less specialist types than the multitude of marques supplied during and after the Korean War. So many specialist types had meant smaller contracts and shorter production runs, thereby raising unit costs, and providing only limited potential for export sales.[23] With far fewer aircraft being sold at home and abroad then the argument for rationalisation of the British aerospace industry became that much stronger. That argument became overwhelming by 1957 and the publication of Duncan Sandys' Defence White Paper, but before that neither the Churchill nor the Eden administrations had been willing to bite the bullet. Consequently, no structural changes took place prior to CRF's death in September 1956, and with an operating profit of over £2 million in the previous financial year there was no feeling in the City or within the industry – nor inside Whitehall – that Fairey Aviation was living on borrowed time.[24]

The company remained in the black partly because modest profits from its subsidiaries helped offset the high level of investment in Fairey's Weapons Division. Work on guided missiles began on a small scale in 1945, funded by the Ministry of Supply, but by the end of the decade attention focused on the 'Blue Sky' project. Blue Sky's design team came up with a highly complex, heavily over-engineered air-to-air missile system. Fired from under the wing of a fighter aircraft, 'Fireflash' would jettison its booster rockets, with the missile then guided to its target along a beam visible to the pilot via his gunsight. Multiple components within each missile, solid-fuel propulsion, and a near impenetrable pre-transistor wiring system, rendered any such weapon highly problematic. Seven years of experimentation saw development costs soar, with the project's managers insisting specialist plant be built in Britain and Australia. Both ministers and the Air Staff were repeatedly assured the Blue Sky project was on track, with all major players inside Whitehall invited to lunch at the Dorchester on 27 July 1955 for a film screening and a progress report from the chairman, making what proved to be his last major speech. It was a convincing performance by Sir Richard, and yet there was no way the paltry contracts awarded for Fireflash, and its companion missiles, could cover the huge amounts allocated to the Weapons Division for its various R&D programmes.[25]

Equally expensive were rocket trials at Aberporth and in Australia, where delta-winged models were fired in simulation of a fighter aircraft's vertical launch off a warship. These tests were both costly and dangerous. They were also a waste of time as the Ministry of Supply soon withdrew its support for the project, but not before Heaton Chapel built a scaled-up turbojet aircraft with provision for booster rockets. The mid-wing delta Type R was a short, squat design vaguely reminiscent of De Havilland's sweptback Swallow. The designers of both aircraft were clearly influenced by Messerschmitt's rocket-powered Komet. But there the comparison ended as Fairey's more rugged machine had a conventional tail, unlike the 'flying wing' of the DH108 and ME163. Nor was the Type R intended to fly at anywhere near supersonic speed, even if its subsequent renaming as the Fairey Delta One suggested a forerunner of the record-breaking FD2. Continuity was evident in the hard data generated by the posthumous FD1 (written off after a crash landing at Boscombe Down in early 1956) and in the experience Fairey's test pilots gained from flying a delta-winged aircraft at transonic speed.[26]

The principal reason Fairey Aviation continued to generate profits throughout Sir Richard's final years was its success in satisfying the Fleet Air Arm's need for a long-range anti-submarine aircraft which could carry powerful search radar *and* the weaponry to destroy an identified target. To power a three-seater reconnaissance/attack aircraft which could stay on patrol for long periods and yet be small enough to fly off fleet carriers, Fairey's design team resurrected an old idea: a coupled pair of engines driving twinned co-axial, counter-rotating propellers through a single transmission, as pioneered prewar in the home-grown P24. In fact, Armstrong Siddeley supplied the Double Mamba coupled-turboprop which powered the production version of the Fairey Gannet. Development of the Gannet proved a long, expensive – and dangerous – process. Chris Staniland had died in 1943 flying a Firefly, and a second test pilot had been fatally injured testing an experimental helicopter. Both deaths, especially Staniland's, had left a lasting impact on Sir Richard. Then, to his horror, a third test pilot only just survived when a prototype Gannet crashed on landing. This was Peter Twiss, of whom more below. A stream of complaints and concerns emanating out of Whitehall took their toll on the chairman and managing director: Sir Richard, already an ill man, aged visibly. Geoffrey Hall recalled his half-brother in the course of 1949 becoming 'ashen grey. Somehow, somewhere his whole world had toppled around him.' Relief for Sir Richard came when a reluctant Hall agreed to assume full responsibility for every aspect of engineering within the company. Once Hall began to exercise full executive power, the Gannet R&D programme turned a corner, in due course regaining the good will of the Ministry of Supply. Slowly, and painfully, Fairey Aviation saw its reputation for building safe and reliable aircraft duly restored.[27]

Beginning in December 1954, Fairey Aviation supplied the Fleet Air Arm with 180 Gannet AS1s, and 171 later versions of the same basic model. Naval air arms in Australia, Indonesia and West Germany ordered a further 70 aircraft. The last Fairey type to fly on operational duties with the Royal Navy remained in service until December 1978. For nearly a quarter of a century the Gannet fulfilled a multitude of roles, and naval air crews rated it highly. The aeroplane was rugged and reliable, with its airborne early warning variant averaging 30 flying hours month on month.[28] Solid and unspectacular, the Gannet attracted little attention, but it made a lot of money for Fairey Aviation, and helped subsidise the company's more glamorous projects – the FD2 record breaker and the turbo-prop 'heli-liner', the Rotodyne.

The Fairey Rotodyne's inaugural flight was on 6 November 1957, nearly 15 months after the death of Sir Richard Fairey. It was a further five months before the hybrid flew in 'winged autogyro' as well as simple helicopter mode. Yet the idea of a large commercially viable 'compound helicopter' dated back to Fairey Aviation's initial experimentation with rotary aircraft a decade or so earlier. In the late 1940s the RAF and the Royal Navy had no option but to buy Sikorsky helicopters from the United States, albeit built under licence by Westland. A handful of British manufacturers saw a gap in the market for both military and commercial helicopters: Westland relied on transatlantic technology, Saunders Roe bought in existing technology, and – encouraged by the Ministry of Supply – Fairey Aviation developed wholly new technology. On 28 June 1948 the FB1 Gyrodyne became the world's fastest helicopter (and more), its subsequent record-breaking achievements abruptly terminated when metal fatigue in the main rotor head caused a fatal crash 14 months later. The death of F.H. Dixon, Fairey's chief test pilot, saw suspension of the White Waltham design team's attempt at fusing the best features of the autogyro and the helicopter.

However, in August 1953 Whitehall offered to share the development costs of realising a generation-jumping rotary aircraft, and less than a year later the Jet Gyrodyne made its first flight. It would take another 14 months before this second prototype was able to switch from a vertical take-off to a level flight cruise, drawing on a theory of propulsion familiar to Wittgenstein before World War I: with a main rotor boosted by dual centrifugal compressors *and* twin variable-pitch pusher propellers mounted on stub wings, the Jet Gyrodyne demonstrated the feasibility of a tip-jet rotor-driven system. This remarkable example of cutting-edge aeronautical engineering proved conclusively that a scaled-up commercial version was possible. Enthusiasm for the project was so great that not enough attention was given to the challenges involved in piloting the machine, not least in-flight re-ignition of the tip burners mounted at the ends of the two-blade rotor, and control of the extraordinarily complex transmission system. The lift and propulsion unit was so big, and thus the power-to-weight ratio so poor, that the Alvis Leonides engine was always operated at full boost, generating serious issues in relation to noise and fuel consumption.[29]

The history of the Rotodyne stretches six years past the end of our story, such that a substantial proportion of the £11 million spent on building prototype XE51 was spent after Sir Richard died.[30] The collective enthusiasm in the early 1950s of the Ministry of Supply, the Ministry of Civil Aviation and British European Airways (BEA) explains why the directors of Fairey Aviation felt confident the company would in due course secure an acceptable return on its sizeable investment. That confidence was boosted by a periodic infusion of state funding, but neither the Treasury nor the Fairey board could have envisaged the size and scale of the experimental facilities built at White Waltham and Boscombe Down, nor the hundreds of hours spent on testing the power plant and rotor-drive: the Jet Gyrodyne validated the use of jet tips, but it gave scant guidance as to the radically different turbine engines, compression system and fuel pump required to power a torque-less rotary aircraft carrying three crew and 40 passengers or a freight equivalent. Having to scale up the means of lift and propulsion to secure a fast vertical take-off and to generate cruise speeds of more than 150 mph compounded all the problems signalled by the Jet Gyrodyne – the jet-tipped main rotor was nearly 28 metres in diameter.[31]

Meeting BEA's strict requirements regarding the noise level inside the aircraft compounded the problem of weight: the original Rotodyne's total load was nearly 15,000 kilograms, but, with still only one machine built, in 1959 the armed forces and American airlines insisted upon a much bigger – and heavier – version (with an estimated development cost of a further £9 million or more). A relatively smooth transition of the much-tested, much-displayed prototype into a production aircraft was now subverted by the demand for a far larger version, powered by hefty Rolls Royce turboprops rather than the current fit-for-purpose Napiers. The hope was that an increase in size would push down unit costs, which both the RAF and BEA saw as disproportionately high for short-haul flights. Ironically, concern was already being expressed over the impact of the original model on an already heavily polluted urban environment – the Rotodyne's critics insisted that anyone living close to a planned inner city 'heli-port' would be subject to a relentless barrage of noise. With BEA increasingly sensitive to hostile criticism, and the airline's enthusiasm rapidly cooling, it became that much harder for Fairey's sales and PR teams to claim that the Rotodyne was no noisier than any comparable fixed-wing aircraft.[32] In February 1960 Fairey Aviation personnel were subsumed within Westland Aircraft, along with all of their company's assets and activities. Yet, with a promise of government support as an inducement to merger, the Rotodyne's future seemed bright. However, a continued delay in commencing production saw the cancellation month by month of all civil and military orders, both American and British, a squeeze on state funding and, finally, on 26 February 1962, an announcement by the Ministry of Aviation that the Rotodyne project had been terminated – the sigh of relief from Westland was palpable.

Yet, for all its commercial failings, the Rotodyne was a remarkable testimony to the inventiveness and can-do mentality of aeronautical engineering in what we now see as the Indian summer of manufacturing industry in late twentieth-century Britain. With cruel irony, at the very moment Sir Richard's company passed into history, its technical staff finally triumphed over adversity – seven years of ceaseless problem-solving had seen the painful realisation of a project unparalleled anywhere in the world. The Fairey Rotodyne looked terrific, an obvious candidate for an *Eagle* cutaway drawing – indeed the fuselage bore a more than passing resemblance to 'Anastasia', Dan Dare's spaceship.[33] This wholly new concept, resplendent in cobalt blue company colours, embodied the spirit of the 'New Elizabethan Age', as prefaced in the Festival of Britain's Dome of Discovery. The 1950s witnessed a panoply of engineering initiatives, from generating nuclear energy at Calder Hall to sending signals into deep space at Jodrell Bank. Specific models, from the XK120 to the Mini, were obvious game-changers, but Britain's car companies rarely generated the same level of excitement and engagement as did Britain's plane makers. Riding on the back of 'a good war' aviation – now aerospace – enjoyed a high profile, with all but the most foresighted sceptics insistent that here was the one industry able to compete with the Americans on equal terms. Manufacturers of both military and civil aircraft saw Britain as still a global player, for most if not all of the

1950s. SBAC stalwarts like CRF and Geoffrey de Havilland simply could not conceive of a time when Britain would not be producing a range of models; they viewed North American sales to the RAF as a temporary measure until their industry was wholly recovered from the disruption of renewed conflict after a mere five years of recovery.[34]

This was the era of test pilots as household names, popular recognition of new aircraft, not least the V-bombers and the stunningly beautiful Hawker Hunter, and mass audience air displays.[35] The biggest air display of all was a fixture in the national calendar, attracting intense newspaper, newsreel and television coverage: the Rotodyne first flew at the Farnborough Air Show in September 1959, and was given star status.[36] Sir Richard Fairey, a pioneer from his early days on Sheppey, saw himself and his company in the vanguard of a national regeneration, rooted in an approach to manufacturing which was anything but complacent and nostalgic for times past. Sir Richard might be scathing about particular aspects of contemporary society, but he never looked back to a non-existent golden age. As he scrutinised Rotodyne blueprints and toured the test rigs at White Waltham, Fairey saw only the future – and a future which he hoped his body would hold out for long enough to see realised. It is impossible to imagine the chairman and champion of Fairey Aviation contemplating the possible failure of such an imaginative enterprise: the Rotodyne would one day be as familiar a presence at the heart of Britain's rebuilt cities as subways and shopping precincts. In the mind of Sir Richard his 'heli-liner' was a powerful manifestation of modernity.

The firm of Fairey embodies the jet age

Public fascination with test pilots in the postwar era saw high-profile individuals become synonymous with 'their' aircraft: at Hawker Siddeley, Neville Duke and the Hunter; at De Havilland, John 'Cat's Eyes' Cunningham and the Comet; at English Electric, 'Roly' Beaumont and both the Canberra and the Lightning; and at Fairey Aviation, Peter Twiss and the Delta Two.[37] Rising from the lower deck, Twiss's credentials rested upon extensive combat and test flying with the wartime Fleet Air Arm. His Fulmar squadron had been attached to *Ark Royal* when the carrier sank in November 1941, after which he remained on active service in the Mediterranean. Having gained the DSC and bar, Twiss was posted home to fly night fighters. He became an ad hoc test pilot, flying his first jet while on a naval mission to the United States; in 1945 he was accepted onto the Empire Test School's super-selective instructional course. A year later he joined Fairey Aviation, assuming responsibility for all deck landing trials. Twiss was number two to Gordon Slade, a former group captain and a future board member. Slade was a formidable figure who oversaw all aspects of the company's testing and experimental programme with a rod of iron: in 1956 the chief test pilot handed over responsibility for the FD2 in recognition that his deputy was better qualified to fly an aircraft which by now he knew intimately.[38]

Within a few months the world would acclaim Peter Twiss as the ice-cool, photogenic record breaker, with Gordon Slade nobly staying hidden behind the headlines. Sir Richard Fairey always enjoyed a close working and personal relationship with his test pilots, most notably Norman Macmillan and Chris Staniland, but even before he broke the world speed record Peter Twiss was seen as someone special. Here was a rare card always welcome at Bossington, and a useful foil to Richard junior, with whom Twiss developed a lasting – and fruitful – friendship. Both men loved flying, knew how to enjoy themselves, had colourful private lives, and were married on more than one occasion. Neither man suffered fools gladly, and neither would back down in the face of adversity – Fairey when faced with losing both legs, and Twiss when told he was not qualified to secure a commission and train as a pilot.[39]

In Twiss's account of flying the FD2 at an average speed of 1,132 mph, and thereby smashing the world air speed record by over 300 mph, Sir Richard was depicted as the only person who could give the go ahead. In this version of events, Richard junior encouraged his friend to secure CRF's permission when a Saturday shoot was adjourned for lunch. Twiss duly asked his host if he had given any thought to record breaking: 'He immediately said, "Yes." It was sharp and short, and definitive. And then, very properly, because he was the head of the firm, he added, "Yes, of course I have. And I think we ought to have a go at it." He said it again. And then changed the subject … What Sir Richard said commanded attention and appropriate action, and it would be a brave man who ever spoke to him out of turn.' Nevertheless Twiss did return to the subject later in the day, and Fairey said he would talk to the board. Nothing further was said, but Twiss remained sanguine across the succeeding weeks, secure in the knowledge that if Sir Richard 'had accepted the proposition. There was nothing more needed to be said. The next thing we knew there would be action.'[40] The great man giving the green light added excitement to the narrative, but Twiss had already made clear that he and Slade had been briefing senior staff on the feasibility of flying at over 1,000 mph since October 1955 when the FD2 first broke the sound barrier.[41]

The Delta 2 was after all designed and built to fly at supersonic speed, Fairey Aviation having responded positively when the Ministry of Supply issued its overdue specification for an experimental aircraft capable of attaining Mach 1.5 (nearly 1,200 mph) at 36,000 feet. Fairey had already been approached informally, as had English Electric, each company securing a development contract in October 1950. Critically, construction of Fairey's flagship aircraft was delayed until 1952, as development work on the money-making Gannet had to take priority.[42] In contrast, English Electric had the capital available to initiate R&D immediately. On 17 November 1954, only five weeks after the FD2 first flew, the flight programme was interrupted for nearly a year owing to damage sustained in a crash landing at Boscombe Down. Approaching Mach 0.9 at 30,000 feet, the first prototype's fuel supply system shut down. With no engine and fast failing hydraulics, Peter Twiss could either eject or attempt to glide his aircraft 30 miles back to the airfield. He chose the latter course of action, his courage and skill earning him the Queen's Commendation for Valuable Service in the Air. Eleven months later Twiss resumed his love affair with the FD2: 'If ever the Delta had had to justify herself to me after her crash, this wonderfully little aircraft did so that morning when she flew as gently as a bird into the hard supersonic October sky. From that moment I knew we had a world-beater.'[43] The following spring, on 10 March 1956, Twiss and the FD2 smashed the world air speed record.

It would be nearly two years before the Americans reclaimed the record. Fairey Aviation could enjoy a brief moment of international acclaim, albeit followed by years of disillusion. Unlike that of the FD2, the flight programme for English Electric's P1 prototype was always a low-key affair. Yet the renamed Lightning would enjoy dazzling commercial success at home and abroad: no fewer than 335 mark F1 to F6 interceptors were sold to the RAF and to the Kuwait and Saudi air forces. Fairey's FD2 was a strikingly graceful aircraft, and with its 'droop' nose and advanced hydraulic system a genuinely revolutionary design. Yet the Delta Two was wholly unsuited to carrying extensive ordnance, and RAF pilots found it difficult to fly – as opposed to the Lightning, with its exceptional performance, range of roles, and reliability in hostile climates. The hammer blow for the FD2 came in April 1957 when Duncan Sandys' white paper insisted that missiles and not aircraft constituted the future face of aerial warfare: the RAF procurement budget allocated funding for only one Mach 2 interceptor aircraft, and that would be English Electric's P1A – the Lightning. In contrast, the Delta Two's twin prototypes were assigned to the Concorde development programme before being signed off as museum exhibits. As Fighter Command's principal contractor, English Electric secured a major stake in BAC when the conglomerate was established in 1960. The same year a humbled Fairey Aviation joined with Westland, very much the junior partner.[44]

To be fair, Fairey's inability to capitalise on the headline-grabbing success of the FD2 was not for want of trying. The mastermind of a brilliant PR campaign was the company's publicity manager, Derek Thurgood, acting on direct instruction from a distant chairman.[45] Once news of the record was announced by the Royal Aero Club on 11 March 1956, Thurgood implemented a carefully planned media blitz, both at home and abroad. Within 24 hours of Fairey Aviation issuing a press release, copies were en route to every air attaché on diplomatic service, with clear guidance re rebuttal of American claims that the United States remained in the vanguard of record breaking. Thurgood placed triumphalist adverts in a host of specialist and non-specialist publications, fast-tracked a company flyer entitled *FD2 The World's Fastest Jet*, and purchased for global distribution 12,000 copies of 'Outpacing the Sun', a supplement hastily compiled by contributors to *The Aeroplane*. Glossy magazines in the Commonwealth, the United States and western Europe received an array of colour photographs, while at home *Picture Post* and *Illustrated* were given a select choice of superior shots. Not that the rest of Fleet Street could feel deprived, nor the newsreel companies: Pathé enjoyed privileged access on the great day, but footage was later released to Movietone. In addition, Fairey Aviation commissioned a promotional short from J. Arthur Rank, and liaised with BP on a film for schools. A sign of the times was that Thurgood prioritised television over radio, devoting as much attention to the fledgling commercial companies as the BBC. Fairey lent a model of the FD2 to Broadcasting House for use as a prop in a discussion programme led by Labour MP Christopher Mayhew. More striking, however, was Peter Twiss's appearance as the mystery guest on BBC TV's popular panel game 'What's My Line?' only six days after breaking the record. Nor was that all: Thurgood informed both the BBC and ITV, as well as the national press, that both Mr and Mrs Twiss were available for interview at any time.[46]

Although the *Daily Sketch* was critical of the news being released on a Sunday (at the insistence of the adjudicating authority, the Royal Aero Club), newspaper coverage was overwhelmingly favourable: correspondents and commentators saw Fairey's success as a great flag-waving event, to be proclaimed loudly on the far side of the Atlantic. In his account of the record-breaking attempt, Peter Twiss recalled how masterly Thurgood orchestrated an ongoing publicity campaign. Both men were strikingly combative in dismissing American claims that Chuck Yeager's X1A, the first machine to fly supersonic machine albeit launched from a mother aircraft, had been a more innovative concept than the Delta Two. So too was Geoffrey Hall the day after the new record was announced: not only did the FD2's success silence any suggestion of superior American technology, but it highlighted yet again the unique status of Fairey and its founder, 'whose pioneer spirit has never ceased to guide and inspire the efforts of the Company.'[47] With Fairey Aviation enjoying front-page coverage, and its share price soaring, the mood in Bermuda was buoyant: on first hearing the news CRF had 'smiled and tears ran down his cheeks.' A telegram from the Queen and Prince Philip – the latter a recent guest of Sir Richard's at the Hayes plant – was especially pleasing.[48] Yet in every respect spring 1956 marked the high point of Fairey Aviation as a global player in the jet age – the firm's fortunes, like those of the Delta Two, had nowhere to go but down.

Living with Labour – Fairey and the contrasting politics of Britain and Bermuda

Evelyn Waugh famously described the six years of Labour government that followed World War II as living in an occupied country.[49] It is hard to imagine Sir Richard Fairey having much in common with the author of *Brideshead Revisited*, but he would certainly have endorsed Waugh's view of life in Mr Attlee's socialist commonwealth. Waugh considered decamping to Ireland but ultimately decided to stay. Fairey could winter in Bermuda. In a speech to the Royal Empire Society in May 1949 CRF gave full voice to the views of his dining companions 5,000 miles away. He conveyed to a sympathetic audience how bemused 'ultra-loyal' Bermudians were by recent events in Britain, not least Aneurin Bevan's unfortunate description of the Conservative opposition: 'the sudden fracture in our national unity and its replacement by a certain rather spiteful disparagement of much that England has stood for they find difficult to understand. To hear Cabinet Ministers referring to their opponents as "vermin" and "not worth a tinker's cuss" does not give the appearance of forward progress in the conduct of public affairs, but seems to reveal the intrusion into Government of a less intelligent and less balanced individual.' Fairey insisted that Bermudians saw 'security from cradle to grave' as a lifelong dependency on state hand-outs, and the benefits of a National Health Service as scant compensation for the continuance of rationing. The same 'ultra-loyal' observers of the British state opposed any expansion of the public sector, and recognised just how much the BBC was still 'tainted by Left Wing propaganda.' CRF acknowledged that these were the views of a white minority, but insisted that black Bermudians were similarly sceptical of top-down reform, from the enforcement of non-discrimination to the provision of free education; thus the latter 'should come about incrementally and not be imposed by left-wing politicians in Britain.' Segregation, official or informal, was clearly indefensible. Fairey implicitly if not explicitly acknowledged this, while at the same time contrasting the majority black population's standard of living with what across the Caribbean constituted the norm.[50]

Fairey continued to rail against Broadcasting House in correspondence with Brabazon. In March 1948 he claimed that everyone in Bermuda believed the BBC to be 'just another socialist propaganda agency.' The Corporation was in cahoots with the Labour Government, with deplorable consequences in Washington and New York. As early as June 1946 CRF was lamenting a complete collapse of British prestige in the United States. Two years later he bemoaned how much the Attlee administration abused the goodwill of even the staunchest Anglophiles: 'they are getting fed up with paying for socialism, as one put it "it's a hell of a job trying to convince the American public that they should work six days a week so that the British can work five", and I don't know the answer to that one.'[51]

No action undertaken by the Government escaped Fairey's censure, but unsurprisingly it was the impoverished state of civil aviation in Britain which provoked the fiercest criticism. Labour was seen as pursuing an already discredited policy of keeping Britain's flagship airlines, notably BOAC, in public ownership. At an SBAC dinner in July 1949 he contrasted the loss-making, still heavily subsidised air operators with the nation's plane makers, collectively endeavouring to exceed the previous year's output of aircraft worth £25 million. Fairey considered aircraft manufacture the ultimate vindication of private enterprise, as tested in two world wars. In an earlier speech to the SBAC he paid tribute to the wartime endeavours of an industry 'free, unfettered, and in keen competition.' This was a remarkable claim by someone so intimately involved in the British war effort. Fairey's rewriting of history went unchallenged by an audience happy to ignore their speaker's wartime service as a senior official in the Churchill coalition's most interventionist department of state.[52] Where Sir Richard was on firmer ground was in suggesting that Labour held back from nationalisation as it considered such a radical move to be counter-productive – prior to being bought out by the state, individual companies would run down their design and development programmes at a critical moment in deteriorating East–West relations.[53]

This was a powerful argument advanced in Whitehall, not least by those in the Cabinet such as Herbert Morrison, who anticipated intense opposition inside and outside Parliament to further nationalisation: a renewed assault upon 'the commanding heights of the economy', focused upon manufacturing industry, could await a renewed mandate and a fresh burst of legislative energy. The 1950 general election would see Labour's Commons majority cut from 146 to a knife-edge five; an exhausted administration left office only 20 months later, after Churchill secured victory at the third time of asking. For all but the most naïve left-winger in the Parliamentary Labour Party, nationalising the aircraft industry was well and truly off the agenda.[54]

In the first six years after the war it felt for Fairey as if the lunatics had taken over the asylum, and both publicly and privately he demonstrated his inability to comprehend why Labour had secured power in July 1945, and his refusal to accept any degree of competence within the Cabinet, or the Government as a whole: 'the socialists' were in office as a consequence of their confidence that 'they can kid people into any required frame of mind, and maybe they are right.'[55] From this perspective Bermuda offered an attractive alternative, its executive operating in harmony with an assembly voted for by a property-qualified electorate. Within a deeply conservative society, where support for the status quo was presumed to cross complex racial barriers, Sir Richard saw an oligarchy operating in the interest of all, not least 'the most prosperous coloured community in the world.' In the spring of 1949, having recently joined the Bermudian Government Board, Fairey described a society based on consensual separate development: 'Although freely intermixing in the work and daily activities socially each race largely keeps to itself but that involves no hint of a superior attitude nor discourtesy or unfriendliness in the relationships between them. There are of course minor discontents and clashes, but in general relationships between the two races are easy and cordial ... both prefer a certain parallelism that has grown up...' If 'left-wing politicians in Britain' left the people of the archipelago well alone then they would 'stay loyal to their old British traditions, and remain a prosperous, contented and cultural oasis of prosperity within the Empire.'[56]

CRF offered a Burkean view of Bermuda, with social and cultural change occurring organically, and incrementally. Sudden change – such as universal franchise – would be deeply damaging, not least in terms of a standard of living far in advance of the archipelago's distant Caribbean neighbours.[57] The Colonial Office, even while Labour was in power, shared a similar view of what was still a vital outpost of western security. During World War II and then into the Cold War, Bermuda fulfilled a crucial role intercepting trans-Atlantic communications, both mail and signals intelligence. Although the U-boats were long gone, the Royal Navy, along with its Canadian and American counterparts, maintained a substantial ASW force in anticipation of Soviet submarine incursion into the western Atlantic. The United States had a large military presence as a consequence of the Destroyers for Bases Agreement signed by Churchill and Roosevelt in September 1940, and from then until 1995 both the Americans and the British operated out of the purpose-built Kindley Air Force

Base. The posting to Bermuda in 1943 of Lord Burghley, a 'safe pair of hands' trusted by the Prime Minister, was a political appointment. The future Marquess of Exeter, a distinguished Olympian, had the requisite charm to sway American VIPs visiting the governor's house. He only stayed two years, but in that time struck up a close friendship with CRF. After 1945 the successors to Burghley were military appointments, first an admiral and then a general.[58] The islands' strategic importance was further demonstrated by the newly crowned Queen's visit in November 1953, and the summit held there a month later – Churchill and his frustrated heir apparent, the Foreign Secretary, Anthony Eden, hosted President Eisenhower and the latest of the Fourth Republic's numerous prime ministers, Joseph Lionel.[59]

Conscious of the need to preserve sound relations with 'the mother country', Sir Richard was temperate in tone regarding Labour's record when addressing an audience in Bermuda: in early 1949 Hamilton's Rotarians were scarcely surprised when hearing that their speaker 'held no brief' for 'the socialistic government of Britain', while at the same time they were pleased to learn that aviation led the way in economic revival and national well-being. Bermuda's great and the good would have been stunned had they heard the speech Fairey intended delivering before common sense kicked in and he anticipated the damage to his company if a truly hair-raising indictment of Attlee's government became common knowledge back home. This would certainly have been the case as attending the Rotary lunch was a local reporter. Unknowingly deprived of a headline-grabbing story, the *Bermuda Mid-Ocean News* found nothing controversial in the views of Sir Richard, a 'towering giant of a Briton who has a voice to match.' The same newspaper, a year earlier, had printed a photograph of CRF receiving the US Medal of Freedom with Silver Palm. The special award ceremony at Kindley Field was a belated acknowledgement by the Truman administration that the former head of the BAC had displayed 'exceptionally meritorious service in the field of scientific research and development.' Fairey would have appreciated the citation's wording, with the emphasis on his credentials as an engineer not a bureaucrat.[60]

1948 and 1949 were the years when Fairey was especially active in what might loosely be termed Bermudian political life. He took his executive appointments seriously, not least his membership of the Airport Board. Eager to improve Bermuda's flying boat service to the United States, he sought and failed to convince the RAF to develop its base on Darrell's Island as a joint civil–military airport. At the same time CRF was energetically lobbying the Colonial Office and BOAC to provide a grant and free travel for those children keen to attend boarding school in Britain. Given the economic and political climate at the time, such a scheme stood little chance of being adopted, not least as selection of the beneficiaries would prove so problematic.[61] These were just two of several attempts by Sir Richard to advance the interests of Bermudians, albeit in practice those drawn from the privileged white community.[62] Fairey liked the old Bermudian families, not least because they did not flaunt their wealth. He had become acquainted with them through sailing, and friends he made on the island encouraged the idea of buying a property. At the time the Fairey family was unusual in that few from Britain became semi-permanent residents. Bermuda was in every respect CRF's second home, but it was never a second country. However congenial life in Paget might be, his extended visits were, above all, a consequence of his chronic susceptibility to bronchitis – in an ideal world he would have spent far more time after the war at Bossington.[63]

Back home at Bossington

Fairey still had status, position and influence in Bermuda, but far less so at home. The Conservatives returned to government in October 1951, but this scarcely affected someone who had never slavishly supported the party, and who surely found Churchill's soft glove approach to the trade unions indefensible. Although still chairman and managing director, Sir Richard was no longer the first person to call if the permanent secretary or the head of procurement needed to get in touch with Fairey Aviation fast. Poor health was now a severe constraint upon CRF's activities, and by the early 1950s his yearly engagements diary showed him spending less and less time away from the Test Valley, albeit compensating with a constant stream of house guests. His closest friends would invariably come to him, with visits rarely reciprocated. The convivial atmosphere at Bossington ensured an invitation to dine or to stay was rarely spurned. Away from the house, company-related appointments became few and far between, such that in 1953 CRF's only regular commitment was attendance at Chatham House's monthly lectures: on 23 April that year the after-dinner speaker for the Royal Institute of International Affairs (RIIA) was the Trotskyist commentator Isaac Deutscher – one can envisage Fairey admiring the man while abhorring his politics.[64]

Attending the RIIA vied with the Royal Aeronautical Society for attracting Sir Richard up to town. In 1950 Fairey was awarded the Society's highest honour, when he became an Honorary Fellow. By this time his involvement with the SBAC was very low key. Although he continued to attend SBAC functions, notably its dinners, his address to Council in June 1947 had a distinctively elegiac note to it: 'I am moved to consider how much I appreciate the good fortune that gave me a place in such a distinguished company as yourselves, and which in all sincerity I have accounted a true privilege; and I wonder sometimes where else, in what age or conditions, I could have found a life which with its ups and downs, its strenuousness and its unbending competitiveness, could have thrown me among a group of associates and friends as stimulating and as altogether interesting as yourselves collectively, and as pleasant as friends individually as the company here tonight.'[65] Remarkably, almost all the great patriarchs of the British aircraft industry were alive and active, and quite possibly listening to Dick Fairey's address. The speaker would have insisted that he still maintained a hands-on role, but among his peers only de Havilland and Handley-Page continued to exercise a level of direction and control comparable to that which they had all enjoyed between the wars.

The future of Fairey Aviation was never far from his mind, but henceforth the priority for CRF was making best use of his river beats without harming the unique eco-system that constitutes the waters of the Test above and below Stockbridge. At the time, 'eco-system' was a term rarely used by non-scientists, but Fairey would have understood instinctively its presumption of pollution-free biodiversity and the defence of environmental interdependency. Thus, extensive development work, ranging from cleaning out river beds to creating a trout hatchery, displayed the owner's sensitivity and keen sense of responsibility. Fairey's priority was improving his estate for the benefit of his family, and to ensure his friends and guests enjoyed a quality of fly fishing unmatched anywhere in the south of England. Geoffrey Hall recalled how in the early 1950s Bossington's trout hatchery – its temperature controlled by a spring of warm water – was nurturing impressively large fish. His half-brother began to use lightweight, bespoke fishing tackle, applying the mind of an engineer to the design of his rod and reel.[66] Bossington retains Sir Richard's collection of flies, all of which he tied himself. One invention popular with visitors to the house was Fathers' Irresistible, a variation of the well-known Adam's Irresistible dry fly, first used in the 1930s on rough water streams by the American designer and dresser Joe Messinger. Fairey's simple to tie pattern was registered with the Flydressers' Guild as a 'purplish grey' hackle and a body of 'yellow hair from a deer's tail.' He was asked to tye a collection of flies for the late Queen Mother, which was presented to her and which probably included Fathers' Irresistible. In management of his beats CRF was clearly influenced by the pioneering work of Houghton Club member Frederic Halford, a prolific writer who before World War I codified the art of upstream dry fly fishing on chalk streams.[67]

Fairey was not, on principle, opposed to new farming methods, but he considered any use of chemical sprays to be indefensible. He was ahead of his time in arguing that large farmers should be subject to stricter control in the deployment of pesticides with the potential to decimate local wildlife.[68] Here lay a dilemma for CRF in that he recognised the damage unregulated farming could inflict upon the countryside, but at the same time he resented any intervention by central or local government, and the agencies acting on their behalf. During the war he had battled successfully with the Catchment Board from the far side of the Atlantic – a radical lowering of the water table to drain water meadows adjacent to the Test and its tributaries had been duly thwarted. After the war Sir Richard believed the best means of countering any attempt to alter the status quo was to join the same boards of which he was so suspicious. This was an astute move, his appointment to the Hampshire River Board ensuring a place on all key policy-making committees. At the same time Fairey became deputy chairman of the Test and Itchen Fishing Association (TIFA) and of the Hampshire Rivers Landowners Committee.[69]

This strategy, although to some extent hindered by CRF's absence for nearly half the year, worked well; only in the last year of his life did it break down. In 1955 Fairey supported the River Board's ban on electric fishing. This triggered a successful attempt by Anthony Tuke, the son of the Barclays Bank chairman, and other local landowners such as the Earls of Inchape and of Perth, to see Fairey voted off the TIFA's executive board. Notably absent from the conspiracy were the Mountbattens, down river at Broadlands, Tommy Sopwith, who by now was living across the valley from Bossington at Compton Manor, and 'riparian owners' north of Stockbridge. Several complainants lived in the Test Valley but were not riparian owners; in other words the river did not flow through their estates. Some members of the Houghton Club might live locally, but they were rarely riparian owners. Tuke and his aristocratic co-signatories drafted a letter, which they hoped would be seen by all proxy voters. This wider constituency included members of TIFA who were paying rods on fisheries of the respective rivers and in consequence knew little about the management of chalk streams, let alone the perceived dangers of electric fishing.[70]

Tuke and his allies argued that their association should have lobbied the River Board to deploy 'an electric fishing association to deal with the vermin in their own water.' Such lobbying did not take place because Sir Richard Fairey had sought to protect eels, 'for the reason that they provide him with a means of remunerating his water-keepers ... we think it quite wrong that the trout-fishing interests of the river as a whole should be prejudiced by views expressed by one who is a director of the Test & Itchen Fishing Association.'[71]

Fairey had cause to feel aggrieved in that the letter, endorsed by hostile members of the Houghton Club, unfairly portrayed him in a bad light: the late autumn trapping of eels, with the Head Water Keeper benefiting from their sale, was a longstanding tradition on local estates. Also, as his daughter pointed out, Fairey did not want to ban the practice of electric fishing, but have it switched from an AC to a DC current, with a licensed system under the control of a bailiff. As an electrical engineer he could in fairness speak with authority, and he viewed the uncontrolled use of electricity as deeply damaging to river life, especially invertebrates.[72] In due course the River Board would adopt his recommendations.

On 9 December 1955 Fairey failed by one vote to secure re-election to the TIFA board. He was incandescent, transferring all his personal accounts from Barclays to Westminster Bank, and urging Richard to do the same. Tuke having ignored his letters of protest, Sir Richard wrote to the Barclays' board demanding their employee's dismissal for his having placed in the public domain 'a petty, mean, and libellous attack' on one of the bank's 'oldest and best customers.' Unsurprisingly, Barclays' directors took no action – A.W. Tuke had been born into one of the bank's founding families, and like his father would in due course become chairman. Clearly being deselected by such a narrow margin was a blow to CRF. Yet, although he had lost the battle, in due course he won the war: Sir Richard would secure his principal objective, namely to prevent uncontrolled electric fishing on both the Test and the Itchen, prior to his death in September 1956.[73]

His daughter, Jane, and before her Geoffrey Hall, testified to Fairey's intense dislike of what he saw as pointless killing, especially the slaughter of wildlife on an industrial scale. He believed the shooting of game birds to be consistent with his commitment to conservation: in Fairey's view cultivating pheasants and partridge was crucial to the sustainability of a rural way of life. As with electric fishing, he believed in a system of organisation and control. The threat of an imbalance in nature was to be avoided at all cost, with the status quo subverted by any threat to the survival of the species, whether that be individual field birds decimated by pesticides or rabbits deliberately infected with myxomatosis. Fairey lobbied energetically to see the latter practice outlawed, his efforts helping secure the presence of a statutory ban in the 1954 Pests Act.[74]

Indebted to Fairey as a mentor, patron and role model, Hall saw his sibling as 'a man deeply absorbed in the worth and beauty of nature and respecting and enjoying the world around him.' He was an adopted countryman, but at the same time he was no son of the city, treasuring his roots in the Fens and never owning a house in London.[75] There was no certainty concerning Sir Richard's mortality, but he was a realist and all too well aware that time was of the essence – life had to be lived to the full each and every day. Bossington, the house and the accompanying estate provided Fairey with a stability, a comforting level of domesticity, and a sense of certainty absent at earlier stages of his life. The certainty lay in the adopted countryman's engagement with the seasons of the year, and the domesticity rested in a marriage and a lifestyle that accommodated both his fragile health and his keen sense of family loyalty and obligation. That obligation was a two-way process in that the now extremely wealthy industrialist took seriously his responsibility for the individual well-being of his extended family, while at the same time anticipating love, deference and a keen sense of duty from those nearest and dearest to him. Close family ties saw CRF feel deep affection for Geoffrey Hall, but sibling love was reinforced by a healthy respect for someone who in terms of personal achievement had more than justified his half-brother's constant support and advice.

As for Esther, Fairey's love was unqualified and unwavering, his only regret being that they had not met earlier. There was no conspiracy of silence surrounding his first marriage – how could there be when Richard was ever present? The first Mrs Fairey simply merited no mention in normal conversation, and so for long periods of time she was forgotten. Esther Fairey had almost certainly never met her predecessor, and the children were scarcely aware of Joan's existence. They were both still young, and Jane Fairey was only in her late teens when her father died.

CRF could take pride in his daughter's intelligence, insight and emerging sophistication. Her privileged background ensured an enviable lifestyle, and yet from an early age she engaged fully with the wider world. Jane kept abreast of contemporary events, and she took a keen interest in the family firm. Mid-fifties photographs of Fairey functions show the chairman's daughter listening politely to middle-aged and elderly gentlemen, soberly suited and suitably opinionated. This was a highly patriarchal and deeply conservative environment, yet despite her youth Jane Fairey was more than able to hold her own. Boarding school had provided Jane with a tolerable education if no great intellectual stimulation, and regrettably the Faireys lived in a world where few considered university the natural destination for bright young women.[76]

Arguably Jane learnt far more at home than she did at school. From both her parents she inherited a clearly definable set of values, with Esther's honesty, good humour, discretion and pragmatism rendering her the ideal role model. Nearly half a century after Fairey's death his daughter recalled that 'one of the last things my father said to me when he had hours to live was: always consult your mother [as] she has great sense.'[77] CRF fostered Jane's zest for knowledge, and the capacity to act in a suitably clear and analytical fashion: 'He was a most wonderful teacher ... and he could instil such a desire to learn.' In Fairey's final years there were far fewer guests at Bossington, and he would take the evening meal as an opportunity to discuss matters of substance with his two children: 'astronomy and the expanding universe, politics, history, the environment, and the increasing world population which particularly worried him ... Religion and the existence of God or not was often discussed. John was always a religious person and I think by that stage of his life my father was an agnostic.' As well as the consequences of rapid population without effective birth control, a prescient Fairey encouraged speculation on what the world would be like once automation had significantly reduced the number of necessary manual jobs. Not that all conversation was of a serious and sombre nature – Fairey's well-honed powers as a raconteur remained intact, however fragile his health.[78]

Like so many fathers and daughters, Sir Richard and Jane shared a powerful bond of love and mutual admiration. Yet at the same time CRF displayed deep paternal affection for his second son, having high hopes for him in terms of scholarly achievement and growing maturity.[79] In due course John progressed from his prep school to Eton, although, surprisingly, he entered a different house from that headed by Charles Routh, Senior Master in History and an acquaintance of CRF.[80] Later, like his father's closest friends in the years before World War I, he read engineering at Cambridge. While studying at Magdalene, Fairey joined the University Air Squadron, having already gained his pilot's licence flying float planes in Canada. John Fairey would go on to have an extraordinarily varied career in aviation, enjoying considerable affection and respect among fellow fliers of commercial and vintage aircraft. He also experienced deep personal tragedy, with his three sons predeceasing him.[81]

By the time of his death Sir Richard must have seen John as very much a chip off the old block, enjoying a simpler, less strained relationship with him than he had with Richard when the latter was of a similar age. In consequence, John Fairey looked back on his childhood with great affection, recalling his father as someone who 'had the ability to stimulate curiosity and a healthy scepticism. He taught us never to take anything at its face value, and I think that this was one of the most valuable gifts that he could hand on to children.'[82] He also passed on a keen sense of fun. Here was a parent whose age and physical condition prevented any direct involvement in his children's more energetic activities, and yet his demonstrable love of the countryside encouraged a shared enthusiasm for every facet of rural life.

That same love of field sports was keenly shared by CRF's oldest child. So too was a deep-rooted love of aeroplanes, although for Richard Fairey that extended to flying the machines as well as overseeing their construction. Here was someone whose eagerness to seize the controls and to embrace the new extended to purchasing a Sikorsky S54 helicopter as a convenient if costly means of visiting factories and commuting from his Maidenhead home. In addition, Richard drove a Lagonda sports car and a Studebaker shooting brake. While petrol was still on ration keeping both cars on the road was a perennial problem, but, as well as his disability allowance, Richard enjoyed access to his employer's commercial allocation. Fairey Aviation generously covered all his travel expenses, helping its general manager to maintain a lifestyle incompatible with his regular income. In addition, Sir Richard continued to support his son well into the 1950s, courtesy of direct subsidies or substantial share transfers. Richard Fairey's personal finances were complicated and chaotic, leaving him heavily dependent upon a parent presumably keen to avoid public scandal as a consequence of bills unpaid – here were strong echoes of CRF tolerating his first wife's extravagances 30 years before.[83]

That tolerance extended to CRF's fellow directors. Richard travelled extensively abroad on behalf of Fairey Aviation, often accompanied by his wife. The firm was suitably generous in funding these trips, some of which were thinly disguised holidays, as on a visit to France in May 1949. Richard felt little obligation to submit a speedy account of his activities while abroad; for example, two months passed before he reported back on a visit in December 1950 to inspect Fairey's Canadian operation. To be fair, when he did submit a report, it was always thorough and professional – here was a highly talented individual, but with a lifestyle unsuited to the attitudes and mores of the postwar boardroom. Broadbent held Richard in high regard, not least for his refusal ever to make any concession to his disability, and his capacity for dealing with a cardiac condition not that dissimilar from his father's. Other members of the board focused on Richard's presumed shortcomings as a husband and a father.[84]

That disapproval extended to the chairman and managing director, with whom Richard increasingly clashed over company policy. A running sore was the airfield taken over in 1944 and by now part of Heathrow: CRF insisted on fighting the Government to the bitter end, while Richard time and again pointed to the firm's mounting legal bills and urged compromise.[85] Clashes between father and son concerning the future direction of Fairey Aviation were fuelled by sharp differences over Richard's complicated private life. Unsurprisingly perhaps, Richard Fairey was notable by his absence from the strategy meetings held at Bossington from the summer of 1954 through to the following spring.[86]

Every week, the deputy chairman, Roland Outen, and the newly appointed managing director, Geoffrey Hall, spent hours debating with Sir Richard about the future structure of Fairey Aviation. CRF was keen to avoid a power vacuum in the event of his early demise. There would be a key liaison role for Richard junior, but no acknowledgement that at some point he would assume control of the company. Apart from any reservations held privately regarding Richard's suitability as managing director and/or chairman, his health was seen to rule him out as the natural successor. CRF's specialist in Boston, Paul White, had examined Richard and declared his alarmingly high blood pressure hereditary – and thus potentially fatal. The patient disregarded this diagnosis, but his father and his colleagues clearly did not. Thus Outen was designated chairman in waiting, with Hall expected to serve as deputy chairman. Geoffrey Hall was therefore the heir presumptive, at some point in the future combining his executive role as managing director with ultimate responsibility for the company as its elected chairman. This plan of succession was finalised in May 1956. At that moment few could have anticipated the death of Sir Richard being followed so swiftly by Roland Outen's, leaving Hall in overall charge less than a year after the firm's founder had acknowledged the suitability of his half-brother and not his son to lead Fairey Aviation into an uncertain future. That uncertainty was rooted in Sir Richard's firm belief that the next Labour government would nationalise the aerospace industry, leaving Fairey dependent upon its ancillary activities. In the light of events that fear seems exaggerated, but prior to the 1955 general election the Shadow Cabinet confidently anticipated an early return to power.[87]

As it transpired, Richard Fairey was in Australia far from the action when Fairey's shareholders voted Geoffrey Hall chairman in March 1957; the political climate post-Suez revived Labour hopes of electoral success, but the City's fears were assuaged by leader Hugh Gaitskell's refusal to force manufacturing industry into public ownership. Hall's election triggered an unanticipated changing of the guard, and an injection of new blood. Maurice Wright and Massey Hilton both died within a few months of each other, and the following year Wilfred Broadbent announced his retirement.[88] It was as if, with their leader gone, the hard-nosed engineers who had made Fairey Aviation the Fleet Air Arm's principal supplier, and built a jet aircraft capable of smashing the world air speed record, had lost the will to carry on. These were old men, too tired to see out their company's Indian summer. The death of Dick Fairey had changed everything – now it was time to go.

Sir Richard's final visit to Bermuda had continued into the late spring of 1956, as before that his family feared he might not survive the rigours of a long journey home. By this time he was in a very delicate state, but not so debilitated that he declined to host shooting parties – invitations for the autumn shoots were sent out as usual. The family's attempt to maintain a degree of normality was aided by Paul White flying over to examine his patient, and advising against further invasive surgery. Dr White's fatalistic attitude contrasted with that of Dr Davey, the family physician, who when visiting Bossington urged Esther to request a more rigorous investigation of her husband's overall condition. Within days CRF had been admitted to the London Clinic. A malign tumour was quickly identified, with consequent surgery. Brabazon came to see his old friend for what both men surely suspected would be a fond, final farewell. Remarkably, given the poor state of his heart, Sir Richard survived the operation. Yet over the next 24 hours there were few if any signs of recovery, and on 30 September 1956 Dick Fairey died. John Fairey sensed that from the moment he entered the hospital his father acknowledged the inevitability of his fate, and in so doing he displayed enormous courage. His mother concurred: her 'wonderful man' had died too soon ('he had so much more in that tremendous brain of his to give the world'), and yet 'he knew he was not going to survive, he was too tired to fight any more.' For Geoffrey Hall the brother to whom he owed so much had 'heard the call of his forebears and the English soil he loved so well claimed him. He lies near the little village church on his estate at Bossington.'[89]

He does indeed lie in what has over time become the Fairey family graveyard. St James's church stands beside the main drive up to the house, halfway between the banks of the River Test and the country lane connecting Mottisfont and Stockbridge. Sir Richard Fairey spent little or no time inside the parish chapel engaged in prayer or quiet contemplation, but he surely treasured this ideal location for his final resting place. This is 'deep England' as idealised by a deeply conservative man whose pastoral vision was rooted firmly in the downs, meadows and river valleys of a broad swathe of southern countryside stretching from the Thames to the Channel – in Angus Calder's words, that 'Green and Pleasant heartland' so familiar from wartime propaganda.[90] Suitably, at his funeral six of the estate's keepers carried Fairey's coffin. His daughter remembered the day of the funeral vividly: 'I can still see Reg Dade, the Head Water Keeper, a great favourite of my father, with tears streaming down his face. Many people were devastated by his death. My mother was distressed by the numbers of men who openly cried when expressing their sympathies to her for her loss in the following months after his death.'[91]

Having overseen arrangements for the funeral, Esther was then approached by Fairey Aviation's public relations chief, Derek Thurgood, regarding Sir Richard's memorial service. She drew up the order of service, while he, with customary thoroughness, briefed the press, drew up an ambitious guest list, and orchestrated proceedings on the day. The service took place in front of a packed congregation at noon in St Martin-in-the-Fields on 19 October 1956. The timing of the event, with the invasion of the Suez Canal Zone imminent, helps explain the absence of cabinet ministers and senior military staff.[92] The same is true of notable absentees from the list of those who sent letters or telegrams of sympathy to Richard Fairey – for example, no First Sea Lord and no chairman of Hawker Siddeley.[93] However, one can assume the likes of Mountbatten and Tommy Sopwith conveyed their condolences in hand-written notes to CRF's widow.

The memorial service, in all its liturgical splendour, exemplified everything Dick Fairey might at one time have questioned, *and* everything he deemed vital to the nation's well-being as an ordered, organised and suitably deferential society – the established church and the liberal democratic state in harmony, and maintaining a uniquely English status quo, as defended so recently in two world wars.[94] Esther Fairey considered the unique combination of Hayes's PR team and the Church of England to have delivered 'a beautiful service' of which 'my darling would have approved.' She thanked John Moore-Brabazon for his 'marvellous address', delivered with characteristic aplomb.[95]

Lord Brabazon's surprisingly short eulogy was well crafted, and no doubt well delivered. He acknowledged his old friend's faults, while at the same time recognising his many excellent qualities. Brabazon summed up the achievements of a lifelong pioneer blessed with an extraordinary breadth of knowledge, 'who never lost his boyish enthusiasm for the fresh and inventive. Here was "a very big man" – both in physique and character' – and 'There was nothing he touched that he did not adorn and embellish.' Brabazon celebrated Fairey's good fortune 'that his work allowed him to taste both success and fame which he so well deserved during his lifetime. And I think we should delight in the fact that the latter part of his life was spent in such domestic felicity and happiness.'[96]

Here surely was the speaker's shrewdest insight: notwithstanding the travails of terminal illness – few, if any, 'die a happy man' – 'dear Dick' experienced a level of contentment he could scarcely have envisaged as a tyro engineer during the Great War or a decade later as a high-achieving industrialist plagued at home by doubt and suspicion. He took great satisfaction from his success as a race-winning yachtsman starting from scratch, and as a wartime envoy realising the potential of the Atlantic alliance, but in his later years the greatest achievement was clearly the lifestyle and the environment his family enjoyed on the banks of the Test. Bermuda was far more than simply a semi-tropical bolthole from lung-testing winter weather, but its attraction and importance could never match all that Bossington had to offer. At Bossington Fairey was for the first time firmly rooted, enjoying a solidity, security and permanence absent from previous houses and estates. Here he left a physical legacy, as becomes clear to anyone visiting the estate six decades later. Crucially, he and Esther created a home, hence Brabazon's celebration of a restless soul finding peace in domesticity and the company of like-minded friends.

Conclusion

A *Flight* photo from April 1954 captures CRF on a visit to HMS *Peregrine*, otherwise known as the Royal Naval Air Station Ford.[1] The West Sussex base had been in continuous service since 1917, with Fairey aircraft a familiar presence. Sir Richard had been driven over from Bossington for the formal handover to the Royal Navy of its first four Gannet anti-submarine aircraft. It is clear from the magazine photograph that his hosts wanted to put on a show, with the occasion overseen by no less a figure of authority than the Flag Officer Air (Home). The crews of No. 703X Flight are kitted up ready to fly, and behind them on the runway ready to roll are a Gannet AS1 and, yes, a Swordfish. Judging by their chiselled features and receding hairlines these men are old enough to have flown the Swordfish and its piston-engined successors in combat, anywhere from the Western Approaches to the Yalu River. It is not surprising that the flight's commanding officer, Lieutenant Commander F.E. Cowtan, was keen to impress. His distinguished guest was the founder of a multi-million pound company which for nigh on 40 years had maintained a unique working relationship with the Fleet Air Arm – in good times and bad Fairey Aviation had facilitated the aggressive exercise of naval air power in home waters, the Atlantic and the Mediterranean; in 1944–5 and 1950–3, strike aircraft built at Hayes and Heaton Chapel were on operational duties in the Pacific and South China Sea. The number of aircraft deployed in RNAS/RAF and Fleet Air Arm colours was staggering – as in the 2,399 Swordfish which rolled off the assembly lines from 1935 to 1944, or the 2,584 Barracudas flown from 1943 to the surrender of Japan. Fairey aircraft – Sea Fox, Barracuda and Swordfish – had contributed to the destruction of Germany's three best-known battleships: the *Graf Spee*, the *Tirpitz* and, most famously, the *Bismarck*. The grim-faced aircrew standing to attention that wintry day in 1954 appeared older, wiser, more cautious versions of the relieved young men filmed on the flight decks of *Victorious* and *Furious* fresh from flying their Barracudas into an Arctic fjord to cripple the *Tirpitz*.[2]

In the Sussex snapshot, Sir Richard, chilled and wind-swept despite his trilby and bespoke calf-length coat, casts an expert eye over both men and machines. He always enjoyed the company of junior and middle-ranking officers, if only because they gave him an honest assessment of his aircraft's performance – between the wars RAF pilots had found themselves royally entertained at weekend parties, or over dinner at the Savoy Grill. Employment on Sheppey before World War I had meant exposure to the founding fathers of the RNAS/Fleet Air Arm, notably Charles Sansom, and an early acquaintance with the reality and the potential of naval air power. Meeting Sansom, Longmore and company left Fairey conscious of the value in cultivating good will with assorted air crew, however humble their rank. Forty years later Sir Richard's weak heart saw fewer visits to Bossington by high-spirited service personnel, and yet he continued to enjoy the company of young men and women, whether in or out of uniform. If invited to dinner, John's male friends could anticipate a lengthy session alone with 'the old man', putting the world to rights as the port decanter was passed speedily around the table.

Fairey's fragile condition ruled out a lengthy inspection of the Fleet Air Arm's most experienced fliers. RNAS Ford's special guest doubtless enjoyed a brief fly-past before the visiting vice-admiral and the station captain whisked him off to the wardroom for a stiff G&T and a good lunch. However delicate his health, CRF felt instantly at home in a convivial atmosphere where men of shared values, not least a deeply conservative disposition, could relax and air their views in suitably forthright fashion. Sir Richard would tower over everyone else – including Lieutenant Commander Cowtan and his fellow fliers. His imposing presence was invariably reinforced by a firmness of view and a mental acuity rooted in continued command of pertinent technical detail. By the time CRF asked for his chauffeur to bring round the Rolls, the officers and men of HMS *Peregrine* would have been quietly impressed by the authority, conviction, expertise and readiness to learn of their VIP guest, without necessarily warming to him. You had to know Dick Fairey for a long time before he let down his guard, relaxed, and treated you as a serious friend and an equal – and in exceptional circumstances, as a confidant.

Fairey formed part of a unique generation. Its achievements are locked into the national mythology of 'this island story' but can also be found in the industrial heritage of France, Germany and the United States. Individual pioneers of manned flight were active elsewhere in continental Europe prior to the Great War, but in size and achievement they could scarcely claim critical mass. Neither Italy, nor Russia, Austria-Hungary and their constituent empires, could boast a similar concentration of aeronauts with the necessary combination of energy, enterprise and expertise. These were engineers boasting a set of skills and a theoretical understanding acquired through intensive study, full or part-time, in academically testing institutions such as Silvanus Thompson's Finsbury Technical College. With Germany in the vanguard, late-nineteenth-century vocational training systems were the hallmark of an advanced industrial society: their curriculum enabled aviation pioneers like Fairey and Handley-Page to grapple with hard science, principally physics and mechanics, and to boast a level of mathematical competence surpassed only by graduates of the major civic universities and Oxbridge. Advanced numeracy plus quiet ambition sowed the seeds of an enduring entrepreneurialism.

These proto-industrialists were driven to succeed by a belief that manned flight could change the world, as indeed it did, and by a keen desire to make their way in that world, and in so doing enjoy, to the full, the fruits of their labour. That drive rarely if ever flagged, and exceptional circumstances – two world wars, and the economic and political upheaval of the intervening years – left them with little opportunity to rest on their laurels, even had they wished to do so. There was a hunger to succeed, and to keep on succeeding. Success could be measured in terms of technical achievement *and* material advancement – the founding fathers of the British aircraft industry enjoyed a quality of life scarcely conceivable at the onset of their careers. One or two boasted comfortable backgrounds, notably Tommy Sopwith, but the majority came from lower-middle-class families with modest incomes and a fragile veneer of gentility. These engineer–entrepreneurs founded their firms in response to an actual or an anticipated demand, and for the following half-century each remained pivotal to the well-being of his creation. In some cases – again Tommy Sopwith – they witnessed the rise and fall of their original enterprise, and had to start all over again. The duration is striking, especially in Britain and the United States: these men lived long lives, for most of which they were synonymous with the firms named after them.

The likes of Donald Douglas and Geoffrey de Havilland, either side of the Atlantic, created company structures which facilitated substantial growth and capital investment without surrender of control: flotation on the stock market meant the advice of institutional shareholders was heard but not necessarily acted upon. In reality this was an era when financial institutions and individual asset managers rarely remarked on company policy so long as year on year the board declared a decent profit margin and a generous dividend.

This passivity worked to the advantage of CRF and his fellow directors after Fairey Aviation became a public joint-stock company in 1929. The governance, structure and share distribution of the company post-flotation ensured a striking degree of continuity with the successive enterprises carrying the Fairey name from 1915 through to 1929. Even before the serial bankruptcies and relaunches of the 1920s, the articles of association were subject to regular revision. Charles Crisp was instrumental in first squeezing out co-founder F.H.C. Rees, and then consolidating control of the company in the hands of the majority shareholder and his closest associates. Dick Fairey benefited enormously from befriending early in his career a company lawyer expert in minimising corporate and personal tax liabilities and maximising income generation – salary, expenses, patent royalties, interest payments and dividends. Complementing Crisp was Morris Myers, a wily operative in the City valued by Fairey for his astute management of a large, diverse and relatively risk-free investment portfolio. Crisp and Myers each helped Dick Fairey become a very rich man, but they were by no means vital to his success – he would have made it without them. Arguably, the long forgotten Frank Rees played a more vital role, in generating the modest funds necessary to launch a plane-making business. Fairey played Rees brilliantly, with careful calculation dropping him at the earliest opportunity.

Always the wheeler-dealer, Frank Rees appeared from the outset untrustworthy and unreliable. A City insider, he had come up trumps when seeking investment capital. Yet soon he was seen as a liability. Rees's speculative plans for diversification proved a distraction from the principal task of progressing beyond subcontracted assembly to the development and manufacture of in-house designs. This was always Fairey's intention, the scale of the Royal Navy's wartime needs providing a rare opportunity to expand at a rate of growth inconceivable in peacetime. Spurred on by Murray Sueter at the Admiralty, the managing director displayed ambition and intent. One year into Britain's first 'industrial war' a prescient single-minded proto-industrialist faced a uniquely advantageous set of circumstances. From initial employment on the north Kent foreshore to establishing his own company, Dick Fairey had benefited from a serendipitous combination of timing and location. With Short Brothers as first employer and then role model, Fairey's mechanical skills merged with a latent management ability, honed in the hangars of Eastchurch and Rochester. J.W. Dunne provided early experience of project management, but Horace Short was a formative influence in demonstrating how developmental progression was inevitably rooted in hard science. Fairey was from the outset a brilliant rigger and stress man, but his talent went beyond fine-tuning other people's designs: he could comprehend the abstract, on the basis of which he approached complex problems with an open mind. That capacity to think outside the box would soon result in the 'wing flap' – the variable camber gear, a major invention and a highly lucrative patent.

CONCLUSION

North Kent may have been the epicentre of Edwardian manned flight, but by the middle of World War I a fast maturing aircraft industry had spread across the south of England, the Midlands and beyond. Fairey established an outpost on the Solent, but like Sopwith and de Havilland he saw the furthest reaches of west and south-west London as ideally suited to establishing a manufacturing plant with good communications, space to expand, and a ready source of labour, whether skilled or unskilled. Across the ensuing 40 years the Fairey enterprise would expand and diversify, its operations undertaken on a genuinely global basis. At home, factories would be established far from the capital's hinterland, principally in the north-west, but Hayes always remained the operational base and the beating heart of the company. Here was the home of Fairey Aviation's longest-serving employees, their loyalty to the firm's founder reflecting a culture of 'personal capitalism' conducive to individual risk-taking and consequent high reward.

Critical to the tyro entrepreneur's early success was a level of trust absent from his dealings with Frank Rees. Fairey's individual and collective relationships with Crisp and their three close friends from Cambridge were crucial to the success of Fairey Aviation at every stage of the company's establishment and exponential wartime growth. That camaraderie, translated into absolute trust, was vital when the company faced the challenge of rapid retrenchment after the Armistice, and during a postwar decade where survival depended upon costly investment in home-grown and trans-Atlantic technology. Respectively test pilot, salesman and seed corn investor, Nicholl, Wright and Dawson each made a critical contribution to the survival and ultimate flowering of Fairey Aviation. Together, the three veterans of varsity and the 'war to end all wars' were much more than the sum of their parts. In tandem with Crisp, their contribution was crucial, not least as each of them was encouraged to provide Fairey with necessary advice.

More often than not that advice was taken on board, however unwelcome. Fairey's circle of loyalists expanded to include lifetime servants of the company, like Broadbent, Barlow, Vinson and Tips. Later acolytes such as Outen, Hilton and Chichester-Smith worked their passage, gaining the trust and approval of the 'big man.' Successive generations of directors and senior managers had to be tough, and blessed with thick skins. Working for Sir Richard Fairey was no easy ride, but the rewards were considerable, extending well beyond hefty dividend shares, sizeable salaries and generous expenses. He may have been careful with his money, but Fairey was exceptionally kind towards those he loved, *and* to those he admired and respected. CRF's fellow directors and senior executives enjoyed the full fruits of a bountiful boss, from box seats at Shaftesbury Avenue shows to champagne suppers at the Savoy, from tennis parties at Woodlands Park to foursomes at Sunningdale. Christmas and special occasions always generated generous gifts, with Miss Burns a formidable operator when it came to ensuring due recognition that yet another year had passed in the service of Fairey Aviation – and of its founder.

It is worth noting here that house servants and estate workers were similarly rewarded for loyalty and long service, with the birthday of Fairey's most valued gamekeeper or waterkeeper an ever present in the desk diary. The boss's secretary was equally adroit in dealing with charitable appeals, sensitive to which requests for money would receive a polite letter of refusal and which would generate a hefty donation: when Fairey gave, he gave generously, and without condition.

Once he joined the company, Richard Fairey was a beneficiary of corporate largesse, and of his father's readiness – however reluctant – to subsidise a lifestyle the wayward yet talented son could scarcely afford. Here was a stormy and complex relationship, which went far beyond the old man signing yet another cheque and shaking his head in exasperation. Sir Richard's reaction to his son's horrific experience in the North Atlantic revealed a depth of love and affection impossible to gauge from damaging newspaper headlines and factory gossip of family quarrels. Young Richard was destined to enjoy the privileges and position of the eldest son in what even after the war remained, to a remarkable degree, a family firm. Yet from the late 1940s those in the know sensed that Richard was no longer the heir apparent, with the founder's half-brother uniquely qualified to take on the top job – as it transpired, Geoffrey Hall found himself heading Fairey Aviation sooner than his late sibling had anticipated. Given the company's unhappy experience after 1957, elevation to chairman and managing director would prove a poisoned chalice.

When Fairey Aviation was in its prime, the rewards of large contracts and large profits were not entirely the preserve of the boardroom and the mess. On a more modest scale the first chairman and managing director rewarded hard work and individual achievement across the workforce. However, between the wars this paternalistic interest in the well-being of employees proved increasingly hard to maintain. The expansion of the company, necessitating new sites in the north of England, saw an inevitable shift in the relationship between managers, especially senior managers, and workers, most of whom felt little or no loyalty to CRF's initial creation. An obvious consequence was a sharp deterioration in industrial relations once rearmament and the waning of economic recession increased job security and the demand for better wages and improved working conditions. Managers close to the shop floor, like Broadbent, appreciated this shift in attitudes, and where appropriate urged accommodation.

CONCLUSION

Negotiation was tantamount to surrender for a company patriarch like CRF, instinctively suspicious of trade unions and reluctant to discriminate between reformist demands and Communist militancy. During and after World War I, Fairey was someone still able to offer a sympathetic ear and a decent rate for the job – a lifelong attachment to Belgium was rooted in his fair and sensitive handling of refugee workers from 1915 through to the end of the war. He may not have been a model employer, but he enjoyed the loyalty of his workforce because he was seen to be a hands-on presence in the assembly sheds and the design office. Yet as the managing director became more distant from the shop floor it is scarcely surprising that attitudes changed, not least because when he was around 'Tiny' Fairey constituted a formidable and for some a frightening presence – here, after all, was someone with a short temper and a reputation for not suffering fools gladly.

The company Richard Fairey ran in the 1930s and into World War II was very different from that forged in conflict a generation earlier. Britain itself was very different from the nation that went to war in 1914, its workers less deferential and more aspirational – profound social, cultural and, above all, economic forces had forged a fast-changing manufacturing landscape, with heavy industry on the wane and southern-oriented 'sunlight industries' such as plane making in the ascendant. This was clearly in the commercial and personal interests of Fairey and his associates, but there were obvious challenges, not least the need to reconcile employees' well-being with a necessary rise in productivity – that requirement was demonstrably not met throughout the years of rearmament and, with CRF absent abroad, proved equally problematic in time of war.

Boardroom turf wars, an absence of effective leadership, and a deeply troubled relationship with the Ministry of Aircraft Production left the question of productivity unresolved, even after Sir Clive Baillieu implemented long overdue reforms. Wartime urgency meant a narrow focus upon maximum output, with any hold-up in production resolved by throwing resource at the problem. The one person ideally equipped to address the issue of poor productivity was absent for the duration of the war, and after 1945 insufficiently robust to push through a radical rethink of assembly methods. Long before World War II, Fairey had shown himself receptive to fresh thinking on the far side of the Atlantic, and his tenure at the British Air Commission proved a steep learning curve – in attitude he shifted from a blinkered view that American plane makers had nothing to teach their British rivals to a grudging acceptance that factory systems adopted by the likes of Boeing and Douglas after 1942 maximised the economies of scale synonymous with big contracts. Fairey also acknowledged that progress from the drawing board to final production could be a more integrated and fluid process, and, much as he still disliked bureaucrats, that vigorous and courteous debate was the most effective means of challenging Whitehall's demand for unduly complex technical specifications.

The Treasury's preference for supposedly cost-effective, multi-purpose aircraft was counter-productive when translated into a specific model's overall performance and operational efficiency. Throughout his career Fairey railed against what he saw as the Air Ministry's unreasonable demands, but ironically the aircraft with which he is most closely associated was designed to fulfil multiple roles (a double irony is that most Swordfish were built by subcontracting their assembly – an arrangement which gave Fairey his initial break, but which in the 1930s he argued vehemently against). Government preference for a highly adaptable combat aircraft, thereby minimising the need to duplicate development and production costs, remained an orthodoxy into the jet age, and it continues to mould procurement planning through to the present day. To a degree Fairey Aviation benefited from this thinking in the early postwar years, through the Fulmar and the Gannet. The commercial success of both aircraft camouflaged the continuing issue of poor productivity, and the full impact of rapid technological advancement on the individual and collective cost of parallel R&D programmes. Cutting-edge technology, rooted in jet propulsion and embryonic avionics, demanded an unprecedented level of capital investment, with the postwar state again looking to private enterprise as the principal funder. Finding large sums to support costly design and development was rendered more difficult because securing the final order was by no means certain: the industry had too many companies competing for too few contracts, with little likelihood of selling abroad in sizeable numbers. Fairey Aviation realised a state-of-the-art design in its delta-winged record breaker, but it was English Electric which provided the RAF with the supersonic fighter most suited to its operational needs. Not only was the Lightning more adaptable, but its prototype flew ahead of the FD2.

Delays in the Fairey Delta's inaugural test flight were partly the consequence of overstretched funding, with work on the Rotodyne and variant versions of the Gannet draining company resources. Sir Richard's enthusiasm for the FD2 was understandable, but a healthier, more hands-on chief executive might have questioned the commercial potential of the Rotodyne project. Whatever his state of health, Fairey was unlikely to question the size of the annual dividend in the first five years after World War II: averaging 64 per cent of the company's earnings ensured an impressive return for shareholders, but it left the level of liquid capital insufficient to support innovative but highly expensive development projects.

CONCLUSION

CRF was unapologetic about the need to maximise profits and to reward risk-takers. Fairey was an engineer, an entrepreneur and a libertarian. (The badge of libertarianism helped explain a troubled relationship with His Majesty's Constabulary, not least in the 1930s when clashes with the police over motoring offences were a regular event. Defending unfettered river rights on the Test was seen as a similar assertion of individual freedom.) CRF's enthusiasm for the free market extended to enthusiasm for a low-tax regime, with minimising his financial obligation to the state deemed an absolute priority, and to a deep distaste for the 'one nation Toryism' advanced by Stanley Baldwin and Neville Chamberlain.[3] He failed to comprehend why a Conservative or Conservative-led administration could facilitate an expansion in the role of the state or maintain only a modest level of imperial preference. Born into a late Victorian middle-class family, Fairey was unsurprisingly an imperialist. One reason why he got on so well with Beaverbrook was that both men felt the assets of empire were scandalously under-utilised.

In the 1920s Dick Fairey shared Sam Hoare's imperial vision, with a new generation of long-haul airliners, whether land- or sea-based, vital to closer communications between the 'mother country' and its furthest colonies and dominions, most especially the Raj. Where CRF differed from Sir Samuel Hoare was in the Secretary of State for Air's readiness to subsidise Imperial Airways and to support pan-national regulation of civil aviation. In reality, Fairey Aviation proved an eager partner in government-sponsored initiatives, not least the pioneering of long-haul flights to India. Unlike after World War II, these speculative ventures were affordable, as the size of the company's profit margins in the late 1920s generated a healthy return for investors *and* the necessary capital to invest. By now a high-profile power broker within the British aircraft industry, Fairey was keen that machines built at Hayes and Hamble be prominent in the much vaunted, much publicised criss-crossing of empire. Large Air Ministry orders, mainly for different versions of the adaptable FIII seaplane, enabled the firm to survive postwar retrenchment when its rivals went to the wall, and to ride recession when Hawker succeeded Fairey as Adastral House's principal contractor after 1927. The Fairey III was critical to the company's survival and prosperity. It was in every sense an aircraft of empire, and Hoare's choice for his final ministerial flight, seeking Sudanese staging posts for the planned service from Cairo to the Cape.

Fairey Aviation's chairman and managing director was instinctively suspicious of the state, and yet he was in the forefront of an industry which more than any other depended upon the goodwill and munificence of government. CRF was well aware of the conundrum, and however strong his private opinions he was invariably cautious and judicious when dealing with ministers and officials, whether representing his company or lobbying bodies such as the SBAC and the Royal Aeronautical Society. Multiple drafts of speeches and articles saw intemperate remarks filtered out, especially after July 1945 when he failed to comprehend how Labour had come to power, and why the Party's programme for a rebalance in the distribution of wealth had attracted such a sizeable vote. Like so many of a conservative inclination, CRF was shocked and surprised that Churchill, the great war leader and hero of 1940, had been ejected from office. Five years out of the country, interrupted only by short and heavily scheduled return visits, had left Fairey less aware of how the home front had witnessed deep-rooted social and political change. Any acknowledgement of a shifting political landscape had translated into a contentious indictment of the 'socialist' BBC and a justified concern for the ecological integrity of the Test Valley. This failure to perceive the real impact of total war was prefaced in autumn 1939 by his mistaken belief that a decentralised system of evacuation undermined social cohesion and endangered lives. On balance, mass evacuation was a force for good, in both preserving life and exposing the middle classes to the harsh reality of inner city life. Sociologist Richard Titmuss was not alone in feeding evacuation into an explanation of Labour's electoral triumph.[4]

Fairey's blanket condemnation of the Attlee administration and its reformist agenda was voiced by all conservative-minded commentators, not least close friends such as Brabazon and Jim Wentworth Day. CRF was similarly suspicious of change, but unlike several of his associates in the 1930s he was wary of political extremism, not least home-grown fascism. Although a product of his class and time, he remained silent when conversation at the nineteenth hole or on the second brandy degenerated into open or scarcely disguised anti-semitism – Myers' friendship and support deserved no less. CRF was an imperialist and an internationalist, witness his affection for Americans at a time when the more affluent sections of English society maintained a singularly unattractive sense of social superiority. Thus, although a patriot, he eschewed crudely nationalist sentiment, let alone the rampant jingoism of 1914–18. He was too intelligent and cosmopolitan to fall prey to far-right populism, his wariness of ideology reinforced by a canny recognition that business was business. Here after all was an industrialist wary of antagonising ministers, even Labour ministers, and someone willing to work with the Bolsheviks when the opportunity arose. Ever the pragmatist, in Washington during the war he enjoyed excellent relations with New Dealers whose views he previously found an anathema. Yet over-riding every other consideration when explaining Dick Fairey's dislike of Sir Oswald Mosley and his ilk was a deep loathing of Germany.

CRF's suspicion of German ambitions and revanchist intent was reinforced by the Nazis' accession to power – his campaigning for a major expansion of the RAF preceded the Führer's arrival in the Reich Chancellery. Fairey's longstanding hostility to disarmament, culminating in his appearance before the 1935 commission, was rooted in a firm belief that the peace settlement signed at Versailles and endorsed at Locarno would not long survive Germany's recovery from the harsh consequences of defeat. Someone so suspicious of propaganda and so keen on rearmament would find scant appeal in a political movement whose black-shirted leader drew inspiration from an unashamedly expansionist Adolf Hitler.

Fairey over-estimated Nazi Germany's military might and industrial performance, and in so doing under-estimated Britain's manufacturing capacity across the winter of 1939–40, not least in aircraft production. Given his senior position within the SBAC, and his roving brief from Chamberlain's government to advise contractors and subcontractors, this is surprising. Yet his pessimism was consistent with the alarmist stance adopted in the early 1930s and conveyed to a wider audience courtesy of the *Daily Mail*. Fairey's brief visit to Germany had left a deep impression, and a belief that the indigenous aircraft industry would experience continuous growth, manufacturing high-quality machines in a multitude of roles. For all Germany's success in the mass production of high-performance fighter aircraft, this proved not to be the case.

Conversely, the visits of CRF and his colleagues to the Soviet Union on the cusp of the first Five Year Plan saw disappointment at the failure to seal a deal translate into dismissal of an aircraft industry sturdy enough to withstand the harsh impact of Stalin's purges. Understandably, Fairey and his lieutenants were so sensitive to the visible consequences of dekulakization and mass starvation that they failed to perceive the strengths of an industrial structure which combined German organisation with Russian genius – the latter extending to a brilliant exploitation of shameless reverse engineering. CRF's correspondence confirms his admiration for Tupolev, but he had more in common with Ilyushin, the great survivor at the heart of the Soviet state apparatus – engineer, technocrat and factory magnate. By 1934 all contact with the Russians, private or official, had ceased. Yet, despite Fairey's unqualified loathing of Communists, he still attended functions at the Soviet Embassy. Although he does not appear in Maisky's diaries, CRF met with the ambassador at various public events. The two men, for example, were photographed together at a Hendon air display. Fairey attended parties and receptions until the eve of the Molotov–Ribbentrop pact. Given the ferocity with which he denounced this unholy alliance, one can only speculate on why throughout the 'thirties CRF made time to visit

Kensington Gardens. The only certainty is that after August 1939 he had no more to do with Maisky; post-Barbarossa he refused to extol Stalin's leadership qualities, choosing only to praise the resilience of Russia's soldiers and airmen.

Mystery surrounding Fairey's continued interest in the Soviet Union contrasts with so many other aspects of his life. He was a very public figure, not least because his physical presence rendered him hard to ignore. As his cuttings albums confirm, CRF appeared regularly in national newspapers and in specialist publications dedicated to flying and sailing. From the final years of World War I, Fairey Aviation attracted generous press coverage, later complemented by newsreel reporting. Inspired and encouraged by its founder, the firm was ahead of the game in terms of self-promotion and favourable PR. Advertisements were striking and contemporary, and Fairey's friendship with Charles Grey ensured a constant stream of good publicity in the columns and supplements of *The Aeroplane*. Other aviation magazines, notably *Flight*, displayed similar goodwill and enthusiasm.

Journalists were always welcome at Hayes, Hamble and Heaton Chapel, with directors and senior management deployed to provide a carefully orchestrated tour and a leisurely lunch. Yet any correspondent critical of the company could expect a testy, even ill-tempered response from the chairman, conveyed courtesy of published and private correspondence with the offending journal's editor – Sir Richard could be courteous, even convivial, in his dealings with the fourth estate, but he found criticism hard to deal with. He was after all famous for a fierce temper and an intolerance of perceived mediocrity. Someone so successful – in the workplace both authoritative and authoritarian – could only be challenged by immediate family, close friends, and counterparts held in similar esteem to himself. He clearly was receptive to painful advice, frank opinions and fresh thinking, but only if articulated by those he respected and/or held in deep affection. If Esther Fairey or 'Wuffy' Dawson suggested that Dick think again then he would do so, but an awkward question from a Fleet Street hack could be contemptuously dismissed.

Sir Richard Fairey was a high-profile plane maker, his reputation in the 1930s boosted by his rapid ascendancy to the most senior level of competitive sailing, and at the same time a very private man. Newspaper profiles are striking as much for what they do not say as for what they do. They rightly portray Fairey as a proud family man – he clearly adored Esther, John and Jane – but beyond reference to Richard junior make no mention of his first marriage. His daughter testified as to how hurt her father was by Joan Fairey's affairs and flagrant displays of wealth: in the 1920s Dick Fairey was a bon viveur who knew how to stage a good party, but a lifelong commitment to securing the best his money could

CONCLUSION

buy became seriously strained once the unforeseen bills arrived on the Grove Cottage doormat. Divorce saw Queenie/Joan written out of his life, notwithstanding a lingering contact with individual members of her family.

Nor was Sir Richard revealing of more intimate details regarding his childhood, beyond an acknowledgement that his father's financial travails and early death gave him the drive and ambition to succeed. The penny-pinching reality of lower middle-class family life, and the full impact of unanticipated tragedy upon a young widow and her offspring, left a deep and lasting impression. Friends and relatives were familiar with Fairey's early life, but he rarely revealed to a wider public how deeply he had been affected by his father's death: tempering the harshest consequences of his family's descent into genteel poverty, CRF always gave the impression that Merchant Taylors' was his only secondary school. To journalists the sacrifices made by his sister Effie, to whom he was forever indebted, and the life-changing impact of Henry Hall, were only hinted at. Yet, for those close to him Dick's unequivocal gratitude was clear to all, witness the financial support given to his sisters throughout their lives. More than any other, the early years of the last century moulded Charles Richard Fairey. With Mrs Fairey yet to remarry, he fulfilled his role as the family's principal provider, and across the following half-century he remained extraordinarily generous to his closest relatives, extending that generosity to mentoring and sponsoring his half-brother's progression from diligent student to designated successor.

Filial duty translated into aspirational endeavour – the critical decision to join John Dunne was driven by that urgent need for personal and professional fulfilment which H.G. Wells recognised as the key factor in the ambition and advancement of Edwardian men and women forced by circumstance to fall back on natural talent. Fairey's lifelong admiration for Wells was rooted in silent acknowledgement that his was a story tailored to the great man's fiction: the black-coated autodidact certain of success, his self-belief rooted in recognition that this is the great age of applied science, when the superstitions of organised religion should be swept away and mankind can achieve stunning technological advancement. With the Wright brothers rendering manned flight a reality, that advancement could most immediately be secured in the field of aeronautics. Model-making in north London was always a means to an end – to the construction of real aircraft, and to a permanent escape from the everyday banality of suburban life. Young Dick Fairey was truly a Wellsian character.

In late middle age, Richard Fairey was more resonant of Henry James: a transatlantic envoy; an intelligent, self-made man comfortable with his wealth; an *arriviste* in the favoured sports of old money; a landowner blessed with a loving family second time around; a generous host happiest in the company of those similar to himself; and a metropolitan figure comfortable in the corridors of power yet deliberately distant from the great and the good. Bossington may recall Gardencourt, but there the analogy ends. No character in *The Portrait of a Lady* engages with the landscape in a manner akin to Sir Richard's warm embrace of the River Test and its immediate environment. In fiction, fishing and shooting might be peripheral activities for bored gentlemen, but back in the real world Fairey saw the beat and the shoot as integral to a rural way of life. That way of life was in his firm opinion threatened by an ill-informed, urban, bureaucratic mindset, rooted in a remote and unfeeling modernity. As a landowner, with England's premier trout stream flowing through his estate, CRF was by no means indifferent to change. Maintaining the status quo was out of the question as the conservation of both land and river necessitated sensitive management and maintenance – the essence of conservatism is controlled change. Fairey's daughter recalled her father's decisiveness and his aversion to half measures. Yet when preserving the most precious features of the Test Valley he duly held back and played the long game, bequeathing a visually stunning inheritance of verdant, unspoilt riverbanks and well-stocked, uninterrupted waterways.

At Bossington, French windows open out onto a close-cut lawn leading down to the river. In early summer this is a still, almost silent scene, the only noise being a blend of birdsong and fast-flowing current. It is fanciful to imagine the spirit of Sir Richard, still protective of his domain. Yet, standing behind the house, staring out over the water, it is easy to understand why he felt so certain that here was *home*, where life must be lived to the full. Here gathered family and friends, old and new, together forging a convivial, congenial, companionable atmosphere in which conversation was lively, even heated, but rarely ever boring. Interesting people convened to discuss interesting topics, with applied science and engineering unusual in their centrality to dinner party debate. The door to the drinks cabinet was always open, the sandwiches were always newly cut and the coffee freshly brewed, the house staff were always on call, and the rod or shotgun was always handy. Home was synonymous with hospitality, and from childhood to maturity John and Jane Fairey relished the readiness of so many visitors, young or old, famous or unknown, to cast aside their respectability and inhibitions, and to enjoy themselves.

Fairey had no dynastic ambitions, witness his preference for Geoffrey Hall and not Richard as chief executive, but he did have an exaggerated sense of family. Shaped by the domestic trauma of his youth he was determined to make his way in the world, to make a great deal of money, and to make certain that his children, and their children, would never experience the upheaval, disadvantage and shame which followed the loss of his father. Bossington fulfilled a unique

CONCLUSION

role in CRF's life, but he was too down-to-earth to foster grandiose ideas that the estate would be handed down from one generation to the next across the centuries – he was a self-made man, not a scion of England's landed classes.

Sadly, the name of Fairey has long since disappeared from what remains of Britain's manufacturing base. Of the firm's flagship aircraft, the Swordfish is a rare presence in the skies above us, and not a single Barracuda survives. Yet Sir Richard can still boast a legacy, albeit an unintentional one. Contrary to his expectations the extended Fairey family remain in the Test Valley, at Bossington and at Pittleworth Manor. Lying at rest in the shadow of his final, favourite country house, 'Tiny' Fairey – inventor, industrialist, technocrat, helmsman, conservationist and late life defender of the rural landscape – can take quiet satisfaction in his descendants' desire to work the land both profitably and sensitively.

Appendix
My Father, Sir Richard Fairey

As his daughter and sole surviving child, indeed one of the few people left alive who knew him, I want to try and explain the man behind the successful aviation pioneer and industrialist. Adrian Smith describes his career as a plane maker and a business man, and I would like to try and convey my father's personality and character. He was endowed with a force of character that I have never encountered in my life again, and he could be very intimidating. He had a will of iron, and once he determined to achieve some objective he was unstoppable. His huge powers of concentration would come into force and nothing would deflect him. I think an example of this would be his driving through the redesign and production of the Swordfish in 14 weeks. He was deeply intelligent, and many who knew him believed him to be a genius. His intellectual interests were very wide, ranging from the sciences and ecology to history, current affairs and politics. His love for the natural world was very profound, and he poured huge amounts of energy in his later years in trying to defeat various forces that he considered destructive to the countryside.

Despite his ferocity, my father had a very soft heart and if he were touched no-one could be kinder. He had a huge love of life with a hedonistic streak, which allied to his generosity made him a wonderful host and a great companion; he loved to enjoy himself and wanted those around him to enjoy themselves too, even though this generally meant dancing to his tune. He had a keen sense of humour, was a good raconteur, and had a rather amusing sense of the ridiculous. He was deeply emotional and had a great and protective love of his family; he liked to think of himself as 'pater familias' and kept in close touch with his sisters all his life. I imagine the tragedies of their childhood caused by the loss of their father and the subsequent descent into poverty and helplessness left a very strong bond between them, and above all reinforced my father's determination to make money. I remember him saying through gritted teeth, 'I swore I would never be poor again.' As the sole surviving male at a very young age of this now destitute family, he felt deeply responsible for them all, and this concern lasted all his life. He seldom talked about these dark days; a veil was drawn across them, and his recounted recollections of childhood were the happy times, skating on the Fens, fishing expeditions with his father, and the acquisition of his first beloved dog. This was achieved when he and his father were lunching at a pub and a terrier appeared, jumped on a seat by the bar and dropped a penny on the counter and was rewarded with a biscuit. This it was explained was the dog's great trick, which he now regularly exploited by begging for pennies from the clientele of the pub. This so fascinated my father that he persuaded his father to buy the dog for him. I suspect there would have been quite a scene had he not got his own way. His wilfulness was already very apparent in childhood. The story goes that at the start of his school life he was carried kicking and screaming to school every day for three weeks until he became interested in the maths lessons. As he put it, 'Life in the nursery with my toys was perfectly pleasant but I could see that once this business of school began, it was never ending.' Fondness for dogs was very much a feature of his life, and he was devoted even to my mother's dachshunds. He was amused by their antics, and, in the case of the dachshund we brought back with us after the war, perfectly prepared to swing rank in order to get the dog into a local army quarantine camp near to our home. There is a letter instructing his agent to point out to the officer in charge of the kennels that he had been an Honorary Colonel in the 27th A.A. Btn. Royal Engineers and therefore qualified to use the kennels for his dog. He described the dog as 'a wonderful companion to us in our semi exile.' As Adrian Smith notes, he was not long a colonel of the 27th, resigning when he discovered his duties were purely ceremonial, and that he would not be used in an advisory capacity come the emergency.

My father was extremely patriotic, and genuinely believed that *Pax Britannica* and the British Empire were together the best possible solution for the world. He also loved the Americans and admired their ruthless 'can-do' attitude, which so beautifully suited his own philosophy, but he also got annoyed with them when their belief in their own superiority clashed with his perception of British superiority. Many of his right-wing and conservative views would ruffle and annoy many people today, but he was a man of his time, and of course believed in self-reliance and minimum state intervention.

Alas, I missed my father's yachting days as I was born in 1937, but I was always aware of his love of the sea. Had my grandfather continued to prosper then my father would have been destined for the Royal Navy. He often said to us, 'To think I might have been an admiral by now.' Yacht racing he loved with a passion. My father told me that in his first racing season with his yacht *Shamrock* he came last in every single race, but nothing daunted and, driven as usual, he carried on and became a highly successful helmsman. He raced a good deal in Scandinavia and used to entertain Crown Prince Olaf on board the motor yacht *Evadne* – a friendly relationship which continued in Washington where the Norwegian royal family spent part of the war. My father's many trophies still reside at Bossington House. After the war my father sold *Evadne* almost by default: he quoted an impossible price to an importuning potential purchaser, who to his horror accepted. It was a move my father deeply regretted. One summer he chartered *Radiant*, a motor yacht of similar size to *Evadne* which belonged to Lord Iliffe: if we thought we were setting off to the south of France then we were very much mistaken as we headed north up the Irish Sea to the west coast of Scotland, and on to the Orkneys where we were marooned for two weeks in the most terrible weather. Rough seas and fierce winds did not deter my father, and you had to be a good sailor to go to sea with him.

My earliest memories of my father were in wartime Washington when as children we saw very little of him, such were the demands of his job and consequent travels back and forth to England, and to the west coast of America where all the major aircraft factories were situated. By the end of the war he had all but worked himself to death and had to resign from his position as Director-General of the British Air Commission during the winter of 1944–5. After undergoing pioneering surgery for his heart condition under the care of the famous Dr Paul White at the Massachusetts General Hospital in Boston, where he was obliged to stay for nearly six months, my father finally settled in Bermuda to convalesce. At first he had to take life very quietly and spent hours at his work bench doing woodwork, at which he excelled, crafting dolls' house furniture for me, mending an eighteenth-century American clock that had wooden cogs, and of course making model aeroplanes and gliders for my brother. As he gathered strength he then partook in the political life of the island, in an advisory capacity, and had many good and dear Bermudian friends. However, his heart always remained in England and he fretted to return to Bossington.

My father excelled at all sports, but by the time I was aware of his activities it was mainly field sports. He was a first-class shot and a superb dry fly fisherman, and through these pursuits he became ever more interested in the ecology of the natural world. This especially applied to the Test river valley, and he fought tooth and nail to protect it from water abstraction, pollution, electric fishing and any attempts to lower the water level of the river ostensibly for the benefit of agriculture. My father understood the interdependency of all nature, and did not wish to see the river cleared of all coarse fish and eels, in whose life cycle he was particularly interested. In one of his speeches he attacked 'those who wanted to see the River Test as a tank for trout.' Of course he was fascinated by the then abundant up-winged fly life of the river, about which he acquired great knowledge. He tyed all his own flies, and his fly tying table stood in the drawing room piled high with boxes of hackles and feathers and all the necessary equipment – a monument to untidiness rather to the annoyance of my mother, but there it stayed. I often spent time on the river with him, and when he was not fishing he would sit quietly watching the bird life and the water voles, totally absorbed and at peace with the world around him. He would be horrified to see our countryside today, devoid as it is of 85 per cent of its birdlife and small mammals, its terrestrial insect life vastly reduced, and the river lacking species that were once common such as crayfish, as well as the variety of aquatic flies that were there then in abundance. He would have felt that his darkest forebodings had been fulfilled.

My father always said that towards the end of your life the things you regretted were not the things you had done but the things you had not done. One of his great regrets was never having taken *Evadne* to the Galapagos Islands as Tommy Sopwith had done with his motor yacht. My father was fascinated by Darwin and his theory of evolution, and would have loved to have seen the place that triggered Darwin's theory. He also had great admiration for Thomas Huxley, and his vital role in seeing that Darwin's work was published.

In the years after the war when Bossington was our home, my parents entertained a certain amount. Frequent visitors were Lew Douglas, the American ambassador, and from time to time his brother-in-law Jack McCloy, US High Commissioner for occupied Germany; Lord Brabazon; Philip Young (then Head of the US Civil Service under Eisenhower's first administration) and his wife Faith; the Ashley Coopers; fellow 'Fen Tiger' Jim Wentworth Day; and Sir Joseph Ball. Various other interesting people who suddenly arrived on the scene included Edwin Hubble, whose 'theory of the expanding universe' revolutionised astronomy. He and his wife Grace were charming and endearing people and became good friends of my parents. I remember a dinner party when among the guests were Sir John Scott, Chief Commissioner of the Metropolitan Police, and his wife. Most unfortunately, that night we were burgled; it was not so lucky for the burglar either, because he was relentlessly pursued, and finally caught about two years later. The family dogs were in disgrace after that episode, having slept peacefully through the whole incident. On one of our shoot days the uncle of King Hussein of Jordan arrived with his ADC; I imagine my father was asked to invite them by the Foreign Office. From time to time Dr White would come to stay, and give my father a medical check-up. By then they had become good friends. At some point Paul White was very keen to study the cardiac system of whales, and I remember my father helped him to procure a harpoon gun which was so light that the whale would not notice being hit and an electrocardiogram could be taken. I believe the experiment was successful, being recounted in the *National Geographic* magazine. Family often came to stay, especially my father's sisters, and my maternal grandfather lived with us for months at a time. It was a warm and hospitable household.

My father was generous to charities of which he approved, among which was the Salvation Army. This came about when he was taken for a day's racing by friends and won a handsome sum on some horse. Airily, he said to the bookie, 'Send it to the Salvation Army.' He was immensely impressed when the Salvation Army returned the cheque thanking him but explaining that they could not accept the proceeds of betting. Their sincerity and integrity so impressed him that he supported them for the rest of his life. God moves in mysterious ways his wonders to perform.

As for my father as a father, I think it was a lot easier to be his daughter than it was to be his son. He loved us all deeply but expected less of his daughter. Richard, he had spoilt, giving him a car and an aeroplane when he was only 17, and allowing him to lead a pleasure-seeking life, which on later reflection I think my father felt contributed to his first son's lack of self-discipline and rather light-hearted view of life. With John, he was much stricter, took a close interest in all his studies, and was demanding of results. He was inclined to over-react, and it was ironic that John, who was serious by nature, should have had to bear the brunt of his father's reformed views on bringing up a son. Having been lavishly generous to Richard in his youth, he was strict with John and myself about money. At the age of 14 I still only had sixpence a week pocket money; this I expect was also my mother's influence. In other ways we were generously treated: the best of schools, voyages on Cunard liners, and holidays in Bermuda. John and I each had a highly interesting youth, and Adrian Smith has already described my father's extraordinary ability to teach and enthuse young people on many subjects. He was a natural teacher and could always explain any problem at a level the listener could understand.

My father's marriage to my mother was a very profound and happy one. He adored her and she him. Her trustworthiness was invaluable to him, as father could tell her anything and know she would never pass it on: a secret was a secret and a promise was a promise with my mother. Of course he could be very difficult and extremely demanding, and someone once described her as being like 'a willow whip which bent before the storm and flipped back into place when it was over.' Fortunately she herself was also a very strong personality. He loved to spoil her, and I can still remember how at Christmas my father would do a stocking for my mother, carefully wrapping every little present and labelling them from her pets – and in the toe there would be his principal present for her, chosen by him personally from Asprey. If she were ill he would hover by her bedside offering remedy after remedy, and showing extreme anxiety for her well-being. They were very seldom apart, and he could not really bear her out his sight. His loss was terrible for her, and my mother never really quite got over his death, despite after eight years of widowhood marrying Philip Young, himself a widower and at that time President of the International Chamber of Commerce. Philip Young was a dear and delightful man, who deserves a book to himself concerning his illustrious career.

To end, just a few words of thanks to Adrian Smith, who worked through mountains of papers and information without the benefit of talking to people who knew my father in the course of his career. All I could do was to reminisce on my father's character and the domestic side of his life. The task of writing his life was infinitely more complicated and challenging than I had imagined it would be at the outset, and I can only say thank you to Professor Smith for his monumental achievement.

Jane Tennant, née Fairey (1937–2017)
Houghton, Hampshire
Summer 2017

Notes

Foreword and Acknowledgements

1 Peter Twiss, *Faster than the Sun: The Compelling Story of a Record Breaking Test Pilot* (London: Grub Street, 2000), p. 28.
2 The RNAS features prominently in Chapters 1–3, but for a concise explanation of why it was so advanced in strategy, tactics and technology when compared to the Army's Royal Flying Corps, see Christina J.M. Goulter, 'The Royal Naval Air Service: a very modern service', in S. Cox and P. Gray (eds), *Air Power History Turning Points from Kitty Hawk to Kosovo* (London: Frank Cass, 2002), pp. 51–65.
3 W.B. Yeats, 'At Stratford', in R.J. Finneran and G. Bornstein (eds), *The Collected Works of W.B. Yeats Vol. IV Early Essays* (New York, NY: Scribner, 2007), p. 82.
4 David Edgerton, *Warfare State Britain, 1920–1970* (Cambridge: Cambridge University Press, 2006), pp. 7–14.
5 Notably via the Arts and Humanities Research Council two-year research network based at the University of Southampton, 'Challenges to biography', 2011–12; subsequently absorbed by the Oxford Centre for Life Writing, headed by another key figure in the study of life writing, Hermione Lee: www.wolfson.ox.ac.uk/oclw.
6 Ray Monk, *Inside the Centre: The Life of J. Robert Oppenheimer* (London: Jonathan Cape, 2012).
7 C.P. Snow, *The Two Cultures* (Cambridge: Cambridge University Press, 1959). For a sustained assault on Snow's so-called 'anti-history', see Edgerton, *Warfare State*, pp. 196–210. Danchev's subjects for biography ranged from pillar of the postwar British state, Oliver Franks, to pioneer of post-impressionism, Paul Cézanne.
8 Robert Innes-Smith, 'James Wentworth Day (1899–1983)', *Oxford Dictionary of National Biography* (Oxford: OUP, 2004), https://doi.org/10.1093/ref:odnb/40790.
9 Sir Richard Fairey to C.C. Vinson, 22 February 1949, FAAM/2010/054/0038, Fleet Air Arm Museum Archives.
10 James Wentworth Day, draft synopsis for book 'Wings over the world: the official life of Sir Richard Fairey, M.B.E., FR., R.Ae.S.', FAAM/32/2010/053/0125.

J. Wentworth Day to Richard Fairey, 5 April 1957 and Richard Fairey to the Board of Fairey Aviation and to J. Wentworth Day, 10 and 15 April 1957, Box 102, AC73/70, RAF Museum.
11 James Wentworth Day to John Fairey, 2 August 1966, in the possession of Mrs Jane Tennant.
12 Evidence that Peter Trippe the Hampstead painter and Peter Trippe the biographer and friend of the Faireys were one and the same courtesy of Jane Wood; Jane Tennant in conversation with the author, 18 August 2017. Under the pseudonym 'Geoffrey Peters', Trippe wrote an epistolary novel *Unfinished Letter* (London: Saturn Press, 1950), its dust-jacket revealing the author's training as an electrical engineer and wartime service with Combined Operations.
13 Peter Trippe, 'The plane maker: the official biography of Sir Richard Fairey': unpublished manuscript held by Mrs Jane Tennant.
14 Adrian Smith, *'Mick' Mannock, Fighter Pilot: Myth, Life and Politics* (London: I.B.Tauris, 2001 and 2015).

Chapter 1

1 'Prelude: contrasting family fortunes' based upon Rev. L. Fairey to C.R. Fairey, 28 November 1931 and 11 April 1932 and C.R. Fairey to Rev. L. Fairey, 18 January 1932; correspondence held by Mrs Jane Tennant.
2 Mrs Jane Tennant, interview, 5 November 2013.
3 Interview with Frances 'Effie' Hulme [CRF's eldest sister] in Peter Trippe, 'The plane maker: the official biography of Sir Richard Fairey', pp. 2–3.
4 *Census of the Commonwealth of Australia 30th June, 1933 Volume I Part VIII. Population and Occupied Dwellings in Localities* (Canberra: Commonwealth of Australia, 1934), p. 630 and *Part XVI. Religion*, ibid., 1020–1.; E.O.G. Scott, *Hagley (A Short History of the Early Days of the Village and District with Notes on the Pioneer Families)* (Launceston: Birchalls, 1985), pp. 40–2: a museum curator in Launceston Eric Scott shared Leslie Fairey's dissenting beliefs, his Methodist-inspired pacifism leading to wartime incarceration. Rhona Hamilton, 'Scott, Eric Oswald (1899–1986)', *Australian Dictionary of National Biography*, vol. 18, 2011: www.adb.anu.edu.au/biography/scott-eric-oswald-15908.
5 C.R. Fairey, 'How I began: in the early days of flying', *The Listener*, 16 February 1938. Mud pattens are better known today as 'splatchers', the name of the boards used for crossing marshland by 'The Mastodon' in Arthur Ransome's 1939 novel *Secret Water*.
6 Conversation with Mrs Jane Tennant, 24 June 2014; family photographs of CRF's holidays in Eaton Socon during the 1890s held by Mrs Tennant.

7   CRF's lifelong love of dogs dated from childhood pets, not least the acquisition of a terrier that could do tricks. The legacy was a keen scientific interest in the behavioural patterns of intelligent animals. Mrs Jane Tennant, interview, 5 November 2013, and conversation, 24 June 2014; J. Wentworth Day, draft synopsis for book 'Wings over the world: the official life of Sir Richard Fairey, M.B.E., FR.R.Ae.S.', FAAM/32/2010/053/0125, Fleet Air Arm Museum Archives; William Wordsworth, 'My heart leaps up when I behold', 1802.

8   Morgan, Gellibrand and Co. was a private joint-stock company that in December 1866 became a partnership; in which form it survived until 1923 when what became a Morgan family business was wound up owing to both directors' retirement: *London Gazette*, 4 January 1867 and 4 January 1924; *Kelly's Directory* [central London], 1912–22.

9   'My father was … a member of the [Hendon] Council and I can remember in the early 1890's the Town being plastered with posters reading "Vote for the solid three – Fairey, Frazer, Hearne."' C.R. Fairey to James E. Walker, 11 June 1956, FAAM/32/2010/054/048.

10 Trippe, 'The plane maker', pp. 3–4 and 6–7; 1891 census return for Ray House, Finchley Lane, Harrow constituency; C.R. Fairey to Rev. L. Fairey, 18 January 1932, Tennant correspondence.

11 Ibid. Frances Fairey's readiness to cross the Atlantic bore echoes of her Elizabethan ancestor John Shrive, who in 1588 sailed his tiny sloop of 20 hands, *The Signett*, to join the fleet facing the Spanish Armada: G.W. Hall, *Sir Richard Fairey: The First Fairey Memorial Lecture* (London, Royal Aeronautical Society, 1959), p. 3; 'Spanish Armada', *The Naval and Military Magazine*, 2 (1827), p. 596.

12 Trippe, 'The plane maker', 15; Frances 'Effie' Hulme quoted in ibid., p. 15.

13 C.R. Fairey to Rev. L. Fairey, 18 January 1932, Tennant correspondence; Trippe, 'The plane maker', p. 18 and pp. 19–20, and interview with Frances 'Effie' Hulme in ibid., pp. 20–1; Julius Dworetsky, 'The diagnosis of tubercular laryngitis', *Journal of the American Medical Association*, 69 (1917), p. 618.

14 Fairey, 'How I began.'

15 Mrs Jane Tennant, interview, 5 November 2013; Frances 'Effie' Hulme quoted in Trippe, 'The plane maker', pp. 10–11 and 39.

16 Hall, *Sir Richard Fairey*, pp. 4–5; interview with Frances 'Effie' Hulme in Trippe, 'The plane maker', p. 12.

17 Hall, *Sir Richard Fairey*, p. 5; Mrs Jane Tennant in conversation, 8 October 2013; Jane Tennant to the author, 8 March 2017.

18 Conversation with Mrs Jane Tennant, 24 June 2014.

19 Trippe, 'The plane maker', p. 16; Hall, *Sir Richard Fairey*, p. 5; pupil records of Merchant Taylors' School, 1897–8.

20 Interview with Frances 'Effie' Hulme in Trippe, 'The plane maker', pp. 21–2 and 26; Fairey, 'How I began.'

21 '7519 Fairey, Charles Richard', St Saviour's School ledger, 1898, Ardingly College archives.
22 Reverend Nathaniel Woodward (1811–91), quoted in Andrea King, 'A short history of Ardingly College', www.ardingly.com; 1901 census return for St Saviour's School, 'Middle Class School', Ardingly, Cuckfield, Sussex.
23 1901 prize volume for Arithmetic and 1902 prize volume for Science held in the Bossington library: Hall, *Sir Richard Fairey*, p. 5; John Fairey to Andrea King, 29 December 2004, Ardingly College archives.
24 Fairey, 'How I began.'
25 Interview with Frances 'Effie' Hulme in Trippe, 'The plane maker', pp. 33–4.
26 Today it is difficult to gain a clear impression of what 27 Station Road looked like in its Edwardian prime, as the front of the house has been extended to the street, and the property converted to commercial use (chartered accountancy).
27 Hall, *Sir Richard Fairey*, p. 5.
28 A premium apprenticeship embraced all aspects of the manufacturing process; as opposed to a craft apprenticeship which, as the name implies, was far more specialist.
29 Arc lamps were an American invention that arrived in Britain via Canada. The Jandus Arc Lamp and Electric Co., named after the Ohio company that held the patent, was set up in 1897 to manufacture shop and street lighting under licence. The company moved from south London to Holloway when rising demand necessitated a larger factory, but a slump in sales saw it wound up in 1912. Norman F. Ball and John N. Vardalas, *Ferranti-Packard: Pioneers in Canadian Electrical Engineering* (Montreal: McGill-Queen's Press, 1994), p. 54.
30 Hall, *Sir Richard Fairey*, p. 5; Fairey, 'How I began'; C.R. Fairey, typescript for radio broadcast, 8 February 1938, 'How I began', FAAM/32/2010/053/0099.
31 As we shall see, G.W. Hall served his apprenticeship with Fairey Aviation, to which he returned in 1932 after study at the Royal College of Science and employment at Rolls Royce. Hall enjoyed no special favours and steady promotion within the company was recognition of his expertise as an engineer and extensive experience as a pilot. He joined the board as director of engineering in 1949 and within a decade had emulated his half-brother by being appointed Fairey's chairman and managing director: H.A. Taylor, *Fairey Aircraft since 1915* (London: Putnam, 1974), p. 27.
32 Hall, *Sir Richard Fairey*, p. 5; interview with G.W. Hall in Trippe, 'The plane maker', p. 31; conversation with Mrs Jane Tennant, 24 June 2014.
33 www.heyshamheritage.org.uk/heysham_harbour; Fairey, 'How I began'; Hall, *Sir Richard Fairey*, p. 5.
34 Ibid.; Fairey, 'How I began'; photograph of CRF outside the Ramsgate home of Bertie Smith [CRF's cousin] *c.*1907 held by Mrs Tennant; Trippe, 'The plane maker', pp. 32, 44.
35 Fairey, transcript for radio broadcast; 'The City and Guilds of London Institute', www.skillsdevelopment.org/aboutus/about_city_guilds.aspx.

36 Contrary to popular assumption, Thompson was not the head of Finsbury Technical College, having been passed over for appointment as principal officer in 1901, perhaps on account of his opposition to the Boer War. Arthur Smithells, 'Thompson, Silvanus Phillips (1851–1916)', Rev. Graeme J.N. Gooday, *Oxford Dictionary of National Biography* (Oxford: Oxford University Press, 2004), www.oxforddnb.com/view/article/36496; Hannah Gay and Anne Barrett, 'Should the cobbler stick to his last? Silvanus Phillips Thompson and the making of a scientific career', *The British Journal for the History of Science*, 32 (2002), pp. 151–86.
37 Ibid., pp. 165–8.
38 'Note to the Committee on the Status and Outlook of the City and Guilds Technical College, Finsbury, by the Principal', 27 December 1901, CLC/211/MS21913/008, and 'Note by the Honorary Secretary on the memorandum of the Principal of the Technical College, Finsbury, as to the grant of an associateship', [?] 1902, CLC/211/MS21912/009, London Metropolitan Archives.
39 Fairey, 'How I began.'
40 Geometry: 26; English: 42: Algebra: 13; Arithmetic 29 = 110 out of 400: result of C.R. Fairey, 1902 Electrical entrance exam, 23 September 1902, CLC/211/MS21913/008, London Metropolitan Archives.
41 Marks lists for Physics and Electronics exams, 15 and 13 May 1903, Mechanical Laboratory Work – Elementary 1903–4, and Mechanics A and Electric Technology exams, 10 and 9 May 1904, CLC/211/MS21980/002, Finsbury Technical College evening students' marks…; ibid.
42 C.R. Fairey, Certificate of the City and Guilds Technical College, Finsbury – Diploma of Fellow of the City and Guilds of London Institute, 1904, in the possession of the late Mrs Jane Tennant. Fairey, 'How I began'; Hall, *Sir Richard Fairey*, p. 5. The previously autonomous Middlesex Technical Education Committee (MTEC) became a subcommittee of the county council's Education Committee from May 1903, so the latter body was in practice Fairey's employer; see the MTEC's history at ref. code MCC/EO/TCOM on the London Metropolitan Archives website.
43 Fairey, 'How I began.' CRF's sister estimated that he contributed between £1 and £1.25 towards the weekly cost of running the house: interview with Frances 'Effie' Hulme in Trippe, 'The plane maker', p. 80. Tottenham Centre constitutes one half of the College of Haringey, Enfield and North East London. Assiduous searching of the relevant Finsbury Technical College records at London Metropolitan Archives failed to uncover C.R. Fairey's final results, but the sheer volume of documentation means the investigation was in no way definitive.
44 Account based on family recollection [Geoffrey Hall, Frances 'Effie' Hulme and John Fairey] in Trippe, 'The plane maker', pp. 154–8, and conversation with Mrs Jane Tennant, 24 June 2014.
45 Geoffrey Hall quoted in Trippe, 'The plane maker', p. 160.
46 John Fairey quoted in ibid., p. 159.
47 Geoffrey Hall quoted in ibid., p. 161.

48 Ibid., p. 159.
49 Hannah Gay and A. Barrett, 'Should the cobbler stick to his last?', *British Journal for the History of Science* 35 (2002), pp. 159–63 and 181–2; Fairey, transcript for radio broadcast.
50 G.W. Hall claimed that H.G. Wells spotted his half-brother playing with model aircraft in the front garden and they conversed at length over the future of manned flight, but this story seems apocryphal – why would the already famous writer be walking down a side street in either Acton or Finchley? A 1927 letter from Fairey to Wells makes no mention of any childhood acquaintance. Re Fairey's admiration for Wells, see Chapter 2. Hall, *Sir Richard Fairey*, p. 6; Fairey, 'How I began'; C.R. Fairey to H.G. Wells, 29 November 1927, H.G. Wells Archive, University of Illinois, Champaign-Urbana.
51 Fairey, 'How I began.'
52 C.R. Fairey, final draft of Walmsley Memorial Lecture – 'The development of the modern aeroplane', FAAM/32/2010/053/0042. For a succinct account of nineteenth-century experimentation in Britain, see Harald Penrose, *British Aviation: The Pioneer Years 1903–1914* (London: Putnam, 1967), pp. 12–42. Strangely, in his 1930 lecture to the Royal Aeronautical Society Fairey made no mention of the mid-Victorian naval architect turned aviator Francis Wenham, whose experiments focused upon wing shape and the role of the tail in ensuring stability. Given CRF's design priorities he must have known the lasting significance of Wenham's aerodynamic research. Ibid., pp. 18–21.
53 Interview with Frances 'Effie' Hulme, and family recollection, in Trippe, 'The plane maker', p. 51. Fairey did recall having previously read a newspaper report, headlined 'Tobogganing on the air', on the Wright brothers' gliding experiments: Fairey, 'How I began.'
54 Ibid.; re Captain J.W. Dunne, see Chapter 2.
55 Fairey, 'How I began'; Hall, *Sir Richard Fairey*, p. 5; family recollection in Trippe, 'The plane maker', pp. 76–7.
56 Interviews with Geoffrey Hall and Frances 'Effie' Hulme in ibid., pp. 80–1 and 86–96.
57 Interview with Frances 'Effie' Hulme in ibid., pp. 38–9.
58 Interview with Geoffrey Hall in ibid., pp. 83–5 and 103–4.
59 Fairey, 'How I began'; Peter King, *Knights of the Air: The Life and Times of the Extraordinary Pioneers Who First Built British Aeroplanes* (London: Constable, 1989), pp. 19–21.
60 Ibid., pp. 62–4, 84–5 and 89–1; H. Penrose, *British Aviation: The Pioneer Years* (London: The Bodley Head, 1967), pp. 155–6.
61 The monoplane's wing was not fixed but held by four light clips, enabling experimentation with the ideal position during trial flights. Acquired by mail order, the band to power the propellers was a stranded form of elastic just under a centimetre wide and less than half a centimetre thick. CRF used first an adapted egg beater and then a modified hand drill to wind up the elastic motor. H.A. Taylor, *Fairey*

Aircraft since 1915 (New York: US Naval Institute Press, 1974), p. 5; Trippe, 'The plane maker', pp. 107–8 and 110; Hall, *Sir Richard Fairey*, p. 6.
62 Ibid.; 'Fairey's miniature aeroplane: Instructions for building and flying. Type 42. Official duration record, February 17th, 1912', FAAM/28/2010/054/0219; Taylor, *Fairey Aircraft since 1915*, pp. 5, 34–5 and 354–78; *Fairey Company Profile 1915–1960*, ed. Martyn Chorlton (Cudham: Kelsey Publishing Group/Aeroplane, 2012), pp. 42–3.
63 Interview with Geoffrey Hall in Trippe, 'The plane maker', pp. 105–7.
64 Geoffrey Hall quoted in ibid., p. 109.
65 Interviews with Geoffrey Hall and Frances 'Effie' Hulme in ibid., pp. 111–16; H.A. Taylor, *Fairey Aircraft since 1915*, p. 5; Hall, *Sir Richard Fairey*, p. 6. The trophies can still be seen at Bossington, CRF's country house.
66 Interviews with Geoffrey Hall and Frances 'Effie' Hulme in Trippe, 'The plane maker', pp. 122–36; Taylor, *Fairey Aircraft since 1915*, p. 5.
67 'Fairey's miniature aeroplane: Instructions for building and flying', FAAM/28/2010/054/0219; Hall, *Sir Richard Fairey*, p. 6. No longer technically an amateur CRF became a competition judge, being elected as a council member of the Kite and Model Aeroplane Association. Twenty years later Fairey Aviation made the 'Heath Row Aerodrome' available to model makers for qualifying rounds of what from 1928 was their hobby's most prestigious competition: 'The Kite and Model Aeroplane Association Official Notices', *Amateur Aviation*, 24 June 1912, and 'The Wakefield Trophy Eliminating Trials', *The Aero-Modeller*, June 1936.
68 Receipt made out to Messrs. Fairey and Witty for £72 18 shillings 4 pence from The Aerial Engineering Company, 7 November 1910, replicated in Trippe, 'The plane maker', pp. 138–9. Given that there are two names on the receipt Fairey's costs extended to having someone handle his accounts.
69 See Chapter 2.

Chapter 2

1 MB2/B4/18-19, 22, 24-25 and MB2/C7/142-147, Broadlands Archives; Margit Fjellman, *Louise Mountbatten Queen of Sweden* (London: George Allen and Unwin, 1968), p. 79; Lt C.R. Sansom, 'Weekly report of instruction to the Officers undergoing Aviation course at Eastchurch', 14 May and 3 June 1911, AIR1/2469, NA; Mrs Jane Tennant, interview, 5 November 2013.
2 Lord Brabazon of Tara, *The Brabazon Story* (London: William Heinemann, 1956), pp. 63–4.
3 Ibid., p. 65; Alan Bramson, *Pure Luck: The Authorised Biography of Sir Thomas Sopwith* (Manchester: Crécy, 2005), pp. 21–2.
4 Hugh Driver, *The Birth of Military Aviation: Britain 1903–1914* (Woodbridge: Boydell and Brewer, 1997), pp. 6, 15–16; P. King, *Knights of the Air* (Constable, 1989),

pp. 32–3 and 37–9. For individual profiles of all Dunne's machines, see Peter Lewis, *British Aircraft 1809–1914* (London: Putnam, 1962), pp. 218–32.

5   Alfred Gollin, *No Longer an Island: Britain and the Wright Brothers* (Standford: Stanford University Press, 1984), pp. 418–21 and 428–36.

6   Sir Geoffrey de Havilland, *Sky Fever: The Autobiography of Sir Geoffrey de Havilland C.B.E.* (Shrewsbury: Airlife, 1979), pp. 70–1.

7   For an assessment of Haldane's record, and the impact of German ideas on his view of developing an indigenous aviation industry in the light of later wartime experience, see H. Driver, *The Birth of Military Aviation* (Woodbridge: Boydell and Brewer, 1997), pp. 230–1, 200–3 and 243–4.

8   Gollin, *No Longer an Island*, pp. 166–78, 230–4 and 269–77; Driver, *The Birth of Military Aviation*, p. 86; Percy B. Walker, *Early Aviation at Farnborough Vol. II The First Aeroplanes* (London: Macdonald, 1974), pp. 193–247, 330–342 and 169–79; Hall, *Sir Richard Fairey*, p. 7.

9   Chapter 9 of H.G. Wells, *The History of Mr Polly* (London: Thomas Nelson and Sons, 1910) commences, 'But when a man has once broken through the paper walls of everyday circumstance, those unsubstantial walls that hold so many of us securely prisoned from the cradle to the grave, he has made a discovery. If the world does not please you *you can change it*.' Given its aviation theme, not surprisingly CRF's favourite Wells novel was the dystopian *When The Sleeper Wakes*, serialised in *The Graphic* in 1903: C.R. Fairey, 'Typescript for radio broadcast, 8 February 1938, "How I began"', FAAM/32/2010/053/0099. Hall, *Sir Richard Fairey*, p. 6.

10 H.G. Wells, *Bealby: A Holiday* (London: Methuen, 1915), pp. vi and 172, and *The War in the Air* (London: George Bell and Sons, 1908); Norman and Jeanne MacKenzie, *The Time Traveller: The Life of H.G. Wells* (London: Weidenfeld and Nicolson, 1974), pp. 222 and 287; Gollin, *No Longer an Island*, pp. 230–2; King, *Knights of the Air*, pp. 38–40. In 1915 Wells was appalled by the opposition newspapers' success in hounding from office an allegedly 'unpatriotic' Lord Chancellor. Haldane, gently satirised in *Bealby* as 'a fluent Hegelian', saw pan-German culture as a counterweight to Prussian militarism. On Wells's changing ideas re air power from the late Victorian era to World War II, see Brett Holman, *The Next War in the Air Britain's Fear of the Bomber, 1908–1941* (Farnham: Ashgate, 2014), pp. 28–30, 33, 35, 74–5, 159–60 and 171.

11 The Aeronautical Society's distinguished mathematician George Hartley Bryan provided a theoretical underpinning for Dunne's experiments, consolidating his calculations in the 1911 treatise *Stability in Aviation*. Walker, *Early Aviation at Farnborough*, pp. 169–70.

12 Driver, *The Birth of Military Aviation*, pp. 188–95 and 209.

13 Walker, *Early Aviation at Farnborough*, pp. 171–9; Gollin, *No Longer an Island*, pp. 170, 230 and 269–72.

14 Hall, *Sir Richard Fairey*, pp. 6–7; 'Fairey's miniature aeroplane: Instructions for building and flying. Type 42. Official duration record, February 17th, 1912', FAAM/28/2010/054/0219.
15 Hall, *Sir Richard Fairey*, p. 7.
16 Ibid.; Geoffrey Hall and Frances 'Effie' Hulme quoted in Trippe, 'The plane maker', pp. 151 and 153; Mrs Jane Tennant, interview, 5 November 2013; 'Pioneers of British Aviation XVI – Mr C.R. Fairey', *Aeronautics*, 22 January 1920.
17 Swale Borough Council, *150 Years of Trains to Sheppey 1860–2010* (Sittingbourne: Swale BC/Kent CC, 2010).
18 Hall, *Sir Richard Fairey*, p. 7; Mrs Jane Tennant, interview, 5 November 2013.
19 CRF identified Everett Edgcumbe's aircraft enthusiast as 'Mr Everett.' H.A. Taylor, *Fairey Aircraft since 1915* (London: Putnam, 1974), p. 6; Fairey, 'How I began.'
20 Ibid., p. 7; de Havilland, *Sky Fever*, pp. 80–1; Driver, *The Birth of Military Aviation*, pp. 24 and 95; Taylor, *Fairey Aircraft since 1915*, p. 6. In 1914 Booth designed a large seaplane, commissioned by the Admiralty from Cowes shipbuilder J. Samuel White. During World War I he headed the Admiralty's Air Technical Branch. Post-war Booth undertook a succession of jobs within the aircraft industry before in the late 1920s joining ICI. At the height of the depression CRF endeavoured to help the now ageing Booth secure a position again building aeroplanes: Harris Booth to C.R. Fairey, 8 and 20 January 1932, and C.R. Fairey to A.E.L. Chorlton, 1 June 1932, FAAM/2010/054/0030.
21 Fairey, 'How I began.'
22 CRF's need for discretion was reinforced by knowledge of Dunne's direct dealings with the Short brothers, as for example when he and Horace joined other plane makers on 5 December 1911 to lobby the calamitous J.E.B. Seeley, Under-Secretary of State for War. Driver, *The Birth of Military Aviation*, pp. 143–4.
23 Hall, *Sir Richard Fairey*, p. 7; Fairey, 'How I began.'
24 www.muswellmanor.co.uk/aviation_history. A further memorial is on the south wall of All Saints, Eastchurch's parish church: a stained glass window commemorates Cecil Grace, who disappeared over the Channel in December 1910, and Charles Rolls, killed at a Bournemouth flying display the previous July. For individual profiles of most of Short Brothers' prewar machines, see P. Lewis, *British Aircraft 1809–1914* (London: Putnam, 1962), pp. 433–65.
25 For images of the Eastchurch aviation memorial, see: www.sheppeywebsite.co.uk/index.php?id=72. Ronald Coleman, *Memorial to Pioneer Airmen* (Eastchurch: Eastchurch Parish Council, 1955).
26 Gordon Bruce, 'Short (Hugh) Oswald (1883–1969)', rev. Robin Higham, *Oxford Dictionary of National Biography* (Oxford: Oxford University Press, 2004), www.oxforddnb.com/view/article/36075; David Edgerton, *England and the Aeroplane Militarism, Modernity and Machines* (London: Penguin, 2013), pp. 125–6; C.H. Barnes, *Shorts Aircraft Since 1900* (London: Putnam, 1967), pp. 29–31.

27 Ibid., pp. 3–9 and 34–5; Driver, *The Birth of Military Aviation*, pp. 43–5 and 64–6.
28 Barnes, *Shorts Aircraft Since 1900*, pp. 5–8, 34–6, 44–8 and 1–2; Driver, *The Birth of Military Aviation*, pp. 66–8 and 41–2; Hall, *Sir Richard Fairey*, p. 7; Fairey, 'How I began'; Lord Brabazon, *The Brabazon Story*, pp. 62 and 51.
29 Barnes, *Shorts Aircraft Since 1900*, pp. 48–53; Driver, *The Birth of Military Aviation*, pp. 68–9; King, *Knights of the Air*, p. 81; Lord Brabazon, *The Brabazon Story*, pp. 66 and 65.
30 Barnes, *Shorts Aircraft Since 1900*, pp. 52–71.
31 Harald Penrose, *British Aviation: The Pioneer Years 1903–1914* (London: Putnam, 1967), pp. 254, 346, 506 and 591.
32 Ibid., pp. 311–12 and 345–6. Peter Fearon, 'The formative years of the British aircraft industry, 1913–1924', *The Business History Review*, 43:4 (1969), p. 487.
33 Penrose, *British Aviation*, p. 286; Fearon, 'The formative years', pp. 482–4.
34 Barnes, *Shorts Aircraft Since 1900*, pp. 76–99; Driver, *The Birth of Military Aviation*, p. 72.
35 Gollin, *No Longer an Island*, pp. 434–5; Eric Grove, 'Seamen or airmen? The early days of British Naval flying' in *British Naval Aviation The First 100 Years*, ed. Tim Benbow (Farnham: Ashgate, 2011), pp. 9–11; profile of Sueter in *Documents Relating to the Naval Air Service Vol. I 1908–1918*, ed. S.W. Roskill (London: Navy Records Society, 1969), pp. 56–8.
36 Grove, 'Seamen or airmen?', pp. 12–13; Penrose, *British Aviation*, p. 293.
37 'Aeroplanes Sheppey Course I to 30 June', Short Brothers to Flag Captain H.J. Langford Clarke, 16 February 1911, AIR1/2469, NA.
38 Correspondence between Lt C.R. Sansom and Capt. M.F. Sueter, March–September 1911, AIR1/2469, NA. David Lee, 'Longmore, Sir Arthur Murray (1885–1970)', rev. Christina J.M. Goulter, *Oxford Dictionary of National Biography* (Oxford: Oxford University Press, 2004), https://doi.org/10.1093/ref:odnb/36368. Sueter enjoyed a tempestuous relationship with Sansom, in 1914–15 his squadron commander at Dunkirk and the Dardanelles, but he deemed his acolyte a fearless aviator, as 'No job was too difficult for him to undertake, and his men would follow him anywhere … Hats off to Sansom, one of the best pioneer naval airmen we ever had!' For Sueter, Longmore was no more than 'a capable flying officer.' Rear-Admiral Murray F. Sueter, *Airmen or Noahs: Fair Play for our Airmen: The Great "Neon" Air Myth Exposed* (London: Sir Isaac Pitman & Sons, 1928), pp. 181 and 46.
39 Lt C.R. Sansom, 'Weekly report of instruction to the Officers undergoing Aviation course at Eastchurch', March–August 1911, AIR1/2469 and 'General statement of the work done by the four officers', 11 August 1911, AIR1/2467, NA.
40 Grove, 'Seamen or airmen?', pp. 13–14; Driver, *The Birth of Military Aviation*, pp. 71–2; Director of Air Department, Admiralty, to CID Air Committee, 29 August 1912 in *Documents Relating to the Naval Air Service*, ed. Roskill, pp. 58–9; Director

of Air Department, Admiralty, 'General progress of the Naval Service up to the end of 1912', January 1913, AIR1/2462, NA; Sueter, *Airmen or Noahs*, p. 95.

41 Winston S. Churchill, *The World Crisis 1911–1914* (London: Thornton Butterworth, 1913), p. 315; Winston S. Churchill to Prince Louis of Battenberg, 7 December 1912, quoted in in Randolph S. Churchill, *Winston S. Churchill Volume II Young Statesman 1901–1914* (London: Heinemann, 1967), p. 689.

42 Christopher M. Bell, *Churchill and Sea Power* (Oxford: Oxford University Press), 15–6 and 28; Churchill, *The World Crisis 1911–1914*, pp. 312 and 337; Winston S. Churchill to Capt. W. Packenham and Capt. M.F. Sueter, 21 and 21 December 1913, and memorandum of First Lord of the Admiralty, 18 May 1914, in *Winston S. Churchill Volume II Companion Part 3 1911–1914*, ed. Randolph S. Churchill (London: Heinemann, 1969), pp. 1898–9 and 1915–17. Churchill's visits to Eastchurch failed to coincide with those of H.G. Wells, although both men engaged in sporadic debate concerning the impact of new technology on the future organisation of advanced capitalist societies: Graham Farmelo, *Churchill's Bomb How the United States Overtook Britain in the First Nuclear Arms Race* (New York: Basic Books, 2013), pp. 15–20.

43 Churchill was 38 when he first flew, and conventional wisdom held 32 to be the maximum age for a pilot. W. Churchill, *Winston S. Churchill Volume II Young Statesman*, pp. 697–705; Winston S. Churchill to Clementine Churchill, 6 June 1914, quoted in ibid., p. 705. A previous death of an officer at Eastchurch, on the ground, had left Wildman-Lushington contemplating resignation until comforted by Sansom, whom he considered, 'in the aviation world as 2nd to none': Mrs D.E. Wildman-Lushington to Winston S. Churchill, 8 December 1913, in *Winston S. Churchill Volume II Companion*, p. 1894.

44 For example a crowded autumn Thursday with parliament in recess, during which Churchill encouraged *The Times*'s Edward Grigg to meet Horace Short and fly in one of his biplanes: Winston S. Churchill to Clementine Churchill, 23 October 1913, ibid., p. 1884. Churchill was using similar tactics to promote the Royal Navy's renewed airships programme, even though after the war he claimed to have prioritised aircraft procurement as the only credible strategic option: W. Churchill, *The World Crisis 1911–1914*, p. 313.

45 King, *Knights of the Air*, p. 123.

46 Ibid., p. 103; Driver, *The Birth of Military Aviation*, pp. 114–17. The terms 'hydro-seaplane' or 'hydro-aeroplane' were soon shortened to 'hydroplane', but in late 1913 Churchill insisted on uniform use of the word 'seaplane' – and that is what such aircraft are labelled here.

47 Malcolm Cooper, *The Birth of Independent Air Power* (London: Allen & Unwin, 1986), pp. 4–8; King, *Knights of the Air* (Woodbridge: Boydell and Brewer, 1997), pp. 129, 102, 97 and 123.

48 Grove, 'Seamen or Airmen?', pp. 14–16; Penrose, *British Aviation*, pp. 369–71.

49 Grove, 'Seamen or Airmen?', pp. 16–20; Barnes, *Shorts Aircraft Since 1900*, pp. 81–4; Penrose, *British Aviation*, p. 372;

50 Grove, 'Seamen or Airmen?', pp. 20–4; Cooper, *The Birth of Independent Air Power*, pp. 4–5 and 8–9; King, *Knights of the Air*, pp. 123–4; Bramson, *Pure Luck* (Manchester, Crecy, 1990), pp. 63–78.
51 Ian Lemco, 'Wittgenstein's aeronautical investigation', *Notes & Records of the Royal Society*, 61 (2007), pp. 39–40 and 49. Both Lamb and Schuster were later knighted for their services to mathematics and physics, respectively.
52 Ibid., p. 41; Driver, *The Birth of Military Aviation*, pp. 73–83; Ray Monk, *Ludwig Wittgenstein: The Duty of Genius* (London: Vintage, 1991), pp. 28–30 and 33–4.
53 Ibid., p. 34.
54 Lemco, 'Wittgenstein's aeronautical investigation', pp. 42–4.
55 Monk, *Ludwig Wittgenstein*, pp. 30–3.
56 Taylor, *Fairey Aircraft since 1915*, pp. 356–78. See chapter 9.
57 Lemco, 'Wittgenstein's aeronautical investigation', pp. 45–6; Taylor, *Fairey Aircraft since 1915*, pp. 392–7 and 405–26. See chapter 9.
58 Lemco, 'Wittgenstein's aeronautical investigation', pp. 48–9; Anne Keynes, 'Introduction' in *A Portrait of Wittgenstein as a Young Man*, ed. G.H. von Wright (Oxford: Basil Blackwell, 1990), pp. iii–iv.
59 Both Cherwell's biographies provide detailed, and complementary, accounts of his experimental flights at the RAF: Earl of Birkenhead, *The Prof in Two Worlds The Official Life of Professor F.A. Lindemann, Viscount Cherwell* (London: Collins, 1961), pp. 58–80; Adrian Fort, *Prof The Life of Frederick Lindemann* (London: Pimlico, 2004), pp. 55–65. CRF was always insistent that, although lacking a pilot's certificate, he too had mastered the art of putting an aircraft into a spin and then successfully pulling out of it at low altitude: conversation with Mrs Jane Tennant, 24 June 2014.
60 Monk, *Ludwig Wittgenstein: The Duty of Genius* (New York: Vintage, 1991), pp. 484–6.

Chapter 3

1 Harald Penrose, *British Aviation: The Pioneer Years 1903–1914* (London: Putnam, 1967), pp. 526–7; Oliver Tapper, *Armstrong Whitworth Aircraft since 1913* (London: Putnam, 1973), pp. 526–7; 4 and 287–96.
2 C.R. Fairey, 'How I began: in the early days of flying', *The Listener*, 16 February 1938
3 C.H. Barnes, *Shorts Aircraft since 1900* (London: Putnam, 1967), pp. 89–90; Peter Lewis, *British Aircraft 1809–1914* (London: Putnam, 1962), pp. 459; Penrose, *British Aviation: The Pioneer Years*, pp. 415–16.
4 Penrose, *British Aviation: The Pioneer Years*, pp. 422–3.
5 Ibid., pp. 417 and 418–19; Alan Bramson, *Pure Luck The Authorised Biography of Sir Thomas Sopwith* (Manchester: Crécy, 2005), pp. 66–9; King, *Knights of*

the Air, *The Life and Times of the Extraordinary Pioneers who First Built British Aeroplanes* (London: Constable, 1989), p. 122; Hugh Driver, *The Birth of Military Aviation: Britain 1903–1914* (Woodbridge: Boydell and Brewer, 1997), pp. 99–100.

6 'When I'm working on a problem, I never think about beauty. I think only how to solve the problem. But when I have finished, if the solution is not beautiful, I know it is wrong': design guru R. Buckminster Fuller, quoted at www.tekgnostics.com/bucky.

7 Barnes, *Shorts Aircraft since 1900*, pp. 102–3.

8 The Short Folder was registered in early 1913. Within months the Sopwith Aviation Company bought rights to the folding wing patent for a mere £15: R.D. Layman, *Naval Aviation in the First World War: Its Impact and Influence* (London: Chatham Publishing, 1996), p. 110.

9 Sir Richard Fairey quoted in Trippe, 'The plane maker', p. 193.

10 Ibid., pp. 94–9; Penrose, *British Aviation: The Pioneer Years*, pp. 450.

11 Barnes, *Shorts Aircraft since 1900*, 12–3, 92 and 16. One unsupported calculation of the original Rochester workforce was 12, but this seems unfeasibly low: King, *Knights of the Air*, p. 120.

12 Interview with Geoffrey Hall in Trippe, 'The plane maker', pp. 175–6.

13 Wright, for example, progressed to Caius College from Marlborough School, renowned for the unusually high quality of its maths teaching. Asa Briggs, *Secret Days Code-Breaking in Bletchley Park* (London: Frontline Books, 2011), pp. 36 and 46–7.

14 Hall, *Sir Richard Fairey*, p. 7; University of Cambridge Department of Engineering, '125 years engineering excellence': www-g.eng.cam.ac.uk/125; Correlli Barnett, *The Audit of War: The Illusion and Reality of Britain as a Great Nation* (London: Papermac, 1987), pp. 218–23. British universities' curricula of both pure and applied science contrasted with the narrow focus upon engineering evident in most *Technische Hoschulen*; as recognised in a succinct riposte of Barnett's argument: David Edgerton, *Science, Technology and the British Industrial 'Decline'* (Cambridge: Cambridge University Press, 1996), pp. 52–6.

15 Fairey, 'In the early days of flying.'

16 Sir Richard Fairey quoted in Trippe, 'The plane maker', pp. 179–80.

17 Maurice Wright quoted in Trippe, 'The plane maker', pp. 177–9.

18 Notably, P.G. Wodehouse, *Psmith in the City* (London: A. & C. Black, 1910).

19 Hall, *Sir Richard Fairey*, pp. 7–8; Trippe, 'The plane maker', pp. 181–2.

20 Nicholl was C-in-C Great Yarmouth; the Air Station's seaplanes proving vital for North Sea reconnaissance. Anon., 'A real Fairey story', *Flight*, 20 February 1941, p. 152; Barnes, *Shorts Aircraft since 1900*, pp. 7, 72–4, 76, 79, 85, 91 and 96; Trippe, 'The plane maker', pp. 186–8.

21 Barnes, *Shorts Aircraft since 1900*, p. 7; 'A real Fairey story', *Flight*, 20 February 1941, p. 152; Trippe, 'The plane maker', pp. 188–9.

22 One secondary source attributes the Dawson family wealth to profitable investment in the Hudson Bay Company. King, *Knights of the Air*, p. 130.
23 Dawson served with Wright at Gallipoli, albeit based on separate ships (the latter flying off the pioneering seaplane depot, HMS *Ben-My-Chree*): Lt Maurice Wright to CRF, 6 August 1915, Box 135, AC73/30, RAF Museum [RAFM].
24 Dawson's later life was spent in Nova Scotia, Canada. For this reason he sold the Heston estate adjacent to Bossington. Barnes, *Shorts Aircraft since 1900*, pp. 1 and 7; Trippe, 'The plane maker', p. 189; Mrs Jane Tennant, interview, 5 November 2013, and to the author, 19 March 2014 and 16 April 2016.
25 'Fairey always used to say that when Shorts told him he had to report to the Admiralty [in autumn 1914], they also told him that they were fed up with him too.' Geoffrey Hall, quoted in Trippe, 'The plane maker', p. 228.
26 Barnes, *Shorts Aircraft since 1900*, pp. 100–5.
27 Ibid.; Peter Fearon, 'The formative years of the British aircraft industry, 1913–1924', *The Business History Review*, 43:4 (1969), pp. 483–5; King, *Knights of the Air*, pp. 138–9, 142–3 and 148–50. From August to December 1914 Britain produced 99 aero-engines and the French 894; and in 1915 respectively 1,721 and 7,096, hence at the start of 1916 France supplying one-third of the RFC's engines. Malcolm Cooper, *The Birth of Independent Air Power* (London: Allen & Unwin, 1986), pp. 15–16.
28 Anon., 'X Aviation Company Limited', 31 July 1914, FAAM/28/2010/054/0121.
29 In 1914 a total of 18 British contractors built 193 machines, of which 60 were experimental types. In 1915 the early impact of subcontracting saw 34 companies construct 1,680 aircraft (710 of them Farnborough's basic scout design, the BE2c); but only five manufacturers produced more than 100: ibid., p. 15.
30 King, *Knights of the Air*, p. 150. Cooper, *The Birth of Independent Air Power*, p. 17. The British aviation industry employed approximately 1,000 workers in August 1914: ibid., p. 16.
31 C.R. Fairey, draft speech for anniversary dinner, 22 December 1949, FAAM /2010/053/0117. Hall, *Sir Richard Fairey*, p. 8; Geoffrey Hall, quoted in Trippe, 'The plane maker', pp. 228–9; Barnes, *Shorts Aircraft since 1900*, p. 1; *Fairey Company Profile 1915–1960*, ed. Martyn Chorlton (Cudham: Kelsey Publishing Group/Aeroplane, 2012), pp. 8–9. Trippe, Barnes and Chorlton all simply replicated what Hall remembered from his youth of how Fairey recalled his company's creation; that is, they drew on a single, yet chances are reliable, source to recycle what was after all an attractive and persuasive story.
32 Barnes, *Shorts Aircraft since 1900*, p. 104; King, *Knights of the Air*, p. 129.
33 Ibid.; Penrose, *British Aviation: The Pioneer Years*, p. 542; Barnes, *Shorts Aircraft since 1900*, pp. 51, 66 and 218.
34 Sueter, *Airmen or Noahs Fair Play for our Airmen: The Great 'Neon' Air Myth Exposed* (London: Sir Isaac Pitman & Sons, 1928), pp. 95, 331–2 and 390; Geoffrey Till, *Air Power and the Royal Navy 1914–1945* (London: Jane's

Publishing Company, 1979), pp. 111–3; Eric Grove, 'Air Force, Fleet Air Arm – or Armoured Corps? The Royal Naval Air Service at war' in *British Naval Aviation: The First 100 Years*, ed. Tim Benbow (Farnham: Ashgate, 2011), pp. 44–5; Barnes, *Shorts Aircraft since 1900*, p. 105.

35 Anon.,'X Aviation Company Limited', 31 July 1914, FAAM/28/2010/054/0121.
36 F.H.C. Rees to C.R. Fairey, 20 August 1914, FAAM/28/2010/054/0122.
37 Ibid.
38 F.H.C. Rees, 'Motor Company', 26 October 1916, FAAM/28/2010/054/0194; D.E.P. Cars Ltd, 'Prospectus' and 'Art Supplement', October 1916, FAAM/28/2010/054/0200 and 0198; F.H.C. Rees and C.R. Fairey 'Facroldaw', 26 October 1916, FAAM/28/2010/054/0194.
39 F.H.C. Rees to C.R. Fairey, 29 April and 11 May 1915, FAAM/28/2010/054/0152 and 0160.
40 F.H.C. Rees to C.R. Fairey, 24 January 1915 and draft prospectus for 'The Fairey Aviation Company Limited', 25 January 1915, FAAM/28/2010/054/0122 and 01225.
41 The RN gave Shorts a guarantee that subcontractors would receive the minimum technical information necessary to assemble the firm's patented machine: Director of Navy Contracts, Admiralty, to Short Brothers, 2 March 1915, Box 135 AC73/30, RAFM.
42 F.H.C. Rees to C.R. Fairey, 21 February and 9 April 1915, FAAM/28/2010/054/0131 and 0144. Ironically, 'The Fairey Aviation Company Limited' was already directly referred to in draft documentation.
43 F.H.C. Rees to C.R. Fairey, 24 and 28 January 1915, and draft prospectus for 'The Fairey Aviation Company Limited', 25 January 1915 , FAAM/28/2010/054/0122 and 01225.
44 C.R. Fairey to F.H.C. Rees, 29 January 1915 [second draft], FAAM/28/2010/054/0120. Fairey made reference to Article 14, but it was in fact Article 16 which banned any financial interest in a naval contractor except via a proxy director of a private joint stock company, as sanctioned by the Admiralty: Royal Navy, *King's Regulations and Admiralty Instructions of His Majesty's Naval Service Volume I* (London: HMSO, 1913), p. 4.
45 Facroldaw Syndicate initial agreement, 7 July 1915, F.H.C. Rees to C.R. Fairey, 11 May 1915 and C.R. Fairey to F.H.C. Rees, 30 March 1915, FAAM/28/2010/054/0185, 0159 and 0140.
46 F.H.C. Rees to C.R. Fairey, 9 April, 24 March and 11 May 1915, FAAM/28/2010/054/0145, 0132 and 0159.
47 Facroldaw Syndicate initial agreement, 7 July 1915, draft Syndicate Agreement, 15 July 1915, Syndicate Agreement, 23 July 1915, and R. Paterson to the Facroldaw Syndicate, 2 July 1915, FAAM/28/2010/054/0185, 0186, 0187 and 0184. The draft agreement is decorated with Fairey's carefully drawn propellors and floats, which constitute the unique doodlings of a highly focused man.

48 Sir Edward Beddington Behrens described himself as the 'Fairey Godfather' to CRF's first biographer. Trippe, 'The plane maker', p. 235.
49 'New companies registered', *The Aeroplane*, 6 August 1915.
50 Fairey Aviation press release, August 1915, and W. Broadbent, speech at Fairey Aviation fortieth anniversary lunch, 1955, quoted in Trippe, 'The plane maker', pp. 235 and 236; *Flight*, 20 February 1941, p. 152.
51 C.R. Fairey to C.J. Ritchie, 23 and 26 June 1937, and C.J. Ritchie to C.R. Fairey, 25 June 1937, FAAM/2010/054/0012; Mrs Jane Tennant to the author, 19 March 2014. Another of Fairey's longstanding associates, Edward Beddington Behrens, was descended from a prominent Jewish family in Manchester.
52 W. Broadbent, quoted in Trippe, 'The plane maker', p. 237; F.H.C. Rees to C.R. Fairey, 11 May and 10 June 1915, FAAM/28/2010/054/0175 and 0177. At its Woolston works in Southampton, the company founded by Noel Pemberton Billing and run by Hubert Scott Paine – from 1916, Supermarine Aviation – also undertook subcontract work for Short Brothers.
53 Barnes, *Shorts Aircraft since 1900*, p. 2; *Flight*, 20 February 1941, pp. 152–3.
54 Not to be confused with the far more influential Major T.M. Barlow, who joined Fairey Aviation as chief engineer in 1924 and joined the board when the company was floated five years later. Harald Penrose, *British Aviation: The Adventuring Years 1920–1929*, pp. 271–2 and 617. Re the recruitment of Tom Barlow, see Chapter 5.
55 Trippe, 'The plane maker', pp. 269 and 271–2; Taylor, *Fairey Aircraft since 1915*, p. 7; CRF's speech at Londonderry House, 22 December 1949, quoted in 'Twenty-five years of service', *The Aeroplane*, 6 January 1950.
56 Ibid. Stress-related ill health forced Barlow to resign from the board in March 1944: CRF to Major T.M. Barlow, 5 April 1944, FAAM/2010/054/0037.
57 Taylor, *Fairey Aircraft since 1915*, p. 2; Trippe, 'The plane maker', pp. 241–2; Mike Collier, 'Crucial role for Hayes in the First World War effort', www.middx.net/articles/munitions.
58 *Fairey Company Profile 1915–1960*, p. 9; Fairey Aviation photograph reproduced in Taylor, *Fairey Aircraft since 1915*, p. 3.
59 Michael Walker, 'Belgium refugees WW1 – Hayes Belgium refugees', 4 February 2014, *Hayes People History*, our history – Hayes.blogspot.com. AMLC's workers were later switched from reconditioning vehicles to building barges for use on Belgian canals behind the Allied front. Broadbent suggested the expatriate company managers were both inefficient and corrupt: Wilfred Broadbent, interviewed in Trippe, 'The plane maker', pp. 246 and 247. Two-thirds of Belgians in Britain were Flemish, which would include refugees from Louvain, and so communication at work must initially have been difficult. Tony Kushner, 'Local heroes: Belgian refugees in Britain during the First World War', *Immigrants & Minorities*, 18:1 (1999), p. 4.
60 King, *Knights of the Air*, pp. 146–7.

61 Ibid., p. 147.
62 Taylor, *Fairey Aircraft since 1915*, p. 51; H.F. King quoted in Trippe, 'The plane maker', pp. 245 and 248; *Fairey Company Profile 1915–1960*, p. 17.
63 Arthur Adams, statement for C.R. Fairey 'of the position regarding various contracts and payments received', 1 February 1917, Box 138 AC73/30, RAFM.
64 Taylor, *Fairey Aircraft since 1915*, pp. 52–3. For a succinct account of RNAS activities 1914–16, see Grove, 'Air Force, Fleet Air Arm – or Armoured Corps? The Royal Naval Air Service at War', pp. 27–40.
65 In 1914–16 the principal Sopwith models flown by the RNAS were the 860, Schneider, and Baby. Ibid., p. 40; Bramson, *Pure Luck*, p. 85; Bruce Robertson, *Sopwith – the Man and his Aircraft* (Bedford: The Sidney Press, 1970), p. 58. In 1915–16 the Sopwith Aviation workforce further impressed the Royal Navy by generously supporting the RNAS Comforts Fund: ibid..
66 Bramson, *Pure Luck*, pp. 93–4, 83–7 and 90–2; Robertson, *Sopwith – the Man and his Aircraft*, pp. 160, 164, 213–5, 67–8 and 81. On Beatty and the Grand Fleet's employment of the Pup and the Strutter for reconnaissance and anti-airship duties in the North Sea, 1917–18, see Layman, *Naval Aviation in the First World War*, pp. 113–15.
67 Ibid., p. 60.
68 'Aeronautical Inspection Directorate from 1939, and brief history 1914–1918', 1945, AVIA46/263, NA; Wilfred Broadbent, interviewed in Trippe, 'The plane maker', pp. 269–70.
69 C.R. Fairey to Wing Commander Randall RN, 17 March 1917, Box 67 AC73/30, RAFM.
70 Taylor, *Fairey Aircraft since 1915*, p. 7; 'De Havilland', *Flight*, 21 November 1930, p. 18.
71 Chief Industrial Commissioner's Department, 'The Fairey Aviation Company. Hayes and London District Committee's Aircraft Industry', 20 December 1916, LAB2/185/IC6452/1916, NA.
72 C.R. Fairey, lectures on 'Trade unionism and industry', Staff College Andover, 18 January and 18 November 1926, FAAM/2010/053/0030 and 2010/053/0029. The Amalgamated Engineering Union (AEU) was founded in 1920 when the Amalgamated Society of Engineers absorbed nine smaller organisations that represented skilled shop floor workers: www.archive.unitetheunion.org/about_us/history/history_of_aeeu.aspx.
73 Hall, 'Sir Richard Fairey', p. 8; Trippe, 'The plane maker', pp. 273–5 and 277; Sir Geoffrey Hall, quoted in ibid., pp. 276.
74 Robertson, *Sopwith – the Man and his Aircraft*, pp. 164, 213 and 60; and 158–9 re the contrasting experience of Sopwith's largest subcontractor, the agricultural machinery specialist Ruston Proctor of Lincoln.
75 Bransom, *Pure Luck*, p. 90; Robertson, *Sopwith – the Man and his Aircraft*, p. 59.

Chapter 4

1. The size of the workforce roughly matched that of the RAF's men under arms. The RAF (RFC/RNAS pre-1 April) flew 484,000 hours of combat in 1918. D. Edgerton, *England and the Aeroplane* (London: Penguin, 2013), p. 24; Peter Dye, 'The Bridge to Air Power – Aviation Engineering on the Western Front 1914–1918', *Air Power Review*, 17/2 (2014), pp. 12–13.
2. Adrian Smith, *'Mick' Mannock, Fighter Pilot: Myth, Life and Politics* (London: I.B.Tauris, 2001 and 2015), pp. 66–126.
3. Harald Penrose, *British Aviation: The Pioneer Years* (London: The Bodley Head, 1967), pp. 544–8; R.D. Layman, *Naval Aviation in the First World War* (Annapolis: US Naval Institute Press, 1996), pp. 110–2 and 72–5. Penrose profiled the Air Department's Technical Department and its senior technician Harris Booth, CRF's foil inside the Admiralty. For succinct overviews of naval air power 1914–18, see Mike Farquharson-Roberts, *A History of the Royal Navy World War I* (London: I.B.Tauris, 2014), pp. 136–54 and Grove, 'Air Force, Fleet Air Arm – or Armoured Corps? The Royal Naval Air Service at War', pp. 27–55.
4. M. Cooper, *The Birth of Independent Air Power* (New York: HarperCollins, 1986), pp. 63–6, 50–3 and 36–7.
5. On Sueter's success in Italy while still trying to influence policy at home, see Grove, 'Air Force, Fleet Air Arm – or Armoured Corps? The Royal Naval Air Service at War', pp. 44–5.
6. In 1936, 59% of 82 large companies surveyed had a 'dominant ownership interest' and 32% 'marginal', and in 1950 46% of 92 companies surveyed were still under family control: data quoted in Brian R. Cheffins, *Corporate Ownership and Control British Business Transformed* (Oxford: Oxford University Press, 2008), p. 12.
7. On Alfred Chandler's influential concept of 'personal capitalism', see ibid., pp. 221–4; and on shareholder passivity, both individual and institutional, see ibid., pp. 123–9, 293–6 and 373–4.
8. See Chapter 9. On the lack of adequate capital to support costly R&D, in the context of all the other problems facing rival manufacturers as Britain entered the jet age, see Keith Hayward, *The British Aircraft Industry* (Manchester: Manchester University Press, 1989), pp. 60–72. Unlike Germany, with financial institutions enjoying boardroom representation, British banks placed ease of liquidity above industrial assets. Cheffins, *Corporate Ownership and Control*, pp. 90–1. The classic account of short-termism's impact upon British manufacturing industry in the last century remains Will Hutton, *The State We're In* (London: Jonathan Cape, 1995), particularly Chapter 6.
9. 'Special Resolution of Fairey Aviation Company Limited', 15 February and 2 March 1917, and C.R. Fairey to Lord Weir, 21 December 1918, Box 67,

AC73/30, RAFM; F.H. Rees to C.R. Fairey, FAAM/2010/01/00? [John Fairey Collection].
10 As late as Christmas 1917, payment for the original Short 827 contract was still not complete. An outstanding debt of £29,122 was blamed for Fairey's suppliers delaying delivery until fully reimbursed: C.R. Fairey to Wing Commander Randall RN, Air Board Offices, 19 December 1917, ibid.
11 Cheffins, *Corporate Ownership and Control*, pp. 256–7; King, *Knights of the Air: The Life and Times of the Extraordinary Pioneers who First Built British Aeroplanes* (London: Constable, 1989), pp. 166–8; H. White Smith, SBAC chairman, to Rt Hon. J. Austen Chamberlain, Chancellor of the Exchequer, 1 March 1919, Box 137, AC73/30. CRF was a founding member of the SBAC's aircraft technical committee, as announced in *Cabinet-Maker*, 7 December 1918.
12 'Special Resolution…', 15 February and 2 March 1917, and anon., handwritten memo and list of weekly salary payments 2 December 1916–10 February 1917, 4 October 1917, ibid.
13 Sir Richard Fairey to C.O. Crisp, 9 November 1944, FAAM/2010/054/0037.
14 Ashurst, Morris & Co., 'Beazley v. The Fairey Aviation Company, Limited. Defence and Counter-Claim, 22 February 1917' and Henry J. Beazley, 'Beazley v. The Fairey Aviation Company Limited. Terms Settlement', 30 July 1917, Box 67, AC73/30, RAFM.
15 Grove, 'Air Force, Fleet Air Arm – or Armoured Corps? The Royal Naval Air Service at War', pp. 35–7.
16 Geoffrey Till, *Air Power and the Royal Navy 1914–45 A Historic Survey* (Colchester: TBS, 1979), pp. 112–15. Commodore Godfrey Paine, 'widely known as an air enthusiast of some standing', was appointed Fifth Sea Lord and Director of Air Services: ibid, p. 115.
17 Commodore Murray Sueter to Fairey Aviation Co. Ltd, 4 and 18 December 1915, ibid.
18 The F2 cost the RN £3,100 plus engine and armament. It would have been truly revolutionary had Fairey pursued his original idea of the two engines operating in tandem from inside the fuselage, driving the propellers via a chain-and-sprocket system. 'The Fairey Aviation Company. Approximate Prices' [1918?], Box 67, AC73/30, RAFM; *Fairey Company Profile 1915–1960*, ed. Chorlton, p. 17.
19 Norman Macmillan, interviewed in Peter Trippe, 'The plane maker: the official biography of Sir Richard Fairey', unpublished manuscript held by Mrs Jane Tennant, pp. 279–80; H.A. Taylor, *Fairey Aircraft since 1915* (London: Putnam, 1974), pp. 51–4. Macmillan was with the RFC on the Western Front flying 1½ Strutters in the spring of 1917 and thus had no direct acquaintance with the F2. Norman Macmillan, *Freelance Pilot* (London: Heinemann, 1937), pp. 27–9 and *Into The Blue* (London: Gerald Duckworth, 1929 and Jarrolds, 1969), pp. 129–31.

20 'Ant-dating the enemy', *The Aeroplane*, 16 January 1918 and 'Built before the Gothas: British bombers with folding wings', *The Graphic*, 23 March 1918.
21 C.R. Fairey, 'The Future of Aeroplane Design for the Services', lecture to RUSI, 11 February 1931, FAAM2010/53/0043.
22 Equally forgotten is a vast experimental bomber, construction of which Fairey Aviation contributed to in the winter of 1916–17. Chronically underpowered and scarcely able to leave the ground, the stranded leviathan was left to rot in a distant corner of Northolt airfield. Taylor, *Fairey Aircraft Since 1915*, p. 11. The so-called 'Kennedy Giant' is possibly the first in a lengthy list of abandoned mega-projects, including Howard Hughes's 'Spruce Goose', the Bristol Brabazon and the Saunders-Roe Princess.
23 Commodore Murray Sueter to Fairey Aviation Co. Ltd, 4 December 1915 and 16 September 1916, and correspondence between C.R. Fairey and Harris Burton, spring and summer 1916, Box 67, AC73/30, RAFM.
24 Director of Navy Contracts, Admiralty, to Fairey Aviation Co. Ltd, 20 January and 1 and 6 February 1917, Box 138 AC73/30, RAFM; *Fairey Company Profile 1915–1960*, ed. Chorlton, p. 17.
25 Taylor, *Fairey Aircraft since 1915*, pp. 7 and 61–6.
26 For bound patent specifications, 1910–39, see Box 07, volumes 1–5, RAFM. From mid-1915 to mid-1918 Fairey sought 15 patents, of which seven were not awarded. Each submission suggests an author confident in his knowledge of mechanics and an ability to engage in sustained experimentation, but there is little or no mathematical calculation to provide a theoretical underpinning. Was there an absence of necessary data, which might explain a near 50% failure rate? C.R. Fairey, 'Provisional Specifications and Letters Patents Vol. 1, 1910–1918', FAAM/2010/055/0037.
27 Patent Number 12,541. *Fairey Company Profile 1915–1960*, ed. Chorlton, pp. 11 and 20; Fairey Aviation claim to the Royal Commission on Awards to Inventors regarding "Hamble Baby" seaplanes, 1921, TS28/61, NA; ibid., quoted in Trippe, 'The plane maker', pp. 285–6.
28 Ibid.; Report of conference of Hamble Baby subcontractors and Commander Seddon, Bristol, 8 May 1917, Box 138, AC73/30, RAFM; Taylor, *Fairey Aircraft since 1915*, pp. 62–3.
29 Ibid., pp. 63 and 9.
30 C.R. Fairey to Director of Patents, Rewards and Royalties, Ministry of Munitions, 20 November 1919 and UK patent transfer agreement re 'Campania', Ministry of Munitions and Fairey Aviation Co. Ltd, [?] 1917, Box 137, AC73/30, RAFM.
31 Taylor, *Fairey Aircraft since 1915*, pp. 63 and 9; C.R. Fairey, draft article 'In defence of the seaplane', 192[?] and draft article 'The organisation of a flying service', 192[?] and proofs for 'Seaplanes' speech and final report, Air Ministry's Air Conference, 6–7 February 1923, FAAM/2010/53/0019 and 2010/53/0018, and FAAM/2010/53/0021.

32 'Fairey Aviation – production', *The Aeroplane*, 6 April 1917.
33 Taylor, *Fairey Aircraft since 1915*, pp. 61–2 and 10.
34 Sueter came to a verbal agreement with CRF and then on 2 June 1916 issued formal instructions to negotiate a contract: C.R. Fairey to Director of Admiralty Contracts, Admiralty, 1 June 1918, Box 137, AC73/30, RAFM.
35 Machines built for tropical use remained with RNAS coastal stations, but five cold weather Campanias saw action in 1919 with the North Russian Expeditionary Force. Taylor, *Fairey Aircraft since 1915*, pp. 54–60.
36 It was in fact Jellicoe who sailed to meet the High Seas Fleet without *Campania*, owing to a signalling error. The C-in-C was not disturbed by *Campania*'s absence as he mistakenly believed she was slower than his battleships. Beatty did have a carrier accompanying his Battle Cruiser Fleet, HMS *Engadine*. The latter, however, had no flight deck, meaning a long delay while spotter seaplanes were lowered into the water for take-off. Thus, throughout the battle, both admirals were handicapped by a lack of reliable, up-to-date intelligence as a consequence of poor or non-existent aerial reconnaissance. Farquharson-Roberts, *A History of the Royal Navy World War 1*, pp. 100–1; Layman, *Naval Aviation in the First World War*, pp. 172–80.
37 The Avro biplane used for the test landings was flown by Maurice Wright. C.O. Port Victoria to the Director of Air Services, Admiralty, 19 September 1916, in *Documents Relating to the Naval Air Service*, ed. Roskill, p. 384.
38 Fairey Aviation was recognised by the RN as a primary supplier of experimental aircraft, along with Sopwith and Handley Page. Preliminary Report of the Admiralty Committee on Deck Landing, December 1917, in ibid., pp. 589–91.
39 Layman, *Naval Aviation in the First World War*, pp. 102–5; Till, *Air Power and the Royal Navy*, pp. 60–2. To the Army's fury the Admiralty was still insisting in 1917 that the RN had first call on every Sopwith aircraft. All ship-borne action in 1917–18 was conducted by Sopwith machines and it was seven Camels that flew off *Furious* on 19 July 1918 to attack the Zeppelin sheds at Tondern. Grove, 'Air Force, Fleet Air Arm – or Armoured Corps? The Royal Naval Air Service at War', p. 40; Bruce Robertson, *Sopwith – The Man and his Aircraft* (Air Review Ltd, 1970), pp. 128–9.
40 'The Fairey Aviation Company. Approximate Prices' [1918?], Box 67, AC73/30, RAFM.
41 In conversation, CRF often expressed his contempt for Lloyd George, but readily conceded that he was an outstanding minister throughout the Great War: Mrs Jane Tennant to the author, 27 October 2016.
42 Cooper, *The Birth of Independent Air Power*, pp. 66–7, 78–81 and 86–95.
43 '…the RFC squadrons which received such a mauling over Arras as to earn the fourth month of 1917 the sobriquet "Bloody April" were arguably at their lowest state of operational effectiveness of the last two years of the war.' Ibid. The RFC saw 275 machines destroyed that month, suffering 421 casualties, of whom

207 were killed in action; from January to the start of June 1917 a total of 708 aircraft were lost. Paul Marr, 'Haig and Trenchard: achieving air superiority on the Western Front', *Air Power Review*, 17/2 (2014), p. 34.

44 Approximately 13,608 kg.
45 Achieving a top speed of 100 mph, the N4 prototypes were powered by four bilateral mounted Rolls Royce Condor engines, each generating 600 horsepower. One wry observation was that together the three planes travelled more miles by road than in the air. Taylor, *Fairey Aircraft since 1915*, pp. 81–4; 'A real Fairey story', *Flight*.
46 Ibid.; Performance graphs and [handwritten] engineering notes, 1917–18, Box 138, AC73/30, RAFM. Barlow may have been in charge of the N4 project but a press release made clear that CRF was 'personally responsible' for the design, 'assisted by the late Major Linton Hope': captions for photographs of Atalanta under construction in *Flight*, *The Aeroplane* and *Aeronautics*, January 1920.
47 Anonymous Fairey Aviation employee, quoted in Harald J. Penrose, *British Aviation The Adventuring Years 1920–1929* (London: Putnam, 1973), p. 94.
48 Penrose, *British Aviation The Adventuring Years 1920–1929*, pp. 94 and 521; Hollis Williams, quoted in ibid., pp. 364 and 209; Charles Lamb, *War in a Stringbag* (London: Cassell, 1977), p. 40. When Lobelle was made head of the design section in 1924, Ralli became head of the technical department. A number of managerial changes took place that year, all emanating from T.M. Barlow's appointment as chief engineer: 'A real Fairey story', *Flight*.
49 Taylor, *Fairey Aircraft since 1915*, pp. 81–4.
50 From the mid-1920s CRF became a regular guest at the London and North Eastern Railway's Felix Hotel and a member of the Felixstowe Golf Club – see Chapter 5.
51 See correspondence re Geoffrey Hall's post-school education at Finsbury Technical College and Imperial College of Science and Technology, FAAM/2010/054/0023.
52 Trippe, 'The plane maker', pp. 267–8; Fairey, 'In the early days of flying'; Mrs Jane Tennant, interview, 5 November 2013; conversation with Charles Fairey, 12 December 2013.
53 CRF's daughter believes he may have died an agnostic, not an atheist. She herself always maintained a close connection with the Church of England. Mrs Jane Tennant to the author, 27 October 2016.
54 CRF's daughter insisted that 'he was a deeply romantic man', as evident in his treatment of her mother, Esther Fairey: 'He loved to pet and spoil her, and for example every Christmas he would do a stocking for her himself, painstakingly wrapping every present and in the bottom there would be something very special from Asprey's.' Mrs Jane Tennant to the author, 27 October 2016.
55 Genealogical research into the Markey family, and profile of George Clarke, by Les Hewett, Marewa, New Zealand: received by the author with thanks, 29 July 2013.

56 Mrs Jane Tennant to the author, 27 October 2017; marriage certificate for C.R. Fairey and Q.H. Markey, 16 August 1915, St Martin Registry Office, London: courtesy of Les Hewett, 29 July 2013.
57 8 Greville Place was an early Victorian villa, on the site of which today is an apartment block.
58 Trippe, 'The plane maker', p. 266.
59 Mrs Jane Tennant to the author, 27 October 2016.
60 Description of Grove Cottage drawn from Trippe, 'The plane maker', pp. 319–23; Nikolaus Pevsner and Elizabeth Williamson, *The Buildings of England: Buckinghamshire* (London: Penguin, 1994), pp. 412–13.
61 Ibid., pp. 415–16.
62 Hall, 'Sir Richard Fairey', p. 8.
63 *Fairey Company Profile 1915–1960*, ed. Chorlton, pp. 18–9; Taylor, *Short Aircraft since 1915*, pp. 67–70. Armstrong Whitworth provided the catapult system: but Fairey patented means of improving upon the system, albeit not translating theory into practice: C.R. Fairey, Patent 4224 and 185821, 11 February 1920 and 8 September 1922, 'Provisional Specifications and Letters Patents Vol. 1, 1910–1918', FAAM/2010/055/0037.
64 C.R. Fairey, remarks in 'New competitor for Atlantic flight', *The Times*, 28 March 1919 and 'Atlantic airman of 21', *Daily Mail*, 21 April 1919 and miscellaneous newspaper cuttings re the Atlantic challenge, March–May 1921, FAAM/2010/050/001.
65 Taylor, *Short Aircraft since 1915*, pp. 71–5; 'The Schneider Cup race', *The Aeroplane*, 27 August 1919.
66 'Seaplanes for civil flying' and C.R. Fairey to the editor, *The Times*, 12 and 13 September 1919; C.G.G., 'Echoes from the Schneider Cup' and 'On the Schneider picnic', *The Aeroplane*, 24 and 17 September 1919.
67 'Seaplanes safe flying', *Manchester Guardian*, 9 July 1919, 'Modern British aeroplanes X. The Fairey Aviation Co. Ltd.', *Aeronautical Engineering* [*The Aeroplane* supplement], 6 August 1919, 'Flight – and the men', *Flight*, 14 August, 1919 and 'Pioneers of British aviation – XXVI Mr C.R. Fairey', *Aeronautics*, 22 January 1920.
68 'Post-Bellum policies', *The Aeroplane*, 16 April 1919 and 'Les premieres ailes variable', *The Aeroplane* [French edition], 24 December 1919; C.R. Fairey quoted in 'Pioneers of British aviation – XXVI Mr C.R. Fairey.'
69 Miscellaneous newspaper cuttings re Pickles's Tower Bridge flight and delivery of the *Evening News*, March and May 1921, FAAM/2010/050/001.
70 Taylor, *Short Aircraft since 1915*, p. 90; C.R. Fairey, remarks in 'Post-Bellum policies', and draft article 'In defence of the seaplane', 192[?], FAAM/2010/053/0019; Brett Holman, *The Next War in the Air: Britian's Fear of the Bomber, 1908–1941* (Abingdon: Routledge, 2014), p. 11. Holman suggests Murray Sueter as the first to use the term 'airmindedness', in 1928: Sueter, *Airmen or Noahs*, p. 296.

Re Geoffrey de Havilland's parallel thinking for land-based machines, see Sir Alan J. Cobham, *A Time To Fly* (London: Shepheard-Walwyn, 1978), pp. 60–2.
71 Taylor, *Short Aircraft since 1915*, pp. 75–6; C.G.G. 'Air Ministry communiqué no. 621' and 'A chance for draughtsman', *The Aeroplane*, 11 October 1920 and 14 September 1921.
72 Taylor, *Short Aircraft since 1915*, pp. 77–80.
73 Ibid., pp. 85–6; interview with Wilfred Broadbent in Trippe, 'The plane maker', p. 371.
74 Three of the civilian IIICs spread word of Fairey Aviation's engineering prowess in Scandinavia and Canada, with the fourth, as described in the text, sent to India. Taylor, *Short Aircraft since 1915*, p. 89.
75 Ibid., p. 98; 'Launch of A.N.A. 1.', *Flight*, 18 August 1921; miscellaneous newspaper cuttings re IIID handover at Hamble, August 1921, FAAM/2010/050/001. 8,568 miles of Australian coastline were circumnavigated in 44 days real time.
76 Norman Macmillan, *Freelance Pilot* (London: Heinemann, 1937), pp. 270–321. As to why 'The Fairey III.C was at that time the best available float-plane,' and how Macmillan's machine was modified before despatch from England, see ibid., pp. 272–3.
77 Norman Macmillan, *Into The Blue* (London: Duckworth, 1929), pp. 250 and 207.
78 Ibid., pp. 28–9; Macmillan, *Freelance Pilot*, pp. 272–3. With masterly understatement Macmillan recalled the Pintail II as 'rather a tricky aeroplane to handle': Macmillan, ibid., pp. 28–9. Taylor, *Short Aircraft since 1915*, pp. 90–3.
79 Interview with Norman Macmillan in Trippe, 'The plane maker', pp. 416–17; C.R. Fairey, draft memo [for SBAC?] re 'the combined opinions of the British designers as to present trend of design and the influence of certain Air Ministry restrictions', 1 January 1924, FAAM/2010/53/0022 and 'The Future of Aeroplane Design for the Services', RUSI illustrated lecture, 11 February 1931, FAAM/2010/53/0043; Penrose, *British Aviation: The Adventuring Years*, pp. 90–3 and 207.
80 Interview with Norman Macmillan in Trippe, 'The plane maker', p. 308. For a list of Fairey Aviation aircraft flown from Northolt, 1917–30 and 1953–4, see Ron Smith, *British Built Aircraft: Greater London* (Stroud: Tempus, 2002), pp. 194–5.
81 Trippe, 'The plane maker', pp. 294–5.
82 Interviews with Norman Macmillan and Wilfred Broadbent, ibid., pp. 309 and 304–5.
83 *Fairey Tales*, January 1918, FAAM/2010/057/0001-2. Around a quarter of female competitors were Belgian, and the men's veteran races were handicapped, with CRF noticeable by his absence from the staff tug of war team: 'Programme for [Fairey Aviation] Garden Party and Sports…', 14 July 1917, FAAM/2010/057/0069.

84 Statement of Fairey Aviation bank balance, 1917–18, Box 67, AC73/30, RAFM. One Friday afternoon in the spring of 1917 the company's bank released cash for the weekly wages only an hour before the workforce clocked off: Wilfred Broadbent quoted in Trippe, 'The plane maker', p. 292.

85 Fairey Aviation and Ministry of Munitions, 'Capital Expenditure on New Building', 30 June 1918, Box 67, AC73/30, RAFM. Captain R.H. Stainforth, Director A.F. (1) to Fairey Aviation, 24 and 30 May and 19 August 1918; Cyril Watts, Asst. Controller Aircraft Finance, to Charles Crisp, 14 December 1918; and C.R. Fairey to B. Holloway, Director of Munitions Finance, 23 August 1918, ibid.

86 C.R. Fairey to Director of Air Contracts, Admiralty, 1 June 1918 and to H.C. Baldwin, Ministry of Munitions, 15 August 1918, and Controller of Aircraft Finance, Ministry of Munitions, to A.G. Hazell, 6 July 1918, Box 137, AC73/30, RAFM.

87 C.R. Fairey to Lord Weir, 21 December 1918, ibid. Dawson's net salary totalled £1,500 and Crisp's post-tax remuneration was £1,000; i.e. the company covered its directors' income tax obligation. A.G. Hazell, 'Overhead charges of the Fairey Aviation Co Ltd', [?] 1920, ibid. Based simply on RPI measured inflation Fairey's salary equivalent in 2017 would be around £85,000, but if measured by economic status or economic power then his income today would be several hundred thousand pounds more.

88 A.G. Hazell, handwritten table of C.R. Fairey's net income and tax paid 1915–20; C.R. Fairey deposit account statement, Barclays Bank Charing Cross Branch, December 1921–June 1922: total outgoings of £56,525 and total income including interest of £63,983 6 shillings 8 pence. Box 143, AC73/30, RAFM.

89 £1,752 8 shillings 2 pence total personal tax paid 1917–20: A.G. Hazell, handwritten table of C.R. Fairey's net income and tax paid 1915–20, ibid.; 'Order of the British Empire': G. Whickhald [Home Office] to C.R. Fairey, 16 January 1920, FAAM/2010/054/0? [held at Bossington].

90 'Statement week ending 16th November 1918' and Statement of Fairey Aviation bank balance, 1917–18, Box 137, AC73/30, RAFM.

91 A.G. Hazell, 'Overhead charges of the Fairey Aviation Co Ltd', [?] 1920, ibid.. W.B. Peat and Co. to H. Buxton, Inspector of Taxes, 25 February 1920; A.G. Hazell to C.O. Crisp, 8 March 1920; and Sir William Peat to C.R. Fairey, 20 May 1920, ibid.. W.B. Peat and Co. served as both company accountants and auditors.

92 C.R. Fairey to Asst. Controller of Supplies (Seaplanes), 22 February 1919; A.G. Hazell, 'Employees engaged directly on aircraft', 13 December 1919', 'Statement week ending 16th November 1918', notes on company finances, 8 January 1919, ibid..

93 'Summary of amount due to the company on 28th January 1920' and 'FAC as at June 30th 1920', 30 June 1920, ibid.; interview with Wilfred Broadbent in Trippe, 'The plane maker', p. 371.

94 Holman, *The Next War in the Air*, pp. 6–7. On the Trenchard–Hoare partnership, see Viscount Templewood [Sir Samuel Hoare], *Empire of the Air: The Advent of the Air Age 1922–1929* (London: Collins, 1957), pp. 35–84. In early 1921 Fairey lobbied energetically for the restoration of a separate Air Ministry, if only to loosen at last the dead hand of the Army: C.R. Fairey to the editor, *The Times*, 28 January 1921.

95 Principal benefactors of the ADC were Handley Page, De Havilland and Bristol – but not Fairey. The disposal abroad of these firms' aircraft provided access to overseas markets, reinforced by HMG transferring surplus machines to the dominions. Fearon, 'The formative years of the British aircraft industry', pp. 492 and 490. Several of Fairey's Type IIIAs and IIBs, without engines were bought by the Dispersal Board and then burnt: interview with Wilfred Broadbent in Trippe, 'The plane maker', pp. 373–4. For the most detailed account of the postwar travails of the British aircraft industry, see Penrose, *British Aviation: The Adventuring Years 1920–1929*, pp. 3–260.

96 Short Brothers constructed only 11 aircraft between 1920 and 1924, surviving courtesy of a contract to build omnibus bodies. Frederick Handley Page floated his company to raise necessary capital for diversifying into vehicle manufacture, with disastrous consequences. Ibid., p. 490; Hayward, *The British Aircraft Industry*, p. 12.

97 Andrew Boyle, *Trenchard* (London: W.W. Norton & Co., 1962), pp. 401–6.

98 Fairey and Charles Ltd was launched with capital of £50,000 in £1 shares, the company's founders being named as joint managing directors. 'New company registered', *Flight*, 22 May 1919; 'A post-war product', *Aeronautical Engineering*, 28 January 1920.

99 Agenda for Fairey Aviation board meeting, 3 December 1919, Box 67, AC73/30, RAFM. Miscellaneous correspondence between C. Crisp and C.R. Fairey and the managers of Lloyd Bank High Holborn Branch and London County Westminster and Parr's Bank Ltd., Bartholomew Lane Branch, London, November 1919–October 1921, Box 135, AC73/30, RAFM. Fairey Aviation borrowed from the second bank on the strength 'of your manufacturing success' without informing its principal lender and in March 1920 paid Fairey and Charles workers via a small clearing bank in Southall, which then drew the sum paid out upon Fairey's principal account at Lloyd's in High Holborn. The latter's manager was clearly incandescent, sending a brusque instruction to CRF on 5 March 1920: 'Please call.' Fairey Aviation Company, 'Notes on statement', [?] 1920, ibid.

100 Taylor, *Short Aircraft since 1915*, p. 96; Wilfred Broadbent, anniversary dinner speech, 22 December 1949, quoted in Trippe, 'The plane maker', p. 420; Penrose, *British Aviation: The Adventuring Years 1920–1929*, p. 93.

101 Lord Brabazon to Sir Richard Fairey, 20 March 1945 and 14 April 1947, AC71/3, LBP, RAFM.

102 Professor S. Clair Thomson, medical certificate for C.R. Fairey, King's College Hospital, 14 December 1920, ibid.
103 A. Bramson, *Pure Luck* (Manchester: Crecy, 1990), pp. 110–17. The pivotal year for Sopwith – and the industry as a whole – was 1935, when his acquisition of Armstrong Siddeley, A.R. Roe and Gloster created the first incarnation of Hawker Siddeley.
104 Peter Fearon, 'The formative years of the British aircraft industry' *The Business History Review* 43, (1969) pp. 493–4; Hayward, *The British Aircraft Industry*, 13; C.R. Fairey to the editor, *Aeroplane*, 8 February 1928. For the suggestion that Fearon and later Correlli Barnett, painted too bleak a picture of the civil–military conundrum following World War I, see Edgerton, *England and the Aeroplane*, pp. 28–32 and 43–4.
105 Hayward, *The British Aircraft Industry*, p. 12; King, *Knights of the Air*, pp. 237–9.
106 Bramson, *Pure Luck*, 111; A.A. Amos, 'Liquidator's statement of account: The Fairey Aviation Company Limited', 9 March 1921, BT34/4112/141133, NA.
107 Ashurst, Morris, Crisp & Co., 'Memorandum and Articles of Association of the Fairey Aviation Company Limited', 9 March 1921, ibid.; Taylor, *Short Aircraft since 1915*, p. 13; Advice to the Air Ministry of F.W. Lawton, on behalf of the Treasury Solicitor, 27 May 1929, TS28/171, NA. Re restructuring of the company in the 1920s, see Chapter 5.
108 'The Fairey Aviation Co., Ltd.', *Flight*, 16 March 1921; Fairey Aviation advertisement, ibid., 15 September 1921; 'A chance for draughtsmen', *The Aeroplane*, 14 September 1921; 'The Fairey Aviation Co., Ltd.', *Aeronautical Engineering* [multi-language supplement to *The Aeroplane*], 9 November 1921.

Chapter 5

1 The Admiralty and Ministry of Information, *Fleet Air Arm* (London: HMSO, 1943), pp. 65–70; John Moffatt with Mike Rossiter, *I Sank the Bismarck Memoirs of a Second World War Navy Pilot* (London: Bantam Press, 2009), pp. 100–1 and 216–35.
2 H.A. Taylor, *Fairey Aircraft since 1915* (London: Putnam, 1974), pp. 98–101; 'He [Coutinho] was really an old man. But what a marvellous navigator. The combined age of himself and his pilot was 120 or something like that! But he was wonderful.' Wilfred Broadbent, quoted in Trippe, 'The plane maker', p. 434. The initial aircraft, prepared by Fairey in secret to a four-month deadline, bore the unfortunate name of 'Lusitania.'
3 Miscellaneous press cuttings, April–June 1922, FAAM/2010/050/002.
4 Taylor, *Fairey Aircraft since 1915*, pp. 101–2.
5 Norman Macmillan, quoted in Penrose, *British Aviation: The Pioneer Years*, pp. 211–12. Letters of condolence to CRF re death of Vincent Nicholl, and replies, FAAM2010/054/0023.g

6 'A Real Fairey Story', *Flight*, 20 February 1941; Penrose, *British Aviation: The Adventuring Years 1920–1929*, pp. 271–2. As well as two Barlows, to confuse matters further Macmillan's deputy was C.R. McMullin. Re the consequences of Major Barlow's arrival for management of the design section and technical department, see Chapter 4.

7 For a detailed reconstruction of the Cape Flight, placing the Fairey IIIDs at the heart of the account, see Julian Lewis, *Racing Ace The Fights and Flights of Samuel 'Kink' Kinkead DSO DSC* DFC** (London: Pen & Sword, 2011), pp. 119–39.

8 Taylor, *Fairey Aircraft since 1915*, pp. 102–3; miscellaneous press cuttings, April–May 1926, FAAM/2010/050/002; 'End of 14,000 miles flight led by Wing-Commander Pulford, the four R.A.F. seaplanes arrive home, having flown from Cairo to the Cape and back', British Pathé, 26 June 1926, www.britishpathe.com/video/end-of-14-000-miles-flight-cuts/query/Wing-Commander. 'Congratulations ... a magnificent show you have put up in our machines': C.R. Fairey to Wing Commander C.W.H. Pulford, 17 May 1926, FAAM/2010/054/0063. Ironically, Pulford was not the firm's preference for CO of the 'Cape Flight': Vincent Nicholl to CRF, 15 October 1925, FAAM/2010/054/0020.

9 See press reception of the Fairey IIID at the 1922 Aero Show, Olympia, for example, 'New type of amphibian aeroplane', *The Times*, 17 January 1922 and predictably, 'A real amphibian' and 'A Fairey triumph', *The Aeroplane*, 11 January 1922.

10 Norman Macmillan, quoted in Trippe, 'The plane maker', p. 452; CRF, draft memo re 'the combined opinions of the British designers as to present trend of design and the influence of certain Air Ministry restrictions', 1 January 1924, FAAM/2010/53/0022. R&D of the Fawn lasted from 1921 to 1924, when the first of 75 production machines came off the assembly line.

11 Ibid.; *Fairey Company Profile 1915–1960*, ed. Chorlton, pp. 30–1; Hall, *Sir Richard Fairey*, p. 8; Taylor, *Fairey Aircraft since 1915*, pp. 113–26.

12 Briefing notes and specification: Fairey Flycatcher, 1923, Box 2, AC73/30, RAFM; Norman Macmillan, quoted in ibid., p. 118; Peter Lewis, *The British Fighter since 1912 Fifty Years of Design and Development* (London: Putnam, 1965), p. 143. Half a century later, John Fairey oversaw construction of a replica Flycatcher.

13 For an argument that interwar FAA aircraft following the Flycatcher scored poorly on both looks *and* performance and an analysis of why, see Geoffrey Till, *Airpower and the Royal Navy 1914–1945, A Historic Survey* (Colchester: TBS, 1979), pp. 86–104.

14 The Sea Fury was too late to see active service during World War II of which over half the FAA's aircraft were American, all of which outperformed their British counterparts; Camm's was a unique carrier design, but technically indebted to

the RAF's Tempest. Ibid., pp. 108–9; A. Bramson, *Pure Luck* (Manchester: Crecy, 1990), pp. 275–6.
15 C.R. Fairey, draft article 'In defence of the seaplane', 192[?], FAAM/2010/53/0019; Taylor, *Fairey Aircraft since 1915*, pp. 127–9.
16 C.R. Fairey to Rita Heelard [*Daily Express*], 5 November 1929, FAAM/2010/054/0028.
17 Proofs for 'Seaplanes' speech and final report, Air Ministry's Air Conference, 6–7 February 1923, FAAM/2010/53/0021; 'Air Conference 1923 programme and Outline of Papers', 6–7 February 1923, Box 27, AC73/30, RAFM.
18 Penrose, *British Aviation: The Adventuring Years*, pp. 211–12.
19 'Cato', *Guilty Men* (London: Victor Gollancz, 1940). Sir Samuel Hoare was ambassador in Madrid from June 1940 until the end of World War II: Viscount Templewood, *Ambassador on Special Mission* (London: Collins, 1946).
20 C.R. Fairey to Sir Samuel Hoare, [?] January 1924, vil (56), Templewood Papers, Cambridge University Library.
21 Templewood [Sir Samuel Hoare], *Empire of the Air*, pp. 39–46 and 52–3.
22 Andrew Boyle, *Trenchard* (London: Collins, 1962), pp. 448–9 and 513; Lords Trenchard and Brabazon, quoted in J.A. Cross, *Sir Samuel Hoare, a Political Biography* (London: Jonathan Cape, 1977), pp. 87 and 88
23 Ibid., pp. 88–91 and Templewood, *Empire of the Air*, pp. 48–52, 179–81, 187–8 and 196–9.
24 Ibid., pp. 196–9. Preceded by professorial appointments, an Aeronautics sub-department was finally established in 1932: '125 Years Engineering Excellence': www-g.eng.cam.ac.uk/125.
25 For example, from late 1925 Fairey Aviation's provision of dinner at The Carlton, *No, No, Nanette* at The Palace, and supper at The Savoy, with none of the RAF guests (with wives) ranked higher than wing commander. Costings and guest list for 18 December 1925 function, Box 143, AC73/30, RAFM.
26 'François Latry', www.cooksinfo.com/francois-latry.
27 C.R. Fairey to H.G. Wells, 29 November 1927, H.G. Wells Archive, University of Illinois, Champaign-Urbana; H.G. Wells to C.R. Fairey, [?] December 1927, FAAM/2010/054/0026.
28 Damian Collins, *Charmed Life: The Phenomenal World of Philip Sassoon* (London: William Collins, 2016), pp. 178–96.
29 Templewood, *Empire of the Air*, pp. 96 and 173–4; M.H., 'Lady Maud Hoare's account of the first flight to India by a woman', ibid., pp. 298–300; S.H., 'The beginnings of the African Air Service: Report by Sir Samuel Hoare', ibid., pp. 301–8. Pathé's extensive coverage of 'our flying Air Minister' aptly concluded in spring 1929 with Hoare (and Brancker) bidding farewell to Imperial Airways' inaugural air mail service to India: '7 Days to India', Pathé News, 4 April 1929, www.britishpathe.com/video/to-india-in-7-days.

30 Taylor, *Fairey Aircraft since 1915*, pp. 144–66. Between the wars more Fairey IIIFs were built pre-rearmament than any British combat aircraft other than the Hawker Hart.

31 Passenger numbers and air freight trebled in the same period. Cross, *Sir Samuel Hoare*, pp. 101–2; Templewood, *Empire of the Air*, pp. 173–6; S.H., 'The beginnings of the African Air Service', ibid., pp. 304–5.

32 *The Aeroplane*, 12 June 1929, and J.T.C. Moore-Brabazon to Sir Samuel Hoare, 13 June 1929, quoted in Cross, *Sir Samuel Hoare*, pp. 103 and 108; ibid., pp. 100 and 108–9. For a later critique of Trenchard's and Hoare's policies, not least re the FAA, see H.R. Allen, *The Legacy of Lord Trenchard* (London: Cassell, 1972), pp. 34, 43 and 46–59. Hoare never succeeding in rendering Fairey less hostile to public subsidy of Imperial Airways: C.R. Fairey to Rita Heelard [*Daily Express*], 5 November 1929, FAAM/2010/054/0028.

33 Alan Cobham and Christopher Derrick, *A Time to Fly: Memoirs* (London: Shepheard-Walwyn, 1978), pp. 126–49.

34 Ibid., pp. 84–94; Penrose, *British Aviation: The Adventuring Years*, p. 349. For Pathé newsreel coverage of Brancker and Cobham leaving for India, and returning home, see www.britishpathe.com/video/blazing-the-trail and www.britishpathe.com/video/record-flight-in-british-plane.

35 Cobham and Derrick, *A Time to Fly*, pp. 94–122 and 1–2. 'Triumph for British aviation', Pathé News, 26 October 1926, www.britishpathe.com/video/triumph-for-british-aviation-version-1-of-2.

36 The RAF Museum's 2014–15 exhibition 'Sir Alan Cobham's flying circus: A life of a pioneering aviator' acknowledged Fairey Aviation's contribution to Cobham's 1926 return flight to Australia, but the posthumous edited version of his memoirs mentions only the contribution of Oswald Short in converting the DH50 into a seaplane.

37 Bramson, *Pure Luck*, p.119; *Fairey Company Profile 1915–1960*, ed. Chorlton, p. 44; Taylor, *Fairey Aircraft since 1915*, pp. 180–1; Penrose, *British Aviation: The Adventuring Years*, pp. 560–1 and 628–9; Trippe, 'The plane maker', 692–4. That Fairey Aviation was the RAF's preference, see Group Captain H.W. Cave to C.R. Fairey, 20 November 1926, FAAM/2010/054/0027.

38 Fairey Aviation agreed with the Air Ministry a price of £15,000 for the airframe, with Napier supplying the 'broad arrow' Lion XIA engine. The monoplane was nicknamed the 'Eversharp' after the maker of slimline pens and propelling pencils. *Fairey Company Profile 1915–1960*, ed. Chorlton, p. 44; Taylor, *Fairey Aircraft since 1915*, pp. 180–1; C.R. Fairey, draft article on streamlining for *The Times* aircraft supplement [published 28 February 1929], 25 February 1929, FAAM/2010/054/0029. For Pathé film footage of the aircraft, see 'Royal Air Force's latest', British Pathé, 24 January 1929, www.britishpathe.com/video/royal-airforces-latest.

39 Penrose, *British Aviation: The Adventuring Years*, pp. 600–3 and 627–9; letters of congratulation to and from C.R. Fairey re the Cranwell-Karachi flight, May 1929, FAAM/2010/054/0029.
40 The most detailed, technically dense, account of the long-range monoplane's design, development and performance flying to India was given by Fairey to the Royal Aeronautical Society in the autumn of 1929. The length of CRF's responses to audience contributions by Handley-Page et al. suggest he took very full notes while listening to his peers. This was a commentary by an engineer for engineers, as confirmed by the accompanying graphs and tables: 'Proceedings: C.R. Fairey, "Range of aircraft", 10 October 1929', *The Journal of the Royal Aeronautical Society*, 34/231 (1930), pp. 220–56.
41 Taylor, *Fairey Aircraft since 1915*, pp. 180–1. Saint-Exupéry flew North African routes for Aéropostale from 1926 to 1929. He survived an air crash in the Libyan desert on 30 December 1935, justifying the unseasonal long-distance flight on the basis of his credentials as a navigator and those of Prévot, his veteran mechanic: Antoine de Saint-Exupéry, *Wind, Sand and Stars* [*Terre des Hommes*, 1939] (London: Penguin, 1995).
42 Crash inquiry papers, 1929–30, in 'Development and trials: Fairey postal aircraft: Design Branch Specification 33/27', 1927–1930, AVIA2/417, NA; Geoffrey Hall interviewed, in Tripp, 'The plane maker', p. 686.
43 Crawnwell to Walvis Bay constituted a distance of 5,309 miles, flown in 57 hours 25 minutes. Plans for a modified aircraft with a diesel engine and retractable undercarriage were abandoned when a joint RAF/RAE concluded a new state-of-the-art machine was more cost effective. Taylor, *Fairey Aircraft since 1915*, pp. 188–92; Penrose, *British Aviation: The Adventuring Years*, pp. 697–8. For composite film footage of both monoplanes, including Pathé's sound report of the 1933 flight to and from southern Africa, see www.youtube.com/watch?v=qGepvIdbad8.
44 Taylor, *Fairey Aircraft since 1915*, pp. 135–6; Penrose, *British Aviation: The Adventuring Years*, p. 240; King, *Knights of the Air*, p. 234; C.R. Fairey, 'The Schneider Cup', lantern lecture in Felixstowe [previously at the Royal Aero Club], FAAM/2010/53/0024.
45 Ibid.
46 Taylor, *Fairey Aircraft since 1915*, pp. 135–6; Hall, *Sir Richard Fairey*, p. 8.
47 Penrose, *British Aviation: The Adventuring Years*, pp. 269–70; 'Glenn Curtiss in Buffalo, New York', www.buffaloah.com/h/aero/curt/#Anchor. Was it a positive signal for CRF that the 'S' in S. Albert Reed (awarded the prestigious Collier Trophy for his propeller design in 1925) stood for 'Sylvanus'?
48 C.R. Fairey, 'Speech at the Keys Lunch', 6 July 1926, FAAM/2010/53/0026. Clem Keys and his wife Indiola became close friends of CRF and his second wife, Esther. Keys lost much of his fortune as a consequence of the Wall Street Crash. Mrs Jane Tennant to the author, 27 October 2016.

49 Wilfred Broadbent recollection of Fairey's public rejection of and apology for the Fawn at a Grove Cottage lunch party in early 1924, in Trippe, 'The plane maker', pp. 465–9.
50 Hall, *Sir Richard Fairey*, p. 8. Lord Thomson and the 1924 Air Estimates: Penrose, *British Aviation: The Adventuring Years*, pp. 274–7.
51 Ian Huntley, 'Fairey's elusive Fox', *Aeroplane Monthly*, January 1979.
52 Norman Macmillan, quoted in Penrose, *British Aviation: The Adventuring Years*, p. 271. C.R. Fairey to Air Vice Marshal Sir John Higgins, 16 March 1926, FAAM2010/054/0063.
53 King, *Knights of the Air*, pp. 234–5.
54 Huntley, 'Fairey's Elusive Fox.'
55 C.R. Fairey to Air Vice Marshal Sir John Higgins, 16 March 1926, FAAM2010/054/0063.
56 Mrs Jane Tennant to the author, 27 October 2016.
57 Passenger lists for *Aquitania*, arrive New York 25 September 1925 and *Berengaria*, arrive Southampton 3 November 1925, National Archives, via www.ancestry.co.uk; Duff Cooper, *The Duff Cooper Diaries*, ed. John Julius Norwich (London: Weidenfeld & Nicolson, 2005), 24 November 1923, p. 182.
58 C.R. Fairey to Vincent Nicholl, 1 and 9 October 1925, Box 2, AC73/30, RAFM; 'The 1925 Schneider Trophy Race', *Flight*, 29 October 1925. Did seeing the 1925 Broadway show spur CRF to enjoy again the West End production of *No, No, Nanette*? See footnote 25.
59 Hall, *Sir Richard Fairey*, p. 9. Sums quoted courtesy of Norman Macmillan, in Penrose, *British Aviation: The Adventuring Years*, p. 271.
60 Wilfred Broadbent, quoted in Trippe, 'The plane maker', pp. 477–8. 'I would say this account is entirely accurate. Mr Broadbent was not a man to invent fables, and I heard the story many times *en famille*. Stateroom cabins were large in those days.': Mrs Jane Tennant to the author, 27 October 2016.
61 Huntley, 'Fairey's elusive Fox.'
62 C.R. Fairey to Air Vice Marshal Sir John Higgins, 10 May 1926, FAAM/2010/054/0027 [letter not sent].
63 Huntley, 'Fairey's elusive Fox.'
64 C.R. Fairey to Sir Vyell Vyvyan, 22 February and 26 March 1926, and to W.L. Fisher, 19 and 25 June and 17 July 1926, FAAM/2010/054/0063.
65 C.R. Fairey to Major (rtd.) C.C. Turner, 16 April 1925, FAAM/2010/054/0020
66 C.R. Fairey to Air Chief Marshal Sir Hugh Trenchard, 29 July 1925 and to Major [?] Innes [Air Ministry], 31 July 1925, ibid. With Fairey Aviation holding an option on Curtiss's new V-1400, Hoare was 'quite horrified at the idea of another American engine coming over but it has got to be done if for no other reason that the question of keeping our people up to the mark.' C.F. Fairey to Air Vice Marshal Sir John Higgins [AOC, Iraq], 16 March 1926, FAAM/2010/054/0063.

67 Taylor, *Fairey Aircraft since 1915*, p. 136; Lt-Col. L.R.F. Fell, quoted in King, *Knights of the Air*, p. 235.
68 Ibid., pp. 234–5; Penrose, *British Aviation: The Adventuring Years*, pp. 362–3; miscellaneous correspondence between C.R. Fairey and H.T. Vane [Napier] and Basil Johnson [Rolls Royce], 1926–7, FAAM/2010/054/0063 and 0026; 'Mr Basil Johnson', *Flight*, 29 April 1926; Basil Johnson to C.R. Fairey, 3 October 1927, FAAM/2010/054/0024; C.R. Fairey to Basil Johnson, 3 January 1928 and 6 and 10 October 1928, FAAM/2010/054/0026.
69 For a full account of the qualification races and Supermarine's success in ultimately winning the 1927 Schneider Trophy, see Lewis, *Racing Ace: The Fights and Flights of Samuel 'Kink' Kinkead DSO, DSC, DFC* (Barnsley: Pen & Sword Aviation, 2011), pp. 140–99.
70 Harold E. Perrin, 'The Royal Aero Club, Schneider Cup Race', Royal Aero Club, 7 September 1927; Maurice Wright to C.R. Fairey, 21 September 1927 and C.R. Fairey to Rt Hon. Sir Philip Sassoon, 7 September 1927 and to R.J. Mitchell, 27 September 1927 [telegrams], FAAM/2010/054/0024; p. 182; Collins, *Charmed Life The Phenomenal World of Philip Sassoon*, p. 182; James Bird [MD, Supermarine] to C.R. Fairey, 30 August 1927, FAAM/2010/054/0023.
71 Templewood, *Empire of the Air*, p. 206; Lewis, *Racing Ace*, pp. 190–1; Air Ministry invitation and table plan for Schneider Trophy luncheon and reception, The Savoy, 4 October 1927, FAAM/2010/054/0023; Speech re Royal Aero Club appeal for £100,000 National Flying Fund, 6 July 1926, FAAM/2010/53/0027.
72 Lewis, *Racing Ace*, pp. 194–7.
73 R.J. Mitchell to C.R. Fairey, 10 August 1928, FAAM/2010/054/0028; 'A real Fairey story', *Flight*. As early as spring 1926 Fairey Aviation were testing more durable, higher tensile alloys to replace duralumin, i.e. produce post-Reed patented propellers: C.R. Fairey to Air Vice Marshal Sir Geoffrey Salmons, 21 May 1926, FAAM2010/054/0063.
74 King, *Knights of the Air*, pp. 250–4.
75 Ibid., pp. 254–69; Templewood, *Empire of the Air*, p. 208; Hayward, *The British Aircraft Industry*, p. 14.
76 'Private and Confidential', C.R. Fairey to C.O. Crisp, 17 March 1926, FAAM2010/054/0063.
77 C.R. Fairey to Major (rtd) C.C. Turner [air correspondent, *Daily Telegraph*] and to Major (rtd) Oliver Stewart [*Morning Post*], 1 and 4 March 1926, FAAM/2010/054/0063. CRF asked a sympathetic but unhelpful C.G. Grey if he could 'muzzle' Geoffrey Dorman, a freelance journalist trenchant in his criticism of Fairey when writing for the *Evening News* if not *The Aeroplane*: C.R. Fairey to C.G. Grey, 10 and 14 December 1925, FAAM/2010/054/0088.
78 Taylor, *Fairey Aircraft since 1915*, pp. 12 and 133; interview with David Hollis Williams, quoted in Penrose, *British Aviation: The Adventuring Years*, pp. 364–5.

79 Norman Macmillan, quoted in ibid., pp. 331–2.
80 Interview with David Hollis Williams, quoted in ibid., pp. 364–5; *Fairey Company Profile 1915–1960*, ed. Chorlton, pp. 37 and 36.
81 Norman Macmillan, *The Aeroplane*, 22 July 1955, quoted in Taylor, *Fairey Aircraft since 1915*, pp. 137–8.
82 Lord Trenchard, quoted in Andrew Boyle, *Trenchard* (London: Collins, 1962), p. 56.
83 C.R. Fairey to Air Vice Marshal Sir John Higgins, 16 March 1926, FAAM/2010/054/0063. Norman Macmillan, quoted in Taylor, *Fairey Aircraft since 1915*, pp. 137–8 and Penrose, *British Aviation: The Adventuring Years*, p. 570.
84 C.R. Fairey to Air Vice Marshal Sir Geoffrey Salmond, 26 March 1926 and to Rt Hon. Sir Samuel Hoare, 30 June 1926, FAAM/2010/054/0063.
85 C.R. Fairey to Air Vice Marshal Sir Geoffrey Salmond, 26 March 1926, ibid.; Norman Macmillan, quoted in Penrose, *British Aviation: The Adventuring Years*, p. 581; Trippe, 'The plane maker', pp. 501–2; Huntley, 'Fairey's elusive Fox.' Worldwide over 2,800 Hart variants were built by Hawker Engineering and its subcontractors at home and abroad: Bramson, *Pure Luck*, pp. 121–2. The Fox's test pilot believed it lost out to the Hart because the latter 'scored heavily on ease of maintenance': Norman Macmillan, quoted in Trippe, 'The plane maker', p. 500.
86 Hall, *Sir Richard Fairey*, p. 9.
87 Taylor, *Fairey Aircraft since 1915*, pp. 140–1. Huntley, 'Fairey's elusive Fox'; Fairey Aviation advertisement cuttings, 1926–31, FAAM/2010/050/002 and Box 130, AC73/30, RAFM, e.g. *The Aeroplane*, 25 June and 23 July 1930; C.R. Fairey, 'The development of the day bomber', 9 August 1927 [draft article commissioned by C.G. Colebrook, aeronautical correspondent of *The Times*], FAAM/2010/054/0023; Rt Hon. Stanley Baldwin, House of Commons, *Hansard*, 10 November 1932.
88 Hall, *Sir Richard Fairey*, p. 9; Mrs Jane Tennant, quoted in an e-mail to the author, 22 December 2014. CRF had commissioned a selection of alternative corporate badges in 1915–16, but dismissed every suggestion: 'Fairey Aviation Company Ld. Badges', Box 2, AC73/30, RAFM. Mrs Needell incorporated the Fairey badge and the Wells quote 'If the world does not please you can change it.' into a very attractive bookplate, to be found still in the volumes held in CRF's library at Bossington. She designed a parallel bookplate for the second Mrs Fairey.
89 Penrose, *British Aviation: The Adventuring Years*, p. 657. For HRH boarding his Royal Bristol Fighter, see 'Our Flying Prince – Prince of Wales, after crowded day fulfilling 8 engagements, flies back to London', 4 June 1928, Pathé News, www.britishpathe.com/video/our-flying-prince-1.
90 466 'The 10th Royal Air Force Display' and 'H.R.H. The Prince of Wales declares the Aero Exhibition open', *Flight*, 18 July 1929.

91 C.G. Grey, *The Aeroplane*, 18 July 1929, quoted in Penrose, *British Aviation: The Adventuring Years*, p. 658; ibid., pp. 664–5; Geoffrey Hall, interviewed in Trippe, 'The plane maker', p. 579; 'H.R.H. The Prince of Wales declares the Aero Exhibition open.' Less visible on the IIIF skeleton was its cadmium plating, a protection against corrosion adopted by Fairey well in advance of the automobile industry.

92 Anonymous visitor to Olympia, quoted in Penrose, *British Aviation: The Adventuring Years*, pp. 665–6; ibid., p. 666.

93 Production data taken from Trippe, 'The plane maker', pp. 557–8 and 580–1 and Taylor, *Fairey Aircraft since 1915*, pp. 159–60 and 165. Re. Fairey Aviation's dealings with the Soviet Union, see Chapter 6.

94 Wilfred Broadbent, quoted in Trippe, 'The plane maker', p. 582.

95 W.G. Halahan [Secretary, Air Ministry] to C.R. Fairey, 27 May 1925, FAAM/2010/054/0020. Vincent Nicholl to C.R. Fairey, 2 and 15 October 1925, ibid.

96 'If there is any other American information or particulars I can give you from my notes please let me know': C.R. Fairey to Major J.S. Buchanan, 9 November 1925, FAAM/2010/054/0020.

97 For example, courtesy of Charles Grey, briefing Higgins on Packard's secret 1250 horse power engine and on Macmillan's invitation to fly with Italy's air strategist General Balbo: C.R. Fairey to Air Vice Marshal Sir John Higgins, 15 February and 21 July 1927, FAAM/2010/054/0021.

98 Major R.E. Penney to C.R. Fairey, 1 February 1927, ibid.

99 C.R. Fairey [at Hotel Continental, Rue de Rivoli, Paris] to Louis Breguet, 1 December 1926, FAAM/2010/054/0027.

100 CRF archived select writings on aeronautics by Fokker: Trippe, 'The plane maker', p. 578; C.R. Fairey to the manager of The Savoy, 17 April 1925, FAAM/2010/054/0020.

101 'Mr Fokker … Please Note!', Pathé News, 7 May 1925, www.britishpathe.com/video/mr-fokker-please-note; Vincent Nicholl to C.R. Fairey, 23 October 1925, FAAM/2010/054/0020.

102 Travel information re CRF's ten days visit to Germany, November 1929, FAAM/2010/054/0029.

103 Geoffrey Hall's account of visiting the Junkers plant, in Trippe, 'The plane maker', pp. 565–6; 'JU-52 History', www.ju52-3m.ch/about. Unlike Junkers, Fairey Aviation's principal use of duralumin in 1929 (by which time more durable alloys were used for forging propellers) was primarily in float construction.

104 [5th Earl of] Onslow, Rev. Robin Higham, 'Thomson, Christopher Birdwood, Baron Thomson of Cardington (1875–1930)', *Oxford Dictionary of National Biography*, www.oxforddnb.com/view/article/36500?docPos=1; Adrian Smith, 'Ramsay MacDonald – aviator and actionman', *The Historian*, 28, 1990, pp. 14–15; Hoare, Thomson's predecessor and successor in 1924, claimed that for ten

months the interim minister undertook only the bare minimum: Templewood, *Empire of the Air*, pp. 103–4.
105 John Fairey and Geoffrey Hall, quoted in Trippe, 'The plane maker', p. 566.
106 See the tribute to recently deceased members by Past President Colonel The Master of Sempill in 'Proceedings: C.R. Fairey, "Growth of Aviation" [Presidential Address], 9 October 1930', *The Journal of the Royal Aeronautical Society*, 35/241 (1931), 4–28; Cobham, *A Time to Fly*, p. 149.
107 Mrs Jane Tennant to the author, 27 October 2016.
108 C.R. Fairey to Sir William Mitchell-Thompson, 7 May 1926, 4 May 1926, FAAM/2010/054/0063; Mrs Jane Tennant to the author, 27 October 2016.
109 C.R. Fairey to The Secretary, Imperial College of Science and Technology, 4 May 1926, FAAM/2010/054/0063.
110 C.R. Fairey to Professor L. Bairstow [Imperial College], 3 October 1927, FAAM/2010/054/0027; C.R. Fairey to Air Vice Marshal Sir John Higgins, 10 May 1926 [letter not sent], FAAM/2010/054/0027. In the year between strike-breaking and beginning his tailored course at Imperial [fee invoiced to CRF: £64 10 shillings], Hall was taken ill in Paris with what appears to have been appendicitis, or even peritonitis: R. Petis [surgeon] to C.R. Fairey, 13 February 1927, FAAM/2010/054/0021.
111 Ibid.
112 C.R. Fairey, 'Trade unionism and industry' lecture notes, Andover Staff College, 18 January and 18 November 1926, FAAM2010/53/0030 and 0029. US trade unions were deemed apolitical and thus the acceptable face of organised labour: Mrs Jane Tennant to the author, 27 October 2016.
113 C.R. Fairey, speech at the House Dinner of the Royal Aero Club, 19 January 1927, FAAM/2010/53/0034 and 0033.
114 C.R. Fairey to Wing Commander Randall RN, 17 March 1917, Box 67, AC73/30, RAFM.
115 CRF was a member of the South Bucks Conservative and Unionist Association, whose MP, Sir Alfred Knox, was a retired general – as was the honorary treasurer. Beaverbrook had a clearer purpose than Rothermere, with whom he soon quarrelled. The owner of Express Newspapers argued that the Conservatives should adopt a policy of 'Empire free trade', whereby Britain turned its back on Europe to create an economic bloc with the Dominions and crown colonies; a view Fairey endorsed. Moore-Brabazon questioned Baldwin's leadership in February 1931 by not seeking selection as Conservative candidate for the Westminster St George's by-election, in which Duff Cooper defeated his Beaverbrook-backed opponent. Stuart Ball, *Baldwin and the Conservative Party: The Crisis of 1929–1931* (New Haven and London: Yale University Press, 1988), pp. 40–1 and 137–8; C.R. Fairey, speech at the House Dinner of the Royal Aero Club, 19 January 1927, FAAM/2010/53/0033.

116 'Selling aircraft abroad', Royal Aero Club after-dinner speech, 1 November 1933, FAAM2010/53/0065.
117 Patrick Zander, '(Right) Wings Over Everest: High Adventure, High Technology and High Nationalism on the Roof of the World, 1932–1934', *Twentieth Century British History* (2010), 21:2, pp. 313–14, 320–2 and 328–9.
118 '…the amazing achievement of the English-speaking races … it is their superior energy and fighting capacity in the past and their trading initiative in the more recent past that have established the position.': C.R. Fairey, 'Trade unionism and industry' lecture notes, Andover Staff College, 18 November 1926, FAAM2010/53/0029.
119 Ian Kershaw, *Making Friends With Hitler: Lord Londonderry and Britain's Road to War* (London: Penguin, 2005), pp. 35, 51, 58–60 and 76–7; C.R. Fairey, opening speech of a Royal Aeronautical Society debate on civil aviation control, 16 January 1933, FAAM/2010/53/0060; C.R. Fairey, 'Mr Fairey's speech', *The Air League Bulletin* (February 1926), pp. 18–20 and 29–33.
120 The *Daily Mail*'s campaign was most vociferous in late 1933 and spring 1935. Kershaw, *Making Friends With Hitler*, pp. 74–7; Richard Griffiths, *Fellow Travellers of the Right: British Enthusiasts for Nazi Germany 1933–9* (Oxford: Oxford University Press, 1983), pp. 140–1. 'November 20th [1937] Göering … A modern Robin Hood: producing on me a composite impression of film-star, gangster, great landowner interested in his property, Prime Minister, party manager, head gamekeeper at Chatsworth.': Edward Frederick Lindley Wood, The Earl of Halifax, *Fulness of Days* (London: Collins, 1957), p. 191. 'He [CRF] genuinely hated fascism and the Germans. "Twice they destroyed our world." Bloody Huns!': Mrs Jane Tennant to the author, 27 October 2016.
121 Robert Innes-Smith, 'James Wentworth Day (1899–1983)', *Oxford Dictionary of National Biography*, https://doi.org/10.1093/ref:odnb/40790; J. Wentworth Day, draft [*Daily Mail*] of 'These men build your war planes (1) Mr C.R. Fairey – Modern Elizabethan' sent for final approval to C.R. Fairey, 21 October 1938, FAAM/2010/053/0124; J. Wentworth Day, draft of 'A five-point plan for a British propaganda campaign' sent for final approval to C.R. Fairey, 3 April 1939, FAAM/2010/053/0045. On Wentworth Day's admiration for 'that splendid old warrior of the air, C.G. Grey', see Zander, '(Right) Wings over Everest', p. 317.
122 Admiral Sir Murray Sueter was invited to Germany as an MP and founding member of the Anglo-German Fellowship. Yet he also chaired meetings of the National Air League, founded by Rothermere in early 1935 with Norman Macmillan as president. Griffiths, *Fellow Travellers of the Right*, pp. 185 and 225; Paul Addison, 'Patriotism under Pressure: Lord Rothermere and British Foreign Policy' in *The Politics of Reappraisal 1818–1929*, eds. Gillian Peele and Chris Cook (London: Macmillan, 1975), pp. 200–1; *Fleet Street, Press Barons and*

Politics: The Journals of Collin Brooks, 1932–1940, ed. N.J. Crowson (London: Royal Historical Society, 1998), 6 December 1935, p. 145.
123 Edgerton, *England the Aeroplane*, pp. 75–6 and 42; Zander, '(Right) wings over Everest', pp. 300–29.
124 C.G. Grey, editorial, *The Aeroplane*, 1 January 1936, quoted in ibid., p. 91. Edgerton highlighted how tributes to Grey in postwar aeronautical literature airbrushed away his appalling views, noting that BEA even named an airliner after him: ibid., pp. 91–2.
125 Correspondence of C.G. Grey and Lt Col. J.T.C. Moore-Brabazon, January–February 1938, Box 11, AC71/3, RAFM.
126 For a less generous judgement of Moore-Brabazon and Mosley's BUF, see Stephen Dorril, *Blackshirt Sir Oswald Mosley and British Fascism* (London: Viking, 2006), pp. 14, 41, 81, 164, 173, 282, 289 and 299. On the two men's early acquaintance, see Sir Oswald Mosley, *My Life* (London: Thomas Nelson, 1968).
127 Grey, like Sueter, promoted the National Air League, several members of which were also in the BUF. Norman Macmillan met Mosley at a Rothermere lunch the same day he was appointed to run the NAL. Robert Skidelsky, *Oswald Mosley* (London: Macmillan, 1985), p. 320; Griffiths, *Fellow Travellers of the Right*, pp. 69–70 and 138–40; *Fleet Street, Press Barons and Politics*, ed. Crowson, 23 January 1935, pp. 80–1.
128 L.J. Dickie [Secretary, Right Book Club] to Miss E.L. Burns, 20 October 1938, FAAM/2010/054/0004. By way of balance, soon after, Claud Cockburn's secretary was instructed to send no further issues of *The Week*, the cyclostyled prewar precursor of *Private Eye*. Clearly a sample issue was sent in the naive expectation that CRF would take out a 16 shillings annual subscription. Miss E.L. Burns to 'The Week', 3 November 1938, and *The Week*, No 288, ibid.
129 Correspondence of C.R. Fairey and Lord Queenborough [Almeric Paget], July–October 1937, FAAM2010/054/0042; Griffiths, *Fellow Travellers of the Right*, pp. 235–6, 238 and 345.
130 C.G. Grey, *The Aeroplane*, 18 July 1929, quoted in Penrose, *British Aviation: The Adventuring Years*, pp 658–9; C.G. Grey to C.R. Fairey, 25 October 1929, FAAM/2010/050/0088.
131 Trippe, 'The plane maker', pp. 561 and 584–5; C.R. Fairey correspondence with Vincent Nicholl and Maurice Wright, FAAM/2010/050/0063 and 0027, and Box 2 and 45, AC73/30, RAFM.
132 Miscellaneous Fairey Aviation restructuring documents, including amendments to articles of association, board resolutions, liquidators' statements and company prospectuses, 1921–7, BT34/4112/141134 and BT34/5027/1205693, NA. On majority shareholders' freedom to amend articles of association before and after enactment of the 1929 Companies Bill, see Cheffins, *Corporate Ownership and Control*, p. 254.

133 Miscellaneous Fairey Aviation Co. Ltd. liquidation documents, 1927; correspondence between C.R. Fairey and Director and Deputy Director of Contracts re a delay in release of £60,000 payment to Fairey Aviation given Air Ministry concern over £150,000 in cash removed from the company as part of its voluntary liquidation, August–September 1927; Treasury Solicitor to F.W. Lawton [Director of Contracts], 27 May 1929, TS28/171, NA.

134 Charles Crisp to C.R. Fairey, 28 June 1927, FAAM2010/054/0021. Crisp feared the Solicitor to the Inland Revenue would focus upon CRF himself given that he was still two years behind in filing his tax returns and only agreed to clear any outstanding amount when threatened with legal action: C.R. Fairey to Charles Crisp, 20 February, and 20 and 30 April 1927, ibid. Receiving a gross dividend was welcome because, under the normal procedure of being taxed 'at source', shareholders 'would typically be liable to pay additional income tax if they were surtax payers since they would only be credited for tax paid on their behalf by the company at the "standard" rate of income…': Cheffins, *Corporate Ownership and Control*, p. 289.

135 Sir Henry Self, response to Treasury Solicitor to F.W. Lawton, 27 May 1929, TS28/171, NA.

136 C.R. Fairey to F.W. Lawton, 8 March 1929, ibid.; Taylor, *Fairey Aircraft since 1915*, pp. 13–14; Penrose, *British Aviation: The Adventuring Years*, pp 617–18; draft trust deed for debenture stock, 1929, Box 137, AC73/30, RAFM. On why in a pre-crash 'bull' market Fairey's tightly controlled floatation was not uncommon, with the 1929 Companies Act tacitly approving pre-prospectus sale of securities, see Cheffins, *Corporate Ownership and Control*, pp. 74 and 274.

137 A.G. Hazell, draft contract for the employment of C.R. Fairey as Managing Director, Expert Adviser and Works Superintendent 1929–34, 1929 and [hand-written] informal spreadsheet for profits on machines, spares and propellers, October 1929, ibid. Hazell estimated a 33% profit on every machine sold: ibid. CRF's 1928 disbursement was £100,001: A.G. Hazell, 'Liquidator's Statement of Account: The Fairey Aviation Company Limited' and 'List of Dividends or Composition', 20 August 1928, BT34/5101/215450, NA.

138 By 1931 one in three of the joint stock companies floated in the 'issue boom' of 1928–9 had been wound up or lacked further investment capital: Cheffins, *Corporate Ownership and Control*, p. 275.

Chapter 6

1 C.R. Fairey quoted in James Wentworth Day, draft synopsis for book 'Wings over the world: the official life of Sir Richard Fairey, M.B.E., F.R.Ae.S.', [?] 1957, FAAM/32/2010/053/0125.

2   C.R. Fairey to James Purdey and Sons Ltd, 3 January 1936, FAAM/2010/054/0002 [CRF's pair of bespoke shotguns cost £281]; cartridges came from Grant and Lang of St James.

3   There is no evidence that CRF caught a tunny during the brief period of fierce competition when blue fish tuna could be found in the North Sea. Miscellaneous correspondence of C.R. Fairey and Harold Hardy [British Tunny Club honorary secretary], 1931–6, FAAM/2010/054/0006. Filey Bay Research Group, 'Big game fishing off the Yorkshire coast', www.fileybay.com/tunnyfish.

4   Miss E.L. Burns to C.R. Fairey, 16 September 1937, FAAM/2010/054/0013.

5   Paragraph based on miscellaneous personal correspondence of CRF, 1924–9, FAAM/2010/054/0020-26 and 0063, including settled invoices from Finnigans of New Bond Street [luggage], Bradleys of Knightsbridge [hotel] and Asprey of Bond Street [jewellery], 1926–7, FAAM/2010/054/0021; Miss E.L. Burns to A. Hughes, 9 February 1927, FAAM/2010/054/0021.

6   For example, inviting CRF to shoot at Martlesham Heath, near Felixstowe, earned his host a case of Chateau Latour 1923: order of CRF to Barry Neame [Hind's Head Hotel], 24 February 1936, FAAM/2010/054/0002.

7   Miscellaneous correspondence of CRF re parties, 1924–9, FAAM/2010/054/0020-26 and 0063; Adrian Smith, *Mountbatten: Apprentice War Lord 1900–1943* (London: I.B.Tauris, 2010), pp. 70–1; Trippe, 'The plane maker', pp. 509–21; Fortnum and Mason invoices, 1927, FAAM/2010/054/0023: the F and M bill for 28 September 1927 was nearly £44, equivalent today to £2,400.

8   The Kodascope Libraries', Robbie's Reels website: www.robbiesreels.com/1920; miscellaneous Kodak-related correspondence, FAAM/2010/054/0022.

9   A particularly memorable theatre visit was seeing a young Brian Aherne star in *White Cargo* at the London Playhouse on 10 March 1927: C.O. Crisp to C.R. Fairey, 8 March 1927, FAAM/2010/054/0066. As well as large productions CRF supported more adventurous small-scale productions, witness his subscription to the Players' Theatre Club: PTC membership card 1938–40, FAAM/2010/054/0004.

10 'Felixstowe Ferry Golf Club', www.felixstowegolf.co.uk/index; miscellaneous correspondence re hotel and travel arrangements of C.R. and Joan Fairey, February–May 1927, FAAM/2010/054/0021; Harold E. Perrin [secretary of the RAeC and Aero Golfing Society] to C.R. Fairey, and match play card for Moor Park tournament, 28 February 1927, 18 February 1927, FAAM/2010/054/0022; ticket application form for Royal Aero Club banquet for Charles Lindbergh, 31 May 1927, ibid.; report of RAeC monthly house dinner, 19 January 1927, *Flight*, 27 January 1927; 'Charles Lindbergh timeline', www.charleslindbergh.com/timeline/index.asp; C.R. Fairey to T.O.M. Sopwith, 28 May 1927, FAAM/2010/054/0022.

11 Paragraph based on miscellaneous personal correspondence of CRF, 1924–9, 1932 and 1938–9, FAAM/2010/054/0020-26, FAAM/2010/054/0063, FAAM/

2010/054/0030 and FAAM/2010/054/0004. Commander James Bird to C.R. Fairey, 13 August 1931 and programme for Royal Aero Club dinner to celebrate British victory in the Schneider Trophy, 9 December 1931, FAAM/2010/34/0090. Claridge's feared the Royal Aero Club would cancel the dinner should Supermarine fail to win again and the RAeC lacked sufficient reserve funds to pay for such a grand occasion up-front. At this point CRF, as President of the Royal Aeronautical Society, led from the front. The RAeC was in a difficult position as in 1930 its advisory committee had concluded that a privately funded attempt to retain the Trophy was not feasible; after which a *Daily Mail* campaign and a gift of £100,000 from Lady Huston had generated sufficient funds to force the Labour Government in January 1931 to reinstate state support for Supermarine/Rolls Royce and sanction the RAF to reform its High Speed Flight. 'R.J. Mitchell: a life in aviation' website, Solent Sky Museum: www.rjmitchell-spitfire.co.uk/schneidertrophy/1931.asp?sectionID=2.

12 Hollis Williams, quoted in Penrose, *British Aviation: The Adventuring Years*, p. 478.

13 'You have always been one of the leading advocates of the superiority of American cars over English cars, and you have been quite right.' C.G. Grey to CRF, 26 October 1929, FAAM/2010/054/0088. CRF even owned a steam car built in the United States: a 1920 Stanley 735A was purchased soon after the war, maintained by Geoffrey Hall and driven on an almost daily basis. Trippe, 'The plane maker', p. 522. When finally Fairey acquired a Bentley (3.5 litre) he was distinctly underwhelmed: Miss E.L. Burns to Bentley Motors (1931) Ltd Service Department, 2 April 1936, FAAM/2010/054/0003; this may explain why in 1945 he had his Pontiac saloon shipped over from America. Conversation with Mrs Jane Tennant, 2 October 2015.

14 Paragraph based on miscellaneous correspondence between CRF and W.C. Gaunt Coy (Pass and Joyce Ltd) and Gerald C. Maxwell (Chrysler Motors), 1925–8, FAAM/2010/054/0020, 0026 and 0088. Trippe, 'The plane maker', p. 522–8.

15 CRF regularly used the Daimler Hire service, as on 30 October 1932 when his driver collided with another car: the passenger's cut head and sprained ankle meant an overnight stay in Kent and Canterbury Hospital. C.R. Fairey, 'Statement handed to Harlington Police', 9 November 1932, FAAM/2010/054/0001.

16 C.R. Fairey to Lord Portarlington, 22 November 1938, FAAM/2010/054/0004. CRF still owned at least one American car – a Cadillac, which he retained after purchasing the company's current limousine in October 1939: correspondence between CRF and Lendrum and Hartman Ltd, October–December 1939, FAAM/2010/054/0007. Miss E.L. Burns to Esther Fairey, 7 September 1939, FAAM/2010/054/0006.

17 For two of several examples where Fairey insisted he was not to blame for an accident or he denied speeding, see C.R. Fairey to Superintendent, Chiswick

Police Station, 25 February 1925 and to Clerk to the Magistrates, Westminster Police Court, 1 December 1926, FAAM/2010/054/0020 and 0024. His car at the time was a six-cylinder coupé, the Chrysler Model 70, its designation signalling a maximum speed of 70 mph.

18 Receipts for fines for speeding in west London of respectively 30 shillings and £3, 26 November 1926 and 5 April 1927, FAAM/2010/054/0024.

19 The 1934 Road Traffic Act introduced a 30 mph limit on roads in built-up areas and CRF saw this as sufficient reason not to donate to his local Conservative association. Ironically, Hore-Belisha was a National Liberal MP and not a Tory. C.R. Fairey to B. Rhodes [NUCUA Wessex Area treasurer], 6 April 1936, FAAM/2010/054/0003.

20 Juliet Gardiner, *The Thirties: An Intimate History* (London: Harper Press, 2011), pp. 682–3 and 679; Moor-Brabazon quoted in ibid., p. 683.

21 C.R. Fairey to Lt Col. J.T.C. Moore-Brabazon, 22 January 1936, FAAM/2010/054/0002. When CRF's chauffeur-driven Rolls Royce was stopped for speeding a month earlier, he allegedly said to the policemen, 'I suppose motor bandits and burglary do not come within the jurisdiction of the police.' The case was dismissed by an Uxbridge magistrate. 'Speed charge fails', *Daily Telegraph* and 'Beat the police on mathematics', *Evening News*, 24 December 1935.

22 After CRF was stopped for speeding in September 1938, the Metropolitan Police chose not to take action: Commissioner of Police to C.R. Fairey, 1 November 1938, FAAM/2010/054/0004. The same was not the case two years earlier when he was charged with dangerous driving in central London and then in Sutton Scotney: ibid., 4 March 1936, FAAM/2010/054/0003 and Superintendent F. Osman [Hants Constabulary] to C.R. Fairey, 14 October 1937, FAAM/2010/054/0010.

23 Geoffrey Hall, quoted in Trippe, 'The plane maker', pp. 550–1.

24 The writer's recollection of CRF's properties in ibid., pp. 590–3; 'Where Jane Austen lived', *Daily Telegraph*, 3 April 1936; Gribble, Booth & Shepherd to C.R. Fairey, 26 February 1936, FAAM/2010/054/0002. In the autumn of 1927 CRF turned down an invitation to purchase Grove Cottage: James Gurney to C.R. Fairey, 3 October 1927, FAAM/2010/054/0023.

25 Particulars of 'Fort Charles', L.A. Page [auctioneer], [?] 1939, FAAM/2010/054/0007. Both the Woodlands and Fort Charles residences were bombed in the course of World War II.

26 Albeit with great reluctance in the case of one hapless elderly cousin whose begging letters secured from Fairey two stern letters and enclosed cheques for modest sums: correspondence of C.R. Fairey and W.J. 'Will' Spencer, July–November 1939, ibid.

27 Correspondence between CRF and C.F. Rowlands [senior partner, Wrentmore & Son solicitors], September–December 1939, ibid. The interest rate of 5% was on a large property in Bournemouth and may have been a consequence of raising the

interest rate when Britain declared war. By this time CRF was regularly inspecting the Hind's Head Hotel's weekly accounts: Barry Neame to Miss E.L. Burns, 14 September 1939, ibid.

28 Geoffrey Hall, quoted in Trippe, 'The plane maker', pp. 553–4.
29 Miscellaneous vacation correspondence, August 1926, FAAM/2010/054/0063.
30 'From all quarters: Fairey managing director', *Flight*, 17 August 1956; Conversation with Mrs Jane Tennant, 2 October 2015.
31 'Richard's problem was that my father adored AND spoilt him.' Mrs Jane Tennant to the author, 2 October 2015.
32 Twyford School website: www.twyfordschool.com; 'Twyford School: notable former pupils', Wikipedia: www.en.wikipedia.org/wiki/Twyford_School#Notable_former_pupils.
33 Interview with Wilfred Broadbent, in Trippe, 'The plane maker', pp. 537–9; 'At the universities', *The Harrovian*, 19 October 1935 [received with thanks, 9 June 2015, from Joanna Badrock, Archivist and Records Manager, Harrow School]; 'Heston', *Aeropilot London*, October 1935; 'University Flying Club Formed', *Cambridge Evening News*, 1 November 1935; C.R. Fairey to S.W. Grose [Tutor, Christ's College], 3 January 1936 and S.W. Grose to C.R. Fairey, 5 January 1936, FAAM/2010/054/0002. Until caution prevailed CRF anticipated his son transferring to the then relatively new Stowe School: see correspondence of C.R. Fairey with J.F. Roxburgh [Headmaster] and C.H. Ivadds [Bursar], February–April 1927, FAAM/2010/054/0021.
34 C.R. Fairey, public school lectures on 'The evolution of the airplane', 1931–2, FAAM/2010/53/0050 and 56; 'Natural Science Club', *The Harrovian*, 15 December 1932 [courtesy of Joanna Badrock]; J.F. Roxburgh to C.R. Fairey, 18 November 1932, FAAM/2010/53/001; President's Address, Royal Aeronautical Society AGM, 31 March 1932, FAAM/2010/53/0051; National Aerospace Library, RAeSoc website: www.aerosociety.com/About-Us/nal. Establishment of the Royal Aeronautical Society's library and archival collection depended heavily upon the Endowment Fund Appeal, which CRF launched in December 1930 and promoted enthusiastically and successfully throughout his presidency: *The Royal Aeronautical Society Monthly Notices*, January 1931.
35 C.R. Fairey, lecture on 'The future of aeroplane design for the Services', Royal United Services Institute (RUSI), 11 February 1931, FAAM/2010/53/0043 and 'The future of aeroplane design for the Services', *RUSI Journal*, 76 (1931), pp. 574–6.
36 HMG to Mr and Mrs C.R. Fairey, invitation to witness the 1931 Schneider Trophy Contest on board HMS *Courageous* and attend the banquet and prize-giving on board SS *Homeric*, 12 September 1931, FAAM/2010/34/0090. 'The [Royal Aeronautical] Society recognises the immense advantages that aeronautical science and engineering have reaped as a result of intensive competition': telegram of thanks and congratulations from RAeSoc President to Lady

Huston, 12 September 1931, ibid.; 'R.J. Mitchell The Science of Speed', Solent Sky Museum: www.solentskymuseum.org/single-post/2017/02/22/RJ-Mitchell---The-Science-of-Speed.

37 C.R. Fairey, lectures to the IDC and RNSC, 7 June 1932 and 15 February 1933, FAAM/2010/53/0055, welcoming address to the Société Française de Navigation Aerienne, [?] April 1933, FAAM/2010/53/0057, opening speech of Royal Society of Arts/RAeSoc debate on 'Civil Aviation Control', 16 January 1933, FAAM/2010/53/0060, and reply to the Director of Scientific Research at the Air Ministry DTD dinner, 18 November 1932, FAAM/2010/053/0053.

38 C.R. Fairey, welcoming address, at joint RAeSoc/RAeC/SBAC dinner for Mrs Amy Mollison, 12 July 1932 and at the Wilbur Wright Memorial Lecture and annual dinner of the RAeSoc Council, 26 May 1932, FAAM/2010/53/0032; photograph of guests, 26 May 1932, FAAM/2010/052/0005; Mary S. Lovell, *Amelia Earhart: The Sound of Wings* (London: Abacus, 2009), p. 221. In May 1936 CRF sent Mrs Mollison his 'sincere congratulations' when unable to attend the *Daily Express*'s late-night reception for her at the Dorchester: C.R. Fairey to *Daily Express* [telegram], 14 May 1936, FAAM/2010/054/0003.

39 In 1950 CRF would be made an Honorary Fellow, the Royal Aeronautical Society's highest award. Hall, *Sir Richard Fairey*, p. 10.

40 Mrs Jane Tennant to the author, 2 October 2015; John Fairey quoted in Hall, *Sir Richard Fairey*, p. 10.

41 Mrs Jane Tennant to the author, 2 October 2015.

42 Miss E.L. Burns to Miss M.M. Howell, 26 July 1927, and Miss M.M. Howell to C.R. Fairey, 22 July 1927, FAAM/2010/054/0024.

43 Documentation re 'The humble petition of Charles Richard Fairey', 7 May 1931, Probate Divorce and Admiralty Division, High Courts of Justice: courtesy of Les Hewett, 29 July 2013.

44 '…he [CRF] simply loathed Joan and never wanted to hear her name again … My mother [Esther Fairey] said a few things about her and these obviously came via my father. He considered Joan a dishonest woman he could not trust.' Mrs Jane Tennant to the author, 2 October 2015.

45 Joan Buxton to Rt Hon. J.T.C. Moore-Brabazon, 16 January 1942, Box 13, AC71/3, Lord Brabazon papers [LBP], RAFM.

46 Correspondence of C.R. Fairey and George Clarke, November 1932, FAAM/2010/054/0001.

47 Correspondence of C.R. Fairey and Mrs Markey, September–November 1932, ibid.

48 That this habit of quietly inspecting the plant and identifying transgressors continued postwar, see Sir Richard Fairey to C.H. Chichester Smith [Assistant Managing Director], 27 June 1947, FAAM/2010/054/038.

49 Conversation with Mrs Jane Tennant, 2 October 2015. Postwar test pilot and world speed record holder, Peter Twiss, had a reputation for being short-tempered, but admitted that he felt intimidated whenever CRF was angry: ibid.

50 The engineers' union facilitated a resolution of the dispute, at the price of Fairey Aviation agreeing to accept AEU shop stewards across the plant. 'Aircraft Strikers Demand Strike Pay', *Evening News*, 8 March 1937 and 'Fairey Strike Ends', *Daily Worker*, 10 March 1937. C.R. Fairey to Lord Winterton [Chancellor of the Duchy of Lancaster], 11 April 1938, Box 124, AC73/30, RAFM.
51 C.R. Fairey to Sir Murray Sueter MP, 17 November 1938, FAAM/2010/054/0004. Maurice Wright to C.R. Fairey, 16 September 1937, FAAM/2010/054/0013: CRF was again cruising on *Evadne* when the strike began.
52 Sebastian Ritchie, *Industry and Air Power: The Expansion of British Aircraft Production 1935–41* (Abingdon: Routledge, 1997), pp. 165 and 162.
53 Ibid., pp. 161 and 165; C.R. Fairey to Sir Murray Sueter MP, 17 November 1938, FAAM/2010/054/0004.
54 Ritchie, *Industry and Air Power*, pp. 85–8 and 44–9; D.E.H. Edgerton, 'Technical innovation, industrial capacity and efficiency: Public ownership and the British military aircraft industry, 1935–48', *Business History*, 26 (1984), p. 254. On armaments manufacturers' genuine fears *c*.1935 re state control, see G.A.H. Gordon, *British Seapower and Procurement Between the Wars* (London: Macmillan, 1988), pp. 164–5. Fairey Aviation's company papers from 1936–8 held at the RAF Museum contain occasional correspondence between CRF and Lord Weir.
55 C.R. Fairey to Lord Weir, 30 September 1937, FAAM/2010/106/0013.
56 T.M. Barlow, 'Malicious damage to "battle" aircraft at The Fairey Aviation Company Limited, Stockport', [6?] April 1938; C.R. Fairey to R.C. Chilver [private secretary to Lord Winterton], 7 April 1938 and to Lord Winterton, 11 April 1938, Box 124, AC73/30, RAFM.
57 C.R. Fairey, Draft speech welcoming Rt Hon. Sir Kingsley Wood MP and Rt Hon. Ernest Brown MP to Fairey Aviation, 24 October 1938, FAAM/2010/53/0100; C.R. Fairey and Rt Hon. Ernest Brown MP, miscellaneous correspondence re sailing, 1937–9, FAAM/2010/53/; Sir Henry Channon, *Chips The Diaries of Sir Henry Channon*, ed. Robert Rhodes James (London: Penguin, 1970), 3 May and 11 July 1939, pp. 243 and 253–4. Brown was even prepared to accept Hitler's ultimatum to Chamberlain at Bad Godesberg: Duff Cooper, *The Duff Cooper Diaries*, 25 September 1938, p. 266.
58 Robert Self, *Neville Chamberlain A Biography* (Farnham: Ashgate, 2006), p. 274. On Ball's role as head of the Conservative Research Department in spying on Churchill and other anti-appeasers, see Lynne Olson, *Troublesome Young Men: The Churchill Conspiracy of 1940* (London: Bloomsbury, 2008), pp. 167–9. John Charmley, *Lord Lloyd and the Decline of the British Empire* (London: Weidenfeld and Nicolson, 1987), pp. 198 and 238; Lord Lloyd to C.R. Fairey, 11 March and 21 February 1940, FAAM/2010/054/0045.
59 'I [Fairey] have been so well satisfied with the results of this policy that I have adhered to it by promoting others from inside the works when vacancies occur,

with the result that employees know that the Board Room door is open to ambition, and none of the Company's funds go to paying directors not working in the concern. Apart from other good results, the shareholders' interests are secured by the loyalty and unanimity of a Board that knows every detail of the business.' C.R. Fairey to Sir William Firth, 19 May 1939, FAAM/2010/54/0005.

60 'Tupolev ANT-9/PS-9', www.aviastar.org/air/russia/ant-9.php; Paul Duffy and Andrei Kandalov, *Tupolev: The Man and His Aircraft* (Shrewsbury: Airlife, 1996), p. 12; L.A. van de Velde, 'Moscow Visit. Diary – 24th October to 7th November 1932. General business impressions, and conclusions. Moscow, 1932', 15 November 1932, entries for 24 October 1932 and 5 November 1932, p. 8, FAAM/2010/106/0010; A.Ozersky to C.R. Fairey, 2 March 1932, FAAM/2010/054/0102.

61 'S.V. Ilyushin', www.ilyushin.org/en/about/history/biography; John T. Greenwood, 'The aviation industry, 1917–1997', in eds. Robin Higham, John T. Greenwood and Von Hardesty, *Russian Aviation and Air Power in the Twentieth Century* (London: Frank Cass, 1998), pp. 132 and 135–6.

62 Ibid., pp. 129–30; Viktor P. Kulikov, 'British aircraft in Russia', *Air Power History*, spring 2004, p. 6.

63 Ibid., p. 9; Von Hardesty, 'Early flight in Russia', in eds. Robin Higham, John T. Greenwood and Von Hardesty, *Russian Aviation and Air Power in the Twentieth Century*, pp. 35–6; Reina Penning, 'From chaos to the eve of the Great Patriotic War, 1922–41' in ibid., pp. 39 and 40.

64 Greenwood, 'The aviation industry, 1917–1997', pp. 134–5; Von Hardesty, 'Early flight in Russia', p. 18.

65 Penning, 'From chaos to the eve of the Great Patriotic War, 1922–41', pp. 40–1; van de Velde, 'Moscow visit…', entry for 2 November 1932, p. 6.

66 Mark Harrison, 'The political economy of a Soviet military R&D failure: Steam power for aviation, 1932 to 1939', *The Journal of Economic History*, 63 (2003), p. 205; Penning, 'From chaos to the eve of the Great Patriotic War, 1922–41', pp. 41 and 44–5; Duffy and Kandalov, *Tupolev: The Man and His Aircraft*, pp. 13–15.

67 C.R. Fairey to Professor A.N. Toupoleff, 27 October 1932 and A. Ozersky to C.R. Fairey, 21 November 1932, FAAM/2010/054/0102.

68 Within the Gulag the NKVD maintained the KOSOS prison aviation group: Robert Conquest, *The Great Terror: A Reassessment* (Oxford: Oxford University Press, 1990), p. 294. Greenwood, 'The aviation industry, 1917–1997', pp. 135–6; van de Velde, 'Moscow visit…', entry for 2 November 1932, p. 6.

69 Miscellaneous correspondence of C.R. Fairey with S.G. Bron [Soviet Trade Delegation], A. Ozersky [Soviet Embassy] and A. Birgenoff and M. Nesteroff [Arcos Ltd] and with Alexis Ignatieff, 1931–2, FAAM/2010/054/0101 and 0102; 'Naturalisation certificate: Alexis Ignatieff. From Russia. Resident in London. Certificate AZ1636 issued 27 June 1932', HO334/129/1636, NA. Unlike his

successor as ambassador, Sokolnikov was hard to cultivate given his puritan lifestyle and poor English: Ivan Maisky, *The Maisky Diaries: Red Ambassador to the Court of St James's 1932–1943*, ed. Gabriel Gorodetsky (London: Yale University Press, 2015), p. xliv.

70 Ibid., pp. 23, 91, 105 and 113; Mrs Jane Tennant to the author, 6 September 2016.

71 C.B. Baker, 'A Fairey for Stalin', 194[?], FAAM/2010/106/0009; van de Velde, 'Moscow visit…', entries for 24 and 25 October 1932, pp. 1 2.

72 For example, gratitude for Russia-related documentation forwarded to the Air Secretary: L.G.S. Reynolds to C.R. Fairey, 6 May 1932 and C. Galpin to C.R. Fairey, 6 May 1932 and 6 March 1933, FAAM/2010/054/0102 and FAAM/2010/106/0010.

73 With no surviving diary of the visit, reconstruction is based upon CRF's recollections in conversations with his daughter and the Air Secretary and photographs held in FAAM/052/0616-0656: conversation with Mrs Jane Tennant, 2 October 2015; 'Note of conversation between Mr. Fairey and Lord Amulree', 10 July 1931, FO371/15614, NA.

74 Ibid.; C.R. Fairey to Air Marshal Sir Robert Brooke-Popham, 29 September 1932, FAAM/2010/054/0102.

75 Miscellaneous correspondence between C.R. Fairey and A. Ozersky, and C.R. Fairey and Alexis Ignatieff, January–June 1932, and C.R. Fairey to V.P. Baranoff [P.I. Baranov], 10 October 1932, ibid.; Ivan Rodionov to the author, 4 September 2015; 'Chronology: 1933', www.warwick.ac.uk/aviaprom.

76 The Fairey party was told its demonstration film for the Firefly could only be screened in Baranov's presence, i.e. not during the visit. van de Velde, 'Moscow visit…', entries for 24 October to 7 November 1932, pp. 1–8.

77 Ibid. Tukhachevsky visited London with Litivinov in January 1936 and could conceivably have met CRF again: Maisky, *The Maisky Diaries*, 30 January 1936, pp. 154.

78 van de Velde, 'Moscow visit…', entry for 5 November 1932, p. 8.

79 Ibid., 'General business impressions', pp. 1–2; ibid., 'Moscow 1932', p. 3.

80 Draft letter of C.R. Fairey's invitation to Professor Toupoleff to speak for an hour at the March 1936 dinner of the Royal Aeronautical Society, 2 January 1936 and the approval of the letter by 'our friends' [Soviet Embassy] confirmed in A. Ignatiev to Miss E.L. Burns, 13 January 1932, FAAM/2010/054/0002. One sign of a recovery in overseas trade by 1936 was Ignatiev's new office in Shell-Mex House off the Strand.

81 C.R. Fairey to A. Ozersky, 7 January 1936, ibid.; H.A. Taylor, *Fairey Aircraft since 1915* (London: Putnam, 1974), p. 262. Note that no Fairey combat aircraft are listed for the air forces of either side in Rafael A. Permuy López, *Air War Over Spain: Aviators, Aircraft and Air Units of the Nationalist and Republican Air Forces 1936–1939* (Hersham: Classic/Ian Allan Publishing, 2009), pp. 26–7.

82 H.A. Taylor, *Fairey Aircraft since 1915* (New York: 1974), pp. 38–9; miscellaneous correspondence of E.O. Tips and C.R. Fairey, 1930s, FAAM/2010/054/0001 and of Baron de Cartier de Marchienne and C.R. Fairey, 1938–9, FAAM/2010/054/0045 and 0046. As a mark of how highly Belgian diplomats rated CRF he was asked to join the Air Secretary and the First Sea Lord at an embassy lunch in the spring of 1937: 'The Belgian Ambassador', *The Times*, 29 May 1937.

83 C.R. Fairey to Lord Marchwood, 20 December 1937, quoted in Trippe, 'The plane maker', p. 610. The award was for services to Belgium's aircraft industry and for fostering good relations with Great Britain; but it could only be worn at an embassy function or if the ambassador was present: ibid., p. 609.

84 Taylor, *Fairey Aircraft since 1915*, pp. 37–44, 172–5 and 197–206. See Chapter 8 re Avions Fairey 1940–6.

85 'Note of conversation between Mr. Fairey and Lord Amulree', 10 July 1931, FO371/15614, NA; Lord Amulree to Arthur Henderson, 10 July 1931 and Foreign Secretary, internal FO memo on the Government's position re Fairey Aviation and the Soviet Union, 21 July 1931, ibid.

86 van de Velde, 'Moscow visit…', entry for 28 October 1932, p. 4. On contrasting views of the Metro-Vickers trial's significance, see *The Diaries of Sir Robert Bruce Lockhart 1915–1938*, ed. Kenneth Young (London: Macmillan, 1973), 19 April 1933, p. 252. CRF told his daughter he saw the trial as reinforcing his reservations re-establishing an industrial presence inside the Soviet Union: Mrs Jane Tennant to the author, 6 September 2015.

87 On Maisky and Litvinov's personal and professional relationships, see Maisky, *The Maisky Diaries*, pp. xxxiii–xxxvi, xliv, 72, 95 and 183–4.

88 Photograph of Moore-Brabazon and Fairey with Ivan Maisky at 1935 RAeSoc garden party, *Flight London*, 9 May 1935.

89 Mrs Jane Tennant to the author, 6 September 2015. Maisky used gifts of caviar and vodka, as well as invitations to a private lunch or dinner, to cultivate good relations, and in the case of journalists even to compromise their professional identity: 'Introduction', Maisky, *The Maisky Diaries*, ed. Gorodetsky, p. xxi. Confirmation of Fairey's absence from Maiskey's diaries: Gabriel Gorodetsky to the author, 14 July 2015.

90 Invitation from HE Monsieur Ivan Maisky to Soviet Embassy reception, 2 May 1935 FAAM/2010/106/0010; C.R. Fairey to Ivan Maisky, 25 February 1936 and to V.K. Putna, 30 April 1936, FAAM/2010/054/0002 and 0003. Like Ozerksy, the military/air attaché Vitovt Putna was executed in 1937.

91 C.R. Fairey to Ivan Maisky, 24 February 1939, FAAM/2010/054/0004; Maisky, *The Maisky Diaries*, ed. Gorodetsky, 2 March 1939, p. 159.

92 'Many thanks for your kind hospitality and a most interesting luncheon': C.R. Fairey to Ivan Maisky, 11 August 1939, FAAM/2010/054/0002. On Maisky's ignorance of negotiations between Moscow and Berlin, see the editor's commentary in Maisky, *The Maisky Diaries*, ed. Gorodetsky, pp. 201–3, 214–15, 217 and 218,

and ibid., 22 and 24 August 1939, and 17 September 1939, pp. 217–18, 219 and 225–6.

93 Harold Nicolson to Vita Sackville-West, 7 April 1936, in Harold Nicolson, *Diaries and Letters 1930–39*, ed. Nigel Nicolson (London: Collins, 1966), pp. 255–6.

94 Among leading Labour politicians Maisky did respect Hugh Dalton, the sort of socialist intellectual CRF would have loathed. On a shared dislike of flying in a civil airliner, see C.R. Fairey to J.T.C. Moore-Brabazon, 30 December 1937, FAAM/2010/106/0013 and Maisky, *The Maisky Diaries*, ed. Gorodetsky, 28 April 1939, p. 176–7. On Maisky's prewar closeness to Beaverbrook, and H.G. Wells, see ibid., pp. xx and 50, and xlii.

95 Sir Alexander Cadogan, *The Diaries of Sir Alexander Cadogan 1938–1945*, ed. David Dilks (London: Cassell, 1971), 14 March 1941, p. 363 and Sir Henry Channon, *Chips The Diaries of Sir Henry Channon*, ed. Robert Rhodes James (London: Penguin, 1967), 28 November 1939, p. 277.

96 Search of intelligence files re C.R. Fairey and Russia/Soviet Union, 1929–40, NA, 28 April 2016.

97 CRF was introduced to Kennedy at the Goldsmiths' Hall on 16 November 1938: Lord Queenborough to C.R. Fairey, 24 October 1938, FAAM/2010/054/0004. Correspondence of Miss E.L. Burns and James E. Brown Jnr. [US Embassy, London], July 1939, FAAM2010/054/0005: the deteriorating international situation prevented the ambassador from joining CRF on the *Evaine* in August 1939.

98 C.R. Fairey to Ernest Brown MP, 12 October 1939 and to Air Commodore Arthur Conningham, 11 October 1939, FAAM/2010/054/0006 and 0045; 'I wrote an article on the two breeds of Bolshevism, to the effect that there is nothing to choose between the philosophies of Moscow and Berlin': Duff Cooper, *The Duff Cooper Diaries*, 23 August 1939, p. 273.

99 C.R. Fairey, draft article on 'Russia', autumn 1939, FAAM/2010/054/0102.

100 Edgerton, 'Technical innovation…', pp. 250–1.

101 King, *Knights of the Air: The Life and Times of the Extraordinary Pioneers who First Built British Aeroplanes* (London: Constable, 1989), p. 299; Ritchie, *Industry and Air Power*, p. 10.

102 Data in AIR2/1322 and *The Statist* [n.d.] quoted in ibid., pp. 10–11; ibid., p. 13.

103 Ibid., pp. 16–17. Within CRF's *bête noir*, the Royal Aircraft Establishment, engineers like Ronald McKinnon Wood (in 1935 a Labour parliamentary candidate) argued the opposite – that public and private experimentation and design should be brought together, which would necessitate nationalisation of the aircraft industry and a consequent avoidance of costly duplication. CRF, Sopwith et al. were vocal in maintaining that competition rendered private design more innovative. Edgerton, 'Technical innovation…', pp. 251–2.

104 Ritchie, *Industry and Air Power*, pp. 16–17.

105 King, *Knights of the Air*, pp. 303 and 262; Taylor, *Fairey Aircraft since 1915*, pp. 33–7. As the Air Ministry assured the SBAC in October 1935 that redundant plant would generate compensation once the expansion programme ended and that the Treasury's 'progress payments' would subsidise investment costs, Fairey had a financial safety-net for expanding its aero-engine division: Ritchie, *Industry and Air Power*, p. 46.

106 Taylor, *Fairey Aircraft since 1915*, pp. 14–15 and 208–30. The Hendon was always a stop gap solution, and when the Whitley and Wellington bombers entered production the Air Ministry cancelled its orders for a further 62. In contrast, 253 Gordon and Seal aeroplanes/seaplanes were sold to the RAF and Fleet Air Arm, and various overseas air forces. *Fairey Company Profile 1915–1960*, ed. Chorlton, pp. 48, 50 and 52.

107 C.R. Mullin's light aeroplane crashed in Belgium in September 1931. On C.S. Staniland's parallel career as an interwar racing driver, see WB, 'The man in white', *Motor Sport*, February 1997, pp. 156–61.

108 Oliver Stewart, 'Air eddies', *The Tatler*, 5 May 1937; 'R.A.S. flying display', *The Times*, 6 May 1935 and 'The garden party of the Royal Aeronautical Society', *The Aeroplane*, 8 May 1935. On 7 June 1935 Londonderry made way for Phillip Cunliffe-Lister [later Lord Swinton] and was effectively demoted to Lord Privy Seal and Leader in the House of Lords.

109 C.R. Fairey, quoted in 'Our air correspondent', 'Big reserves of warplanes needed – Mr Fairey's warning', *Daily Mail*, 5 January 1934. Rothermere made his demand for a 5,000-aircraft RAF in the *Daily Mail*, 7 November 1933: Brett Holman, *The Next War in the Air: Britain's Fear of the Bomber, 1908–1941* (Farnham: Ashgate, 2014), p. 189.

110 C.R. Fairey, draft briefing for National Press Agency, 22 January 1934, FAAM/2010/53/0071 and lecture to Imperial Defence College and RAF Staff College parties visiting Hayes plant, 14 February and 2 March 1934, FAAM/2010/53/0072; introduction of new Air League chairman at RSA debate on civil aviation control, 16 January 1933, FAAM/2010/53/0060.

111 Fenner Brockway, *The Bloody Traffic* (London: Victor Gollancz, 1934), Philip Noel-Baker, *Hawkers of Death: The Private Manufacture of Armaments* (London: Labour Party, 1934) [note the more than coincidental title – the author advocated an International Air Police to enforce global peace] and W.H. Williams, *Who's Who in Arms* (London: Labour Research Department, 1935); C.R. Fairey, speech to the AID annual dinner, 27 April 1934, FAAM/2010/53/0074. As well as being prominent in the ILP, from 1932 disaffiliated from the Labour Party, Brockway – a strict pacifist imprisoned as a CO – was chairman of the No More War Movement and thus active in the War Resisters International: Richard Overy, *The Morbid Age: Britain and the Crisis of Civilization, 1919–1939* (London: Penguin, 2010), pp. 192 and 239. Re Baldwin's openly expressed pessimism with regard to any future war, even after he succeeded MacDonald as PM and rearmament was under way, see ibid., pp. 176–7.

112 'Thanks to Pacifist Governments of the past this country ... was left more or less at the mercy of the dictatorships of Europe...': C.R. Fairey, after-dinner speech marking production of 100 Battle airframes, Stockport plant, 28 January 1938, FAAM/2010/53/089; Holman, *The Next War in the Air*, p. 189; 'The new air programme', *Daily Mail*, 23 May 1935, quoted in ibid. Re Rothermere's and Londonderry's contacts with Hitler, and their increasingly acrimonious dealings with each other, see Kershaw, *Making Friends with Hitler: Lord Londonderry and Britain's Road to War* (London: Penguin, 2005), pp. 35, 51, 58 60, 74 7, 84–5 and 102–11.

113 Addison, 'Patriotism under pressure', pp. 200–1; C.R. Fairey, introduction of new Air League chairman at RSA debate on civil aviation control, 16 January 1933, FAAM/2010/53/0060 and motion of appreciation of HRH Prince of Wales, Municipal Airports Conference, 8 December 1933, FAAM/2010/53/0068. The Air League's respectability extended to exclusion of dissident Tory MPs, including Churchill: Edgerton, *England and the Aeroplane*, p. 78.

114 Fairey Aviation, 'Report of the directors', December 1934, FAAM/2010/054/0001.; ibid., December 1933: 5% dividend paid out of £56,575 in 1934, as opposed to a 10% dividend paid out of £98,541 in 1933.

115 Marquess of Londonderry to C.R. Fairey, 12 October 1932, FAAM/2010/054/0030; Fairey Aviation Sales Department, 'Memorandum on export transactions', 21 January 1936, Box 07, [?], RAFM; C.R. Fairey, 'Selling aircraft abroad', Royal Aero Club after-dinner speech, 1 November 1933, FAAM/2010/53/0065. 'We make the finest aircraft in the world and they [Canadians] don't want to admit it. They don't want to buy them because they are British, that's all.': C.R. Fairey to Charles G. Grey, 1 February 1932, FAAM/2010/050/0088.

116 *The Statist*, 30 March 1935, quoted in Ritchie, *Industry and Air Power*, p. 190. 'After 1936, firms often succeeded in fulfilling their financial requirements by capitalising reserves and by borrowing, and the need to float new issues was, in any case, reduced after the government began to finance capital extensions in 1938 ... By the outbreak of war ... the share capital of the principal aircraft firms had been raised to more than £19 million.': ibid., p. 90.

117 'Redeemable preference share issue': F. Goetz [of Myers & Co share brokers] to C.R. Fairey, 1 February 1934 and annual Fairey Aviation, 'Report of the directors', December 1934–9, FAAM/2010/054/0001. 'Wuffy' Dawson left the board in 1934 and was replaced by one-time RNAS test pilot L. Massey Hilton. He joined CRF, Maurice Wright, Wilf Broadbent, T.M. Barlow and successive company secretaries: A.G. Hazell, Charles Crisp (temporary following Hazell's death) and from 1937 C.C. Vinson. Other than Vinson, they were all still board members when CRF died in 1956.

118 'Shares purchased through Myers & Co [and Barclays Bank]', April 1936 and E.L. Burns to C.R. Fairey, 23 April 1936, FAAM/2010/054/0003. An opportunity lost was CRF's repeated refusal to become the first chairman of London and Coastal

Oil Wharves Ltd, which went on to build a highly profitable oil and chemical bulk storage facility on Canvey Island [today's Oikos Storage]: C.R. Fairey to W.H. Botsford, 6 April 1936, ibid. W.L. Duke to C.R. Fairey, and C.R. Fairey to F.G.T. Dawson [in Montreal], 2 November 1939, FAAM/2010/054/0006.

119 'Fairey Aviation', *Sunday Times*, 1937.
120 C.R. Fairey, draft letter to the editor of *The Times*, 27 April 1936 and Charles Vinson to C.R. Fairey, 2 May 1936, FAAM/2010/054/0003.
121 Edgerton, 'Technical innovation...' pp. 251–6; David G. Anderson, 'British rearmament and the "merchants of death": The 1935–6 Royal Commission on the Manufacture and Trade in Armaments', *Journal of Contemporary History*, 29 January (1994), pp. 5–37.
122 *Minutes of Evidence Taken Before the Royal Commission on the Private Manufacture of and Trading in Arms, Seventeenth Day, Friday 7th February, 1936* (London: HMSO, 1936), pp. 518–30.
123 Correspondence of C.R. Fairey and Charles Allen, November 1935–February 1936, drafts of 'Memorandum by the SBAC Limited, January 1936, annotated Royal Commission evidence and commission member profiles, Box 07, [?], RAFM; *Minutes of Evidence Taken Before the Royal Commission on the Private Manufacture of and Trading in Arms*, pp. 518–30; 'Arms Commission told of the RAF's needs', 7 February 1936 and miscellaneous national press, 8 February 1936; C.G.G., 'On the Royal Commission on the private manufacturing of and trading in arms and its relations with the British aircraft industry', *The Aeroplane*, 14 February 1936; *Baltimore Sun, New York Herald Tribune* and *New York Times*, 8 February 1936; Commander Sir Charles Craven to C.R. Fairey, 11 February 1936 and C.R. Fairey to Commander Sir Charles Craven, 18 February, 1936, FAAM/2010/054/0002.
124 Anderson, 'British rearmament and the "merchants of death"', pp. 28–9. News of the Rhineland's remilitarisation prompted a brief suspension of the Guildhall sittings and the crisis's debilitating impact on the Commission was compounded by the announcement of a second defence white paper and the parallel appointment of Inskip: ibid., p. 24.

Chapter 7

1 W.H. Auden, 'September 1, 1939.'
2 Sung to the tune of 'My bonny lies over the ocean' and included in *The Fleet Air Arm Songbook*, quoted in David Wragg, *Stringbag: The Fairey Swordfish at War* (Barnsley: Pen & Sword Aviation, 2004), p. 171.
3 Ibid., pp. 5–11. On flying a Swordfish, see ibid., pp. 12–23.
4 Ibid., pp. 116–21, 42–83 and 103–164. On the largely forgotten role of Swordfish spotters and dive-bombers in the Norwegian campaign, spring 1940, see Geoffrey

Till, *Air Power and the Royal Navy 1914–45: A Historic Survey* (Colchester: TBS, 1979), pp. 17–19 and 178.
5. Wragg, *Stringbag*, pp. 130–69.
6. Installing a 93-gallon tank in the middle cockpit, as for torpedo squadrons taking part in the Taranto raid, doubled the Swordfish's normal range of 450 nautical miles. Maximum speed was 138 mph, although aircrew believed 100 knots to be more normal when the aircraft was fully loaded. Ibid., pp. 170–1.
7. Ibid., pp. 42–62; David Wragg, *Swordfish: The Story of the Taranto Raid* (London: Weidenfeld and Nicolson, 2003), pp. 16–18; Charles Lamb, *War in a Stringbag* (London: Leo Cooper, 1987), pp. 40–4.
8. John Moffatt with Mike Rossiter, *I Sank the Bismarck: Memoirs of a Second World War Navy Pilot* (London: Bantam Press, 2009), p. 247.
9. Ibid., pp. 64–5; Lamb, *War in a Stringbag*, pp. 78, 168–9 and 199; H.A. Taylor, *Fairey Aircraft since 1915* (New York: Putnam, 1974), pp. 236–60 and 288–95; *Fairey Company Profile 1915–1960*, ed. Chorlton, pp. 58–9, 76–7, 64–5 and 68–9; Till, *Air Power and the Royal Navy*, pp. 95–6 and 101–2.
10. For details of all FAA Swordfish squadrons, see Wragg, *Stringbag*, pp. 182–210.
11. G. Till, *Air Power and the Royal Navy*, pp. 101–2 and 125.
12. Wilfred Broadbent, interviewed in Trippe, 'The plane maker', pp. 720–7.
13. H.A. Taylor, *Fairey Aircraft since 1915*, pp. 231–60; Matthew Willis, 'A recipe for obsolescence? British naval torpedo bomber development in the 1930s', Naval Air History, 17 September 2015: www.navalairhistory.com/2012/09/17/a-recipe-for-obsolescence-british-naval-torpedo-development-in-the-1930s/.
14. G. Till, *Air Power and the Royal Navy*, pp. 104–10; Wragg, *Swordfish*, pp. 18–19; *Fleet Air Arm Aircraft of World War 2: British-built Types Used by the Royal Navy*, ed. M. Hooks, (Cudham: Kelsey Publishing Goup, 2013), p. 63; *Fairey Company Profile 1915–1960*, pp. 58–9 and 72–3 and 78–9.
15. For the Royal Navy's own Swordfish-related propaganda, see Admiralty and Ministry of Information, *Fleet Air Arm: The Admiralty Account of Naval Air Operations* (London: HMSO, 1943) and *East of Malta West of Suez: The Admiralty Account of the Naval War in the Eastern Mediterranean: September 1939 to March 1941* (London: HMSO, 1943).
16. McClintock's reputation rested on his financial nous and his forensic abilities as an auditor: his highly publicised inspection of the BBC's finances in 1934 gave a clean bill of health to Sir John Reith's much criticised corporation. Asa Briggs, *The History of Broadcasting in the United Kingdom Volume II: The Golden Age of Wireless* (Oxford: OUP, 1995), p. 392.
17. Chairman's report to Fairey Aviation OGM, 31 December 1936, read by Maurice Wright as CRF absent with influenza: 'Aircraft profit limit criticism', *Financial Times*, 1 January 1937. Report of 11th OGM in 'The Fairey Aviation Co.', *Financial News*, 30 December 1939.

18 N.H. Gibbs, *History of the Second World War: Grand Strategy. Volume 1 Rearmament Policy* (London: HMSO, 1976), pp. 106–9 and 560.
19 Ritchie, *Industry and Air Power*, pp. 41 and 32–3; C.R. Fairey [on behalf of SBAC] to The Secretary, Air Ministry, 12 February 1935, FAAM/2010/054/0041.
20 Gibbs, *History of the Second World War: Grand Strategy Volume 1*, pp. 561, 177 and 564–5. 'By October 1936 no fewer than 55 different experimental designs were in progress and six more were under consideration.' A positive view is that multiple projects spread the risk of failure and that delayed product specialisation facilitated volume production of proven designs. Ritchie, *Industry and Air Power*, pp. 54, 49 and 258.
21 Ibid., pp. 41–3 and 257; G.A.H. Gordon, *British Seapower and Procurement Between the Wars* (London: Macmillan, 1988), p. 287.
22 'When I tell you that the German output exceeds 600 aeroplanes a month you will realise the scope of the task before us. We see no remedy for the situation other than to make England strong and too dangerous to interfere with. We have tried the other scheme of disarmament and sweet reason and it did not work … The Government is very earnest in its intentions that that security shall be achieved. We do not know how much time we have. We only know that it is too short.' C.R. Fairey, after-dinner speech marking production of 100 Battle airframes, 28 January 1938, FAAM/2010/053/0089.
23 Gibbs, *History of the Second World War: Grand Strategy Volume 1*, pp. 562–5 and 362–70; Hayward, *The British Aircraft Industry*, p. 20; King, *Knights of the Air*, pp. 305 and 312.
24 Whitehall perception of his company as the exception to the rule is seen in Fairey's invitation to a discreet briefing for key industrialists by Sir Thomas Inskip, Minister for the Coordination of Defence, at The Savoy on 14 April 1937: Sir Charles Craven [MD, Vickers] to C.R. Fairey, 24 March 1937, FAAM/2010/106/0013.
25 Participant firms received an annual management fee of £50,000 and a royalty for each unit produced [£75 for aero-engines and £200 for aircraft], as well as bonuses for costs kept below estimates. The contribution of motor manufacturers was recognised once the war began, as previously there were severe problems with Rootes' shadow factory at Speke and, despite eventually running the giant Spitfire production plant at Castle Bromwich, Lord Nuffield was initially averse to involving Morris Motors. Ibid., p. 304; Hayward, *The British Aircraft Industry*, p. 20; Edgerton, 'Technical innovation, industrial capacity and efficiency: Public ownership and the British military aircraft industry, 1935–48', pp. 257–9.
26 Ibid.; Taylor, *Fairey Aircraft since 1915*, pp. 20–1; J.T.C. Moore-Brabazon's Commons speech, 17 November 1938, quoted in Edgerton, 'Technical innovation…', p. 261.

27 King, *Knights of the Air*, p. 304; Taylor, *Fairey Aircraft since 1915*, pp. 18–19; C.R. Fairey, Reply to speech of Lord Mayor of Manchester, opening of Ringway Airport, 8 June 1937, FAAM/2010/57/087.
28 Gibbs, *History of the Second World War: Grand Strategy Volume 1*, pp. 567–71, 302–3 and 574–89.
29 C.R. Fairey to Sir Charles Bruce-Gardner, 14 April 1938, FAAM/2010/054/0043; Ritchie, *Industry and Air Power*, pp. 54–9, 83, 43 and 259–61.
30 J.T.C. Moore-Brabazon to C.R. Fairey, 30 May 1938 and C.R. Fairey to J.T.C. Moore-Brabazon, 2, 8 and 9 June 1938, AC71/3, LBP, RAFM; Robert Self, *Neville Chamberlain: A Biography* (Farnham: Ashgate, 2006), pp. 312–13.
31 In June 1938 Moore-Brabazon showed his gratitude for the Fairey family's hospitality by despatching a television to Woodlands, which 'performed perfectly in the midst of an admiring family circle.' Ibid. He had previously booked premier class accommodation for the liner *Queen Elizabeth*'s inaugural voyage, which because of the war never materialised: J.T.C. Moore-Brabazon correspondence with Cunard Ltd, February 1938, AC71/3, LBP, RAFM.
32 C.R. Fairey, draft letter to the editor of *Truth* re issue of 24 May 1940, 27 May 1940, ibid. R.B. Cockett, 'Communication: Ball, Chamberlain and *Truth*', *The Historical Journal*, 33/1 (1990), pp. 131–42. CRF ignored earlier requests from the Conservative Party's National Publicity Bureau to generate financial support for *Truth* among his fellow executives: Cyril Potter [NPB treasurer] to C.R. Fairey, 23 November 1938 and 9 January 1939, FAAM/2010/054/0004.
33 C.R. Fairey to J.T.C. Moore-Brabazon, 21 July 1940, ibid. Moore-Brabazon expressed similar sentiments in a hand-written reply the following day.
34 Freeman's director generals for R&D and for production were the RAF's rising star Air Vice-Marshal Arthur Tedder and the brilliant railway engineer/industrialist Ernest Lemon, separately or together more than a match for CRF and the SBAC's other senior members. Ibid., pp. 257 and 50–2; miscellaneous correspondence of C.R. Fairey and Air Marshal Sir Wilfred Freeman [Air Member for Development and Production], 1938–9, FAAM/2010/53/.
35 Edgerton, 'Technical innovation…', pp. 261–2. Between postings as AMDP and MAP Chief Executive, Fedden served as Deputy Chief of the Air Staff, 1940–2.
36 Ritchie, *Industry and Air Power*, pp. 152 and 150–1; Edgerton, 'Technical innovation, industrial capacity and efficiency: Public ownership and the British military aircraft industry, 1935–48', p. 274, fn 51.
37 Ibid., pp. 261–2; Edgerton, *England and the Aeroplane*, pp. 47–8.
38 Ritchie, *Industry and Air Power*, pp. 88–9, 94, 153, 104 and 220–1. CRF remained certain that maintaining a parent plant at maximum efficiency was more conducive to higher output than subcontracting: C.R. Fairey to Sir Charles Bruce-Gardner, 27 October 1938, quoted in ibid., p. 104.
39 Wilfred Broadbent, interviewed in Trippe, 'The plane maker', pp. 658–62.

40 Instead of Barlow the Deputy Air Member for Development and Production visited the shadow factory for two days in autumn 1938 and consequently imposed a lengthy action plan: correspondence re Austin of Air Marshal Sir Wilfred Freeman and C.R. Fairey, September–October 1938 and October 1939, Box 136, AC73/30, RAFM.

41 In 1937 only 80 of the anticipated 220 Battles were built. Ritchie, *Industry and Air Power*, pp. 77 and 79; 'Building the Battle' and 'Designed for mass-production', *Flight*, 17 June and 19 August 1937, p. 184 and pp. 593–9.

42 C.R. Fairey to Air Ministry Supply Officer, 24 February 1936, quoted in Ritchie, *Industry and Air Power*, p. 83; ibid., pp. 83, 77, 79, 84, 85 and 90.

43 'The Battle was particularly vulnerable with its slow speed, poor defensive armament, lack of armoured protection for the crews, small bombload, and peculiar tactical characteristics. Inexplicably, the Battle's bombsight was connected to the Merlin Engine manifold, thus becoming white hot in flight preventing readjustment. Thus, Battle crews were constrained to level predictable attacks at 2,000 feet – easy targets for German Flak.': Stuart W. Peach, 'A neglected turning point in air power history: Air power and the fall of France' in Sebastian Cox and Peter Gray (eds), *Air Power History: Turning Points from Kitty Hawk to Kosovo* (London: Frank Cass, 2002), p. 155.

44 The Battle was largely confined to daylight operations as at night sparks from the Merlin's exhaust blinded the pilot: ibid., p. 170, fn. 59.

45 Taylor, *Fairey Aircraft since 1915*, pp. 264–83; *Fairey Company Profile 1915–1960*, pp. 62–3.

46 Official war historian M.M. Postan, cited in King, *Knights of the Air*, pp. 314–15; Taylor, *Fairey Aircraft since 1915*, pp. 271, 275 and 277; W.L. Scott [Air Ministry procurement] to C.R. Fairey, 13 April 1939 and C.R. Fairey to Air Marshal Sir Wilfred R. Freeman, 6 October 1939, Box 136, AC73/30, RAFM.

47 C.R. Fairey to Lt Col. J.T.C. Moore-Brabazon, 30 December 1937, FAAM/2010/106/0013; C.R. Fairey, draft of statement to the Civil Aviation Planning Committee, Whitehall, 27 January 1938 [drafted 26 January 1938], FAAM2010/53/088; draft of oral evidence given to the sixth meeting of the CAPC in Sir Charles Bruce-Gardner's office, 28 March 1929, Box 136, AC73/30, RAFM. In November 1939 Imperial Airways was merged with the much smaller British Airways to create the British Overseas Airways Corporation (BOAC), operational from 1 April 1940.

48 *Fairey Company Profile 1915–1960*, p. 70.

49 Taylor, *Fairey Aircraft since 1915*, pp. 296–300; W. Hildred [Deputy D.-G. of Civil Aviation] to C.R. Fairey, 13 June 1938, C.R. Fairey to Air Ministry Secretary, 19 August 1938 and E.A. Pearse to C.R. Fairey, 17 October 1939, Box 136, AC73/30, RAFM; confirmation of order agreed with BOAC at meeting on 23 January 1939 in C.R. Fairey to Sir John Reith, 24 January 1939 [not sent], FAAM/2010/054/0004.

50 Ibid., pp. 300–3. The intention was that the initial Bristol Taurus radial engine would soon make way for the powerful sleeve-valve Rolls Royce Exe.
51 Ibid., pp. 301–2; 'A real Fairey story', *Flight*, 20 February 1941, p. 162. There was speculation at Hayes in 1945 that CRF might on his return from Washington sanction three new prototypes of the FC1, boasting even more powerful engines and a production target of 200 aircraft: *Fairey Company Profile 1915–1960*, p. 70.
52 Anthony Heckstall-Smith, quoted in Ian Dear, *Enterprise to Endeavour The J-Class Yachts* (London: Adlard Coles Nautical, 1999), p. 69.
53 Passenger lists for Atlantic Ferry Organisation/USAAF C54, arrive New York, 7 June 1943, *Olympic*, arrive New York 12 September 1934 and *Aquitania*, arrive Southampton 9 October 1934 and 7 April 1937, National Archives, via www.ancestry.co.uk [outward journey March 1937, on *Queen Mary*]. On the superior performance and comfort of the Skymaster over its British equivalent in the mid-1940s, the sparsely equipped Avro York, see Air Commodore John Mitchell LVO DFC AFC with Sean Feast, *Churchill's Navigator* (London: Grub Street, 2010), pp. 115–19.
54 Lord Brabazon, *The Brabazon Story* (London: Heinmann, 1956), pp. 172–3. With Parliament in recess Moore-Brabazon was in Newport as the *Morning Post*'s sailing correspondent.
55 Contrary to popular assumption the size of a 12-metre class yacht is *not* determined by its length, but a succession of periodically revised complex formulae known as regulation rules.
56 Bramson, *Pure Luck*, pp. 139–41; Camper & Nicholsons, 'Our Heritage', www.camperandnicholsons.com/about/company-heritage.htm; Dear, *Enterprise to Endeavour*, pp. 84–6.
57 C.R. Fairey, 'On some applications of aeronautical science to yacht design', draft article for September 1939 issue of *Aeronautics*, 12 July 1939, FAAM/2010/106/0101.
58 Major B. Heckstall-Smith, 'The America's Cup Challenge "The right man to sail her"', *Yachting World and Motor Boating Journal*, 3 November 1933.
59 The RTYC clubhouse's 'Coffee Room' is in fact a restaurant. Correspondence of C.R. Fairey with T.O.M. Sopwith and Brice Slater, March–May 1927, FAAM/2010/054/0022; C.R. Fairey, correspondence re hotel bookings and moorings, July–October 1927, FAAM/2020/054/0023. In February 1936 Fairey joined the Royal Dart Club, reflecting the amount of time he now spent in Devon.
60 'Yachting and other engagements May to August 1929', 'Yacht racing – entries and results. 1930' and 'Yacht racing – entries and results. 1931', FAAM/2010/106/0017-189; correspondence re renewed rental of Pier House, IOW, 1936, FAAM/2020/054/0003.
61 Miscellaneous business correspondence of E.L. Burns and C.R. Fairey, April–May 1936 and August–October 1937, including arrangements for Maurice Wright to meet *Evadne* at Cannes, 18 April 1936, ibid.

62 'Cover Picture', *Yachting World*, February 1956; www.superyachts.com/motor-yacht-3192/marala.htm; Bramson, *Pure Luck*, pp. 140 and 147–54. *Evadne* is today registered as M/Y *Marala*, with basic specification the same as when she was launched in 1931. Early in 1937 she underwent an extensive – and expensive – refit and rebuild, with Heal's commissioned to refurbish the interior. Correspondence between C.R. Fairey and Barry Neame, January–November 1932, FAAM/01/1932-11/1932; www.hindsheadbray.com.

63 Louis Breguet to C.R. Fairey, 6 May 1936, FAAM/2010/054/0003; 'Whitsun and Deauville', *Morning Post*, 20 May 1936.

64 C.R. Fairey to A.-H. van Scherpenberg, 7 January 1936 and Wolfgang Gans Edler zu Putlitz [First Secretary] to C.R. Fairey, 11 January 1936, FAAM/2010/054/0002. CRF's daughter believes he was blackballed by RYS members because he was divorced: conversation with Mrs Jane Tennant, 15 January 2016. In the 1960s Peter Trippe claimed that Fairey felt harshly treated by the RYS and 'Nothing would satisfy him now until he became a leader of yachting society.' Trippe, 'The plane maker', p. 608.

65 Director, National Maritime Museum to C.R. Fairey, 3 March 1937, Joseph Mears Launches and Motors Ltd to E.L. Burns, 24 April 1937, invitation of C.R. Fairey to 'Coronation Naval review. 20 May 1937', correspondence of C.R. Fairey and Sir Malcolm Campbell, May 1937 and miscellaneous correspondence re the Coronation Regatta and *Evadne* guests, 18 June–3 July 1937, FAAM/2010/106/0013; C.R. Fairey, draft talk for 25 minutes broadcast from the Marine Spa, Torquay, Saturday 19 June 1937, 2 June 1937, FAAM/2010/053/0083; E.L. Burns to G.R. Barnes [Assistant Director of Talks, BBC], 27 April 1937, FAAM/2010/054/0010.

66 Hall, *Sir Richard Fairey*, p. 10; 12 Metre Class, 'Flica', www.12mrclass.com/yachts/detail/273-itemId.511707028; R.A. Smith, 'The Flica Project', www.americascupmasters.com/_/The_Flica_Project; 'Yacht racing – entries and results. 1932' and 'Yacht racing – entries and results. 1933', FAAM2010/106/0020-1.

67 Ibid.; C.R. Fairey, 'On some applications of aeronautical science to yacht design'; 'Twelve metre yacht "Flica" – 1/8th scale model sails research at Hayes, Middlesex. Model Sails – 1st series. Interim report nos 1–8', February–May 1933, FAAM/2010/106/0011-16; Dear, *Enterprise to Endeavour*, pp. 90–1. The quadrilateral jib, 'had amongst its several advantages the ability not to backwind a yacht's staysail as its luff was cut parallel to the latter's leach': ibid., p. 90.

68 With Breguet a fellow competitor in the Baltic, CRF sailed *Flica* to three firsts, a second and a third in seven races: 'K.N.S. Jubileumsregatta ved Hanko 5–9 juli 1933', 'Yacht racing – entries and results. 1932' and 'Yacht racing – entries and results. 1933', FAAM/2010/106/0020-22; 'Prize winning yachts in 1933', *Yachting World and Motor Boating Journal*, 8 December 1933.

69 R.A. Smith, 'The Flica Project.' The name *Flica* was revived by the Fairey family when Richard entered a Huntress 23 yacht in the 1960 Miami-Nassau Race.

Author's conversation with America's Cup historian Bob Fisher, 7 April 2016. Heckstall-Smith, 'The America's Cup Challenge "The right man to sail her"'; 'Another challenge to U.S.', *The Times*, 15 December 1933.
70 12 Metre Class, 'Westra', www.12mrclass.com/yachts/detail/273-itemId.511707080.html; correspondence of C.R. Fairey and Arthur C. Connell, February 1936, FAAM/2010/106/0002.
71 *Evaine*, 70 feet in length and 23.84 tons in weight, was launched at Gosport on 20 May 1936. Sir William Burton [President International Yacht Racing Union] to C.R. Fairey, 21 May 1936: '…may I say how pleased I am that you have bought [C.E.A.] Hartridge's boat and come back into the Class?', FAAM/2010/054/0003.
72 Miscellaneous correspondence re *Evaine* of C.R. Fairey and Laurent Giles and Partners, February–April 1939, and C.R. Fairey's assistant private secretary to C.B. Bond, 13 April 1939, FAAM/2010/054/0005. C.R. Fairey to Ernest Brown, 21 March 1939 and to Sir William Burton, 16 May 1939, ibid.
73 12 Metre Class, 'Evaine' and 'Trivia', www.12mrclass.com/yachts/detail/273-itemId.511707058 and www.12mrclass.com/yachts/detail/273-itemId.511707076.
74 Miscellaneous correspondence of C.R. Fairey and Director of Sea Transport, Mercantile Marine Department, April and August–September 1939 and Ernest Brown to C.R. Fairey, 30 October 1939 [including Winston Churchill to Sir John Simon, 16 October 1939], FAAM/2010/054/0005-6. Re a coordinated claim for compensation, see C.R. Fairey to T.O.M. Sopwith, 5 October 1939, FAAM/2010/054/0007.
75 C.R. Fairey to Vice-Admiral Sir Noel Lawrence, 12 October 1939 and to Director of Sea Transport, Mercantile Marine Department, 7 December 1939, FAAM/2010/054/0006; miscellaneous correspondence of C.R. Fairey with Captain J.C.J. Soutter, September–November 1939 and with Lieutenant H.N. Taylor, September–December 1939, ibid., and April–June 1940, FAAM/2010/054/0075.
76 C.R. Fairey to Lieutenant H.N. Taylor, 13 December 1939, ibid.
77 Interview with Noel Raymond Pugh, 28 April 1999, Imperial War Museum oral archive, www.iwm.org.uk/collections/item/object/80017910.
78 'Precis of Attacks by EVADNE, RECRUIT, and PINCHER', 23 April 1945, Anti-U-Boat Division, ADM1/1760, NA.
79 V.R., 'Who's who in the air: Charles Fairey', *Daily Express*, 10 February 1936.
80 Bramson, *Pure Luck*, pp. 140–2; 'Shamrock V. Model sails: testing and research at Hayes, Middlesex. Progress report, 17 February to 31 May 1934', 31 May 1934, FAAM/2010/106/00?? [uncatalogued]; Arthur Lamsley, 'A season with the international classes', *Yachting World and Motor Boating Journal*, 22 September 1934. For the 1934 season CRF had *Shamrock*'s hull painted RAF Blue.
81 Arthur Lamsley, 'Endeavour's crew strike' and 'Mr. Sopwith completes his new crew', *Yachtsman*, 21 and 24 July 1934; Dear, *Enterprise to Endeavour*, pp. 91–3

and 104–5. Another wife regularly raced on her husband's yacht: Esther Fairey served as time-keeper, 'for which you had to have nerves of iron!': Mrs Jane Tennant to the author, 15 June 2017.
82 J Class Association, 'History: 1929–1937 The Golden Years', www.jclassyachts.com/history/1919-1937; Bramson, *Pure Luck*, pp. 144–7; Hall, *Sir Richard Fairey*, p. 10. On the ungainly, overweight *Endeavour II* and its ill-fated challenge, see Dear, *Enterprise to Endeavour*, pp. 135 and 144–50.
83 Hall, *Sir Richard Fairey*, p. 10.
84 Summary of after-dinner speech by C.R. Fairey, autumn 1935, in Dear, *Enterprise to Endeavour*, p. 118.
85 Beecher Moore interview in *Yachts and Yachting*, February 1976, quoted in Dear, *Enterprise to Endeavour*, p. 92.
86 Dear, *Enterprise to Endeavour*, p. 133–4; 'America's Cup withdrawal. Challenger's friendly gesture', *News Chronicle*, 27 November 1935; Special Correspondent, 'America's Cup – Mr Fairey talks of future contests', *Evening Standard*, 23 December 1935; C.R. Fairey to R.A. Eccleston and to [?] Morgan [NYYC], 1 February 1936, FAAM/2010/054/0002. Travel arrangements re meeting of C.R. Fairey with New York Yacht Club, March-April 1937, FAAM/2010/054/0003.
87 C.R. Fairey to T.O.M. Sopwith, 3 January 1936, and to Sydney Walton, 17 January 1936, FAAM/2010/054/0002.0.
88 C.R. Fairey to R.A. Eccleston, 1 February 1936, ibid.; 'America's Cup withdrawal. Challenger's friendly gesture', *News Chronicle*, 27 November 1935; Special Correspondent, 'America's Cup – Mr Fairey talks of future contests', *Evening Standard*, 23 December 1935.
89 Invitation to shooting party on 8 December 1939: C.R. Fairey to T.O.M. Sopwith, 29 November 1939, FAAM/2010/0007.
90 The variety of Fairey's investment portfolio is shown in his ownership of the Hind's Head Hotel, near Maidenhead and a Butlins-style holiday camp, which in its first year received 17,000 bookings and made a profit, prior to being taken over by the War Office at the start of the war: 'British Holiday Estates Limited: Report of the Directors for the Year ended 31st October 1939', 23 May 1940, FAAM/054/0006.
91 Mrs Jane Tennant to the author, 2 October 2015.
92 A. Camden [Goldsmiths and Silversmiths Company] to Miss E.L. Burns, 16 May, 22 June and 1 July 1939, FAAM/2010/0005.
93 Miscellaneous newspaper reports, 23 and 24 April 1935; garden portrait of the Fairey family with pet dog, *Sketch London*, 14 August 1935 and 'Mr and Mrs Charles Fairey at Home', *Tatler*, 21 August 1935. Angus, the much photographed west highland terrier and Dan his companion dachshund, were stolen over an Easter weekend; a forlorn CRF, despite his low opinion of the Metropolitan Police, found it necessary to request the capital's constabulary assist their Uxbridge colleagues and publicise a £20 reward for the dogs' safe return: Miss E.L. Burns to Secretary, New Scotland Yard, 23 April 1937, FAAM/2010/054/0013.

94 Conversation with Mrs Jane Tennant, 15 January 2016.
95 Miscellaneous correspondence re residency of Mrs E. Fairey and staff at St Mildred's Hotel, Westgate-on-Sea, 4–30 June 1937, ibid. Mrs Jane Tennant to the author, 2 October 2015.
96 John Fairey and the now Mrs Jane Tennant, quoted in Trippe, 'The plane maker', p. 66.
97 Trippe, 'The plane maker', pp. 64–5 and 66–7. The ghost story is set at dusk on the bank of the River Test, where the narrator contrasts fly fishing methods with his counterpart from the late 21st century and it appeared in the *Journal of the Fly Fishers' Club*: C.R. Fairey, manuscript of 'A mid-summer's eve encounter', [n.d.], FAAM/2010/053/0018.
98 Details re Richard Fairey's 21st birthday party in Wentworth Day, draft synopsis for book 'Wings over the world: the official life of Sir Richard Fairey, M.B.E., F.R.Ae.S.'
99 Trippe, 'The plane maker', pp. 621–2; Mrs Jane Tennant to the author, 2 October 2015.
100 Miss E.L. Burns to C.R. Fairey, 7 April 1936 and Park Ward & Co to Miss E.L. Burns, 28 April 1936, FAAM/2010/054/0003; miscellaneous correspondence re Fairey family cruise of the Caribbean, February 1937, FAAM/2010/106/0013; Miss E.L. Burns to W. Langton, 22 June 1937, re visit of Richard Fairey to RAF Display, 26 June 1937 [Langton, the chauffeur, and an accompanying maid to provide RF and his seven guests with six bottles of gin, two of whiskey and one of vermouth, as well as two dozen lagers!], FAAM/2010/106/0012; Max Miller booking via Fosters Agency Entertainment Bureau, 23 December 1937, FAAM/2010/106/0010.
101 Mrs Jane Tennant to the author, 2 October 2015.
102 Trippe, 'The plane maker', pp. 976–7; *Thistledown* (Arthur B. Woods, Warner Bros, UK, 1938); 'Mrs L.A. Fairey [neé Berger]', obituary, *The Times*, 11 July 1944.
103 Mrs Jane Tennant to the author, 2 October 2015. Trippe, 'The plane maker', pp. 987–92 and 980–4.
104 Invitation of C.R. Fairey to dinner of the Derby Club, 27 May 1937, FAAM/2010/050/0003; 'The Shipwrights' Company', *Shipping World*, 21 April 1937; 'Speed of air raid held London peril', *New York Times*, 28 May 1937.
105 Miscellaneous correspondence re the Savage Club and the British Sportsman's Club, 1937–9, FAAM/2010/054/0005; 'Bradman's Verbal Innings', British Movietone, 28 March 1938, www.youtube.com/watch?v=4UtNRBBtxWM.
106 London *Gazette*, 31 December 1937.
107 CRF felt the pressure of the rearmament drive meant he had no time to fulfil his obligations as a TA honorary colonel: Correspondence of C.R. Fairey and Colonel H. Leigh, March–April 1938, FAAM/2010/054/0073.
108 Schedule for 7 June 1939, in Miss E.L. Burns to C.R. Fairey, 5 June 1939, FAAM/2010/054/0005.

109 Fairey's Lagonda dealer drove his saloon to and from the Beau Site Hotel in Cannes: Richard Watney [Lagonda] to C.R. Fairey 1939, FAAM/2010/054/0004; C.R. Fairey to Mrs E.L. Haydon, 3 and 12 April 1939, FAAM/2010/054/0005. Conversation with Mrs Jane Tennant, 15 January 2016.

110 C.R. Fairey to B. Rhodes [Wessex area treasurer, National Union of Conservative and Unionist Associations], 6 April 1936, FAAM/2010/054/0003; miscellaneous correspondence re shoot and room bookings at respectively Flixton Hall, Bungay, and Magpie Hotel, Harleston, November–December 1938 and Lord Marchwood to C.R. Fairey, 29 December 1938, FAAM/2010/054/0004.

111 Ibid.; re meeting ambitious young MPs on the butts, see W.D. 'Toby' Thelleson to C.R. Fairey, 22 January 1939, FAAM/2010/054/0004.

112 Ernest Brown, dedication in memory of 3 September 1939 on the reverse of a photograph the Minister of Labour gave to Fairey later that year; in the possession of Mrs Jane Tennant.

113 Chamberlain's last visit was on Sunday, 24 March 1940: Lord Lloyd to C.R. Fairey, 11 March 1940, FAAM/2010/054/0044.

114 Algernon E. Maudslay to C.R. Fairey, 10 May 1937, FAAM/2010/106/0013; C.R. Fairey to R.T. Outen, 29 April 1937 and to Duncan B. Gray, 9 June 1937, FAAM/2010/106/0010.

115 Horsebridge Lake on the Bossington estate is aquifer-fed. 'Chalkstream Fly Fishing in Hampshire', www.fishingbreaks.co.uk/hampshire; 'Bossington – fly fishing', www.bossingtonestate.com/flyfishing?gclid=EAIaIQobChMI4bbNs_y-2AIVbrHtCh 30VQXWEAAYASAAEgJ6yPD_BwE; John Waller Hills, *A Summer on the Test*, 1930 second edn reprint (London: André Deutsch, 1983), pp. xv and 34; C.R. Fairey, manuscript of 'A mid-summer's eve encounter.' Forty years later fishing the Test prompted Ted Hughes to celebrate 'Where a body loves to be/Rapt in the river of its own music': 'Be a dry-fly purist', Ted Hughes, *Collected Poems*, ed. Paul Keegan (London: Faber & Faber, 2003), p. 845. Thank you to Dr Mark Womald for confirming that the poet laureate found inspiration from fishing upriver of Stockbridge.

116 Lewis Douglas was a southern Democrat who broke with Roosevelt over the New Deal. He was appointed to the wartime administration as deputy head of the War Shipping Administration, where he became acquainted with Fairey, in Washington with the British Air Commission. Douglas was the American ambassador in London from 1947 to 1950, during which time he lost an eye when fishing at Bossington and hit by a miscast line. He remained a keen fly fisherman and wore an eye patch. CRF correspondence re Douglas's accident, August 1949, FAAM/2010/054/0044.

117 C.R. Fairey quoted in J. Wentworth Day, draft of 'These men build your war planes (1) Mr C.R. Fairey – Modern Elizabethan.' The annual wage bill for Bossington's servants in 1939 was only £290, meaning there was no financial

pressure to speed up a permanent move from Woodlands: Maud Hudd to Miss E.L. Burns, 15 June 1939, FAAM/2010/054/0005.
118 Geoffrey Hall, 'Sir Richard Fairey', p. 11 and quoted in Trippe, 'The plane maker', pp. 595–6.
119 Correspondence of C.R. Fairey and Sir Francis Peek, July–September 1939 and Fred Sigrist to C.R. Fairey, 14 July 1939, FAAM/2010/054/0007.
120 C.R. Fairey to Bertram T. Rumble, 19 August 1939, FAAM/2010/054/0007.
121 C.R. Faircy to Dr Maurice Lees [Torquay GP], 26 October 1939, FAAM/2010/054/0007.
122 C.R. Fairey to Lord Beaverbrook, 31 March 1938, FAAM/2010/054/0043.
123 C.R. Fairey to Ernest Brown, 21 March 1939 and to Lord Kemsley, 27 March 1939, FAAM/2010/054/0005.
124 'Land Forces Committee: minutes, 23 October 1939' CAB92/111, in Winston S. Churchill, *The Churchill War Papers Vol. I September 1939–May 1940*, ed. Martin Gilbert (London: Heinemann, 1993), pp. 284–5.
125 C.R. Fairey to Lord Kemsley, 30 October 1939 and 30 November 1939 [not sent], FAAM/2010/054/0006.
126 C.R. Fairey to H.C. Knapman [Chairman, Romsey and Stockbridge UDC], 26 January 1939, FAAM/2010/054/0006.
127 A.E. Payne [Eton RDC Officer for Evacuation] to C.R. Fairey, 25 August 1939, Miss E.L. Burns to A.E. Payne, 31 August 1939, Heal & Son to C.R. Fairey, 29 August, 4 September and 28 October 1939, and Miss E.L. Burns to C.R. Fairey, 27 September 1939, ibid.
128 Geoffrey Hall, interviewed in Trippe, 'The plane maker', p. 980.
129 C.R. Fairey to Air Commodore Arthur 'Mary' Conningham 11 October 1939, and to Ernest Brown, 12 October 1939, FAAM/2010/054/0006.
130 Philip Morgan, *Italian Fascism 1919–1945* (Basingstoke: Macmillan, 1995), p. 157.
131 Graham Stewart, *Burying Caesar Churchill, Chamberlain and the Battle for the Tory Party* (London: Weidenfeld & Nicolson, 1999), p. 245; John Charmley, *Lord Lloyd and the Decline of the British Empire* (London: Weidenfeld & Nicolson, 1987), pp. 198–9, 200, 202, 204, 211, 212–13; Count Ciano, *Ciano's Diary 1939–1943*, ed. Malcolm Muggeridge (William Heinemann: 1947), 14 January 1939, p. 14.
132 Lord Lloyd to Count Grandi, [?] September 1939, quoted in Charmley, *Lord Lloyd*, p. 237.
133 Ibid., pp. 198 and 238.
134 Ibid., pp. 205 and 239–40; Maurice Cowling, *The Impact of Hitler: British Politics and British Policy 1933–1940* (Cambridge: Cambridge University Press, 1975), p. 247; Lord Lloyd to C.R. Fairey, 21 February 1940, FAAM/2010/054/0045.

135 Churchill's first choice to oversee FAA procurement was Moore-Brabazon, someone even closer to Fairey than Lloyd: Winston S. Churchill to Neville Chamberlain, 29 September 1939, in Churchill, *The Churchill War Papers Vol. I*, pp. 174–6.
136 Ernest Brown to C.R. Fairey, 13 October 1939, FAAM/2010/054/0006; C.R. Fairey to Lord Lloyd, 12 October 1939, FAAM/2010/054/0007.
137 Ciano, *Ciano's Diary 1939–1943*, 1 December 1939, p. 178.
138 Morgan, *Italian Fascism 1919–1945*, pp. 182–5; Ciano, *Ciano's Diary 1939–1943*, 23 December 1943, pp. 559–60.
139 The inhaler was found to be unexpectedly 'efficacious': C.R. Fairey to Dr Neilson Davie [Golders Green GP], 25 May 1937, FAAM/2010/106/0013.
140 C.R. Fairey to Dr Neilson Davie, 26 September 1939, and to Ernest Brown, 26 and 30 October 1939, and to Frederick Handley Page, 23 October 1939, FAAM/2010/054/0006; Dr B.T. Parsons-Smith to C.R. Fairey, 29 October 1939, FAAM/2010/054/0007.
141 C.R. Fairey to Admiral H.C. [Clive] Rawlings, 22 December 1939, ibid.

Chapter 8

1 W.P. Crozier, *Off The Record Political Interviews 1933–1943*, ed. A.J.P. Taylor (London: Hutchinson, 1973), 24 August 1940, p. 195.
2 Fuelling the fear of longstanding anti-appeasers, that Beaverbrook might exploit his newly acquired power base should defeatism prevail: Harold Nicolson, *Diaries and Letters 1939–1945*, ed. Nigel Nicolson (London: Collins, 1967), 11 August 1940, p. 106.
3 Hoare enjoyed sharing political gossip with Crozier, who was so well connected that after seeing Sinclair he then spent 45 minutes with the Prime Minister. Crozier, *Off The Record*, 3 April 1940 and 23 August 1940, pp. 158–61, 189–90 and 191–2. The *MG*'s editor had previously noted the contrast between Sinclair's scepticism over how much had changed and Churchill's belief that, 'when you've got the aircraft industry working like this with a never-ceasing dynamo like Beaverbrook behind it, things move at a great pace.' Ibid., 26 July 1940, pp. 174 and 177.
4 Ibid., 24 August 1940, pp. 196 and 199–201. Crozier, a consummate note-taker, anonymised the names of all MAP personnel, leading his editor to advance an alternative 'Mr. X', Morris Wilson, a buying agent for Beaverbrook in the United States.
5 Ibid.; Anne Chisholm and Michael Davie, *Lord Beaverbrook: A Life* (New York: Alfred A. Knopf, 1993), p. 377.
6 Ibid., pp. 376, 387 and 385.
7 Re the mutual respect of Beaverbrook and Sir Hugh Dowding, C-in-C Fighter Command, see Crozier, *Off The Record*, 23 August 1940, p. 199 and A.J.P. Taylor,

Beaverbrook (London: Penguin, 1974), p. 592. Air Chief Marshal Lord Dowding and Sir Archibald Rowlands, and Sir Maurice Dean and Air Chief Marshal Lord Tedder, all referenced in Chisholm and Davie, *Lord Beaverbrook*, pp. 384–5 and 396–7. Beaverbrook's record was stoutly defended by his official biographer: Taylor, *Beaverbrook*, pp. 556–8.

8   Lord Brabazon, *The Brabazon Story* (London: Heinemann, 1956), pp. 201–3. 'I bear him [Beaverbrook] much affection, but I don't think he ever liked me much.': ibid., p. 201.

9   Lord Beaverbrook to Sir Henry Self, [?] August 1940, quoted in Hall, *Sir Richard Fairey*, p. 11; Sir Richard Fairey, speech to Vancouver aircraft workers, [?] 1944, quoted in Peter Trippe, 'The plane maker: the official biography of Sir Richard Fairey', unpublished manuscript held by Mrs Jane Tennant, pp. 997–8.

10 Chisholm and Davie, *Lord Beaverbrook*, pp. 376 and 387; Sir Archibald Rowlands [PUS, MAP] to C.R. Fairey, 27 May 1940, FAAM/2010/054/0066.

11 Geoffrey Hall, interviewed in Trippe, 'The plane maker', pp. 999–1001.

12 Pugh, *Hurrah for the Blackshirts!*, p. 279.

13 Lt Col J.T.C. Moore-Brabazon, correspondence concerning Briggs Manufacturing Company, July–September 1940, AC71/3, LBP, RAFM.

14 Lord Beaverbrook to Lt Col. J.T.C. Moore-Brabazon, 21 September 1940, AC71/3, LBP, RAFM.

15 Lt Col J.T.C. Moore-Brabazon to Lord Beaverbrook, 9 July 1940 and Lord Beaverbrook to Lt Col. J.T.C. Moore-Brabazon, 10 July 1940, AC71/3, LBP, RAFM.

16 K.J. Meekcoms, *The British Air Commission and Lend-Lease: The Role, Organisation, and Work of the BAC (and its Antecedents) in the United States and Canada, 1938–1945* (Tunbridge Wells: Air-Britain, 2000), p. 29; James L. Darroch, *Canadian Banks and Global Competitiveness* (Montreal: McGill-Queen's Press, 1999), pp. 129–31; Gavin J. Bailey, *The Arsenal of Democracy Aircraft Supply and the Anglo-American Alliance 1938–1942* (Edinburgh: Edinburgh University Press, 2013), p. 88. '[Morris Wilson] works without salary, he gives his services free, and they are worth while. The whole business of the Ministry is in his hands, he can take decisions, he can spend money subject only to the responsibility of the Minister.': Lord Beaverbrook, statement in the House of Lords, [?] 1940, quoted in David Farrer, *The Sky's The Limit: The Story of Beaverbrook at M.A.P.* (London: Hutchinson, 1943), p. 42. Farrer claimed Beaverbrook 'gives Wilson the major credit for what the American programme achieved', and that his first choice for the United States had been the head of Bristol's aero-engine division, Roy Fedden, but the firm vehemently objected: ibid., pp. 45 and 41.

17 Meekcoms, *The British Air Commission and Lend-Lease*, pp. 26–7; Bailey, *The Arsenal of Democracy Aircraft*, pp. 18–19, 40–3 and 54–6. Re the significant

consequences for USAAF strategic bombing of placing a Merlin engine in the previously poorly performing P-51 Mustang (and adding drop-tanks), see Paul Kennedy, *Engineers of Victory The Problem Solvers who Turned the Tide in the Second World War* (London: Penguin, 2014), pp. 116–35.

18   Meekcoms, *The British Air Commission and Lend-Lease*, pp. 26–7; Bailey, *The Arsenal of Democracy Aircraft*, pp. 54, 86, 156–9 and 200–1; Winston S. Churchill to Franklin D. Roosevelt, 1 June 1940, in *Churchill & Roosevelt The Complete Correspondence I. Alliance Emerging October 1933–November 1942*, ed. Warren F. Kimball (London: Collins, 1984), pp. 41–2.

19   H.H. Arnold, *Global Mission* (London: Hutchinson, 1951), pp. 125 and 134–6; Bailey, *The Arsenal of Democracy*, pp. 18–9, 22, 42–3 and 57–8; Meekcoms, *The British Air Commission and Lend-Lease*, pp. 26–7.

20   Ibid., pp. 28–9; Lord Lothian, NBC broadcast, 22 July 1940, in *Lord Lothian Speaks to America*, ed. H.U. Hudson (Oxford: Oxford University Press, 1941), p. 23. Variants of the Curtiss P40 flown by the RAF were renamed the Kittyhawk and then the Tomahawk. The latter was inferior in performance to the Spitfire in tests, and outgunned by the Bf109 in combat, but this embarrassing information was withheld from the Americans so as not to compromise future supplies of aircraft. Bailey, *The Arsenal of Democracy*, pp. 114 and 116.

21   Ibid., pp. 90–1 and 87–9.

22   Ibid., pp. 91–3; Taylor, *Beaverbrook*, pp. 554–5.

23   Lothian famously told American reporters, 'Well boys, Britain's broke. It's your money we want.': quoted in Bailey, *The Arsenal of Democracy*, pp. 93–4; Jean Edward Smith, *FDR* (New York: Random House, 2008), pp. 482–6; Lord Halifax to Sir Richard Fairey, 7 April 1945, FAAM/054/0076.

24   Meekcoms, *The British Air Commission and Lend-Lease*, pp. 14–21; H. Duncan Hall, *North American Supply*, History of the Second World War UK Civil Series (London: HMSO, 1955), pp. 118–19.

25   'My father thought very highly of Henry Self and thought he was phenomenally intelligent.' Mrs Jane Tennant to the author, 10 February 2017.

26   C.R. Fairey to Rt Hon. J.T.C. Moore-Brabazon, 27 April 1941, AC71/3, LBP, RAFM. 'But he [CRF] and Self were *such a contrast* in terms of personality ... that they were able to get along with each other.' Philip Young, quoted in Trippe, 'The plane maker', pp. 1030–1.

27   Between 1939 and 1945 Canada manufactured 16,431 and assembled 3,200 aircraft, and overhauled or repaired 6,530 machines: Meekcoms, *The British Air Commission and Lend-Lease*, p. 54.

28   N.M. Munro, 'History of the British Air Commission', p. 11, 1946, CAB102/123, NA.

29   Philip Young, quoted in Trippe, p. 1058.

30   Smith, *Mountbatten Apprentice War Lord*, pp. 304–5. 'I don't think he was their [US businessmen] idea of a typical Englishman, so he had no problems': Philip

Young, quoted in Trippe, p. 1059. Mrs Jane Tennant to the author, 10 February 2017.

31 'It was Lord Beaverbrook of course who sent him out … who said he was the only person who could handle it.': Lady Fairey quoted in Trippe, p. 1019. Seemingly there were damaging rumours that CRF joined the BAC solely so he could leave England: ibid., pp. 1017–18. C.R. Fairey to Lt Col. J.T.C. Moore-Brabazon, 21 September 1940, AC71/3, LBP, RAFM.

32 Criticism of the absence of adequate provision to cover CRF's absence presumably came from institutional shareholders: Taylor, *Fairey Aircraft since 1915*, p. 24.

33 Poole Flying Boats Celebration, 'Friends Newsletter', spring 2010, www.pooleflyingboats.com/pdfs_newsletters/Friends%20Newsletter%20Spring%202010.pdf; Trippe, 'The plane maker', pp. 1026–8. In New York the Fairey family took a suite at the Volney Hotel, but Esther was warned to look out for spies lurking in the corridors: Mrs Jane Tennant to the author, 10 February 2017.

34 *Baltimore Sun*, 5 August 1940; Smith, *FDR*, p. 450; Bailey, *The Arsenal of Democracy*, p. 86. Donovan, the future founder of the OSS (forerunner of the CIA), was FDR's classmate at Columbia Law School. The third passenger was the second Lord Cunliffe's younger brother, the Hon. Geoffrey Cunliffe, one-time of Eton and Cambridge. Although he had little in common with either Fairey or Donovan, Cunliffe's presence suggests he was charged with an important task upon his arrival in Canada.

35 'Wardman Tower – History': www.wardmantower.com/history/#founder; C.R. Fairey to Lt Col. J.T.C. Moore-Brabazon, 21 September 1940, AC71/3, LBP, RAFM; Hall, *Sir Richard Fairey*, p. 11.

36 Ibid.; Lord Beaverbrook to Winston S. Churchill, 26 December 1940, quoted in Taylor, *Beaverbrook*, p. 568; Bailey, *The Arsenal of Democracy*, pp. 108–9.

37 Sir Richard Fairey to Air Marshal Arthur 'Mary' Coningham, 20 April 1943, FAAM/2010/054/0037. From the outset Arnold was considered an efficient and effective chair of the JAC: C.R. Fairey to Air Vice Marshal George Baker, 7 January 1942, ibid.

38 David Reynolds, 'The President and the King: The diplomacy of the British Royal visit of 1939', in *From World War to Cold War Churchill, Roosevelt, and the International History of the 1940s* (Oxford: Oxford University Press, 2007), pp. 137–47.

39 Meekcoms, *The British Air Commission and Lend-Lease*, pp. 33 and 37. The RAF rejected its version of the P38 Lockheed Lightning, and suggested the P39 Bell Airacobra was better suited to the eastern front: ibid.

40 Mary Benington's recollection of CRF's work practices, in Trippe, 'The plane maker', p. 1071.

41 Meekcoms, *The British Air Commission and Lend-Lease*, pp. 57–8, 38–9 and 33–4; C.R. Fairey to Rt Hon. J.T.C. Moore-Brabazon, 6 June 1941, AC71/3, LBP, RAFM.

42 Mary Benington, quoted in Trippe, 'The plane maker', p. 1082.
43 Meekcoms, *The British Air Commission and Lend-Lease*, pp. 40–1; C.R. Fairey, private correspondence with Sir Archibald Rowlands and Lord Halifax, 1941–44, FAAM2010/054/0037.
44 Reynolds, 'The wheelchair president and his special relationships' in *From World War to Cold War*, 173–4.
45 Meekcoms, *The British Air Commission and Lend-Lease*, pp. 36–7, 32–3 and 42; Munro, 'History of the British Air Commission', chapter 2, appendix B, pp. 1–5.
46 'U.S. aircraft production held key to Allied victory' and 'British air leader opens exhibit', *Dayton News*, 9 and 11 July 1942.
47 Taylor, *Fairey Aircraft since 1915*, p. 24; C.R. Fairey to Air Vice Marshal George Baker, 7 January 1942, FAAM/2010/054/0037: Philip Young, interview in Trippe, 'The plane maker', p. 1046.
48 Ibid., pp. 1047 and 1049; 'Roosevelt and Hopkins shape Allied set-up', 'Hopkins set up as head of Lend-Lease program', 'Economy of lend-lease sets model for capital' and 'Philip Young is dead at 76; Eisenhower's personnel chief', *New York Times*, 30 March 1941, 15 April 1941, 14 March 1943 and 19 January 1987; Mrs Jane Tennant to the author, 10 February 2017.
49 'Jack Towers interpolates brief and pithy remarks, always very much to the point.': C.R. Fairey to Air Vice Marshal George Baker, 7 January 1942, FAAM/2010/054/0037.
50 Clark G. Reynolds, *Admiral John H. Towers: The Struggle for Naval Air Supremacy* (Annapolis, MD: Naval Institute Press, 1991), p. 375.
51 Bailey, *The Arsenal of Democracy*, pp. 119–21; Arnold, *Global Command*, pp. 141–52; General Henry H. Arnold, *American Air Power Comes of Age General Henry H. 'Hap' Arnold's World War II Diaries Volume I*, ed. Major General John W. Huston USAF Rtd (Maxwell, AL: Air University Press, 2002), 12–27 April 1941, pp. 139–65.
52 General Henry H. Arnold to Sir Richard Fairey, 27 December 1945, FAAM/054/0076.
53 Philip Young, account of CRF and the JAC, in Trippe, 'The plane maker', p. 1051. On the USAAF chief's growing disenchantment with the British, including the RAF, see Arnold, *American Air Power Comes of Age*, ed. Major General John W. Huston USAF Rtd.
54 Arnold, *American Air Power Comes of Age*, 11–12 August 1941, pp. 229–2; Bailey, *The Arsenal of Democracy*, pp. 178–82; Sir Henry Self, 18 May 1942 and Lord Beaverbrook, 16 August 1941, quoted in ibid., pp. 249 and 182. Beaverbrook continued to correspond directly with FDR and his vice president, Henry Wallace, holding face-to-face meetings with them and Arnold during the Arcadia Conference in Washington: Duncan Hall, *North American Supply*, p. 342.
55 Bailey, *The Arsenal of Democracy*, pp. 137 and 184–92; Arnold, *Global Mission*, pp. 157–60, 166–7, 176–7 and 178–80. The Fairey Fulmar's failings became evident

as early as 1940, hence the value placed on Supermarine's Spitfire carrier version, the Seafire; and the Albacore was deemed obsolete as a successor to the Swordfish as it entered service with the FAA: Ben Jones, 'The Fleet Air Arm and the Struggle for the Mediterranean, 1940–44' in *British Naval Aviation: The First 100 Years*, ed. Tim Benbow (Farnham: Ashgate, 2011), pp. 81–2, and Taylor, *Fairey Aircraft since 1915*, pp. 293–4.

56 Fairey occasionally acted as an emissary in three-way negotiations between Roosevelt, Churchill and Arnold, as when instructed in March 1942 to lobby the USAAF Chief over the scant delivery of bombers: Duncan Hall, *North American Supply*, p. 362.

57 Bailey, *The Arsenal of Democracy*, pp. 148–9; The Earl of Halifax, *Fulness of Days* (London: St James' Place, 1957), pp. 262–3.

58 Mrs Mary Benington (neé Bell), recollection of D-G personal communications, in Trippe, 'The plane maker', pp. 1066–7 and 1068–9; Duncan Hall, *North American Supply*, pp. xiv–xv. For the suggestion that Freeman's return to MAP was at Churchill's behest, see Kennedy, *Engineers of Victory*, pp. 120–1 and 368.

59 Brabazon, *The Brabazon Story*, pp. 202–3; C.R. Fairey to Rt Hon. J.T.C. Moore-Brabazon, 2 May and 27 April 1941, AC71/3, LBP, RAFM; C.R. Fairey to Air Vice Marshal George Baker, 7 January 1942, FAAM/2010/054/0037.

60 Rt Hon. J.T.C. Moore-Brabazon to C.R. Fairey, 23 September 1941, and C.R. Fairey to Rt Hon. J.T.C. Moore-Brabazon, 16 October 1941, AC71/3, LBP, RAFM.

61 Angus Calder, *The People's War: Britain 1939–45* (London: Jonathan Cape, 1969), p. 263; Brabazon, *The Brabazon Story*, pp. 207–10; ibid., p. 211. Cripps may have demanded Moore-Brabazon's dismissal as one part of the price for accepting office on his return from serving as ambassador in Russia: Gabriel Gorodetsky, *Stafford Cripps' Mission to Moscow 1940–42* (Cambridge: Cambridge University Press, 1984), p. 295.

62 C.R. Fairey to Rt Hon. J.T.C. Moore-Brabazon, 9 March 1942, FAAM/2010/054/0037.

63 Munro, 'History of the British Air Commission', p. 41; Meekcoms, *The British Air Commission and Lend-Lease*, pp. 43–4; Bailey, *The Arsenal of Democracy*, pp. 148–9.

64 Ibid., pp. 160–3, 199–26, 272, 274, 260 and 241. The USAAF first flew Spitfires, followed by Beaufighters, and then Mosquitoes, i.e. Supermarine, Bristol and De Havilland marques.

65 Richard Overy, *Why The Allies Won* (London: Jonathan Cape, 1995), pp. 195–8.

66 'Weapons production of the major powers 1939–45' in ibid., p. 331. 'Whereas we told the President of the United States we wanted a total of approximately 60,000 planes in the Air Force, in 1944 we actually had 72,700', Arnold, *Global Mission*, p. 168.

67 David Edgerton, *Britain's War Machine Weapons, Resources and Experts in the Second World War* (London: Allen Lane, 2011), pp. 208–9.

68 On Towers' relaxed view of RN demands in May 1942, see Arnold, *Global Mission*, p. 177.
69 Ibid., pp. 257–8, 254 and 270.
70 The United Kingdom had produced around 50 per cent more aircraft than the United States in 1940 and 1941. USAAF and Air Ministry data courtesy of ibid., pp. 251–3 and 255.
71 Ibid., pp. 255–6 and 264–7; Arnold, *Global Mission*, pp. 204 and 218; Arnold, *American Air Power Comes of Age Volume II*, 6 September 1943, p. 40 and editorial commentary, ibid., pp. 44–5; Field Marshal Lord Alanbrooke, *War Diaries 1939–1945*, eds. Alex Danchev and Dan Todman (London: Weidenfeld and Nicolson, 2001), 20 January 1943, p. 364.
72 C.R. Fairey to Air Marshal Arthur 'Mary' Coningham, 29 November 1941, FAAM2010/054/0037. For all Coningham's wartime failings, CRF remained a loyal friend and was bereft when the Avro Tudor carrying 'Mary' disappeared en route from the Azores to Bermuda on 30 January 1948. Mrs Jane Tennant to the author, 10 February 2017; Aviation Safety Network, www.aviation-safety.net/database/record.php?id=19480130-0.
73 Lady Fairey, recollection CRF's health and lifestyle, and Mary Benington's recollection of CRF's work practices, in Trippe, 'The plane maker', pp. 1077, 1071 and 1079–80.
74 Lady Fairey, quoted in ibid., p. 1077.
75 CRF endeavoured each night to read his children a bedtime story, introducing them both to Kipling's *Just So* stories and *Puck of Pook's Hill*: Mrs Jane Tennant to the author, 10 February 2017.
76 Lead Belly, 'The bourgeois blues', 1937.
77 Smithsonian, 'Separate not equal: Washington D.C.', www.iwm.org.uk/collections/search?query=INTERVIEW+NOEL+RAYMOND+PUGH&=Search&items_per_page=10. '…some of the liftmen had only one arm, presumably this was to give a job to the disabled … as children we were on friendly terms with one liftman called Holly, my recollection is that the blacks were very kind to children': Mrs Jane Tennant to the author, 10 February 2017.
78 Previously CIGS, in Washington Dill served as Chief of the British Staff Mission and then Senior British Representative on the Combined Chiefs of Staff, 1942–4.
79 Franklin D. Roosevelt quoted in Smith, *FDR*, p. 108.
80 Halifax, *Fulness of Days*, p. 262.
81 'The Post presents Mrs C.L. [sic.] Fairey…', *Washington Post*, 14 February 1942.
82 Correspondence re knighthood of C.R. Fairey, June–August 1942, FAAM/2010/054/0076.
83 Report of a meeting between CRF's solicitor and Treasury officials facilitated by Moore-Brabazon: Roland Outen [partner of Charles Crisp] to C.R. Fairey, 22 December 1941, AC71/3, LBP, RAFM; Lady Fairey, recollection of CRF's

dispute over travel expenses with the Treasury, in Trippe, 'The plane maker', pp. 1071–2.
84 C.R. Fairey to Flying Officer later Squadron Leader Archie McClellan, 25 September and 28 November 1941, and to Rev. Mr W.S. Daubeney, 7 November 1941, FAAM2010/054/0037. Daubeney had christened John and Jane Fairey in a joint ceremony prior to their departure for America: Mrs Jane Tennant to the author, 10 February 2017.
85 Lady Fairey, recollection of 7 8 December 1941, quoted in Trippe, 'The plane maker', pp. 1036–7; C.R. Fairey to Commander Norman Holbrook VC [veteran submariner, returned to service at the Admiralty], 20 December 1941, FAAM2010/054/0037. Andrew Roberts, *The Storm of War: A New History of the Second World War* (London: Allen Lane, 2009), pp. 191 and 193. A classic account of the Taranto raid, 11 November 1940, is in a best-selling wartime booklet which boosted public awareness of Fairey Aviation's importance to the FAA: Ministry of Information, *Fleet Air Arm The Admiralty Account of Naval Air Operations* (London: HMSO, 1943/2001), pp. 54–60.
86 '…a grateful Esther sends her love': C.R. Fairey to Rt Hon. J.T.C. Moore-Brabazon, 7 January 1942, FAAM2010/054/0037.
87 C.R. Fairey to Rear Admiral H.C. Rawlings, 29 January 1942, ibid.
88 C.R. Fairey, miscellaneous private correspondence, 1941–4, ibid.
89 C.R. Fairey to Squadron Leader Archie McClellan, 28 November 1941, ibid.
90 Sir Richard Fairey to Gerald Penny [Nursling fly fisherman], 6 October 1942, and correspondence of C.R. Fairey and Reverend W.S. Daubeney, 1941–4, ibid.
91 C.R. Fairey to Reverend W.S. Daubeney, 7 November 1941, ibid.; Sir Richard Fairey to Air Marshal Arthur 'Mary' Coningham, 20 April 1943, and to Lord Brabazon, 27 April 1943, ibid.
92 Lord Brabazon, legal testimony, 1960, replicated in Trippe, 'The plane maker', pp. 1090–2.
93 As strongly suggested by the content of Richard Fairey to Rt Hon. J.T.C. Moore-Brabazon, [?] February 1942, AC71/3, LBP, RAFM.
94 This account of the last voyage of DS *Ringstand* is based upon the astonishing volume of information to be found at: www.uboat.net/allies/merchants/ships/1286 and www.warsailors.com/singleships/ringstad.
95 Mrs Jane Tennant to the author, 10 February 2017.
96 Ibid.
97 'Korvettenkapitän Peter-Erich Cremer', www.uboat.net/men/cremer.htm. CRF maintained a blanket condemnation of all Germans, always insisting that, 'Twice they have wrecked our world': Mrs Jane Tennant to the author, 10 February 2017.
98 Ibid.; Richard Fairey, recollection of the *Rinstand* sinking, in Trippe, 'The plane maker', p. 1100.

99 Mary Benington, quoted in ibid., p. 1082; Mrs Jane Tennant to the author, 10 February 2017. 'Eight days of anxiety': C.R. Fairey to Wing Commander J.L. Lawson, 10 February 1942, FAAM/2010/054/0037.

100 'News of his condition was bad, but nothing in comparison with what I had thought to face, and I have no words with which to thank you both for your sympathy and assistance,': C.R. Fairey to Rt Hon. J.T.C. Moore-Brabazon, 9 March 1942, FAAM/2010/054/0037.

101 Today's Douglas Bader Rehabilitation Centre, the name reflecting Queen Mary's 90-year history of enabling military and auxiliary personnel who have lost limbs to regain their independence.

102 Charles Desoutter [pioneer prosthetic limb builder, Roehampton], recollection in Trippe, 'The plane maker', pp. 1106–7. Richard Fairey's post-amputation golfing ability was paralleled in his prowess at shooting and fly fishing.

103 Fairey Aviation, press release on the retirement of Richard Fairey, 1960, replicated in ibid., pp. 1085–6; Geoffrey Hall, recollection of Richard Fairey at the AAEE in ibid., p. 1111.

104 Peter Twiss, *Faster than the Sun: The Compelling Story of a Record Breaking Test Pilot* (London: Grub Street, 2000), p. 29; recollection of associates collected in Trippe, 'The plane maker', pp. 1109–10.

105 'Her dress bill was £1,600 in 1940' and 'Spent £1,600 on clothes in 1940', *Daily Express* and *Daily Worker*, 6 November 1940; 'Mrs L.A. Fairey [neé Berger]', obituary, *The Times*, 11 July 1944; Trippe, 'The plane maker', p. 1102.

106 Sir Stafford Cripps' Cabinet report on the British aircraft industry, CAB87/63, November 1943, quoted in King, *Knights of the Air: The Life and Times of the Extraordinary Pioneers who First Built British Aeroplanes* (London: Constable, 1989), p. 342; Ministry of Information, *What Britain Has Done 1939–1945: A Selection of Outstanding Facts and Figures* (London: HMSO, 1945/Atlantic Books, 2007), p. 104.

107 Ministry of Aircraft Production, *The Aircraft Builders* (London: HMSO, 1947), p. 80, quoted in Edgerton, *England and the Aeroplane*, p. 124.

108 Sir Richard Fairey to Sir Frederick Handley-Page, 8 June 1943, FAAM/2010/054/0037.

109 King, *Knights of the Air*, pp. 346–7; Hayward, *The British Aircraft Industry*, pp. 27–8.

110 Sir Richard Fairey to Air Marshal Arthur 'Mary' Coningham, 29 December 1943, FAAM/2010/054/0037. On the unique status within MAP of Air Marshal Sir Wilfred Freeman between 1942 and 1945, see David Edgerton, *Warfare State Britain, 1920–1970* (Cambridge: Cambridge University Press, 2006), p. 152. For an overview of the power and position enjoyed by other service personnel, plus seconded industrialists like CRF, inside the wartime supply departments, see ibid., pp. 150–6.

111 Sir Richard Fairey to Sir Frederick Handley-Page, 30 November 1944, ibid.

112 Correlli Barnett, *The Audit of War: The Illusion and Reality of Britain as a Great Nation* (London: Papermac, 1987), pp. 143–58; Edgerton, *England and the Aeroplane*, pp. 126–31 [a critique more fully developed in David Edgerton, 'The prophet militant and industrial: the peculiarities of Corelli Barnett', *Twentieth Century British History*, 2/3 (1991), pp. 360–79].
113 Barnett, *The Audit of War*, pp. 154-6.
114 Taylor, *Fairey Aircraft since 1915*, pp. 21–2.
115 Staniland was succeeded by his deputy test pilot since the mid-1930s, Duncan Menzies. Menzies was no pushover if critical of a Fairey prototype: his refusal to sign off the Mk1 Barracuda delayed production and infuriated directors: Matthew Willis, 'From Firefly to Swordfish: The distinguished aviation career of Duncan Menzies, part two' *Aviation Historian*, 12 (July 2015), pp. 56–64.
116 Ibid., pp. 22, 326–8, 312–14 and 319–20; Martyn Chorlton, *Fairey Company Profile 1915–1960* (Cudham: Kelsey Publishing Group, 2012), p. 86. 1,700 Barracudas seems a sizeable number, but to put it in perspective Westland alone delivered 2,115 Supermarine Seafires to the FAA: Hooks, *Fleet Air Arm Aircraft of World War 2*, p. 36.
117 Broadbent was especially critical of a 'daughter firm' for the Hayes plant, General Aircraft at Feltham, which attracted fierce criticism from MAP for its record building the Firefly deck fighter: Wilfred Broadbent, interview in Trippe, 'The plane maker', pp. 1128.
118 Ibid., pp. 1114–23. Trippe's account, written when wartime personnel were available to interview, is an unexpectedly clear and succinct exposition of the company's production problems. Predictably, the writer shoots himself in the foot by blaming MAP and lambasting Cripps.
119 Peter Clarke, *The Cripps Version: The Life of Sir Stafford Cripps 1889–1952* (London: Penguin, 2002), pp. 330–5. On Cripps' high poll ratings in 1942 and Churchill's calculated response, see Steven Fielding, 'The Second World War and popular radicalism: the significance of the "movement away from party"', *History*, 80:258 (1995), pp. 38–58.
120 '…Cripps – a freak and not a person who will ever be a controlling force in political life – he is too aloof and impersonal … what a bad advertisement he is – and looks – for non-drinking, nut-eating, etc. – he has a perfectly blue nose': Sir Cuthbert Headlam, *Parliament and Politics in the Age of Churchill and Attlee: The Headlam Diaries 1935–1951*, ed. Stuart Ball (Cambridge: Cambridge University Press, 1999), 25 September 1943, p. 385; Crozier, *Off The Record*, 20 February 1942, p. 293.
121 Correspondence of Sir Richard Fairey and Sir Stafford Cripps, 1943–5, FAAM/2010/054/0037.
122 Sir Stafford Cripps to Sir Richard Fairey, [?] 1943, quoted in Trippe, 'The plane maker', p. 1201; John Fairey, recollection of CRF's angry reference to Cripps re his gift, ibid., p. 1202.

123 Clarke, *The Cripps Version*, pp. 377–8; Edgerton, *England and the Aeroplane*, pp. 125–6.
124 Sir Stafford Cripps, address to the Fairey Aviation workforce at Hayes, 19 December 1942, replicated in Trippe, 'The plane maker', pp. 1155–60.
125 Sir Stafford Cripps and MAP officials, 'Production problems in Fairey Aviation due to bad planning', 1943, ADM1/15003, and 'Fairey Aviation: internal administration', 1942–3, AVIA9/38, NA; Maurice Wright, policy paper, 6 September 1943, Sir Richard Fairey to Sir Archibald Rowlands, 13 September 1943 and Sir Archibald Rowlands to Sir Richard Fairey, 1 October 1943, replicated in Tripp, 'The plane maker', pp. 1174–86.
126 Taylor, *Fairey Aircraft since 1915*, p. 22; Tripp, 'The plane maker', pp. 1167–73.
127 Taylor, *Fairey Aircraft since 1915*, pp. 22–3; Hayward, *The British Aircraft Industry*, p. 24.
128 Re the British left's support for 'scientific management', as promoted by Cripps, see the work of Michael Weatherburn [www.imperial.academia.edu/MichaelWeatherburn], including his case study of MAP's use of management consultants (UOR – Urwick, Orr & Partners) at Fairey's northern plants: 'Scientific management at work: the Bedaux system, management consulting and worker efficiency in British industry, 1914–48', PhD, Imperial College, 2014, pp. 193–5. On Marden's efforts to cultivate the goodwill of the Communist-dominated Joint Planning Committee at Hayes, see ibid., p. 195.
129 Geoffrey Hall, recollection of CRF convincing Sir Clive Baillieu to join Fairey Aviation, in Trippe, 'The plane maker', pp. 1189–90.
130 Taylor, *Fairey Aircraft since 1915*, pp. 22 and 25; J.P. Poynter, 'Sir Clive Lathan Baillieu (1889–1967)', *Australian Dictionary of Biography*, volume 7, 1979: http://adb.anu.edu.au/biography/baillieu-sir-clive-latham-5629. 'They [the workforce] understood his [Bailleu's] reasons because he went out of his way to gain their confidence and make them understand. They trusted him. And they respected him.': Wilfred Broadbent, quoted in Trippe, 'The plane maker', p. 1191.
131 Sir Clive Bailleu, correspondence re Fairey Aviation organisational change and budgetary control, 1944–5, FAAM/2010/054/0037; Sir Richard Fairey to Sir Archibald Rowlands and to Sir Harold Scott [PUS MAP], 14 July and 19 September 1944, ibid.
132 C.R. Fairey to Commander Norman D. Holbrook VC, 20 December 1941, FAAM2010/054/0037.
133 Sir Richard Fairey to Lord Brabazon, 8 October 1945, Box 13, AC71/3, LBP, RAFM.
134 C.R. Fairey to Jim Wentworth Day, 16 October 1941, and to Reverend Mr W.S. Daubeney, 7 November 1941, FAAM2010/054/0037.
135 C.R. Fairey to Jim Wentworth Day, 16 October 1941, ibid.; Ministry of Agriculture and Fisheries, 'Rivers Avon and Stour Catchment Board', *London Gazette*, 8 September 1933; 'Sir Richard Treherne', obituary, *Daily Telegraph*,

22 December 2001. The 1948 River Boards Act reduced the 47 catchment boards created in 1930 to 32 river authorities.
136 C.R. Fairey to Reverend Mr W.S. Daubeney, 7 November 1941, Squadron Leader Archie McClellan, 28 November 1941 and Commander Norman D. Holbrook VC, 20 December 1941, FAAM2010/054/0037.
137 Sir Richard Fairey to Captain Rt Hon. Harold H. Balfour, [?] August, 1942, ibid.
138 Sir Richard Fairey to Lord Brabazon, 27 April 1943, Jim Wentworth Day, 4 December 1943 and Lord Brabazon, 13 July 1944, ibid.
139 C.R. Fairey to Richard Fairey, 2 April 1942, FAAM2010/054/0037.
140 C.R. Fairey to Jim Wentworth Day, 16 October 1941, ibid.
141 C.R. Fairey to Reverend Mr W.S. Daubeney, 7 November 1941, Commander Norman D. Holbrook VC, 20 December 1941 and Richard Fairey, 2 April 1942, ibid. Re prewar 'socialistic whinings about merchants of death and war mongering,' see C.R. Fairey to Squadron Leader Archie McClellan, 16 March 1942, ibid.
142 Sir Richard Fairey to Air Vice Marshal George Baker, 10 August 1942, ibid.; Chisholm and Davie, *Lord Beaverbrook*, pp. 430–5; Sir Richard Fairey to Major R. Croilshan, 21 August 1942, FAAM2010/054/0037; Hugh Purcell, *The Last English Revolutionary: Tom Wintringham 1898–1949* (Stroud: Sutton, 2004), pp. 189–230.
143 Sir Richard Fairey to Captain Rt Hon. Harry Crookshank, 2 February 1943, FAAM/2010/054/0077; Sir Richard Fairey to Lord Brabazon, 27 April 1943 and 3 January 1944, FAAM/2010/054/0037.
144 Sir Richard Fairey to Lord Brabazon, 24 and 25 November 1953 and 3 January 1944, FAAM/2010/054/0085.
145 C.R. Fairey to Richard Fairey, 2 April 1942, FAAM/2010/054/0037.
146 Sir Richard Fairey to Major R. Croilshan, 21 August 1942 and to Squadron Leader Archie McClellan, 6 June 1942, ibid.
147 Sir Richard Fairey to Air Marshal Arthur 'Mary' Coningham, 20 April 1943 and C.R. Fairey to Richard Fairey, 2 April 1942, ibid.
148 Sir Richard Fairey to Air Marshal Arthur 'Mary' Coningham and to Lord Brabazon, 20 and 27 April 1943, ibid.
149 Sir Richard Fairey to Charles Crisp and Sir Frederick Handley-Page, 19 July and 8 June 1943, ibid.
150 Royal Aeronautical Society, Annual Council Dinner List of Guests and Table Plan, 27 May 1943, FAAM/2010/054/0085. On the American envoy's popularity within the Labour movement, see David Reynolds, 'The President and the British Left: The Appointment of John Winant as US Ambassador in 1941' in *From World War to Cold War*, pp. 148–64.
151 Smith, *FDR*, pp. 570 and 573–5.
152 Sir Richard Fairey, 'Destroying the enemy with his own weapon', *Philadelphia Inquirer*, 3 January 1943.

153 Lester D. Gardner [IAS secretary] to Sir Richard Fairey, 29 December 1943 and programme for the Institute of the Aeronautical Sciences awards ceremony, 24 January 1944, FAAM/2010/054/0037.
154 Wilfred Broadbent, recollection of 1943 compulsory purchase of 'Great West aerodrome', in Trippe, 'The plane maker', p. 1192; Sir Richard Fairey to Lord Brabazon and to Sir Archibald Rowlands [not sent], 19 January and 5 April 1944, FAAM/2010/054/0037.
155 'Cripps fails to satisfy Fairey workers', *Daily Worker*, 21 January 1944.
156 '…we were all on holiday at a place called Big Wynne Inn on an island in a big lake … my father joined us but was very unwell … as it was extremely hot he slept out on the veranda of the cottage at night.' Mrs Jane Tennant to the author, 10 February 2017.
157 Lady Fairey, recollection of CRF's 1943 thrombosis, in Trippe, 'The plane maker', pp. 1196–7; Sir Richard Fairey to Sir Harold Scott, 4 July, 19 September and 24 October 1944, FAAM/2010/054/0037.
158 Mary Bell, quoted in Trippe, 'The plane maker', p. 1082.
159 Sir Richard Fairey to Sir Harold Scott, 14 and 28 December 1944, FAAM/2010/054/0037. 'We left Washington to Baltimore from where we flew on a seaplane (a BOAC Clipper) to Bermuda and landed by Darrel's Island and I am certain it was the last day of 1944. We stayed at the Belmont Hotel for a while before moving into our house, Lyndham.': Mrs Jane Tennant to the author, 15 June 2017.
160 Lady Fairey, recollection of CRF's 1945 heart attack and treatment, in Trippe, 'The plane maker', p. 1197; Kenneth A. Evelyn et al., 'Effect of sympathectomy on blood pressure hypertension: a review of thirteen years' experience at the Massachusetts General Hospital', *Journal of the American Medical Association*, 140:7 (1949), pp. 592–602; Sir Richard Fairey to Lord Brabazon, 8 October 1945, Box 13, AC71/3, LBP, RAFM; Mrs Jane Tennant, interview, 5 November 2013; Mrs Jane Tennant to the author, 10 February 2017.
161 Ibid., 15 June 2017.
162 Lord Halifax to Sir Richard Fairey, 7 April 1945, FAAM/054/0076.

Chapter 9

1 During and immediately after the war, with Furness Withy's *Queen of Bermuda* still in use as a troop ship, the Faireys travelled on Canadian cargo ships from the Great Lakes, *Fort Townsend* and *Fort Amherst*. The ships carried only a few passengers but a large quantity of forage, as the islands' agriculture could not support Bermuda's system of horse-drawn transport.
2 Correspondence re Sir Richard Fairey's absences in Bermuda, the United States and Canada, 1946–50, FAAM/2010/054/0038; Sir Richard Fairey to Lord Brabazon,

10 June 1946, Box 13, AC71/3, BP, RAFM; Sir Richard Fairey to Richard Fairey, 1 February 1950, and appointments diary, 1953, Boxes 102 and 77, AC73/30, ibid.; passenger lists re Charles Richard Fairey, Southampton/New York/Hamilton, 1945–55, National Archives, via www.ancestry.co.uk.

3   Minutes of Fairey Aviation board meetings, 1945–56, Boxes 35, 40 and 37, AC73/30, RAFM.

4   See, for example, 'Expenses incurred by Sir Richard Fairey on matters directly connected with the Fairey Aviation Co. Ltd., Hayes, Middlesex', 23 October 1946, FAAM/2010/054/0038 and claims for US and Canada travelling expenses, 1945–51, Boxes 35, 40, 37 and 38, AC73/30, RAFM.

5   Roland Outen, weekly reports to Sir Richard Fairey, 1946–51 and Sir Richard Fairey to C.C. Vinson, 5 January and 15 March 1949, FAAM/2010/054/0038.

6   Outen became a serial non-executive director, often serving as deputy chairman or – for the boards of International Harvester and Quaker Oats Ltd, as well as Fairey Aviation – as chairman.

7   Sir Richard Fairey, draft designation of board members' executive duties and principal functions, 21 May 1951, Box 102, AC73/30, RAFM; C.C. Vinson to Sir Richard Fairey, 27 January 1949, FAAM/2010/054/0038. Fighting Hayes and Harlington in the 1950 general election Vinson – 'the company man' – secured just half of the successful Labour candidate's votes. He worked as a business consultant, from 1952 to 1960, when he joined the board of the restructured Fairey Company.

8   Mrs Jane Tennant to the author, 8 March 2017.

9   Correspondence, June–September 1946, and invitation/acceptance lists re Fairey Aviation cocktail party, Dorchester Hotel, 17 September 1946, FAAM/2010/054/0038.

10 Taylor, *Fairey Aircraft since 1915*, pp. 22–3 and 27; Charles Lawrence, *Fairey Marine Boats, Raceboats, Rivals and Revivals* (Chiswick: Charles Lawrence, 2014), p. 14; Major C.H. Chichester-Smith, 1920 photographs of Japan, personal archives of P.G. Currey [PAPGC].

11 Taylor, *Fairey Aircraft since 1915*, pp. 42–4: minutes of Burtonwood Repair Depot management meetings, 1940–2, PAPGC. Fairey SA/SONACA built 220 F16s for the Danish and Belgian air forces under licence from General Dynamics, following which the Wallonia state enterprise became a partner in building the Airbus: Sonaca Group, 'History', www.sonaca.com/content/belgium/entreprise.

12 One unhappy tour of the Hayes plant left CRF complaining bitterly of untidiness and too many skiving workers: Sir Richard Fairey to Major C.H. Chichester-Smith, 27 June 1947, FAAM/2010/054/0038.

13 Also, CRF encouraged Richard Fairey's global expansion of Fairey Air Surveys, created as far back as 1929 when Fairey Aviation took over Aerial Surveys, an India-based mapping and aerial photography company founded six years earlier: 'History of Fairey surveys', June 2013, www.faireysurveys.co.uk/category/history.

14　Lawrence, *Fairey Marine Boats*, pp. 14, 12–13 and 16; HLP/FMC, '12-FT. BOAT', 5 June 1946, PAPGC; 'Uffa Fox, 1898–1972', http://www.uffafox.com/uffabiog.htm; correspondence of Sir Richard Fairey and Major C.H. Chichester-Smith re Fairey Marine, 1946–51, FAAM/2010/054/0038.
15　Lawrence, *Fairey Marine Boats*, pp. 14–16.
16　Taylor, *Fairey Aircraft since 1915*, pp. 326–55 and 45.
17　Ibid., p. 7; Fairey Aviation, menu and programme, Connaught Rooms, 22 December 1949, Box 102, AC73/30, RAFM; Sir Richard Fairey to Air Marshal Lord Trenchard, 5 December 1949, Add. 9429/1B/51, Trenchard Papers, Cambridge University Library; Air Marshal Lord Trenchard, acceptance of invitation, 7 December 1949, and 'Speech for Fairey Dinner', 22 December 1949, Add. 9429/1B/52 (i) and (ii), ibid.; 'Twenty-five years of service', *The Aeroplane*, 6 January 1950 and 'Sir Richard gives a party', *The Tatler*, 4 January 1950; Sir Richard Fairey, 'Draft speech for anniversary dinner, 22 December 1949', [?] December 1949, FAAM/2010/053/0117.
18　J.C. Macpherson [company secretary] to Sir Richard Fairey, 19 August and 3 September 1947, 'Proposed heads of agreement between The Martin Aircraft Corporation and The Fairey Aviation Co. Ltd.', 21 September 1947 and C.C. Vinson to Sir Richard Fairey, 28 July 1950, FAAM/2010/054/0038. Hawker Siddeley bought Blackburn and revived its fortunes with the success of the Buccaneer S2 strike aircraft.
19　Fairey Aviation's aggregate earnings, 1946–52 based on net profits for the first two financial years and pre-tax profits for the following three; dividends totalled £1,343,729: Sir Richard Fairey, board folders 1948–56, Boxes 40, 37, 38 and 50, AC73/30, RAFM. Sir Richard Fairey, statement to shareholders, Fairey Aviation AGM, 24 November 1955, quoted in Trippe, 'The plane maker', pp. 1323–4.
20　Taylor, *Fairey Aircraft since 1915*, pp. 326–55 and 45. UTC Aerospace Systems closed Claverham Ltd, formerly Fairey Hydraulics, in 2014.
21　Keith Hayward, *The British Aircraft Industry* (Manchester: Manchester University Press, 1989), pp. 52 and 62–4.
22　Ibid., pp. 54, 55 and 57. On just how much a nation's hopes rested on the success of Comets 1 to 3 and were subsequently dashed, see Adrian Smith, 'The dawn of the jet age in austerity Britain: David Lean's *The Sound Barrier* (1952)', *Historical Journal of Film, Radio and Television*, 30:4 (2010), pp. 487–514.
23　Hayward, *The British Aircraft Industry*, pp. 64–5.
24　1954–55 pre-tax profit of £2,159,568: Fairey Aviation, 'Draft report on accounts for the year ended 31st March 1955', 1 October 1955, Box 50, AC73/30, RAFM.
25　Taylor, *Fairey Aircraft since 1915*, pp. 45–50. Guest list and approved draft of Sir Richard Fairey's speech, 'Blue sky luncheon', Dorchester Hotel, 27 July 1955, FAAM/2010/054/0038. The speech was drafted and redrafted by Fairey Aviation's PR team, with three iterations couriered between Hayes and Bossington. Detailed

briefing notes identified guests 'of particular interest to us': 'Brief for F.A.C. directors only. Blue sky luncheon – Dorchester Hotel, 27th July, 1955', Box 102, AC73/30, RAFM.

26 James Hamilton-Paterson, *Empire of the Clouds: When Britain's Aircraft Ruled the World* (London: Faber & Faber, 2010), pp. 65–6; *Fairey Company Profile 1915–1960*, ed. Chorlton, pp. 104–5; Taylor, *Fairey Aircraft since 1915*, pp. 382–7.

27 Ibid., pp. 356–68; Geoffrey Hall memoir re the GR17/Gannet, drawn on in Trippe, 'The plane maker', pp. 1275–89 and 1308–14 and quoted, p. 1289.

28 Taylor, *Fairey Aircraft since 1915*, pp. 356–68.

29 *Fairey Company Profile 1915–1960*, p. 120; Taylor, *Fairey Aircraft since 1915*, pp. 388–97.

30 The following account of the Rotodyne's unhappy history draws heavily upon ibid., pp. 405–26.

31 The tip jets on the main rotor relied on fuel and compressed air (with compressors fitted to the rear of each of the two Napier Eland engines). These were 'fed via the leading edge of the main plane up to the rotor head and then to opposing tip jets. Once a speed of 60 mph was reached, the tip jets were extinguished and the Eland engines took over while the 46ft 6in [1.38 meters] span main wing took over 50% of the lift away from the rotor blades. This effectively turned the aircraft into a large high-speed autogyro in level flight.' *Fairey Company Profile 1915–1960*, p. 120.

32 'Government monitors worked out that noise levels within 500 ft from the pad during take-off and landing were "intolerable" and that those within 1,000ft (305 m) of the Rotodyne in mid-flight were "unpleasantly noisy" – the same as hearing a raised voice from 2 ft (60 cm) away.' Justin Parkinson, 'Why did the half-plane, half-helicopter not work?', *BBC News Magazine*, 12 February 2016, www.bbc.co.uk/news/magazine-35521040.

33 *Eagle's* cutaway drawings of the Rotodyne and of 'Anastasia' featured in the Science Museum exhibition, 'Dan Dare & the birth of hi-tech Britain', that ran from May 2008 to October 2009.

34 Hamilton-Paterson, *Empire of the Clouds*, p. 122.

35 Ibid., pp. 4–13, and 125. Especially evocative of this period is the documentary 'Jet! When England ruled the skies: Episode 1/2 Military marvels', BBC4, 2012: www.bbc.co.uk/programmes/b01m81f5.

36 'Farnborough Air Display', Pathé News, 10 September 1959: www.britishpathe.com/video/farnborough-air-display-10.

37 Smith, 'The dawn of the jet age in austerity Britain: David Lean's *The Sound Barrier* (1952)', pp. 492–4.

38 Twiss, *Faster than the Sun*, pp. 67–71, 23–4 and 88–9.

39 Ibid., pp. 28–33 and 67.

40 Ibid., pp. 29–30.

41 Ibid., pp. 25–7.

42 Taylor, *Fairey Aircraft since 1915*, p. 429. Excluding the Rotodyne's huge development costs, by 1955 the Gyrodyne had cost Fairey Aviation £338,000 in R&D, the FD1 £382,000, and the Gannet no less than £539,500: Sir Richard Fairey, statement to shareholders, Fairey Aviation AGM, 24 November 1955, quoted in Trippe, 'The plane maker', pp. 1323–4.

43 Twiss, *Faster Than the Sun*, pp. 13–18 and 19.

44 'English Electric Lightning history', *Thunder and Lightnings*, 20 November 2016, www.thunder-and-lightnings.co.uk/lightning/history.php; Taylor, *Fairey Aircraft since 1915*, pp. 434–7; Hamilton-Paterson, *Empire of the Clouds*, pp. 182–4; Stuart Croft, Andrew Dorman, Wyn Rees and Matthew Uttley, *Britain and Defence 1945–2000: A Policy Re-evaluation* (London: 2001), pp. 12–4; Twiss, *Faster Than the Sun*, pp. 34–8, 104–5 and 149–50.

45 Derek Thurgood [?], 'List of actions suggested by Sir Richard Fairey in his letter of 2nd March 1956', Box 102, AC73/30, RAFM.

46 Correspondence of Derek Thurgood re FD2 publicity campaign, January–December 1956 and Derek Turgood to L. Massey Hilton, 'Preliminary Memorandum on F.D.2 Speed Record Publicity, 9 April 1956, Box 116, AC73/30, RAFM. 'Air speed record smashed', Pathé News, 15 March 1956, www.britishpathe.com/video/air-speed-record-smashed; 'We the British' and 'What's my line', BBC TV, 24 April and 16 March 1956. Thurgood was later sent a detailed treatment for a feature film about a pilot who is based on Twiss but haunted by wartime memories and witnessing the death at Farnborough of De Havilland test pilot John Derry, until breaking Mach 1.5 speeds his recovery and enables him to test Britain's 'sleek new civil airliner of revolutionary design.' The similarity to David Lean's 1952 *The Sound Barrier* must have been obvious to Fairey's chief publicist. Franklin Collings, 'The flying buccaneer or (beating the sun)', Box 118, AC73/30, RAFM.

47 Twiss, *Faster Than the Sun*, pp. 117–24; newspaper cuttings re FD2, March-June 1956, G.W. Hall, 'Speech at Claridge's – F.D.2. World air speed record', 12 March 1956, Box 116, AC73/30, RAFM. Robert Hotz, 'Bring back the world speed record now' and letter to the editor from Derek Thurgood, *Aviation Week*, 26 March and 18 April 1956.

48 Lady Fairey quoted in Trippe, 'The plane maker', p. 1325; Secretary, *Britannia*, to Secretary-General, The Royal Aero Club [12?] March 1956, op. cit.

49 For example, Evelyn Waugh, *The Diaries of Evelyn Waugh*, ed. Michael Davie (London: Phoenix, 1995), 23 November 1946, p. 663.

50 Sir Richard Fairey, 'Lunch-time speech on Bermuda to the Royal Empire Society', 11 May 1949 – speaker introduced by Lord Brabazon of Tara and Admiral of the Fleet Lord Chatfield, FAAM/2010/054/0114.

51 Sir Richard Fairey to Lord Brabazon, 14 March 1948, 8 October 1945, 10 June 1945 and 14 March 1948, Box 13, AC71,3, BP, RAFM.

52 Sir Richard Fairey, 'Second draft of speech for S.B.A.C. Dinner, Wednesday 26th July 1949', 7 July 1949, and 'Second draft of speech proposing the toast of the

Society at the S.B.A.C. Council Dinner, Wednesday 25th June 1947', [?] June 1947, FAAM/2010/054/0112 and 0109.
53 '...whatever they may say in public these people know that we are not really a closed shop of profiteers but a highly competitive industry, and without us the technical ascendancy on which the country depends would be lost.': Sir Richard Fairey, 'Second draft of speech for S.B.A.C. Dinner, Wednesday 26th July 1949.'
54 On the Cabinet clash between Morrison and Hugh Dalton on nationalisation of iron and steel giving the green light to extensive public ownership of manufacturing industry, see Ben Pimlott, *Hugh Dalton* (London: Jonathan Cape, 1985), pp. 496–8, John Bew, *Citizen Clem: A Biography of Attlee* (London: riverrun, 2016), pp. 464–6 and Peter Hennessy, *Never Again Britain 1945–1951* (London: Jonathan Cape, 1992), pp. 337–8.
55 Sir Richard Fairey to Lord Brabazon, 10 June 1946, Box 13, AC71/3, BP, RAFM.
56 Sir Richard Fairey, 'Lunch-time speech on Bermuda to the Royal Empire Society.'
57 Ibid. Bermuda did not join the West Indies Federation in 1957, and in 1995 a referendum of its 58,000 inhabitants saw a clear majority eschew independence and vote in favour of the archipelago remaining a British Overseas Territory.
58 Admiral Sir Ralph Leatham and Lieutenant General Sir Alexander Hood, followed by two further generals, 1955–64: 'Previous governors of Bermuda', www.gov.bm/previous-governors-bermuda.
59 Lionel, in Bermuda on American insistence, spent most of his stay in bed with pneumonia: Lord Moran, *Churchill: The Struggle for Survival 1940–65* (London: Constable, 1966), 3–10 December 1953, pp. 503–12.
60 'First draft speech for the Bermuda Rotary Club, 25 January 1949', [?] January 1949, FAAM/2010/054/0108; 'Fairey has no criticism for British airline detractors' and photograph of award ceremony at USAF Bermuda Base Command, *Bermuda Mid-Ocean News*, 26 January 1949 and 23 February 1948; citation quoted in Colonel John B. Ackerman, USAF, to Sir Richard Fairey, 13 November 1947, FAAM/2010/054/0080.
61 Correspondence of Sir Richard Fairey and Air Marshal Lord Tedder [CAS] re Darrell's Island project, January–August 1949 and of Sir Richard Fairey and Sir Thomas Lloyd [PUS, Colonial Office] re free UK education for Bermudian children, July 1949, FAAM/2010/054/009.
62 Bermuda's two annual Rhodes Scholarships were technically open to both white and black. The most prominent members of the 60 per cent black population were the island's lawyers, beneficiaries of local schemes to support the brightest non-white children in an education system superior to almost all Britain's other colonies.
63 Mrs Jane Tennant to the author, 15 June 2017.
64 Sir Richard Fairey, 1953 appointments diary, Box 77, AC73/70, RAFM.
65 'Second draft of speech proposing the toast of the Society at the S.B.A.C. Council Dinner, Wednesday 25th June 1947.'

66 Hall, *Sir Richard Fairey*, pp. 11–12.
67 'Fly fishing in the Sierra', http://stevenojai.tripod.com; Chris Reeves [Flydressers' Guild membership secretary] to the author, 19 April 2017; Alastair Robjent [Robjent's Fine Country Pursuits Ltd, Stockbridge] in telephone conversation with the author, 18 April 2017. John Fairey's second wife, Beverley, was a descendant of Frederic Halford.
68 Sir Richard Fairey to the editor, *The Times*, 27 December 1952; Hall, *Sir Richard Fairey*, p. 12.
69 Sir Richard Fairey, papers re the Fisheries, Pollution, and General Finance Committees of the Hampshire River Board, the Test and Itchen Fishing Association, and the Hampshire Rivers Landowners Committee, 1949–56, Box 102, AC73/30, RAFM.
70 A.W. Tuke and 12 co-signatories to Major T.T. Phelps [TIFA board hon. secretary], 30 November 1955, ibid. David Tennant in conversation with the author, 20 July 2017; Jane Tennant to the author, 29 July 2017.
71 A.W. Tuke and 12 co-signatories to Major T.T. Phelps [TIFA board hon. secretary], 30 November 1955, Box 102, AC73/30, RAFM.
72 Sir Richard Fairey to Major The Hon. A.J. Ashley Cooper, 2 November 1953, Fairey family papers, Pittleworth Manor; Mrs Jane Tennant to the author, 15 June 2017. 'He pitted himself against the introduction of electrical fishing which in its early days was viewed by those who had to maintain the river as housewives might have viewed the introduction of the vacuum cleaner, purely a blessing … His concern was for the fish themselves when the electrical jolt could literally break their backs and then for all the smaller beings from eel elvers to aquatic fly larvae. In a speech to the River Board he said that there were some who wished to turn the Test into a tank for trout': Jane Tennant, 'The Environmentalist', June 2017, Pittleworth Manor.
73 Sir Richard Fairey to Richard Fairey, 2 January 1956, and to Directors of Barclays Bank, 30 December 1955, Box 102, AC73/30, RAFM.
74 Hall, *Sir Richard Fairey*, p. 12; Mrs Jane Tennant to the author, 15 June 2017 and in conversation with the author, 10 February and 15 June 2017; Jane Tennant, 'The Environmentalist.'
75 Hall, *Sir Richard Fairey*, p. 12; Mrs Jane Tennant to the author, 15 June 2017.
76 Mrs Jane Tennant in conversation with the author, 15 June 2017.
77 Mrs Jane Tennant to the author, 2 October 2015.
78 Mrs Jane Tennant, quoted in Trippe, 'The plane maker', p. 66 and to the author, 15 June 2017; Jane Tennant, 'The Environmentalist.'
79 For example, 'The children are in grand form, Jane is beyond description, and Johnny's last school report was a wow': Sir Richard Fairey to Lord Brabazon, 8 October 1945, Box 13, AC71/3, BP, RAFM.
80 Sir Richard Fairey to C.R.N. Routh, 30 September 1946, FAAM/2010/054/0038.

81 'Captain John Fairey', obituary, *Daily Telegraph*, 23 July 2009 and *The Times*, 7 August 2009.
82 John Fairey quoted in Trippe, 'The plane maker', p. 67.
83 Papers re Richard Fairey's personal finances, 1940–60, Box 142, AC73/30, RAFM.
84 Papers re Richard Fairey's visits abroad, 1945–53, Box 13 AC73/30, ibid.; Richard Fairey, 'Visit to Halifax – 6th December to 10th December 1951', ibid.; Wilfred Broadbent, interviewed in Trippe, 'The plane maker', pp. 1293 and 1320–1.
85 Ibid., pp. 1321–2.
86 The account of Hall, Outen, and Fairey planning the latter's succession based on Geoffrey Hall's version of events in Trippe, 'The plane maker', pp. 1410–3.
87 Labour had secured a greater percentage of the popular vote in the 1951 general election and the party maintained higher poll ratings than the Conservatives throughout the early 1950s: in an essentially two party contest a net swing of three votes in a hundred would ensure a Labour victory: Michael Pinto-Duchinsky, 'Bread and Circuses? The Conservatives in Office, 1951–1964', in *The Age of Affluence 1951–1964*, eds. Vernon Bogdanor and Robert Skidelsky (London: Macmillan, 1970), pp. 70–1.
88 Trippe, 'The plane maker', pp. 1416–9.
89 John Fairey's account of his father's death, in Trippe, 'The plane maker', pp. 1325–7; Esther Fairey to Lord Brabazon, 21 October 1956, Box 13, AC71/3, BP, RAFM; Hall, *Sir Richard Fairey*, p. 12.
90 Angus Calder, *The Myth of the Blitz* (London: Pimlico, 1992), p. 182.
91 Mrs Jane Tennant to the author, 15 June 2017.
92 Documents and correspondence of Derek Thurgood re the memorial service of Sir Richard Fairey, October 1956, Box 120, AC73/30, RAFM.
93 'List of names of people from whom letters and telegrams of sympathy have been received by Mr. Fairey', 16 October 1956, ibid. The only members of the Eden Government on the list were two ministers of state, John Profumo and Reginald Maudling, the latter a family friend who gave the address at Richard Fairey's memorial service on 26 October 1960.
94 'Memorial Service for Sir Richard Fairey: Order of Service', 19 October 1956, ibid.
95 Esther Fairfax to Lord Brabazon, 21 October 1956, Box 13, AC71/3, BP, RAFM.
96 Lord Brabazon, 'Funeral address for Sir Richard Fairey, 16 October 1956', ibid.

Conclusion

1 'Gannet, R.N.', *Flight*, 9 April 1954; magazine photograph of CRF inspection No 703X Flight replicated in H.A. Taylor, *Fairey Aircraft since 1915* (New York: Putnam, 1974), p. 27.

2 'The Tirpitz bombed', Pathé News, 17 March 1944, www.britishpathe.com/video/the-tirpitz-bombed.
3 'Note he [CRF] stayed resident and domiciled in Britain when taxes were at their very highest (top rate 19/6 in the pound) when many fled to tax havens.' Jane Tennant to the author, 20 July 2017.
4 Richard Titmuss, *Problems of Social Policy* (London: HMSO, 1950), pp. 137–50 and 166–82.

Bibliography and Filmography

Sir Richard Fairey and family and Fairey Aviation archives

Papers of Sir Richard Fairey and family, Fleet Air Arm Museum, Yeovilton, National Museum of the Royal Navy
Papers of Sir Richard Fairey and family, Pittleworth Manor
Papers of Sir Richard Fairey and family, Bossington
Papers of Fairey Aviation, RAF Museum, Hendon
Fairey, C.R., 'How I began: In the early days of flying', *The Listener*, 16 February 1938.
——— 'Proceedings: C.R. Fairey, "Growth of aviation" [Presidential Address], 9 October 1930', *The Journal of the Royal Aeronautical Society*, 35/241 (1931).
——— 'The future of aeroplane design for the Services', *RUSI Journal*, 76 (1931).
Fairey, Sir Richard, 'Destroying the enemy with his own weapon', *Philadelphia Inquirer*, 3 January 1943.
Hall, G.W., *Sir Richard Fairey: The First Fairey Memorial Lecture* (London: Royal Aeronautical Society, 1959).
Tennant, Jane, 'The environmentalist', June 2017, Pittleworth Manor.

Family interviews and informal conversations

Jane Tennant (née Fairey)
Charles Fairey
David Tennant

Primary source material – archival and reference

Admiralty papers, National Archives
Air Ministry papers, National Archives
Board of Trade papers, National Archives

Prime Ministerial papers, National Archives
Ministry of Labour papers, National Archives
Ardingly College archives
Brabazon papers, RAF Museum, Hendon
Finsbury Technical College papers, London Metropolitan Archives
Middlesex Technical Education Committee papers, London Metropolitan Archives
Mountbatten papers, Broadlands Archives, University of Southampton
Templewood papers, Cambridge University Library
Trenchard papers, Cambridge University Library
H.G. Wells archive, University of Illinois, Champaign-Urbana

Primary source material – newspaper, magazine and journal articles

[anon. unless otherwise indicated]

'Spanish Armada', *The Naval and Military Magazine*, 2 (1827).
'The Kite and Model Aeroplane Association Official Notices', *Amateur Aviation*, 24 June 1912.
'New companies registered', *The Aeroplane*, 6 August 1915.
Dworetsky, Julius, 'The diagnosis of tubercular laryngitis', *Journal of the American Medical Association*, 69 (1917).
'Ant-dating the enemy', *The Aeroplane*, 16 January 1918.
'Built before the Gothas: British bombers with folding wings', *The Graphic*, 23 March 1918.
'New competitor for Atlantic flight', *The Times*, 28 March 1919.
'Post-Bellum policies', *The Aeroplane*, 16 April 1919.
'Atlantic airman of 21', *Daily Mail*, 21 April 1919.
'New company registered', *Flight*, 22 May 1919.
'Seaplanes safe flying', *Manchester Guardian*, 9 July 1919.
'Modern British aeroplanes X. The Fairey Aviation Co. Ltd', *Aeronautical Engineering* [*The Aeroplane* supplement], 6 August 1919.
'Flight – and the men', *Flight*, 14 August, 1919.
'The Schneider Cup Race', *The Aeroplane*, 27 August 1919.
'Seaplanes for civil flying', 13 September 1919.
'Les premieres ailes variable', *The Aeroplane* [French edition], 24 December 1919.
'Pioneers of British aviation – XXVI Mr C.R. Fairey', *Aeronautics*, 22 January 1920.
'A post-war product', *Aeronautical Engineering*, 28 January 1920.
C.G.G., 'On the Schneider picnic', *The Aeroplane*, 17 September 1919.
C.G.G., 'Echoes from the Schneider Cup', *The Aeroplane*, 24 September 1919.
C.G.G. 'Air Ministry Communiqué No. 621', *The Aeroplane*, 11 October 1920.
'The Fairey Aviation Co., Ltd.', *Flight*, 16 March 1921.

BIBLIOGRAPHY AND FILMOGRAPHY

'Launch of A.N.A. 1.', *Flight*, 18 August 1921.
C.G.G., 'A chance for draughtsman', *The Aeroplane*, 14 September 1921.
'The Fairey Aviation Co., Ltd.', *Aeronautical Engineering* [multi-language supplement to *The Aeroplane*], 9 November 1921.
'New type of amphibian aeroplane', *The Times*, 17 January 1922.
'A real amphibian', *The Aeroplane*, 11 January 1922.
'A Fairey triumph', *The Aeroplane*, 11 January 1922.
'The 1925 Schneider Trophy Race', *Flight*, 29 October 1925.
Report of RAeC monthly house dinner, 19 January 1927, *Flight*, 27 January 1927.
'The 10th Royal Air Force Display', *Flight*, 18 July 1929.
'H.R.H. The Prince of Wales declares the Aero Exhibition open', *Flight*, 18 July 1929.
'De Havilland', *Flight*, 21 November 1930.
'Natural science club', *The Harrovian*, 15 December 1932.
Major B. Heckstall-Smith, 'The America's Cup Challenge "The right man to sail her"', *Yachting World and Motor Boating Journal*, 3 November 1933.
'Whitsun and Deauville', *Morning Post*, 20 May 1936.
'Prize winning yachts in 1933', *Yachting World and Motor Boating Journal*, 8 December 1933.
'Another challenge to U.S.', *The Times*, 15 December 1933.
'Our air correspondent', 'Big reserves of warplanes needed – Mr Fairey's warning', *Daily Mail*, 5 January 1934.
Lamsley, Arthur, 'A season with the international classes', *Yachting World and Motor Boating Journal*, 22 September 1934.
'R.A.S. Flying Display', *The Times*, 6 May 1935.
'Mr and Mrs Charles Fairey at home', *Tatler*, 21 August 1935.
'The garden party of the Royal Aeronautical Society', *The Aeroplane*, 8 May 1935.
'Heston', *Aeropilot London*, October 1935.
'At the universities', *The Harrovian*, 19 October 1935.
'University flying club formed', *Cambridge Evening News*, 1 November 1935.
'America's Cup withdrawal. Challenger's friendly gesture', *News Chronicle*, 27 November 1935.
Special Correspondent, 'America's Cup – Mr Fairey talks of future contests', *Evening Standard*, 23 December 1935.
'Aircraft profit limit criticism', *Financial Times*, 1 January 1937.
'Aircraft strikers demand strike pay', *Evening News*, 8 March 1937.
'Fairey strike ends', *Daily Worker*, 10 March 1937.
'The shipwrights' company', *Shipping World*, 21 April 1937.
'Speed of air raid held london peril', *New York Times*, 28 May 1937.
'The Belgian ambassador', *The Times*, 29 May 1937.
'Building the Battle', *Flight*, 17 June 1937.
'Designed for mass-production', *Flight*, 19 August 1937.
'Speed charge fails', *Daily Telegraph*, 24 December 1935.

'Beat the police on mathematics', *Evening News*, 24 December 1935.
V.R., 'Who's who in the air: Charles Fairey', *Daily Express*, 10 February 1936.
C.G.G., 'On the Royal Commission on the Private Manufacturing of and Trading in Arms and its relations with the British aircraft industry', *The Aeroplane*, 14 February 1936.
'Where Jane Austen lived', *Daily Telegraph*, 3 April 1936.
'The Wakefield Trophy eliminating trials', *The Aero-Modeller*, June 1936.
Stewart, Oliver, 'Air eddies', *The Tatler*, 5 May 1937.
Report of 11th OGM in 'The Fairey Aviation Co.', *Financial News*, 30 December 1939.
'A real Fairey story', *Flight*, 20 February 1941.
'Roosevelt and Hopkins shape Allied set-up', *New York Times*, 30 March 1941.
'Hopkins set up as head of Lend-Lease program', *New York Times*, 15 April 1941.
'U.S. aircraft production held key to Allied victory', *Dayton News*, 9 July 1942.
'British air leader opens exhibit', *Dayton News*, 11 July 1942.
'Economy of Lend-Lease sets model for capital', *New York Times*, 14 March 1943.
'Cripps fails to satisfy Fairey workers', *Daily Worker*, 21 January 1944.
Evelyn, Kenneth A. et al., 'Effect of sympathectomy on blood pressure hypertension: a review of thirteen years' experience at the Massachusetts General Hospital', *Journal of the American Medical Association*, 140 (1949).
'Sir Richard gives a party', *The Tatler*, 4 January 1950.
'Twenty-five years of service', *The Aeroplane*, 6 January 1950.
'Gannet, R.N.', *Flight*, 9 April 1954.
Hotz, Robert, 'Bring back the world speed record now', *Aviation Week*, 26 March 1956.
'From all quarters: Fairey managing director', *Flight*, 17 August 1956.
Anne Matheson, 'Fairey gran'ma left with the baby', *The Australian Women's Weekly*, 2 October 1957.
'Philip Young is dead at 76; Eisenhower's personnel chief', *New York Times*, 19 January 1987.
'Sir Richard Treherne', obituary, *Daily Telegraph*, 22 December 2001.

Primary source material – diaries, correspondence and memoirs

Alanbrooke, Field Marshal Lord, *War Diaries 1939–1945*, Alex Danchev and Dan Todman, eds (London: Weidenfeld & Nicolson, 2001).
Arnold, H.H., *Global Mission* (London: Hutchinson, 1951).
——— *American Air Power Comes of Age: General Henry H. 'Hap' Arnold's World War II Diaries Volume I*, Major General John W. Huston USAF Rtd, ed. (Maxwell, AL: Air University Press, 2002).
——— *American Air Power Comes of Age: General Henry H. 'Hap' Arnold's World War II Diaries Volume II*, Major General John W. Huston USAF Rtd, ed. (Maxwell, AL: Air University Press, 2002).

Barnato Walker, Diana, *Spreading My Wings* (London: Grub Street, 2003).
Brabazon of Tara, Lord, *The Brabazon Story* (London: William Heinemann, 1956).
Briggs, Asa, *Secret Days Code-Breaking in Bletchley Park* (London: Frontline Books, 2011).
Brockway, Fenner, *The Bloody Traffic* (London: Victor Gollancz, 1934).
Brooks, Colin, *Fleet Street, Press Barons and Politics: The Journals of Collin Brooks, 1932–1940*, N.J. Crowson, ed. (London: Royal Historical Society, 1998).
Bruce Lockhart, Sir Robert, *The Diaries of Sir Robert Bruce Lockhart 1915–1938*, Kenneth Young, ed. (London: Macmillan, 1973).
Cadogan, Sir Alexander, *The Diaries of Sir Alexander Cadogan 1938–1945*, David Dilks, ed. (London: Cassell, 1971).
'Cato', *Guilty Men* (London: Victor Gollancz, 1940).
Channon, Sir Henry, *Chips The Diaries of Sir Henry Channon*, Robert Rhodes James, ed. (London: Penguin, 1967).
Churchill, Winston S., *The World Crisis Volume 1 1911–1914* (Thornton Butterworth, London, 1923).
―――― *Winston S. Churchill Volume II Companion Part 3 1911–1914*, Randolph S. Churchill, ed. (London: Heinemann, 1969).
―――― *The Churchill War Papers Vol. I September 1939–May 1940*, Martin Gilbert, ed. (London: Heinemann, 1993).
Churchill, Winston S. and F.D. Roosevelt, *Churchill and Roosevelt: The Complete Correspondence I. Alliance Emerging*, Warren F. Kimball, ed. (London: Collins, 1984).
Ciano, Count, *Ciano's Diary 1939–1943*, Malcolm Muggeridge, ed. (William Heinemann, 1947).
Cobham, Sir Alan J., *A Time To Fly* (London: Shepheard-Walwyn, 1978).
Crozier, W.P., *Off The Record Political Interviews 1933–1943* A.J.P. Taylor, ed. (London: Hutchinson, 1973).
Cooper, Duff, *The Duff Cooper Diaries*, John Julius Norwich, ed. (London: Weidenfeld & Nicolson, 2005).
Cox, Sebastian and Peter Gray, eds., *Air Power History Turning Points from Kitty Hawk to Kosovo* (London: Frank Cass, 2002).
Farrer, David, *The Sky's The Limit: The Story of Beaverbrook at M.A.P.* (London: Hutchinson, 1943).
Halifax, Earl of, *Fulness of Days* (London: Collins, 1957).
de Havilland, Sir Geoffrey, *Sky Fever: The Autobiography of Sir Geoffrey de Havilland C.B.E.* (Shrewsbury: Airlife, 1979).
Headlam, Sir Cuthbert, *Parliament and Politics in the Age of Churchill and Attlee: The Headlam Diaries 1935–1951*, Stuart Ball, ed. (Cambridge: Cambridge University Press, 1999).
Hudson, H.U., ed., *Lord Lothian Speaks to America* (Oxford: Oxford University Press, 1941).
Lamb, Charles, *War in a Stringbag* (London: Cassell, 1977).

Macmillan, Norman, *Freelance Pilot* (London: Heinemann, 1937).
────── *Into The Blue* (London: Gerald Duckworth, 1929).
Maisky, Ivan, *The Maisky Diaries: Red Ambassador to the Court*, Gabriel Gorodetsky, ed. (London, Yale University Press, 2015).
Mitchell, Air Commodore John, LVO DFC AFC with Sean Feast, *Churchill's Navigator* (London: Grub Street, 2010).
Moran, Lord, *Churchill: The Struggle for Survival 1940–65* (London: Constable: 1966).
Nicolson, Harold, *Diaries and Letters 1930–39*, Nigel Nicolson, ed. (London: Collins, 1966).
────── *Diaries and Letters 1939–45* Nigel Nicolson, ed. (London: Collins, 1967).
Noel-Baker, Philip, *Hawkers of Death: the Private Manufacture of Armaments* (London: Labour Party, 1934).
Moffatt, John, with Mike Rossiter, *I Sank the Bismarck Memoirs of a Second World War Navy Pilot* (London: Bantam Press, 2009).
Mosley, Sir Oswald, *My Life* (London: Thomas Nelson, 1968).
Snow, C.P. *The Two Cultures* (Cambridge: Cambridge University Press, 1959).
Sueter, Rear-Admiral Murray F., *Airmen or Noahs Fair Play for our Airmen: The Great 'Neon' Air Myth Exposed* (London, Sir Isaac Pitman & Sons, 1928).
Templewood, Viscount, *Empire of the Air: The Advent of the Air Age 1922–1929* (London: Collins, 1957).
────── *Ambassador on Special Mission* (London: Collins, 1946).
Twiss, Peter, *Faster than the Sun: The Compelling Story of a Record Breaking Test Pilot* (London: Grub Street, 2000).
Waller Hills, John, *A Summer on the Test*, 1930 2nd edn. reprint (London: André Deutsch, 1983).
Waugh, Evelyn, *The Diaries of Evelyn Waugh*, ed. Michael Davie (London: Phoenix, 1995).

Primary source material – official publications

Hansard
London Gazette

The Admiralty and Ministry of Information, *Fleet Air Arm* (London: HMSO, 1943/2001).
────── *Fleet Air Arm: The Admiralty Account of Naval Air Operations* (London: HMSO, 1943/2001).
────── *East of Malta West of Suez: The Admiralty Account of the Naval War in the Eastern Mediterranean: September 1939 to March 1941* (London: HMSO, 1943/2001).
Commonwealth of Australia, *Census of the Commonwealth of Australia 30th June, 1933 Volume I Part VIII. Population and Occupied Dwellings in Localities* and *Part XVI. Religion* (Canberra: Commonwealth of Australia, 1934).

HMSO, *Minutes of Evidence Taken Before the Royal Commission on the Private Manufacture of and Trading in Arms, Seventeenth Day, Friday 7th February, 1936* (London: HMSO, 1936).
Ministry of Information, *What Britain Has Done 1939–1945: A Selection of Outstanding Facts and Figures* (London: HMSO, 1945/Atlantic Books, 2007).
Royal Navy, *King's Regulations and Admiralty Instructions of His Majesty's Naval Service Volume I* (London: HMSO, 1913).
―――― *Documents Relating to the Naval Air Service Vol. I 1908–1918*, S.W. Roskill, ed. (London: Navy Records Society, 1969).
Titmuss, Richard, *Problems of Social Policy* (London: HMSO, 1950).

Secondary source material – books

Kelly's Directory

Bailey, Gavin J., *The Arsenal of Democracy Aircraft Supply and the Anglo-American Alliance 1938–1942* (Edinburgh: Edinburgh University Press, 2013).
Ball, Norman F. and John N. Vardalas, *Ferranti-Packard: Pioneers in Canadian Electrical Engineering* (Montreal: McGill-Queen's Press, 1994).
Ball, Stuart, *Baldwin and the Conservative Party: The Crisis of 1929–1931* (New Haven and London: Yale University Press, 1988).
Barnes, C.H., *Shorts Aircraft since 1900* (London: Putnam, 1967).
Barnett, Correlli, *The Audit of War: The Illusion and Reality of Britain as a Great Nation* (London: Papermac, 1987).
Bell, Christopher M., *Churchill and Sea Power* (Oxford: Oxford University Press, 2013).
Benbow, Tim, ed., *British Naval Aviation The First 100 Years*, ed. Tim Benbow (Farnham: Ashgate, 2011).
Bew, John, *Citizen Clem: A Biography of Attlee* (London: riverrun, 2016).
Birkenhead, Earl of, *The Prof in Two Worlds: The Official Life of Professor F.A. Lindemann, Viscount Cherwell* (London: Collins, 1961).
Bogdanor, Vernon and Robert Skidelsky, eds., *The Age of Affluence 1951–1964* (London: Macmillan, 1970).
Boyle, Andrew, *Trenchard* (London: Collins, 1962).
Bramson, Alan, *Pure Luck: The Authorised Biography of Sir Thomas Sopwith* (Manchester: Crécy, 2005).
Briggs, Asa, *The History of Broadcasting in the United Kingdom Volume II: The Golden Age of Wireless* (Oxford: Oxford University Press, 1995).
Bruce, Gordon, 'Short (Hugh) Oswald (1883–1969)', Rev. Robin Higham, *Oxford Dictionary of National Biography* (Oxford: Oxford University Press, 2004).
Calder, Angus, *The People's War Britain 1939–45* (London: Jonathan Cape, 1969).
―――― *The Myth of the Blitz* (London: Pimlico, 1992).

Charmley, John, *Lord Lloyd and the Decline of the British Empire* (London: Weidenfeld & Nicolson, 1987).

Cheffins, Brian R., *Corporate Ownership and Control British Business Transformed* (Oxford: Oxford University Press, 2008).

Chisholm, Anne and Michael Davie, *Lord Beaverbrook: A Life* (New York: Alfred A. Knopf, 1993).

Chorlton, Martyn, ed., *Fairey Company Profile 1915–1960* (Cudham: Kelsey Publishing Group/Aeroplane, 2012).

Churchill, Randolph S., *Winston S. Churchill Vol. II Young Statesman 1901–1914* (London: Heinemann, 1967).

Clarke, Peter, *The Cripps Version: The Life of Sir Stafford Cripps 1889–1952* (London: Allen Lane, 2002/London: Penguin, 2003).

Coleman, Ronald, *Memorial to Pioneer Airmen* (Eastchurch: Eastchurch Parish Council, 1955).

Collins, Damian, *Charmed Life: The Phenomenal World of Philip Sassoon* (London: William Collins, 2016).

Conquest, Robert, *The Great Terror: A Reassessment* (Oxford: Oxford University Press, 1990).

Cooper, Malcolm *The Birth of Independent Air Power* (London: Allen & Unwin, 1986).

Cowling, Maurice, *The Impact of Hitler: British Politics and British Policy 1933–1940* (Cambridge: Cambridge University Press, 1975).

Cox, Sebastian and Peter Gray, eds, *Air Power History: Turning Points from Kitty Hawk to Kosovo* (London: Frank Cass, 2002).

Croft, Stuart, Andrew Dorman, Wyn Rees and Matthew Uttley, *Britain and Defence 1945–2000 A Policy Re-evaluation* (London: 2001).

Cross, J.A. *Sir Samuel Hoare, a Political Biography* (London: Jonathan Cape, 1977).

Darroch, James L., *Canadian Banks and Global Competitiveness* (Montreal: McGill-Queen's Press, 1999).

Dear, Ian, *Enterprise to Endeavour The J-Class Yachts* (London: Adlard Coles Nautical, 1999).

Dorril, Stephen, *Blackshirt Sir Oswald Mosley and British Fascism* (London: Viking, 2006).

Driver, Hugh, *The Birth of Military Aviation Britain 1903–1914* (Woodbridge: Royal Historical Society/Boydell Press, 1997).

Duffy, Paul and Andrei Kandalov, *Tupolev: The Man and His Aircraft* (Shrewsbury: Airlife, 1996).

Edgerton, David, *Science, Technology and the British Industrial 'Decline'* (Cambridge: Cambridge University Press, 1996).

―――― *Warfare State Britain, 1920–1970* (Cambridge: Cambridge University Press, 2006).

―――― *Britain's War Machine Weapons, Resources and Experts in the Second World War* (London: Allen Lane, 2011).

BIBLIOGRAPHY AND FILMOGRAPHY

—— *England and the Aeroplane Militarism, Modernity and Machines* (London: Penguin, 2013).

Farmelo, Graham, *Churchill's Bomb How the United States Overtook Britain in the First Nuclear Arms Race* (New York: Basic Books, 2013).

Farquharson-Roberts, Mike, *A History of the Royal Navy: World War I* (London: I.B.Tauris, 2014).

Finneran, Richard J. and George Bornstein, eds, *The Collected Works of W.B. Yeats Vol. IV Early Essays* (New York: Scribner, 2007).

Fjellman, Margit, *Louise Mountbatten Queen of Sweden* (London: George Allen and Unwin, 1968).

Fort, Adrian, *Prof: The Life of Frederick Lindemann* (London: Pimlico, 2004).

Gardiner, Juliet, *The Thirties: An Intimate History* (London: Harper Press, 2011).

Gibbs, N.H., *History of the Second World War: Grand Strategy Volume 1 Rearmament Policy* (London: HMSO, 1976).

Gilbert, Martin, *Winston S. Churchill Vol. III 1914–16* (London: Heinemann, 1971).

Gollin, Alfred, *No Longer an Island Britain and the Wright Brothers* (Standford: Stanford University Press, 1984).

Gordon, G.A.H., *British Seapower and Procurement Between the Wars* (London: Macmillan, 1998).

Gorodetsky, Gabriel, *Stafford Cripps' Mission to Moscow 1940–42* (Cambridge: Cambridge University Press, 1984).

Griffiths, Richard, *Fellow Traveller of the Right British Enthusiasts for Nazi Germany 1933–39* (Oxford: Oxford University Press, 1983).

Hamilton-Paterson, James, *Empire of the Clouds: When Britain's Aircraft Ruled the World* (London: Faber and Faber, 2010).

Hayward, Keith, *The British Aircraft Industry* (Manchester: Manchester University Press, 1989).

Hennessy, Peter, *Never Again Britain 1945–1951* (London: Jonathan Cape, 1992).

Higham, Robin, John T. Greenwood and Von Hardesty, eds., *Russian Aviation and Air Power in the Twentieth Century* (London: Frank Cass, 1998).

Holman, Brett, *The Next War in the Air: Britain's Fear of the Bomber, 1908–1941* (Farnham: Ashgate, 2014).

Hooks, Mike, *Fleet Air Arm Aircraft of World War 2* (Cudham: Kelsey Publishing Group, 2013).

Hutton, Will, *The State We're In* (London: Jonathan Cape, 1995).

Innes-Smith, Robert, 'James Wentworth Day (1899–1983)', *Oxford Dictionary of National Biography* (Oxford, 2004).

Kennedy, Paul, *Engineers of Victory: The Problem Solvers who Turned the Tide in the Second World War* (London: Penguin, 2014).

Kershaw, Ian, *Making Friends With Hitler: Lord Londonderry and Britain's Road to War* (London: Penguin, 2005).

King, Peter, *Knights of the Air: The Life and Times of the Extraordinary Pioneers who First Built British Aeroplanes* (London: Constable, 1989).
Lawrence, Charles, *Fairey Marine Boats, Raceboats, Rivals and Revivals* (Chiswick: Charles Lawrence, 2014).
Layman, R.D., *Naval Aviation in the First World War: Its Impact and Influence* (London: Chatham Publishing, 1996).
Lee, David, 'Longmore, Sir Arthur Murray (1885–1970)', Rev. Christina J.M. Goulter, *Oxford Dictionary of National Biography* (Oxford: Oxford University Press, 2004).
Lewis, Julian, *Racing Ace: The Fights and Flights of Samuel 'Kink' Kinkead DSO DSC* DFC** (London: Pen & Sword, 2011).
Lewis, Peter, *British Aircraft 1809–1914* (London: Putnam, 1962).
—— *The British Fighter Since 1912 Fifty Years of Design and Development* (London: Putnam, 1965).
Lovell, Mary S., *Amelia Earhart: The Sound of Wings* (London: Abacus, 2009).
MacKenzie, Norman and Jeanne, *The Time Traveller: The Life of H.G. Wells* (London: Weidenfeld and Nicolson, 1974).
Meekcoms, K.J., *The British Air Commission and Lend-Lease: The Role, Organisation, and Work of the BAC (and its Antecedents) in the United States and Canada, 1938–1945* (Tunbridge Wells: Air-Britain, 2000).
Monk, Ray, *Ludwig Wittgenstein The Duty of Genius* (London: Vintage, 1991).
—— *Inside the Centre: The Life of J. Robert Oppenheimer* (London: Jonathan Cape, 2012).
Morgan, Philip, *Italian Fascism 1919–1945* (Basingstoke: Macmillan, 1995).
Olson, Lynne, *Troublesome Young Men: The Churchill Conspiracy of 1940* (London: Bloomsbury, 2008).
Onslow, 5th Earl of, Rev. Robin Higham, 'Thomson, Christopher Birdwood, Baron Thomson of Cardington (1875–1930)', *Oxford Dictionary of National Biography* (Oxford: Oxford University Press, 2004).
Overy, Richard, *Why The Allies Won* (London: Jonathan Cape, 1995).
—— *The Morbid Age Britain and the Crisis of Civilization, 1919–1939* (London: Penguin, 2010).
Peele, Gillian, and Chris Cook, eds., *The Politics of Reappraisal 1818–1929* (London: Macmillan, 1975).
Penrose, Harald, *British Aviation: The Pioneer Years 1903–1914* (London: Putnam, 1967).
—— *British Aviation: The Adventuring Years 1920–1929* (London: Putnam, 1973).
Permuy López, Rafael A., *Air War Over Spain: Aviators, Aircraft and Air Units of the Nationalist and Republican Air Forces 1936–1939* (Hersham: Classic/Ian Allan Publishing, 2009).
Pevsner, Nikolaus, and Elizabeth Williamson, *The Buildings of England: Buckinghamshire* (London: Penguin, 1994).

Pimlott, Ben, *Hugh Dalton* (London: Jonathan Cape, 1985).
Pugh, Martin, *Hurrah for the Blackshirts! Fascists and Fascism in Britain Between the Wars* (London: Jonathan Cape, 2005).
Purcell, Hugh, *The Last English Revolutionary: Tom Wintringham 1898–1949* (Stroud: Sutton, 2004).
Reynolds, Clark G., *Admiral John H. Towers: The Struggle for Naval Air Supremacy* (Annapolis: Naval Institute Press, 1991).
Roberts, Andrew, *The Storm of War: A New History of the Second World War* (London: Allen Lane, 2009).
Robertson, Bruce, *Sopwith – the Man and his Aircraft* (Bedford: The Sidney Press, 1970).
Reynolds, David, *From World War to Cold War: Churchill, Roosevelt, and the International History of the 1940s* (Oxford: Oxford University Press, 2007).
Ritchie, Sebastian, *Industry and Air Power: The Expansion of British Aircraft Production 1935–41* (London: Frank Cass, 1997).
Scott, E.O.G., *Hagley (A Short History of the Early Days of the Village and District with Notes on the Pioneer Families)* (Launceston: Birchalls, 1985).
Self, Robert, *Neville Chamberlain: A Biography* (Farnham: Ashgate, 2006).
Skidelsky, Robert, *Oswald Mosley* (London: Macmillan, 1985).
Smith, Adrian, *Mick Mannock, Fighter Pilot: Myth, Life and Politics* (London: I.B.Tauris, 2001/2015).
Smith, Jean Edward, *FDR* (New York: Random House, 2008).
Smith, Ron, *British Built Aircraft: Greater London* (Stroud: Tempus, 2002).
Smithells, Arthur, 'Thompson, Silvanus Phillips (1851–1916)', Rev. Graeme J.N. Gooday, *Oxford Dictionary of National Biography* (Oxford: Oxford University Press, 2004).
Stewart, Graham, *Burying Caesar Churchill, Chamberlain and the Battle for the Tory Party* (London: Weidenfeld & Nicolson, 1999).
Swale Borough Council, *150 Years of Trains to Sheppey 1860–2010* (Sittingbourne: Swale BC/Kent CC, 2010).
Taylor, A.J.P., *Beaverbrook* (London: Hamish Hamilton, 1972/London: Penguin, 1974).
Taylor, H.A., *Fairey Aircraft since 1915* (London: Putnam, 1974).
Tapper, Oliver, *Armstrong Whitworth Aircraft since 1913* (London: Putnam, 1973).
Till, Geoffrey, *Air Power and the Royal Navy 1914–1945* (London: Jane's Publishing Company, 1979).
von Wright, G.H., ed., *A Portrait of Wittgenstein as a Young Man* (Oxford: Basil Blackwell, 1990).
Walker, Percy B. *Early Aviation at Farnborough Vol. II The First Aeroplanes* (London: Macdonald, 1974).
Wragg, David, *Swordfish: The Story of the Taranto Raid* (London: Weidenfeld and Nicolson, 2003).
―――― *Stringbag: The Fairey Swordfish at War* (Barnsley: Pen & Sword Aviation, 2004).

Secondary source material – magazine and journal articles

Anderson, David G., 'British rearmament and the "merchants of death": The 1935–6 Royal Commission on the Manufacture and Trade in Armaments', *Journal of Contemporary History*, 29 (1994).

Cockett, R.B., 'Communication: Ball, Chamberlain and *Truth*', *The Historical Journal*, 33:1 (1990).

Dye, Peter, 'The bridge to air power – aviation engineering on the Western Front 1914–1918', *Air Power Review*, 17 (2014).

Edgerton, D.E.H., 'Technical innovation, industrial capacity and efficiency: public ownership and the British military aircraft industry, 1935–48', *Business History*, 26 (1984).

―――― 'The prophet militant and industrial: the peculiarities of Corelli Barnett', *Twentieth Century British History*, 2 (1991).

Fearon, Peter, 'The formative years of the British aircraft industry, 1913–1924', *The Business History Review*, 43 (1969).

Fielding, Steven, 'The Second World War and popular radicalism: the significance of the "movement away from party"', *History*, 80 (1995).

Fountain, Nigel 'Lettice Curtis', obituary, *Guardian*, 28 July 2014.

Gay, Hannah, and Anne Barrett, 'Should the cobbler stick to his last? Silvanus Phillips Thompson and the making of a scientific career', *The British Journal for the History of Science*, 32 (2002).

Harrison, Mark, 'The political economy of a Soviet military R&D failure: Steam power for aviation, 1932 to 1939', *The Journal of Economic History*, 63 (2003).

Huntley, Ian, 'Fairey's elusive Fox', *Aeroplane Monthly*, January 1979.

Kushner, Tony, 'Local heroes: Belgian refugees in Britain during the First World War', *Immigrants & Minorities*, 18 (1999).

Lemco, Ian, 'Wittgenstein's aeronautical investigation', *Notes & Records of the Royal Society*, 61 (2007).

Marr, Paul, 'Haig and Trenchard: Achieving air superiority on the Western Front', *Air Power Review*, 17 (2014).

Smith, Adrian, 'Ramsay MacDonald – aviator and actionman', *The Historian*, 28 (1990).

―――― 'The dawn of the jet age in austerity Britain: David Lean's *The Sound Barrier* (1952)', *Historical Journal of Film, Radio and Television* 30/4 (2010).

WB, 'The Man in White', *Motor Sport*, February 1997.

Willis, Matthew, 'From Firefly to Swordfish: The distinguished aviation career of Duncan Menzies, part two' *Aviation Historian* 12, July 2015.

Secondary source material – website articles

Anon., 'Charles Lindbergh Timeline', www.charleslindbergh.com/timeline/index.asp

―――― 'English Electric Lightning history', Thunder and Lightnings, 20 November 2016, www.thunder-and-lightnings.co.uk/lightning/history.php

BIBLIOGRAPHY AND FILMOGRAPHY

——— 'Felixstowe Ferry Golf Club', www.felixstowegolf.co.uk/index
——— 'Fly fishing in the Sierra', www.stevenojai.tripod.com/irresistible
——— 'Glenn Curtiss in Buffalo, New York', www.buffaloah.com/h/aero/curt/#Anchor
——— 'Heysham Harbour', www.heyshamheritage.org.uk/heysham_harbour
——— 'S.V. Ilyushin', www.ilyushin.org/en/about/history/biography
——— 'François Latry', www.cooksinfo.com/francois-latry
——— 'Korvettenkapitän Peter-Erich Cremer', www.uboat.net/men/cremer
——— 'R.J. Mitchell: a life in aviation' website, Solent Sky Museum, www rjmitchell-spitfire.co.uk/schneidertrophy/1931.asp?sectionID=2
——— 'Muswell Manor', www.muswellmanor.co.uk/aviation_history
——— 'The City and Guilds of London Institute', www.skillsdevelopment.org/aboutus/about_city_guilds.aspx
——— 'The Eastchurch aviation memorial, www.sheppeywebsite.co.uk/index.php?id=72
——— 'The Kodascope Libraries', Robbie's Reels, www.robbiesreels.com/1920
——— 'Tupolev ANT-9/PS-9', www.aviastar.org/air/russia/ant-9.php
——— 'Twyford School', www.twyfordschool.com
——— 'Uffa Fox, 1898–1972', www.uffafox.com/uffabiog
——— 'Wardman Tower – History', www.wardmantower.com/history/#founder
AEEU, 'Amalgamated Society of Engineers', www.archive.unitetheunion.org/about_us/history/history_of_aeeu.aspx
Barclays Bank, 'Anthony Tuke (Chairman 1951–1962)', www.archive.barclays.com/items/show/19496
Bossington Estate, 'Bossington – Fly Fishing', www.bossingtonestate.com/field-sports/fly-fishing
Camper & Nicholsons, 'Our Heritage', www.camperandnicholsons.com/about/company-heritage
Collier, Mike, 'Crucial role for Hayes in the First World War effort', www.middx.net/articles/munitions
Fairey Surveys, 'History of Fairey Surveys', June 2013, www.faireysurveys.co.uk/category/history
Filey Bay Research Group, 'Big Game Fishing off the Yorkshire Coast', www.fileybay.com/tunnyfish
Fishing Breaks, 'Chalkstream Fly Fishing in Hampshire', www.fishingbreaks.co.uk/hampshire
Hamilton, Rhona, 'Scott, Eric Oswald (1899–1986)', *Australian Dictionary of National Biography*, volume 18, 2011, www.adb.anu.edu.au/biography/scott-eric-oswald-15908.
Hinds Head Hotel, home page, www.hindsheadbray.com
King, Andrea, 'A Short History of Ardingly College', www.ardingly.com
'National Aerospace Library' Royal Aeronautical Society, www.aerosociety.com/About-Us/nal
J Class Association, 'History: 1929–1937 The golden years', www.jclassyachts.com/history/1919-1937

Parkinson, Justin, 'Why did the half-plane, half-helicopter not work?', *BBC News Magazine*, 12 February 2016, www.bbc.co.uk/news/magazine-35521040
Poole Flying Boats Celebration, 'Friends Newsletter', spring 2010, www.pooleflyingboats.com/pdfs_newsletters/Friends%20Newsletter%20Spring%202010.pdf
Poynter, J.P., 'Sir Clive Lathan Baillieu (1889–1967), *Australian Dictionary of Biography*, volume 7, 1979, www.adb.anu.edu.au/biography/bailleu-sir-clive-latham-5629/text8517
Smith, R.A., 'The Flica Project', www.americascupmasters.com/_/The_Flica_Project
Smithsonian, 'Separate Not Equal: Washington D.C.', www.americanhistory.si.edu/brown/history/4-five/washington-dc-1
Sonaca Group, 'History', www.sonaca.com/content/belgium/entreprise
University of Cambridge Department of Engineering, '125 years engineering excellence', www-g.eng.cam.ac.uk/125
Walker, Michael, 'Belgium Refugees WW1 – Hayes Belgium refugees', 4 February 2014, *Hayes People History*, ourhistory – Hayes.blogspot.com
Willis, Matthew, 'A recipe for obsolescence? British naval torpedo bomber development in the 1930s', Naval Air History, 17 September 2015, www.navalairhistory.com/2012/09/17/a-recipe-for-obsolescence-british-naval-torpedo-development-in-the-1930s/
12 Metre Class, 'Evaine', www.12mrclass.com/yachts/detail/273-itemId.511707058
12 Metre Class, 'Flica', www.12mrclass.com/yachts/detail/273-itemId.511707028
12 Metre Class, 'Trivia', www.12mrclass.com/yachts/detail/273-itemId.511707076
12 Metre Class, 'Westra', www.12mrclass.com/yachts/detail/273-itemId.511707080

Secondary source material – unpublished papers and theses

Trippe, Peter, 'The plane maker: The official biography of Sir Richard Fairey', Pittleworth Manor.
Wentworth Day, J., draft synopsis for book 'Wings over the world: the official life of Sir Richard Fairey, M.B.E., FR.R.Ae.S.', Fairey papers, Fleet Air Arm Museum.
Weatherburn, Michael, 'Scientific management at work: the Bedaux system, management consulting and worker efficiency in British industry, 1914–48', PhD, Imperial College, 2014.

Fiction and verse

Auden, W.H., *Selected Poems*, Edward Mendelson, ed. (London: Faber & Faber, 1979).
de Saint-Exupéry, Antoine, *Wind, Sand and Stars* [*Terre des Hommes*, 1939] (London: Penguin, 1995).
Hughes, Ted, *Collected Poems*, Paul Keegan, ed. (London: Faber & Faber, 2003).

Wells, H.G., *The War in the Air* (London: George Bell & Sons, 1908).
―― *The History of Mr Polly* (London: Thomas Nelson & Sons, 1910).
―― *Bealby: A Holiday* (London: Methuen, 1915).
Wodehouse, P.G., *Psmith in the City* (London: A. & C. Black, 1910).
Wordsworth, William, *The Selected Poetry and Prose of William Wordsworth*, G.H. Hartman, ed. (London: Signet Classic, 1970).

Film and television

'Blazing the trail', Pathé News, 24 November 1924, www.britishpathe.com/video/blazing-the-trail

'Record flight in British plane', Pathé News, 19 March 1925, www.britishpathe.com/video/record-flight-in-british-plane

'Mr Fokker … Please note!', Pathé News, 7 May 1925, www.britishpathe.com/video/mr-fokker-please-note

'End of 14,000 miles flight led by Wing-Commander Pulford, the four R.A.F. seaplanes arrive home, having flown from Cairo to the Cape and back', British Pathé, 26 June 1926, www.britishpathe.com/video/end-of-14-000-miles-flight-cuts

'Triumph for British aviation', Pathé News, 26 October 1926, www.britishpathe.com/video/triumph-for-british-aviation-version-1-of-2

'Our Flying Prince – Prince of Wales, after crowded day fulfilling 8 engagements, flies back to London', 4 June 1928, Pathé News, www.britishpathe.com/video/our-flying-prince-1

'Royal Air Force's Latest', British Pathé, 24 January 1929, www.britishpathe.com/video/royal-airforces-latest

'7 Days to India', Pathé News, 4 April 1929, www.britishpathe.com/video/to-india-in-7-days

'Cheers all the way', Pathé News, 6 April 1933, www.britishpathe.com/video/cheers-all-the-way-1

Thistledown (Arthur B. Woods, Warner Bros, UK, 1938).

'Bradman's verbal innings', British Movietone, 28 March 1938, www.youtube.com/watch?v=4UtNRBBtxWM

'The Tirpitz bombed', Pathé News, 17 March 1944, www.britishpathe.com/video/the-tirpitz-bombed

'Air speed record smashed', Pathé News, 15 March 1956, www.britishpathe.com/video/air-speed-record-smashed

'Farnborough Air Display', Pathé News, 10 September 1959, www.britishpathe.com/video/farnborough-air-display-10

'Jet! When England ruled the skies: Episode 1/2 Military Marvels', BBC4, 2012: www.bbc.co.uk/programmes/b01m81f5

Index

AASF (Advanced Air Striking Force), 194, 195
accident, 40, 45, 95, 115, 122–3, 150, 175
   crash landing, 44, 99, 282, 287
   Dixon, F.H., 282
   Gannet, 281–2
   Grace, Cecil, 40, 42
   Pinsent, David, 49, 50
   R101: 115, 138
   Rolls, Charles, 39
   Staniland, Chris, 281–2
   Twiss, Peter, 282
   Wildman-Lushington, Gilbert, 44
Adastral House, 99, 113, 116, 117, 119, 128, 132, 144, 190, 313
Admiralty, 41, 119
   Fairey Aviation and, xii, 70, 71, 86, 186, 308
   modernisation of, 28
   Short Brothers, 71
   see also Air Department; Royal Navy
The Aerial Engineering Company, 25
Aero Exhibition, 122, 134–6
   see also Olympia
Aero Models Association, 23

Aero Show, 52, 59
   see also Olympia
aerodynamics, 38, 47, 48, 49, 139
Aeronautical Research Council, 108, 114
aeronautics, 29, 31, 36, 53, 75, 200, 317
   Short Brothers, 42, 78
Aeronautics, 96
The Aeroplane, 52, 66, 82, 96, 107–108, 142, 143, 180, 288, 316
   see also Grey, Charles G.
AEU (Amalgamated Engineering Union), 159
AID (Aeronautical Inspection Division), 72
Air Committee, 43
Air Department (Admiralty), 30, 60, 61, 62, 70, 71, 81, 84, 86, 87
   Sueter, Murray, 43, 62, 81
Air League, 130, 141, 177
Air Ministry, 180, 189, 192
   contracts, 104, 105, 172–4, 225, 280, 312
   CRF and, 85, 103, 107
   Fairey Aviation and, 97, 105, 125, 131, 138, 159, 186, 187, 189, 194, 195, 196, 197, 312, 313
   rationalisation, 173, 187
   Shadow Scheme, 189, 190
   United States and, 225

INDEX

aircraft design, 21, 25, 29, 38
   American technology and design, 131, 132, 133
   CRF and, 34, 47, 69, 82–9
   De Havilland, Geoffrey, 47
   disconnect between aircraft design and engine efficiency, 78
   Dunne, John William, 32, 84
   engine design, 39, 59
   Fairey Aviation, design team, 88–9
   innovation, 30
   Short, Horace, 47, 52, 53, 60
   Short Brothers, 30
   Sopwith, Thomas, 30, 53, 71, 83, 106, 107, 114
   wing design, 22, 32, 47, 83, 198
Aircraft Disposal Company, 104
'airmindedness', 97, 154–5, 177
Aitken, Max, 141, 221, 223
Alford, Kay, 273
Allen, Charles, 179
AMLC (Army Motor Lorries Company), 68, 69, 73–4
Amos, Arthur A., 66, 108, 144
amphibian, 95, 97
   see also seaplane
Anderson, John, Sir, 215–16
Aquitania, ship, 126, 199
Argus, aircraft carrier, 86
Armistice, 2, 54, 56, 77, 86, 97, 102, 103, 105, 218, 309
Armstrong-Siddeley, 48, 281
Armstrong Whitworth, 51
   AW52 'flying wing', 51
Army, 32, 45, 53, 63, 78, 275
   aircraft production, 30, 46
   World War I, 46, 71
Arnold, Henry 'Hap', General, 225–6, 227, 232, 236, 237–8, 240, 241, 242–3
ASW (anti-submarine warfare), 205, 291
   anti-submarine aircraft, 205, 237, 281, 305

Attlee, Clement Richard, xiii, 257, 260, 261, 289, 290, 292, 314
Austin, Herbert, Lord, 194, 222
Austin Motors, 190, 194, 195
Australia, 1, 2, 98, 120, 275
Australian Naval Air Service, 98
Avions Fairey (Belgium), 168, 169, 180, 274–5
   Fairey SA, 275
   Hall, Geoffrey, 275
   reconstruction of, 275
Avro, 21, 48, 60, 95, 163, 257, 266
   see also Roe, Alliott Verdon

BAC (British Air Commission), 224, 228, 232, 236, 240, 287
   1785 Massachusetts Avenue, 233
   Planning Department, 234
   staff, 233–4
   Washington, 232
   see also CRF and the United States
Baillieu, Clive, Sir, 259–60, 274, 311
Balbo, Italo, Air Marshal, 217–18
Baldwin, Stanley, xiv, 115, 133, 139, 141, 175, 176, 213, 313
Ball, Joseph, Sir, 161, 192
balloon, 29
   Short Brothers, 37, 39, 54
Balloon Factory, Farnborough, 30
Baranov, P.I., 162, 166, 167–8, 169
Barlow, A.C., 67
   Fairey Aviation and, 67, 68, 83, 88, 100, 108, 121, 125, 145, 190, 196, 211, 254, 309
Barlow, Tom M., Major, 112, 160, 194, 230
Barnett, Corelli, 55, 254
Battenberg, Louis, Vice-Admiral Prince, 28, 42, 43, 47, 62
Battenberg family, 28, 42
Bauhaus School, 137

INDEX

BBC (British Broadcasting Corporation), 139, 179
   Broadcasting House, 202, 263, 288, 290
   Home Service broadcast, 7, 13, 15–16, 202
   left-wing bias, 263–4, 289, 290, 314
BEA (British European Airways), 283, 284
Beatty (Lord), David, Admiral of the Fleet Sir, 86
Beaverbrook, Lord, 171, 215, 227, 230, 232, 237, 238, 246, 256, 257, 263
   Aitken, Maxwell, 141, 221, 223
   CRF/Beaverbrook relationship, 223, 229–30, 235, 239–40, 246, 313
   MAP and, 221, 222, 223
Beazley, Henry, 81
Beddington Behrens, Edward, 66
Belgium, 1, 68–9, 73, 88
   exiled Belgians as Fairey Aviation workforce, 1, 68–9, 73, 88, 101, 168, 311
   see also Avions Fairey
Bell, Gordon, 52, 54
Bell, Mary, 234, 251, 267
Bermuda, 58, 206, 289–90, 291, 293
   CRF and politics at, 291–3
   CRF's final visit to, 300–301
   Darrell's Island civil–military airport, 292
   Lyndham, 268, 272
   as winter retreat, xii, xiii, 267, 271, 272, 289, 302
   World War II, 291–2
Bermuda Mid-Ocean News, 292
Bevin, Ernest, 246, 256
Blackburn, 83, 89, 130, 185, 189, 242, 254, 255
BOAC (British Overseas Airways Corporation), 115, 197, 198, 230, 290, 292

Boeing, 198, 311
   Boeing B-29 bomber, 163
Boeing, William E., 162
Boer War, 17, 19, 29, 212
bomber, 78, 163, 188, 194
   1½ Strutter, 71
   Bermuda/Buffalo, scout-bombers, 241
   Fairey Aviation, 70, 82, 125, 132–3, 175, 195
   Hawker Hart, 132–3, 173, 177
   strategic bombing doctrine, 104
   V-bomber, 285
   see also fighter aircraft; Swordfish
Booth, Harris, 35
Bossington, xv, 58, 138, 152, 161, 192, 213–15, 237, 248, 273, 293, 303, 306, 318–19
   Catchment Board and, 261–2
   convivial atmosphere at, 293
   CRF's final resting place, xvi, 301, 303, 319
   CRF's library at, 31
   fishing at, 213, 214–15, 219, 223, 248, 265, 294
   move to, 152
   principal home of Fairey family, xii, 318
   returning home, 268
   Slater, Brice G., 68
   St James and St Mary, 91–2, 213–14, 301
   World War II, 247, 267, 268
   see also River Test; Test Valley
Brabazon, Lord, *see* Moore-Brabazon, John
Brancker, Sefton, Sir, 116, 117–18, 119–20, 138
Bray, C.F., 88, 100
Breguet, Louis Charles, 136, 202, *illustration 9*
Briggs Manufacturing Co., Michigan, 224

INDEX

British aerospace industry, 280, 284–5, 300
British aircraft industry, xiii–xiv, 26, 34, 100, 161, 188, 279–80, 312
    1950s, 284–5
    aeronauts, concentration of, 306–307
    Anglo-American collaboration, 225
    birth of, 37
    Edwardian aviation, 27, 45, 309
    founding fathers of, 2, 306–307
    interservice rivalry, 78
    interwar period, 138, 172–3, 192
    Labour Government and, 290
    lagging behind, 113, 191, 254
    manpower shortage, 193
    modern aircraft, 40
    peacetime collaboration, 120–1
    rationalisation, 173, 187, 279, 280
    Victorian era, 29–30
    World War I, 2, 32, 45, 46, 59, 69–70, 73, 74, 77, 87, 309
    World War II, 191, 222, 225, 253, 279, 315
    see also R&D
British/Allied Purchasing Mission, 224–8, 232, 234–5, 259
British Empire, 1, 3–4, 17–18, 98, 115, 141, 313
    CRF, an imperial story, 4
British manufacturing industry, xiv
Broadbent, Wilfred, 66, 100, 105, 108, 127, 135, 185, 230, 277, 299, 309, 310
    retirement, 300
Brockway, Fenner: *The Bloody Traffic*, 176
Brown, Ernest, 160–1, 202, 205, 213, 217–18, 246
Bruce-Gardner, Charles, Sir, 192, 193
BSC (British Sportsman's Club), 212
BSC (British Supply Council), 239
Burghley, Lord, 292
Burns, Miss (CRF's secretary), 148, 157, 201, 202, 217, 219, 230, 234, 310
    CRF's dependence on, 212–13, 273
    as Mrs Haydon, 212
Busk, Edward, 35

Café Royal, 117, 129
Calshot, 54
Cambridge University, 35, 29, 88, 117, 154, 298, 307
    CRF's Cambridge companions, 55, 57, 58, 60, 309
Camm, Sydney, 106, 114, 121, 132, 155
Campbell, Bruce, 276
Campbell, Malcolm, Sir, 202
Campbell Orde, A.C., 198
Camper and Nicholsons, 89, 199, 201, 204, 206
Canada, 5, 58, 215, 226, 275
    CRF: family, 229–30
    as the Empire and Commonwealth's principal weapons supplier, 228
Carlton Grill, 117
CAS (Chief of the Air Staff), 116, 117, 132, 188, 232
CAT (Commonwealth Air Training Plan), 195, 196
Cayley, George, Sir, 18
Central Air Office, 45, 46, 52
Chamberlain, Joseph, 17–18
Chamberlain, Neville, 161, 171, 180, 188, 190, 191–2, 213–14, 217, 221, 256, 313, 315
Chance Vought, 242
Chaplin, Charlie, 148, 247
Charles, F.M., 88, 100, 101, 102–103, 104–105, 108
Chichester-Smith, C.H., Major, 273–5, 276, 309
Churchill, Winston, Sir, 37, 42–3, 104, 161, 256, 260, 262, 291
    CRF and, 161, 235, 293, 314
    Eastchurch, 44
    First Lord of the Admiralty, 43

430

INDEX

MAP and, 221, 222, 240
as pilot, 44
RNAS and, 43–4, 45, 81
Short Brothers and, 44
Sueter, Murray and, 45
United States and, 225, 227, 238
World War II, 225, 291
Ciano, Count, 217–18
CID (Committee of Imperial Defence), 30, 41, 188
City and Guilds Institute for the Advancement of Technical Education, 13
civil aviation, 98, 107, 115, 116, 279–80
civilian pilot, 49
see also commercial aviation; passenger aircraft
Clare, S30 C-class flying boat, 230–1, 264
Clarke, George, 92, 158
Clyde, Scotland, 86, 87
Cobham, Alan, Sir, 112, 119–21, 230
Cockburn, George, 42
Cody, Samuel, 30
Coker, E.G., 14
Cold War, 163, 275, 280, 291
commercial aviation, 119, 280
seaplane, 85
see also civil aviation; passenger aircraft
Commonwealth, 3, 228, 239, 288, 289
Communism, 159, 160, 172, 179, 180, 217, 256, 263, 264, 311, 315
CRF's anti-communism, 166
Coningham, Arthur 'Mary', Air Marshal Sir, 243
Conservative Party/Government, 44, 141, 161, 169, 176, 213, 218, 256, 289, 293
'One Nation Toryism', xiv, 141, 313
Coutinho, Gago, 111
Cowtan, F.E., Lieutenant Commander, 305, 306

Cox, Lieutenant and Admiralty Inspector, 72
Cremer, Peter-Erich, Kapitänleutnant, 250, 251
CRF (Charles Richard Fairey), xi, 1, 317–18, *illustration 2, 3, 8, 9, 10, 13, 14, 16*
atheism/agnosticism, 3, 91, 93, 298
biographies, xiv, xv–xvi, 13, 16–17
birth, 3, 4
childhood, 7, 317
a civilian, 61
criticism of, 130, 140, 178, 180, 254, 316
'Dick' Fairey, xi, 1
environmental preservation, xii, 4, 261, 294–6, 313, 314, 318
generosity, 149–50, 153, 158, 229, 245, 248, 309–10, 317
leading ideas of, 3, 13, 297
legacy, 319
letters by, 2–3, 10, 58, 66, 81, 83, 201, 248, 263, 264, 290
a libertarian, 158, 313
personality, xi, xiii, 13, 57, 67, 74, 90, 93, 96, 140–1, 151–2, 229, 277–8, 306
physical appearance, xi, 13, 24, 57, 96, 141, 316
a self-made man, xii, 206, 317, 319
temper, 157, 158, 179, 311, 316, 316
'Tiny' Fairey, 108, 140, 162, 311, 319
a typical weekday of, 212
see also the entries below for CRF
CRF: achievements, awards and titles, xi, xii, 212, 213, 246, 302, 306
competitive flying, 23
Fellow of the City and Guilds of London Institute, 15
governor of Westminster Hospital, 212
honorary colonel in the Territorial Army, 212

INDEX

IAS Honorary Fellow, 266–7
knighthood/MBE, xiii, 102, 106, 114, 213, 246
material success/wealth, 12, 13, 55, 75, 89, 90, 102, 143, 146, 178, 209, 245, 317
Royal Aeronautical Society, Honorary Fellow, 293
Royal Aeronautical Society, Silver Medal, 130
Royal Aeronautical Society, Wakefield Gold Medal, 85, 156
US Medal of Freedom with Silver Palm, 292
CRF: death, xiii, 249, 276, 282, 297, 298, 300
  30 September 1956: 4, 301
  Bossington as CRF's final resting place, xvi, 301, 303, 319
  funeral, 301
  Lord Brabazon's eulogy, 302–303
  memorial service, 301–302
  Suez crisis, 4, 302
CRF: education, 30, 47, 96
  apprenticeship, 10, 11–18, 36
  autodidact, 17, 19, 21
  boarding school, Sussex, 8, 9
  engineering, 11, 13, 15, 52
  Hendon Preparatory School, 7
  mathematics, 7–8, 9, 15, 30, 47
  Merchant Taylors' School, 8, 96, 317
  night-school tuition, 10, 13
  performance as student, 7, 15
  science and technology, 11, 15
  St Saviour's School, 9–11, 15, 21, 96
  technical drawing, 9, 12
  technical education and expertise, xiii, 9, 11, 12, 13, 17, 21, 25, 47
  see also Finsbury Technical College
CRF, the entrepreneur, xiii, 2, 23, 25, 28, 52, 78–9, 108, 109
  capitalism, 140
  model aircraft, selling of, 23–5, 52
  neo-liberal free-market mentality, xiii, 179, 313
  see also CRF and Fairey Aviation; Fairey Aviation
CRF: estates, houses, properties, xii, 93, 152, 217
  8 Greville Place, Maida Vale, 92–3
  Fort Charles, Salcombe, 152–3, 215, 247
  Hind's Head Hotel, Maidenhead, 153
  Home Counties, 92, 151, 209
  Oakley Hall, north Hampshire, 152, 214
  see also Bossington; Grove Cottage; Woodlands Park
CRF and Fairey Aviation, xi, xii–xiii, 1, 108, 144, 153, 201, 309
  chairing from US, 272–3
  chairing *in absentia*, 272, 273
  creating his own company, 53, 57–8, 60, 61
  as manager, 158–62, 234
  nominal control only, 230
  see also Fairey Aviation; Fairey Aviation: staff
CRF: family, xii, xvi, 91–3, 108, 156–7, 210, 212, 234, 297, 302, 316, 317, 318
  children, xiv, 19, 58, 91, 93, 148, 210, 268, 298
  christening, 91, 93
  family first, 271
  first marriage, 92–3, 147–8, 297, 316
  second marriage, 58, 91, 209–10, 271
  World War II, 215, 217, 243–4, 247
  see also Fairey family
CRF: health issues, 2
  amnesia, 16–17
  bronchitis, 218, 267, 293
  exhaustion, 106, 131, 152, 246–7

432

INDEX

heart condition/attack, 58, 218–19, 243, 267, 268, 272, 301, 306
hernia, 219
malign tumour, 301
motorcycle accident and coma, 16–17
peritonitis, 34
pneumonia, 272
poor health, xi, xiii, 204, 218–19, 223, 259, 262, 267, 282, 293, 297
second heart operation, Boston, 272
sympathectomy, Boston, 268, 271
thrombosis, 267
CRF: leisure and sports, 148–9, 201, 212
cars, 90, 150–1, 212, 299
eating, 117, 171
field sports, xiv, 200, 212, 298
films, 148
golf, 149, 152, 169, 191, 199, 200, 212, 243
Hinds Head Hotel, 201
musicals, 92, 126, 148
sailing, 4, 144, 148–9, 153, 200, 212
shooting, xv, 4, 152, 168–9, 191, 209, 213, 266, 296, 318, *illustration 13*
at United States, 243, 245
see also fishing; yachting
CRF: plane making and aircraft industry, xi, xiii, 1, 22, 161, 187, 316
aircraft design, 34, 47, 69, 82–9
American technology and design, 131, 132, 133
career as aircraft manufacturer, 2, 22, 29, 34, 51, 69, 306
engineering expertise, xiii, 60, 88
as figurehead of the aircraft industry, xii
Short Brothers and, 51
see also Fairey Aviation: aircraft; model aircraft
CRF and politics, xiii, 119, 138–43, 313–15
anti-communism, 166

Bermuda, 291–3
Conservative Party, 141, 213, 293
fascism, xiii, xiv, 142
Labour Government, 138–42, 290–1, 292
National Government, dislike and suspiciousness of, xiv, 160, 176, 290, 313–14
shooting and, 213
CRF: teaching/lectures/speeches, 17, 19, 25, 82–3, 85, 154–5, 212, 289, 290, 293–4, 314
1923 Air Conference, 115–16
1927 House Dinner of the Royal Aero Club, 140
Technical Education Committee, Middlesex County Council, 15
Tottenham Polytechnic, 15–16
CRF and the United States, 131, 132, 133, 140, 228–9, 235, 245, 278, 314
15 Broad Street, 231
CRF and BAC, 57, 58, 193, 222, 228–9, 232–40, 241–2, 243, 259–60, 311
CRF as BAC Director-General, xiii, 234, 239–40, 246, 263, 267–9, 292
CRF's trips to the States, 126, 127, 136, 144, 199, 224, 230–1
Fairey Aviation, chaired from US, 272–3
'Fairland', 239
fishing, 248
Hollywood, 247
leisure and sports, 243, 245
New York, 231
transatlantic crossings, 264–5, 268, 272
Wardman Park Hotel, 231, 232
Washington, 231–2, 243, 244, 269
CRF: work, 25, 33–4
Finchley Power Station, 19–20, 34

433

INDEX

RAF, 41
Short Brothers, xii, 28, 36, 40, 41, 47, 51–4, 59, 64, 308
see also Eastchurch airfield; Jandus Electric Company
Cripps, Stafford, Sir, 138, 192, 240, 253, 254, 256, 259, 260, 265, 266, 267
CRF/Cripps relationship, 256–7
Crisp, Charles, 56, 66, 74, 81, 107, 128, 130, 139, 145, 265, 273, 308, 309
sailing, 148–9
Crookshank, Harry, 263–4
Crozier, W.P., 221–2
Cunard liner, 62, 86, 126, 127, 268, 272
Currey, Charles, 275–6
Curry, Mrs., 8
Curtiss, 124, 150, 225, 229, 233
Curtiss engine, 124
Curtiss P36 and P40 fighters, 225
Curtiss V-1400: 126
D12 engine, 124, 127, 128, 132
Fairey Aviation and, 70, 127–8, 129, 131, 132
Curtiss, Glen, 124–5, 236

Daily Express, 206
Daily Mail, 141, 165, 175, 176, 315
Daily Mail prize, 36, 38, 95
Daily Sketch, 288
Daily Telegraph, 130
Daimler, 100, 104–105, 106, 150
Danchev, Alex, xiv
Darwin, Charles, 10
On the Origin of Species, 18
Daubeney, Reverend, 248, 265
Davey, Dr, 301
David, Prince of Wales (Edward VIII), 118, 134, 177
Dawson, F.G.T. ('Wuffy'), 55–6, 57–8, 63
CRF/Dawson relationship, 58, 67, 246, 316

Fairey Aviation and, 58, 64, 65, 66, 80, 100, 107, 126, 127, 144, 309
De Cartier de Marchienne, Baron, 168–9
De Havilland, Geoffrey, Sir, 30, 46, 73, 107, 118, 120, 122, 285, 294, 307, 309
aircraft design, 47
Comet 3, transcontinental airliner, 280
DH9A, 113
DH50: 120
Mosquito, 237, 275
Sea Vixen, 114
Dean, Maurice, Sir, 223
Deperdussin, Armand, 46
deployment of aircraft, 46, 133, 194, 239
Depression, xii, 2, 94, 115, 123, 146, 149, 204, 236
Derby, Lord, 212
Derby Club, 212
Deutscher, Isaac, 293
DFC (Distinguished Flying Cross), 56, 97, 274
Diaper, Herbert 'Dutch', 203
Dickens, Charles, 26, 55
Great Expectations, 27
Dill, John, Field Marshal Sir, 244
disarmament, 155, 169, 172–81, 199, 315
Disarmament Conference, 169, 176
Disney, Henry, 160
Disney, Walt, 247
Dixon, F.H., 282
DNAD (Director of the Naval Air Division), 185
Donovan, William J., Colonel, 231, 235
Doolittle, James, General, 126, 267
Douglas, airliner/carrier aircraft, 197, 198, 242, 265, 311
Douglas, Donald, 247, 307
Douglas, Lew, *illustration 14*

434

INDEX

Dowding, Hugh, Air Chief Marshal Sir, 123, 148, 179, 188, 192, 193, 222–3
DS *Ringstadt*, 249, 250, 251
Duncanson, Fred, 83, 87–9, 94, 108, 130
Dunne, John William, Captain, 25, 28, 30, 308
  aeroplane lateral stability and pilot control, 19, 31, 32
  aircraft design, 32, 84
  Blair Atholl Syndicate, 31, 51
  D5 prototype, 32–3
  D8 prototype, 35, *illustration 2*
  D9 prototype, 51
  Dunne/CRF relationship, 30, 31, 32–3, 36, 51, 317
  Eastchurch airfield, 29, 31, 32
  gliders experimental flights, 30–1, 32
  Wells, H.G. and, 31, 33
duraluminium, 124, 127, 135, 137, 185, 203

Earhart, Amelia, 156, 229
Eastchurch airfield, 27–9, 39, 54
  aviation memorial, 37
  civil–military partnership, 43
  CRF at, 33–6, 308
  Dunne, John William, 29, 31, 32
  see also Isle of Sheppey
Eastchurch Naval Flying School, 43
Eden, Anthony, Sir (Lord Avon), 246, 256, 280, 292
Edgerton, David, xiv, 254
Edwardian era, 1, 11, 12, 13, 56
  British aviation in, 27, 45, 309
Eisenhower, Dwight D., 236, 292
Elizabeth, Queen Mother, 294
Ellington, Edward, Marshal of the RAF Sir, 188
engine, 128–9
  Admiralty Type 184: 59

Curtiss engine, 124
D12 engine, 124, 127, 128, 132
demand of, 59
disconnect between aircraft design and engine efficiency, 78
engine design, 39, 59
Fairey Aviation as aero-engine manufacturer, 127–9, 174
French engine, 39, 45, 48, 52, 59, 87
jet engine, 280
light, high-speed engine, 113
multi-engine machine, 40
Napier engine, 95, 122, 126, 129, 284
P24: 22, 174, 281
piston engine, 134, 196
PV12: 174
Rolls Royce, 59, 98, 128, 129, 153
Rolls Royce's Merlin, 130, 155, 174, 195, 241
Short, Horace, 40
single engine, 21
Sunbeam engine, 59
tip-jet technology, 48, 49, 50, 283
turboprop, 22, 49, 281, 284
water-cooled engine, 87
Whittle's model, 48
World War I, 59, 87
see also propeller
English Electric, 285, 287, 312
  Lightning, 241, 285, 287, 312
entente, 142, 155, 226
EPD (Excess Profits Duty), 80, 101, 106
Everett Edgcumbe, 35

FAAM (Fleet Air Arm Museum), xvi, 84
Fairey III (Type III), 56, 85, 86–7, 94–100, 104, 313
  all-metal IIIF, 135
  civilian use, 98
  competition, 95–6, 97
  Fairey IIIA, 97
  Fairey IIIB, 97, 103

INDEX

Fairey IIIC, 96, 97, 98–9, 103, 163
Fairey IIID, 98, 105, 111, 112, 113, 114, 118, 135
Fairey IIIF, 118, 129, 131, 134–5, 165, 175, *illustration 7*
   first aerial crossing of the south Atlantic, 111
   importance of/success, 94, 98, 107–108, 113, 135, 145, 313
   N9/N10 experimental seaplanes, 94–5, 97
   origins, 94–5
   passenger-carrying IIIF, 118, 129
   promotion of, 96–7
   see also Fairey Aviation: aircraft
Fairey, Atalanta (Richard Fairey Jr's third wife and widow), xv
Fairey, Benjamin (CRF's uncle), 5
Fairey (Bergö), Aino (Richard Fairey Jr's first wife), 211, 252–3, 299
Fairey, Charles (CRF's grandfather), 3, 4
Fairey, Charles (CRF's grandson), xvi–xvii, 252
Fairey, Charles Richard, Sir, *see* CRF
Fairey (Craig), Diana (Richard Fairey Jr's second wife), 252
Fairey, Esther, Lady (CRF's second wife), 58, 91, 151, 152–3, 154, 199, 219, 243, 247, 268, 316, *illustration 10*
   Canada and United States, 229–30, 231, 245–6
   the ideal companion, 209–10, 245, 297
   second marriage, 236
   Whitney, Esther Sarah, 209
   as widow, 19, 209, 301–302
   Young, Esther, 236
Fairey, Ethel (CRF's aunt), 4, 5
Fairey, Frances Ethel ('Effie', CRF's sister), 7, 8, 19, 55, 90, 317
   as CRF's technical assistant, 7, 20, 23
   Effie/CRF relationship, 7, 20
   jobs, 8–9, 34
   Mrs Hulme, 20, 153
   Roman Catholicism, 91
Fairey, Frederick (CRF's brother), 7, 10
   Wanstead Orphanage School, 9
Fairey (Jackson), Frances (CRF's mother), 5–6, 7, 33, 55
   death, 91
   as needlewoman, 8
   Roman Catholicism, 91
   see also Hall, Mrs
Fairey, James (CRF's uncle), 5
Fairey, Jane (CRF's daughter, second marriage), 210, 248, 249, 268, 296, 297–8, 316, 318
   CRF/Jane relationship, 244, 298
   education, 297
   see also Tennant, Jane
Fairey, Joan (CRF's first wife), 92–3, 94, 108, 149, 210, 297
   adultery, 157, 316
   CRF/Joan tension, 147–8, 316
   divorce, 157, 252, 317
   remarriage/Mrs Buxton, 158, 317
   see also Markey, Queenie
Fairey, John (CRF's son, second marriage), xvi, 16, 91, 157, 306, 316
   CRF/John relationship, 210, 244, 298
   education, 248–9, 298
   as pilot, 298
Fairey, Leslie, Reverend (CRF's cousin), 1, 2–4, 5
Fairey, Margery (CRF's sister), 7, 9, 33, 55, 90, 134
Fairey, Phyllis (CRF's sister), 7, 9, 33, 90, 153
Fairey, Richard Jr. (CRF's son, first marriage), xiv, xv, 2, 93, 108, 148, 158, 202, 211, 298, *illustration 13*
   Air Transport Auxiliary, 211
   amputation of both legs, 251, 286

INDEX

CRF/Richard Jr relationship, xvi, 157, 210–11, 249–50, 252–3, 298, 299, 310
education, 154
Fairey Aviation and, 211, 273, 274, 299–300
Fairey Marine and, 276
first marriage, 211, 299
high blood pressure, 299
as pilot, 211, 249, 251–2, 286, 298–9
second marriage, 252
United States, trip to, 249–51, 252, 253
Fairey, Richard Sr (CRF's father), 3, 4, 223
bankruptcy and poverty, 6, 8, 317
Bishopsgate, 6, 8
Canada, 5–6
death, 4, 6–7, 8, 317, 318
family, 5
Morgan, Gellibrand and Co., 4–5
New Brunswick, 5, 6
personality and physical appearance, 10
Fairey, Sarah-Jane (CRF's granddaughter), xvi
Fairey Aviation, xi, 29
1921: 56–7
1929 flotation, xii, 2, 63, 79, 107, 144, 145, 146, 307
advertising and press coverage, 85, 95, 96–7, 98, 107–108, 111, 112, 133, 190, 288–9, 316
as aero-engine manufacturer, 127–9, 174
as aerospace company, 79, 271, 277
archives, xvi, 7, 84
backlog, 255
beginnings of, 58, 60, 61–6, 79–80, 308, 309
capital, 66, 108, 127, 144–5, 177–8
cash-flow problem, 80, 101, 102–104, 273, 308
challenges, 70, 87, 108

Cold War, 275
Curtiss and, 70, 127–8, 129, 131, 132
demise of, 279, 280
diversification, 104–105, 174, 275, 308, 309
downturn, 173–4, 175, 177, 253–5, 289, 308
as dual company, 63, 65
first contract, 61–2
interwar period, 57, 90, 104–105, 108–109, 114–15, 172, 310
logo, 134, 185
patronage, 66, 80
postwar period, 267, 274, 278, 309
restructuring, 107, 108, 143–6, 257–8, 274, 311
shareholders, 64, 79, 80, 107, 144, 153, 174, 177–8, 308
Sopwith Aviation and, 70, 74, 83–4
subcontracting, 87–8, 135, 184–5, 186, 193, 195, 254, 255, 308, 312
Westland, merger with, 49, 59, 78, 284, 287
World War I, 69, 70, 75, 80, 118, 277, 311
World War II, 37, 58, 171, 185, 190, 192, 194–5, 253, 274, 305, 311
see also the entries below for Fairey Aviation; CRF and Fairey Aviation
Fairey Aviation: achievements and rewards, xi, 108, 183, 209, 309
1924 Britannia Trophy, 98
as American aircraft supplier, xiii
Britain's largest airframe manufacturer, xii–xiii
expansion, 78, 87, 100, 114, 152, 190, 193, 308, 309, 310
FD2, speed record, xi, 183, 281, 282, 286, 287, 288
global status, 1, 3, 57, 94, 135, 136, 275, 282, 289, 309

437

INDEX

profit, 80–1, 101, 103, 105, 113, 114, 146, 187, 209, 278, 280, 313
rotary technology, xiii, 277, 278, 282
volume production, 52, 74, 114, 184, 186, 187, 188, 194, 196
Fairey Aviation: aircraft, 70, 74, 143
1½ Strutter, 71–2, 73, 74, 83, 97
aero-engine manufacturing, 127–9, 174
airliner FC1: 197–8
Albacore, 184, 185, 186, 195, 255
all-metal aircraft, 131
ANA1/ANA3, 98
Atalanta, 88
Barracuda, 185, 196, 198, 237, 255, 279, 305, 319
Campania seaplane, 85, 86, 101–102
Fairey Felix, 128, 132
Fairey Hamble Baby, 83–4
Fawn, 113, 114, 125, 131
Ferret, 131
Flycatcher, 113–14
flying boat, 87–8, 89
Fulmar, 184, 185, 186, 195, 237, 255, 285, 312
functionality over good looks, 114
G.4/31 light bomber, 175
Hendon bomber, 175
N4 series, 89
operational model, 78–9
P24: 22, 174, 281
patent, 84, 85, 101, 102, 178
Pintail, 99
prototype F1: 82, 83
prototype F2: 70, 82, 83
PV12: 174
R&D, 196–7, 278, 281, 312
seaplane, 70, 84, 85, 97, 109, 115
Spearfish, 22, 255, 279
tip-jet technology, 49
Titania, 88
Type 827: 66, 70, 72

variable camber gear, 84, 85, 95, 96, 113, 308
see also Fairey III; Fairey Battle; FD2; Firefly; Gannet; Gyrodyne; Rotodyne; Swordfish
Fairey Aviation: facilities:
design team, 88–9
drawing office, 66, 68, 72, 193
Elmdon Aerodrome, 194
Errwood Park, 190, 255
factories, 67, 68, 72–4, 100
'Great West' Aerodrome, 174, 267, 276
Hamble Spit, xi, 67, 68, 70, 99, 100, 145, 177, 276
Harmondsworth, 174–5, 276–5
headquarters at 175 Piccadilly, 66, 67, 72
Heath Row Aerodrome, 175
Heston Aerodrome, 267, 276
Kingsbury, 73, 83
Northolt aerodrome, 99, 100, 131, 145
Ringway Aerodrome, 190, 194
shadow factory, 159, 193, 194
Stockport, 159, 160, 190, 193, 194, 195, 219, 255
White Waltham, 267, 276, 282, 283, 285
see also Harlington; Hayes; Heaton Chapel
Fairey Aviation: staff, 67–9, 150, 158–9, 161–2, 254, 309–11
1949 dinner, 67–8, 277
Belgium/exiled Belgians, 1, 68–9, 73, 88, 101, 168, 311
board, 66, 79, 100, 107, 108, 144, 178, 272, 274, 275, 309
directors at, xv, 66, 79, 80–1, 90, 102, 107, 126, 145, 178, 230, 257, 309
engineering team, 143
Hall, Geoffrey, as chairman successor, 300, 310, 317, 318

438

INDEX

joint chairmanship, 259
management team, 100
'project team', 258
salaries, 65, 80–1, 102, 145, 178, 309
test pilot, 70, 95, 99, 112, 175, 281–2, 285, 286
trade union and strike, 73, 139–40, 159, 194, 254, 259, 310–11
Fairey Aviation: state and military institutions:
Admiralty, xii, 70, 71, 86, 186, 308
Air Ministry, 97, 105, 125, 131, 138, 159, 186, 187, 189, 194, 195, 196, 197, 312, 313
Fleet Air Arm, 1, 22, 113, 114, 118, 185, 195–6, 239, 255, 278, 281, 282, 300, 305
MAP, 255–6, 257, 311
RAF, xi, xii, 1, 117, 118, 131, 132–3, 184, 189, 190, 305
RNAS, 87, 113, 305
Royal Navy, xi, 67, 69, 87, 113, 184, 185, 282, 308
state patronage, xiii, 69, 74, 78, 87, 100, 101, 173
tax-related issues, 79, 80, 102–104, 105, 106, 108, 145, 178
Treasury, 68, 80, 100, 101–102, 145, 177, 186
Whitehall, 74, 101, 145, 160, 173, 276–7
see also Ministry of Munitions
Fairey Battle, 160, 184, 185, 186, 190, 194–6, 237
see also Fairey Aviation: aircraft
Fairey family, 5, 33, 90–1, *illustration 1*
27 Station Road, Finchley Central, 11, 12
52 Station Road, 20, 55, 90
78 Goldsmith Avenue, Acton, 8, 9, 10, 12

CRF's filial and family loyalty, 55, 90, 210, 297, 317
Eaton Socon, 4
fall into poverty, xii, 6, 8–9, 317
grandmother, 5
Hendon, 5, 7, 10
income and occupations, 3, 4–5
New Brunswick, 5
Ray House, 5, 12
religious beliefs, 3
St Neots, 3, 4, 8
Victorian era, 3, 4, 7, 313
see also CRF: family; Hall/Fairey family
Fairey Fox, 1, 121, 128, 129, 152
12 Squadron, 133, 134
all-metal Fox II, 131, 132
Fox I, 129, 131–2, 133
Fox IIM, 133, 169
high-performance combat aircraft, 121
Sea Fox, 184, 185, 305
Fairey Gannet, *see* Gannet
Fairey Hydraulics, 279
Fairey Marine, xiii, 275–6
Fairey Swordfish, *see* Swordfish
Fairey Tales, 101
Fairey's Weapons Division, 280
anti-submarine aircraft, 281
'Blue Sky' project, 280–1
drone, 277
Fireflash, 280, 281
mid-wing delta Type R, 281
missile, 277, 280
Farman, Henri, 39, 40
Farnborough, 30, 35
fascism, xiii, xiv, 142, 217
FD1 (Fairey Delta 1), 281
FD2 (Fairey Delta 2), xii, xiii, 278, 288–9, *illustration 15*
delta-winged, 183, 312
as Fairey Aviation's crowning achievement, 183

439

INDEX

Fairey's best known aircraft, 113
  limitations, 287
  supersonic speed, 183, 286–7
  world air speed record, xi, 183, 281, 282, 286, 287, 288
  see also Fairey Aviation: aircraft
FD2 The World's Fastest Jet, 288
FDR (Franklin Delano Roosevelt), 225–6, 227, 238, 244, 245, 291
  re-election, 227, 232–3
Fedden, Roy, Sir, 253–4
Fernandez-Llorente, Esther (CRF's granddaughter), xvi
Fernandez-Llorente, Fran (Jane Tennant's son-in-law), xvi
fighter aircraft, 71, 104, 186, 191, 280–1
  Curtiss P36 and P40 fighters, 225
  Fairey Aviation, 74, 82, 83, 99, 113, 125–6, 128, 131, 155, 239, 255, 305
  Germany, 315
  Hurricane, 191, 195, 225
  Supermarine, 239
  United States, 125–6, 233
  see also bomber; Fairey Fox; Firefly
Finsbury Technical College, 13–15, 17, 18, 21, 35, 47, 88, 139, 307
  CRF's Electrical Engineering certificate, 15, 16, 17
  Electrical Engineering programme, 15
  see also CRF, education
Firefly, 1, 128, 133, 135, 185, 186, 196, 198, 255, 258, 281, 282
  Firefly I, 131
  Firefly II, 169
  FR5: 278
  Russia and, 165, 167, 168, 169, 170
  see also Fairey Aviation: aircraft
fishing, xii, xv, 4, 147, 191, 318
  Bossington, 213, 214–15, 219, 223, 248, 265, 294
  CRF's collection of flies, 294

dry fly fishing, 213, 248, 294
electric fishing, 295, 296
United States, 248
  see also River Test
Fleet Air Arm, 189, 237, 280
  CRF inspection of Flight703X, *illustration 16*
  Fairey Aviation and, 1, 22, 113, 114, 118, 185, 195–6, 239, 255, 278, 281, 282, 300, 305
  Swordfish and, xii, 183, 185, 186, 187
  United States and, 233, 242
flight, 18, 19
  manned flight, 1, 10, 28, 47, 50, 306, 307, 317
  powered flight, 18, 21, 25
  see also British aircraft industry
Flight, 96, 107, 149, 194, 305, 316
float plane, 20–1, 53
  asymmetrical floats, 21, 22
  single-float 'Bat Boat', 53
  see also seaplane
flying boat, 89, 95, 115
  biggest flying boats in the world, 87
  C-Class Empire flying boat, 197
  Catalina, 241
  Dornier Wal flying boat, 136
  Fairey Aviation, 87–8, 89
  see also Clare; seaplane
flying club, 97, 154
Fokker, Anthony, 71, 83, 136–7, 162, 163
Forsyth, A. Graham, Captain, 174
France, 71
  aircraft industry, 225, 226
  British aviation and reliance upon France, 59
  French engine, 39, 45, 48, 52, 59, 87
  United States and, 226
  World War II, 225, 226
Franco, Francisco, Generalissimo, 116, 136, 266

INDEX

Freeman, Wilfred, Air Chief Marshal Sir, 192–3, 222, 223, 238, 239–40, 254, 258
fuselage, 33, 52, 124, 185–6, 284
  aluminium sheeting, 40, 131

Gamage's, 23–4, 25, 33, 34
Gannet, 281, 287, 305, 312
  accident, 281–2
  Armstrong-Siddeley's dual propeller system, 48, 281
  Gannet AS1: 282
  R&D programme, 282
  reconnaissance aircraft, 48
  turboprop, 22, 281
Garnham, Bob, 204
Gayford, O.R., 123
Geddes, Eric, Sir, 104
General Strike, 139–40
George V, King of the United Kingdom and Northern Ireland, 46, 52, 53, 120, 175
George VI, King of the United Kingdom and Northern Ireland, 202
German battleship, 184
  Bismarck, 111, 184, 233, 305
  Graf Spee, 184, 305
  Tirpitz, 255, 305
Germany, 137, 141–2, 314–15
  aircraft production, 46, 138, 191, 216, 254, 307, 315
  appeasement of, 161, 189, 191
  Deruluft airline, 163
  expansionism, 171, 315
  remilitarisation of, 181
  Russia and, 163
  Sueter, bombing Germany, 70, 78
  threat from, 30
  World War I, 46
  World War II, 191, 216
  see also Nazism; Zeppelin
Gibbs, Philip, Sir 180

gliding/glider, 18–19, 20, 22, 30–1, 32, 55
Grace, Cecil, 40, 41, 42
Graham-Wright, Claude, 41
Grandi, Count, 217–18
Great Crash, 2, 137, 178
Great War, *see* World War I
Grey, Charles G., 52, 82, 96, 108, 119, 134, 141, 143, 316
  anti-semitism, 142
  see also The Aeroplane
Grove Cottage (Iver) 93–4, 102, 108, 147, 148, 152, 317
Grumman, 242
  Grumman Avenger, 239, 255
Gyrodyne:
  FB1 Gyrodyne, 282
  Jet Gyrodyne, 49, 283

Haldane, Richard B., Lord, 30, 31
Halifax, Lady, 244
Halifax, Lord, 205, 213, 217, 225, 227, 234, 239, 244–5, 263, 268–9, *illustration 12*
Hall/Fairey family, 33–4, 90–1
  52 Station Road, 20
  see also Fairey family
Hall, Geoffrey (CRF's half brother), 12, 16, 20, 22, 23, 24, 34, 55, 61, 90, 123–4, 152, 153–4, 214–15, 296
  apprenticeship at Hayes, 139
  as chairman successor to CRF at Fairey Aviation, 300, 310, 317, 318
  CRF/Geoffrey relationship, 12, 91, 210, 296, 297, 317
  Fairey Aviation and, 153–4, 274, 282, 288, 300
  Fairey SA, 275
  Gannet programme, 282
  Grove Cottage, 94
  Woodlands Park and, 154, 210, 217
Hall, Henry Clayton (CRF's stepfather), 10–11, 12, 90, 317

441

INDEX

Hall, Mrs (CRF's mother, second marriage), 12, 23, 55, 90
  see also Fairey (Jackson), Frances
Handley Page, Frederick, Sir, 21, 52, 78, 79, 82, 161, 179, 209, 265, 294, 307
Handley Page, Lady, 265
Harlington (Fairey Aviation), 68, 70, 72–3, 100, 101, 273
Harlington New Building, 103
Harrap, George G., xv
Harrow, 2, 154
Hawker Siddeley Group, 106, 133, 143, 172, 173, 177, 194, 237, 279, 313
  Hawker, 106
  Hawker Hart, 132–3, 173, 177
  Hawker Hunter, 285
  Hawker Sea Fury, 114
  Hurricane, 191, 195, 225
  long-distance flight, 121
  see also Sopwith, Thomas
Hayes (Fairey Aviation), xi, 68–9, 73, 86, 97, 127, 145, 159, 177, 186, 254, 257, 272
  beating heart of the company, 309
  expansion of, 70, 130, 135, 193
  General Strike, 139
  wind tunnel at, 198, 203, 204
Hazell, Archie G., 103, 105, 108, 144, 145, 146
Heathrow, 276, 299
Heaton Chapel (Fairey Aviation), xi, 160, 186, 190, 194, 195, 254, 255, 275, 276, 281, 305, 316
  Errwood Park, 190
helicopter, 281, 282
  FB1 Gyrodyne, 282
  Jet Gyrodyne, 283
  Rotodyne, 49
  Sikorsky S54: 299
Henderson, Arthur, 139, 169
Hendon aerodrome, 7, 21, 35
  Hendon Air Display, 114, 315

Heysham Harbour, Midland Railway Company, 12–13, 14, 19
H.G. Hawker Engineering Company, 106
  see also Hawker Siddeley Group
Higgins, John, Sir, 125, 136
Hilton, Massey, 97–8, 275, 300, 309
Hitler, Adolf (Führer), xiv, 141, 142, 143, 156, 170, 172, 175, 176, 177, 181, 217, 243, 315
HMS *Ark Royal*, 46, 53
HMS *Campania*, 62, 86
HMS *Evadne*, 206
HMS *Furious*, 86, 305
HMS *Glorious*, 186
HMS *Hermes*, 46
HMS *Pegasus*, 97
HMS *Peregrine* (Royal Naval Air Station Ford), 305, 306
HMS *Slinger*, 95
HMS *Victorious*, 111, 305
Hoare, Samuel, Sir (Lord Templewood), 104, 116–17, 128, 129, 130, 132, 221, 313
  CRF/Hoare relationship, 116
  long-distance flight, 118–19, 121
Hope, Linton, 89
Hopkins, Harry, 235, 236, 237, 238
Hopkinson, Bertram, 55
Hore-Belisha, Leslie, 213
Houghton Club, 214, 294, 295, 296
Houston Mount Everest Expedition, 141
Hughes, Billy, 98
Hughes, Mary, 98

IAS (Institute of Aeronautical Sciences), 266–7
Ignatiev, Alexis, 165, 166, 168
Ilyushin, Sergei, xiii, 162–3, 164, 167, 315
Imperial Airways, 118–19, 128, 197, 198, 313

INDEX

Imperial College, 13, 14, 47, 139, 153
Imperial Japanese Navy, 99
industrialisation, 11, 18
Inland Revenue, 102, 103, 105, 106, 145
Inskip, Thomas, Sir, 180
interwar period, xii, xiii
   British aircraft industry, 138, 172–3, 192
   British Empire, 155
   Fairey Aviation, 57, 90, 104–105, 108–109, 114–15, 172, 310
   peacetime collaboration, 120
   RAF, 106, 173
Isle of Grain, 53, 89
Isle of Sheppey, 27
   CRF's career as aircraft manufacturer, 2, 22, 29, 306
   see also Eastchurch airfield
Isle of Wight, 35, 124, 200, 201
Italy, 95, 99, 123, 129–30
   Regia Aeronautica, 123
   World War II, 190, 217–18
Itchen, 89

JAC (Joint Aircraft Committee), 235–6, 238, 240
Jackson, Frances, *see* Fairey (Jackson), Frances
Jacques Schneider International Trophy, *see* Schneider Trophy
James, Henry, 317
Jandus Electric Company, 11–12, 15, 19
Japan, 190, 237, 279, 305
Jellicoe (Lord), John, Admiral of the Fleet Sir, 82, 86
Jenkins, Flight Lieutenant, 1122–3
jet age, 37, 50, 114, 279, 312
   CRF/Fairey Aviation and, 271, 289
   see also FD2
Johnson, Amy (Mrs Mollison), 156
Johnson, Basil, 129, 130
Jones, Adrian, 11–12

Jones-Williams, Squadron Leader, 122–3
Junkers, Hugo, 137, 163, 168

Kennedy, Joseph, 172, 231
Keys, Clement, 124, 125, 126
Kingsley Wood, Sir, 160, 212, 246
Kipling, Joseph Rudyard, 17–18, 141
Kite and Model Aeroplane Association, 23
Knudstad, Jakob, Captain, 250, 251, 252
Korean War, 278, 279, 280

Labour Party/Government, 116, 119, 135, 161, 164, 165, 169, 173, 256, 260, 261, 262, 264, 300, 314
   British aircraft industry, 290
   CRF and, 138–42, 290–1, 292
   postwar period, 289, 290–1
   see also Socialism
Lagonda, 105, 151, 299
Lamb, Horace, 47, 48
Lancastria, liner, 275
land-based aircraft, 60, 97, 115, 197
landing, 84
   crash landing, 44, 99, 282, 287
   on water, 40, 43
   seaplane, 47, 85, 86
Lansbury, George, 140
Laski, Harold, 260, 263
League of Nations, 115, 140, 170, 176
Lindbergh, Charles, 121, 149, 156, 266–7
Lindemann, Frederick, Viscount Cherwell, 49–50
Lionel, Joseph, 292
Litvinov, Maxim, 170
Lloyd George, David, Lord, 59, 69, 87, 103, 106, 161, 179, 217–18, 222
Lobelle, Marcel, 88–9, 99, 121, 124, 125, 131, 185, 194, 196, 211
   Swordfish, 89, 185
Lockheed, 197, 198, 225, 233

INDEX

Londonderry, Lady, 139, 175
Londonderry, Lord, 123, 141, 156, 160, 167, 175, 176–7, 188
long-distance flight, 39–40, 98–9, 115, 118, 121
   Cobham, Alan, 119–20
   Fairey Aviation, 111, 113, 115, 118–19, 120, 121–3
   first aerial crossing of the south Atlantic, 111
   Hoare, Samuel, Sir, 118–19, 121
   long-distance record breaker, xiii, 121–2, 123
   monoplane, 122, 123, 135
   RAF, 112–13, 121, 123
   seaplane/flying boat, 115, 121
   United States, 121, 122
Longmore, Arthur, Air Chief Marshal Sir, 42–3, 45, 306
Lothian, Lord (Philip Kerr), 226, 227
Ludlow-Hewitt, Edgar, 133
Luftwaffe, 169, 186, 191, 192, 215, 216, 223

MacDonald, Ramsay, 116, 138–9, 140, 165, 169, 175–6, 179
Macmillan, Norman, 98–9, 111–12, 119, 165, 175
   test pilot, 99, 112, 113, 131–2, 286
MAEE (Marine Aircraft Experimental Establishment), 89
Maisky, Ivan, 170–2, 265–6, 315–16
Manchester Guardian, 96, 221
Manchester University, 47–8
Mannock, Edward 'Mick', Major, 77
MAP (Ministry of Aircraft Production), 37, 192, 221–3, 239, 240–1, 253, 311
   aircraft supply from United States, 227, 242
   Beaverbrook, Lord, 221, 222, 223
   Churchill, Winston, 221, 222, 240

   Fairey Aviation and, 255–6, 257, 311
   Short Brothers and, 257
Marden, G.E., 257–9, 260, 267, 274
Markey, John (CRF's father-in-law, first marriage), 92
Markey, Mrs (CRF's mother-in-law, first marriage), 158
Markey, Queenie, 92
   see also Fairey, Joan
Martin, Glenn L., 229
Martin Aircraft Corporation, 278
Maud, Lady, 118
Maxim, Hiram, Sir, 31
MBE (Most Excellent Order of the British Empire), xiii, 102, 106, 114, 213, 246
McClean, Francis, Sir, 29, 37, 38–9, 41, 43, 52, 62
   long-distance flight, 39–40
McClintock Agreement, 190
McCudden, James, Major, 77
McKinnon, Ronald, 179
Meldola, Raphael, 14
metal, 48, 159, 282
   all-metal aircraft, 118, 131, 132, 135, 175
   all-metal propeller, 124
   fuselage, aluminium sheeting, 40, 131
   see also duralumin
Milligan, Clifford, 16
Ministry of Munitions, 69, 72, 74, 80, 82, 85, 87, 100, 101, 103
Ministry of Supply, 280, 281, 282, 283, 286–7
missile, 277, 287
Mitchell, Reginald, 121, 130, 155
Model Aeroplane Competition, Crystal Palace, 23
model aircraft (by CRF), 17, 19, 20, 21, 317
   cane and brown paper aeroplane model, 10

INDEX

competitive flying, 22–3
contra-rotating propellers in, 22
Fairey's Miniature Aeroplane, 24–5, 33, *illustration 3*
Fairey's Miniature Aeroplane, manual for the assembly of, 24, 25
float plane, 20–1
Glider model advert, *illustration 4*
gliding, 20, 22
production model, 25–6, 34
powered model, 22
selling of, 23–4
talent at, 22
see also CRF: plane making and aircraft industry
Molotov, Vyacheslav Mikhailovich, 166, 170
Molotov–Ribbentrop Pact, 171, 218, 315
Monk, Ray, xiv
monoplane, 1, 22, 35, 121, *illustration 4*
all-metal monoplane, 175
Deperdussin Monocoque, 46
Fulmar, 184, 186, 237, 255, 285, 312
long-distance flight, 122, 123, 135
see also Fairey Battle
Moore-Brabazon, John, Lord, 29, 37, 105, 116–17, 119, 151, 175, 190, 240, 277, 290, *illustration 5, 13*
competitive flying, 39
CRF/Moore-Brabazon relationship, 39, 240, 249–50, 302–303
Daily Mail prize, 36, 38
politics, 142, 223–4, 314
Short No. 2: 36
Short, Horace and, 38
World War II, 191
Morgenthau, Henry, 234, 236
Morning Post, 130, 178
Morrison, Herbert, Lord, 290
Mosley, Oswald, Sir, 142, 143, 177, 233, 263, 314

Mountbatten, Admiral Lord (Prince Louis of Battenberg), 28, 148, 229, 302
Mountbatten, Edwina, Lady, 148
Murdoch, Frank, 199, 206, 207
Mussell Manor, 27, 29, 36, 38
Mussolini, Benito, xiv, 129–30, 142, 217, 218
Myers, Maurice, 66, 308, 314

Napier, 125, 129, 130, 174
Napier engine, 95, 122, 126, 129, 284
National Aerospace Library, 155
National League of Airmen, 177
National Physical Laboratory, 35, 49
wind tunnel, 47, 84, 198
nationalisation, 160, 179, 275, 290–1, 300
Navy Wing, 43–4, 45, 46, 53
Nazism, xiv, 137, 142–3, 170, 172, 217–18, 315
Neame, Barry, 201
Nicholetts, G.E., 123
Nicholl, Vincent, Colonel, 55–7, 64, 67, 97, 98, 99, 131
death, 112, 144
as test pilot, 56, 57, 70, 95, 112, 309
Nicholson, Charles, 89, 199, 201, 203, 204, 207

Ogilvie, Alec, 62
long-distance flight, 39–40
Short Brothers and, 38, 39, 62
Olaf, Crown Prince of Norway, 203
Olympia, 52, 53, 54, 134–6, 137, 143
Outen, Roland, 273, 300, 309
Overy, Richard, 254
Oxford University, 259, 307
Ozersky, Aleksander, 164, 166

P24: 22, 174, 281
Page, Ernest, 73

INDEX

Pan Am, 115, 126, 198, 265, 266
Parnall, 62, 83
Parsons, Charles, 38
passenger aircraft, 197
    Britain, 197, 199, 265
    CRF as passenger, 35, 264–5
    Deruluft airline, 163
    DH50: 120
    Fairey IIIF, passenger-carrying, 118, 129
    Fairey Aviation, FC1: 197–8
    land-based passenger aircraft, 197
    Rotodyne, 49
    Short Brothers, 39
    United States, 197, 198–9, 265
    see also civil aviation; commercial aviation
Peat, William, Sir, 107
    W.B. Peat and Co., 107
Penny, George (Lord Marchwood), 213
'personal capitalism', 79, 309
Petavel, Joseph Ernest, 47
Pethick-Lawrence, Frederick, 178
Philadelphia Inquirer, 266
Pickles, Mrs, 95
Pickles, Sydney, 70, 82, 86, 95, 96
Pilcher, Percy, 18
pilot:
    autopilot, 123, 197
    certificate as, 41, 43, 49
    Churchill, Winston, 44
    civilian as, 49
    CRF as, 35
    Fairey, John, 298
    Fairey, Richard Jr., 211, 249, 251–2, 286, 298–9
    fighter pilot, 186
    instructors, 41, 44
    RNAS pilots, 28, 56
    Royal Aero Club pilot's certificate, 41
    Short Brothers and, 28, 39, 42, 47
    training, 28, 41–2, 47
    see also test pilot

Pinsent, David, 49, 50
Pittleworth Manor, xvi, 319
Polikarpov, Nikolai, 164
    Polikarpov I-5: 168
Portal, Charles, Marshal of the RAF Sir, 237, 238, 242
Pott Brothers, 63
Pott cars, 63, 64
propeller, 22, 35, 88
    Armstrong-Siddeley's dual propeller system, 48
    contra-rotating propellers, 21–2, 281
    CRF and, 48, 49
    Fairey Aviation and, 127–8
    propeller design, 48
    'pusher' propeller, 21, 49
    Reed propeller, 127
    Short Brothers, 39, 40
    single-propeller float plane, 20–1
    Wittgenstein: 'Improvements in Propellers applicable for Aerial Machines', 48
    see also engine
Purvis, Arthur, 234

Queen Mary, 126, 268
    Verandah Grill, 126
Queenborough, Lord, 143
Queenborough Pier, 54

R&D (Research and Development), xiii, 129, 187, 279, 287
    aircraft R&D, 30
    collaborative R&D, 45
    Fairey Aviation, 196–7, 278, 281, 282, 312
    non-military R&D programme, 196–7
    Short, Horace, 51
    state-sponsored R&D, 30, 32, 69, 78, 107, 189

INDEX

RAF (Royal Air Force), 129, 280
  CRF and, 41, 117
  defensive role, 104
  effective air defence and strategic bombing, 190
  expansion of, 175–6, 178, 188
  Fairey Aviation and, xi, xii, 1, 117, 118, 131, 132–3, 184, 189, 190, 305
  long-distance flight, 112–13, 121, 123
  interwar period, 106, 173
  Scheme C, 188, 190
  Scheme F, 188, 189, 190
  Scheme J, 191
  Scheme M, 192
  Supermarine Spitfire, 183
  United States and, 226, 237, 238–9, 241, 242, 285
  World War I, 104
  World War II, 191, 226, 237, 238–9, 242
RAF Museum: Fairey Aviation archives, xvi, 7
Ralli, P.A., 88, 121, 125, 130
Rawlings, H.C., Captain, 185
rearmament, 177, 179, 186, 310, 311
  aircraft and agenda for rearmament, 187–94
  CRF and, xii, 159, 171, 204
record breaking, 99, 121, 130
  Campania non-stop flight, 86
  FD2 and world air speed record, xi, 281, 282, 286, 287
  long-distance record breaker, xiii, 121–2, 123
  United States, 288
Reed Propeller Company, 124, 127, 128
Rees, Frank, 62–3
  Fairey Aviation and, 63–5, 80, 81, 100, 308
religion, 298
  Anglicanism, 91

  christening, 91, 93
  Church of England, 302
  CRF's atheism/agnosticism, 3, 91, 93, 298
  Roman Catholicism, 91
RFC (Royal Flying Corps), 39, 43, 60, 61, 71, 77, 78
RIIA (Royal Institute of International Affairs), 293
Rittenhouse, David, Lieutenant, 125
River Test, xv, 152, 161, 213, 247, 301, 318
  Catchment Board and, 260–2, 264, 295
  CRF and environmental preservation, xii, 4, 261, 294–6, 313, 314, 318
  England's finest, most eminent chalk stream, 214
  A Summer on the Test, 214
  see also fishing; TIFA
RNAS (Royal Naval Air Service):
  capability, 46
  Churchill, Winston, 43–4, 45, 81
  CRF at, 61
  Fairey Aviation and, 87, 113, 305
  pilots, 28, 56
  Short Brothers and, 28, 39, 54, 59
  Sueter, Murray, 45
  World War I, 77, 78
Roe, Alliott Verdon, 21, 47–8, 52
  see also Avro
Roe, Saunders, 282
Rolls, Charles, 29
  death, 39
  Short Brothers and, 37, 38
Rolls Royce, 87, 104, 121, 125
  Comet 3, transcontinental airliner, 280
  CRF and, 123, 130, 174
  dominant force within British aero-engine industry, 130
  Eagle, 59, 98

INDEX

engine, 59, 98, 128, 129, 153
Merlin, 130, 155, 174, 195, 241
P51 Mustang, 225, 241
turboprop, 284
Roosevelt, Franklin Delano, *see* FDR
Rose, 72, 100
Rothermere, Lord, 96, 141, 175, 176, 177
Rotodyne, xiii, 282, 283–4, 285, 312
   passenger-carrying, 49
   turbo-prop 'heli-liner', 282
   vertical take-off, xiii, 49
   see also Fairey Aviation: aircraft
Rowlands, Archibald, Sir, 223, 234, 239–40, 258, 265, 267
Royal Aero Club, 27, 29, 43, 56, 130
   expansion of, 74
   James Gordon Bennett Cup, 27
   Royal Navy and, 43
Royal Aeronautical Society, 37, 130, 265–6
   CRF and, 1, 19, 85, 114–15, 154, 156, 162, 314
Royal Aircraft Establishment, 51, 128, 132
Royal Aircraft Factory, 30, 45, 46, 49, 84
Royal Commission on the Private Manufacture of and Trading in Arms, 178–81, 187
Royal Navy, 8, 28, 41, 44
   air defence and, 70
   aircraft, offensive capabilities, 46, 70–1
   CRF at, 60–1
   Fairey Aviation and, xi, 67, 69, 87, 113, 184, 185, 282, 308
   maritime air power, 86
   Royal Aero Club and, 43
   Short Brothers and, 43, 46, 51, 54
   Sopwith Aviation, 71
   Swordfish, 183
   see also Admiralty

Russell, Bertrand: *The Principles of Mathematics*, 48
Russia (Soviet Union), xiii, 98, 104, 140, 162–72, 180, 233, 242, 263, 306, 316
   Anglo-Soviet relations, 165, 167, 168, 169, 170
   ANT-9 'Wing of the Soviets', 162
   apparatchiks, xiii, 167, 169
   Arcos, 135, 164, 165, 168
   CRF and, 162, 164–8, 169, 170–2, 264, 314, 315–16
   CRF, trip to, 165–6, 315
   Deruluft airline, 163
   Fairey Firefly, 165, 167, 168, 169, 170
   Five-Year Plan, xiii, 163, 166, 315
   Germany and, 163
   Hitler's invasion of, 242
   Red Workers-Peasants Air Fleet, 163
   Soviet aircraft, 163–4, 166–8
   Soviet Embassy, xiii, 164–5, 171, 315
   TsAGI (Ventral State Aero-Hydrodynamic Institute), 162, 164, 167
   VVS, 163, 164, 165, 168, 169, 171

Saint-Exupéry, Antoine de, 122–3
Salmond, Geoffrey, Air Chief Marshal Sir, 117, 131, 133, 136
Salmond, Ian, 117
Sandys, Duncan: Defence White Paper, 280, 287
Sansom, Charles, Lieutenant, 42, 43, 44, 45–6, 52, 53, 74, 78, 236, 306
   Eastchurch Naval Flying School, 43
Sassoon, Philip, Sir, 117–18, 129, 217
Savage Club, 212
Savoy Grill, 117, 306
SBAC (Society of British Aircraft Constructors), 104, 130, 173, 179, 187, 253, 265, 279

INDEX

CRF and, 107, 108, 114, 116, 120, 156, 175, 187, 209, 253, 285, 290, 293–4, 314
  Fairey Aviation and, 74, 80
  foundation of, 74
  Harrogate programme, 193–4
  Royal Commission, 179–81
  Sopwith, Thomas, 74, 106
Schneider Trophy, 56, 94, 95–6, 121, 123, 126
  Supermarine, 88, 128, 129, 149, 155
Schuster, Arthur, 47
science and technology, 18
  aircraft technology, 30
  British tuition in universities, 55
  CRF and, 11, 15, 18, 25, 28, 30, 174, 279, 312
  design technology, 38
  new technology, 25, 28, 44, 133, 134, 141, 199, 282
  technological imperative of 'industrial war', 49
seaplane, 54
  commercial operation, 85
  Fairey Aviation, 70, 84, 85, 97, 109, 115
  'hydro-aeroplane', 52
  martial credentials, 52
  rivers, 85
  RNAS, 46, 52
  Short Brothers, 40, 52, 53, 54, 59, 78
  Sopwith Aviation, 78
  take-off and landing, 47, 85, 86
  torpedo gear, 54, 59
  twin floats and, 40
  World War I, 78
  see also amphibian; Fairey III; float plane; flying boat
Seddon, John, 84, 87
Self, Henry, Sir, 224–5, 228, 232, 238, 240
Sempill, Master of, 141, 142

ship-based aircraft, 46, 62, 86
Short, Eustace, 28, 36, 37, 54
  see also Short Brothers
Short, Horace, 28, 36, 37, 38, 40, 44, 46, 54
  aircraft design, 47, 52, 53, 60
  brachycephaly, 38
  CRF/Short relationship, 51, 58, 67, 308
  death, 54
  Moore-Brabazon, John and, 38
  pilot training, 42, 47
  R&D, 51
  Rolls, Charles and, 38
  see also Short Brothers
Short, Oswald, 28, 36, 37, 54, 120, 257
  CRF/Short relationship, 58, 61
  flotation bags, 45
  see also Short Brothers
Short aircraft:
  biplanes, 39, 41, 45–6
  C-Class Empire flying boat, 197
  glider, 38
  passenger aircraft, 39
  seaplane, 40, 52, 53, 54, 59, 78
  Short Admiralty No. 2: 52
  Short No. 1: 38
  Short No. 2: 36
  Short No S.26: 39
  Short Nos S.26–81: 39
  Short No S.27: 39
  Short No S.28: 39
  Short No S.38: 39, 46
  Short No S.41: 46
  Short No S.80: 39
  Short–Wright biplanes, 37, 38
  Sunbeam V-12: 59
  'tractor' aircraft, 39, 40, 46, 52
  Type 166: 59
  Type 827/184: 59, 62, 66
  Type 830: 59
  see also Short Brothers

INDEX

Short Brothers, 29, 36–41
   Admiralty and, 71
   aeronautics, 42, 78
   aircraft design, 30
   as aircraft manufacturer, 39
   balloon-making, 37, 39, 54
   Churchill, Winston and, 44
   CRF at, xii, 28, 36, 40, 41, 47, 51–4, 59, 64, 308
   a family business, 38
   MAP and, 257
   pilot, 28, 39, 42, 47
   RNAS and, 39, 54, 59
   RNAS pilots, 28
   Rochester/Medway factory, 54–5, 59–60, 62, 308
   Royal Aero Club and, 37, 39
   Royal Navy, 43, 46, 51, 54
   Sheppey factory, 54, 59
   subcontracting, 60
   test pilots at, 39, 40, 45
   World War I, 40, 51–2, 54, 59–60
   World War II, 37
   see also Short aircraft
Shute, Nerina, xv
Siddeley, John (Lord Kenilworth), 79
Sigrist, Fred, 71, 83–4, 143, 215, 234
Simon, John, Sir, 161, 179, 205
Sinclair, Archibald, Sir (Lord Thurso), 221–2, 266
Slade, Gordon, 286
Slater, Brice G., 67–8, 200, 203, 277
   Bossington estate, 68
Slessor, John, Marshal of the RAF Sir, 232, 237, 242
Socialism, 138, 180, 191, 260–4, 289, 290, 291
   see also Labour Party/Government
Sokolnikov, Grigori, 164
solo flight, 156, 252
Sopwith, Thomas ('Tommy'), xii, 1, 37, 45, 74, 83–4, 133, 199, 307, 309

aircraft design, 30, 53, 71, 83, 106, 107, 114
Baron de Forest prize, 29
'Bat Boat', 53
CRF/Sopwith relationship, 209, 302
death, 37
yachting, 199, 200, 202, 203, 205, 206–207
see also Hawker Siddeley Group
Sopwith Aviation, 46–7, 53, 60, 86, 106
   1½ Strutter, 71–2, 78
   Baby, 83
   Fairey Aviation and, 70, 74, 83–4
   Kingston factories, 71, 74, 83
   Pup, 71, 86, 150
   Royal Navy, 71
   seaplane, 78
   Sopwith Camel, 77, 83, 86
   triplane, 71
   two-seaters, 71–2
Soviet Union, *see* Russia
speed:
   cruising speed, 124
   FD2 and world air speed record, xi, 183, 281, 282, 286, 287, 288
   High Speed Flight, 130, 155
   S6B, record-breaking, 155
   transonic speed, 281
Spitfire, 130, 183, 186, 191, 195, 218, 225, 231
   see also Supermarine
Stalin, Joseph, 135, 162, 170, 172, 316
   Five-Year Plan, 163, 166
   purges, 162, 164, 315
Staniland, Chris, 175, 196, 281–2, 286
Stettinius, Edward, 235, 236
Stringfellow, John, 18
Sueter, Murray, Captain/Commodore/Rear Admiral Sir, 41, 44, 46, 52, 61, 142, *illustration 6*
   Air Department, 43, 62
   bombing Germany, 70, 78

INDEX

Churchill, Winston and, 45
Fairey Aviation and, 61–2, 64, 69, 78, 80, 81–2, 83, 277, 308
RNAS and, 45
Sunbeam, 59, 62, 87
Sunday Times, 178
Supermarine, 86, 95, 97, 149, 130, 239
  CRF and, 123
  S6/S6B, 121, 130, 153, 155
  Schneider Trophy, 88, 128, 129, 149, 155
  Supermarine Seafire, 114
  Supermarine Spitfire, 130, 183, 186, 191, 195, 218, 225, 231
Swinton, Lord (Philip Cunliffe-Lister, Sir), 160, 187, 191, 225
Swordfish, 46, 111, 183, 305, 319, *illustration 11*
  825 Squadron's destruction, 196
  diving, 184
  Fairey's best known aircraft, xii, 94, 113
  first blueprints of, 46
  Fleet Air Arm, xii, 183, 185, 186, 187
  Lobelle, Marcel, 89, 185
  multiple roles, 184, 186, 312
  origins 185–6
  production of, 184
  Royal Navy, 183
  success, 184, 186, 187, 196
  take-off, 184
  Taranto, 183, 247
  a torpedo-bomber, 184
  TSR prototype, 185–6, 193
  World War II, 183, 184, 239
Sykes, Frederick, Sir, 104

take-off, 46, 84, 122, 184
  from water, 21, 40
  seaplane, 47, 85, 86
  vertical take-off, xiii, 49, 174, 281
Tasmania, 2
  Hagley, 1, 3

The Tatler, 175
Taylor, H.N., Lieutenant, 205
technology, *see* science and technology
Tennant, David, xvi
Tennant, Jane (née Fairey, CRF's daughter, second marriage), xvi, 91, 93, 321–6
  see also Fairey, Jane
test flight, 23, 42, 49, 54, 123, 124, 167, 279, 312
test pilot, 50, 106, 285
  Bell, Gordon, 52
  Fairey Aviation, 70, 95, 99, 112, 175, 281–2, 285, 286
  McClean, Francis, 40
  postwar era, 285
  Short Brothers, 39, 40, 45
  Slade, Gordon, 286
  Wright, Maurice, 57, 86
  see also Macmillan, Norman; Nicholl, Vincent; Pickles, Sydney; pilot; Sansom, Charles; Staniland, Chris; Twiss, Peter
Test Valley, xii, xvi, 58, 91, 209, 243, 272, 293, 295, 314, 318, 319
  see also Bossington; Pittleworth Manor; River Test
Thompson, Silvanus, 13–14, 15, 16, 17, 47, 88, 307
Thomson, Lord, 138
Three-Seater, 53
thrust, 18, 21, 32, 48
Thurgood, Derek, 288, 301–302
TIFA (Test and Itchen Fishing Association), 295–6
The Times, 95, 122, 140, 178
Tips, Ernest Oscar, 168, 169, 274–5, 309
  Tipsy light monoplane, 169
Titmuss, Richard, 314
Tizard, Henry, Sir, 49, 50
torque, 21, 22, 30, 283
Towers, Jack, 236–7, 238, 240, 242

INDEX

'tractor' aircraft, 39, 40, 46, 52, 82
Treasury, 41, 115, 181, 189, 312
  CRF and, 246
  Fairey Aviation and, 68, 80, 100, 101–102, 145, 177, 186
Trenchard (Lord), Hugh, Marshal of the RAF Sir, 104, 116, 119, 123, 128, 130, 131–2, 148, 277
Trippe, Peter: CRF's biography, xv–xvi, 13, 16–17
Truman, Harry S., 292
Truth, 191–2
Tuke, Anthony, 295, 296
Tukhachevsky, M.N., 167
Tupolev, Andrei, xiii, 162–4, 168, 315
  TU-4: 163
Twiss, Peter, 282, 285–6
  FD2 and world air speed record, xi, xv, 286, 287, 288
  Queen's Commendation for Valuable Service in the Air, 287
  test pilot, xi, 282, 285
Type III, *see* Fairey III

U-boat, 206, 250, 251, 253, 274, 291
United States, 121, 124, 128
  Air Ministry and, 225
  aircraft production, 242, 253, 254, 255, 307, 311–12
  American neutrality, 227
  Anglo-American collaboration, 225–7
  Anglo-American diplomatic relationship, 245
  Anglo-American rivalry, xiii, 233
  British community in 244–5
  Churchill, Winston and, 225, 227, 238
  fighter aircraft, 125–6, 233
  Fleet Air Arm and, 233, 242
  isolationism, 227, 231, 233

Lend-Lease policy, 226, 227, 232, 235–6, 238, 241, 242
long-distance flight, 121, 122
MAP and, 227, 242
New Deal, xiii, 235, 245, 314
passenger aircraft, 197, 198–9, 265
RAF and, 226, 237, 238–9, 241, 242, 285
record breaking, 288
US Office of War Mobilization, 266
US Treasury, 226, 227, 234, 236
Washington, 244
White House, 225, 226–7, 232, 233, 235, 236, 245
World War II, 224–9, 238, 242
see also CRF and the United States
University College Bristol, 14
USAAC (US Army Air Corps), 126, 227, 228, 232
USAAF (US Army Air Forces), 226, 235, 237, 241, 242–3, 275, 276
USN (US Navy), 115, 125, 228, 235
USS *Swanson*, 250, 251

Van de Velde, L.A., 166, 167–8, 169
Vane, Herbert, 129
Varlars, Benjamin, 23–4
Vickers, 41, 95, 97, 170, 173
  'Mayfly', 41
Victorian era, 3, 4, 17, 29–30, 47, 313
  Victorian sentimentality, 13
Vines, Alan, 275, 276
Vinson, Charles, 230, 273, 309
volume production, 17, 28, 59, 191
  absence of, 45
  Fairey Aviation, 52, 74, 114, 184, 186, 187, 188, 194, 196
  World War I, 46, 77

Wakeling, Florence, 268
War Office, 30, 32, 45, 46, 69, 72, 78, 213

INDEX

'warfare state', 50, 87
Washington Post, 245–6
Waugh, Evelyn, 143, 289
Weir, Lord, 102, 160, 179
Wells, H.G., 11, 18, 62, 117, 317
  Bealby, 31, 33
  Dunne, John William and, 31, 33
  The History of Mr Polly, 31
  The War in the Air, 31
Wentworth Day, James (Jim), xiv–xv, 141–2, 263, 314
Westland Aircraft, xv, xvi, 78, 282
  Fairey Aviation, merger with, 49, 59, 78, 284, 287
  operational model, 78–9
White, Paul, 268, 272, 300, 301
Whitehall, 30, 43, 80, 106, 113, 119, 179, 181, 189, 194, 274
  CRF and, 246
  Fairey Aviation and, 74, 101, 145, 160, 173, 276–7
  United States and, 227
  Whitehall inertia, 216
Whitney, Esther Sarah, *see* Fairey, Esther, Lady
Whittle, Frank, Air Commodore Sir, 48, 49
Wildman-Lushington, Gilbert, Captain, 44
Williams, David Hollis, 122, 150, 258
Wilson, Morris, 224
wing, 135
  delta-winged aircraft, 183, 281, 312
  duralumin, 135, 137, 185
  Fairey's Miniature Aeroplane, 33
  fixed-wing aircraft, 277
  flexible wing surface, 84
  high-lift wings, 84, 85, 87
  power-folding wings, 278
  Short No S.41: 46

wing design, 22, 32, 47, 83, 198
wing-tips, 32, 33
Wittgenstein, Ludwig, 47–50, 124, 283
  English translation of *Tractatus Logico-Philosophicus*, 49
  'Improvements in Propellers applicable for Aerial Machines', 48
Wood, Kingsley, Sir, 160, 212, 246
Woodlands Park, 152, 188, 215, 216–17, 247, 310
  as evacuees' residence, 216
  Hall, Geoffrey and, 154, 210, 217
Woodward, Nathaniel, Reverend, 9
World War I (Great War), xii, 18, 59, 87, 271
  aerial warfare, 77
  Battle of Jutland, 86
  Battle of the Somme, 72
  British aircraft industry, 2, 32, 45, 46, 59, 69–70, 73, 74, 77, 87, 309
  expedition against the Bolsheviks, 97, 98
  Fairey Aviation, 69, 70, 75, 80, 118, 277, 311
  Gallipoli Campaign, 2, 54, 58
  Germany, 46
  'industrial war', 49, 70, 77, 308
  interservice rivalry, 78
  onset of, 61
  RAF, 104
  RFC, 77, 78
  RNAS, 77, 78
  seaplane, 78
  Short Brothers, 40, 51–2, 54, 59–60
  volume production, 46, 77
World War II, xii, 50, 184, 190, 195
  Allies, 83, 97, 105, 218, 226, 241
  American entry into the war, 238, 246
  American neutrality, 227
  Anglo-American collaboration, 225–9, 232, 235–6, 238–9, 242, 265–6

INDEX

Axis, 197, 217, 266
Battle of Britain, 183, 192, 228
Bermuda, 291–2
the Blitz, 233, 238, 288
Bomber Command, 104, 237, 243
Bossington and, 247, 267, 268
British aircraft industry, 191, 222, 225, 253, 279, 315
British/Allied Purchasing Mission, 224–8, 232
Churchill, Winston, 225, 291
CRF and, 216, 217–18, 228–9, 247–8, 262, 314
CRF: family, 215, 217, 243–4, 247
D-Day, 253
declaration of war, 194, 246
evacuation programme, 215–16, 314
Fairey Aviation, 37, 58, 171, 185, 190, 192, 194–5, 253, 274, 305, 311
Fighter Command, 133, 223, 225, 226, 237, 241, 287
France, 225, 226
Germany, 191, 216
'industrial war', 190
Italy, 190, 217–18
Lend-Lease policy, 226, 227, 232, 235–6, 238, 241, 242
London, bombing of, 216, 223, 253
Norwegian campaign, 186, 225
Pearl Harbor, 233, 235, 247, 274
RAF, 191, 226, 237, 238–9, 242
Short Brothers, 37
Swordfish, 183, 184, 239
United States, 224–9, 238, 242
Worshipful Company of Shipwrights, 212
Wright, Fred, 67

Wright, Maurice, 55–6, 57, 64, 67, 86, 126, 129, 135–6, 137, 145, 152, 166, 258, 309
as CRF's successor, 230, 254
death, 300
Wright brothers (Orville and Wilbur Wright), 18–19, 21, 27, 29, 31, 38, 237, 317
the Flyer, 37
front elevator, 32
Short–Wright biplanes, 37, 38

yachting, 153
12-metre yacht, xii, 2, 204, 205, 275
America's Cup, xii, 200, 203–204, 205, 206, 207, 210
CRF and competitive sailing, xii, 2, 200–201, 202, 203–204, 208–209, 212, 271, 302, 316
Endeavour, 199, 203, 204, 206, 207
Endeavour II, 208
Evadne, 156, 191, 201–202, 205–206, 230, 323, 324
Evaine, 172, 202, 204–205, 207
Firefly, 276
Flica, 2, 203, 204
IYRC (International Yacht Racing Union), 204
J-class yacht, xii, 199, 203, 208
K-class yacht, 208
Modesty, 149, 200, 202–203
motor yacht, xii, 156, 191, 201, 202, 323, 324
NYYC (New York Yacht Club), 199, 204, 208
Radiant, 323
RLYC (Royal London Yacht Club), 66, 200, 208
RTYC (Royal Thames Yacht Club), 143, 164, 200

INDEX

RYS (Royal Yacht Squadron), 199, 202, 204, 208
Shamrock V, 203, 206, 208
Sopwith, Thomas, 199, 200, 202, 203, 205, 206–207
Westra, 204
see also Fairey Marine
Yachting World, 200, 203, 204
Yeats, William Butler, xii

Young, Esther (CRF's widow, second marriage), 236
see also Fairey, Esther, Lady
Young, Faith, 236
Young, Philip, 236, 238

Zeppelin, 52, 54, 66, 70, 78, 83, 274
Zindel, Ernst, 137